MW00761164

STRUCTURE OF VACUUM
ELEMENTARY MATTER

To Marshall,
for his freindship over
half a lifetime. Thank
you!
        Valter

Kona , July 2000

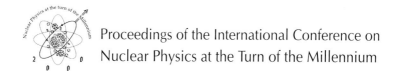

Proceedings of the International Conference on
Nuclear Physics at the Turn of the Millennium

STRUCTURE OF VACUUM & ELEMENTARY MATTER

Wilderness, South Africa   *March 10–16, 1996*

editors

Horst Stöcker
Goethe Universität Frankfurt

André Gallmann
Université Louis Pasteur, Strasbourg

Joseph H. Hamilton
Vanderbilt University, Nashville

World Scientific
Singapore • New Jersey • London • Hong Kong

Published by

World Scientific Publishing Co. Pte. Ltd.

P O Box 128, Farrer Road, Singapore 912805

USA office: Suite 1B, 1060 Main Street, River Edge, NJ 07661

UK office: 57 Shelton Street, Covent Garden, London WC2H 9HE

British Library Cataloguing-in-Publication Data
A catalogue record for this book is available from the British Library.

ISBN 981-02-2789-2

Printed in Singapore by Uto-Print

International Conference on
Nuclear Physics
at the Turn of the Millennium:

Structure of
Vacuum and Elementary
Matter

March 10 to March 16, 1996

Wilderness, South Africa

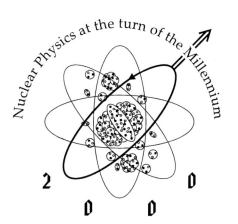

Topics:

- **Supercritical QED Fields** and Pair Production in Heavy-Ion Collisions

- **Superheavy Nuclei**, Exotic Nuclear States and Decays

- **Superdense Matter** in Relativistic Heavy-Ion Collisions: Collective Flow, Particle Production and the Nuclear Equation of State, Phase Transitions in QCD, Strange Matter and Signatures of the Quark Gluon Plasma.

Organizing Commitee:

A. Gallmann, Université Louis Pasteur, Strasbourg;
F.J.W. Hahne, University of Stellenbosch;
J.H. Hamilton, Vanderbilt University, Nashville;
J.P.F. Sellschop, University of the Witwatersrand, Johannesburg;
G. Soff, TU Dresden;
E. Stein, Goethe Universität Frankfurt;
H. Stöcker, Goethe Universität Frankfurt;
M.R. Strayer, Oak Ridge National Laboratory.

Editor's Preface

The International Conference on Nuclear Physics at the Turn of the Millennium was held at Wilderness, South Africa, from March 10 - 15, 1996.

The conference brought together physicists from all over the world to discuss the most important recent advances and future directions, experimental and theoretical, in four active fields of nuclear and elementary particle physics research:

1. superheavy elements and exotic nuclei

2. supercritical fields and atomic physics

3. superdense matter, particle production and quark matter

4. special theoretical topics

Highlights from the experimental side include first reports on the discovery of element 112, newly discovered radioactive decay modes (cluster radioactivity, cold fission, triple fission), inspiring data from the new heavy exotic beam facilities, and atomic physics. The intriguing analyses of the new high energy gold- and lead-beam experiments at the Gesellschaft für Schwerionenforschung (GSI), Lawrence Berkeley Laboratory (LBL), Brookhaven National Laboratory, and Conseil Europeen pour la Recherche Nucleaire (CERN), have given strong evidence for the formation of high density very hot matter by collective compression.

The newly presented discovery of collective flow, even at the ultrarelativistic energies, as well as the formation of Resonance Matter, further the excitement in the community about the formation of new, hitherto unknown forms of elementary matter in heavy-ion collisions. Also, first results on the search for Antimatter and Hypermatter have been presented at this meeting.

A stimulating physics program can be expected on these topics which will extend well over the turn of the millennium: future perspectives are being offered by the Relativistic Heavy Ion Collider (RHIC), the Large Hadron Collider (LHC), the Japanese Hadron Facility as well as a future GSI machine.

These exciting new experimental findings were matched by a wealth of recent theoretical investigations, covering areas both tightly related to these experiments as well as more formal recent developments.

Many theoretical talks were devoted to the structure and dynamics of nuclear and elementary matter under extreme conditions, a fruitful and many-faceted area of research.

Walter Greiner, teacher, colleague, and friend, has played a pioneering role in defining, developing and nourishing - over the last three decades - this wide range of topical fields covered in this conference - we have summarized Walter's research interests in the epilogue. More than 120 Ph.D. students have evolved from Walter's school, most of them highly successful in their career, and more than 30 of his former students now have university positions, world-wide. The livelihood of the research presented here is proof for Walter's intuition and scientific taste.

Walter loves South Africa, its beauty and climate. For nearly three decades he has very good friends here. So it was decided, on the occasion of his 60th birthday, to invite Walter to this country, where he received his first Doctor Honoris Causa (from the University of the Witwatersrand).

This was a fortunate decision as the participants of the conference will bear witness: its setting, both in geographical as well as in the scientific sense, was superb.

Indeed, special thanks go to Sue Sellschop for selecting the – most beautiful! – Wilderness site and for initiating many of the activities for the accompanying persons.

Sanel de Swardt's tireless and friendly help is gratefully acknowledged – she took care of most of the organisational matters here.

Many colleagues, students, and friends, theorists and experimentalists alike, who are actively engaged into research in these areas, have come to this scientific conference, to present their views on the most recent progress in these fields, to discuss the future activities, and, also, to celebrate Walter's prominent role in our science. The scientific papers contained in these proceedings, therefore, also serve as the "Festschrift", dedicated to Walter Greiner, on the occasion of his 60th birthday...

Therefore, from your colleagues, friends and students, both those present in Wilderness as well as those at home, best wishes for your and your family's future, remain as vigorous and productive as we know you, and watch the lions, elephants and wildebeasts in Kruger park, but carefully!

Finally a special bow to Bärbel Greiner, who encourages and guides her husband Walter, as Walter's teacher Hans Marschall expressed it so most appropriately: "Bärbel ist die Hüterin Deiner Bequemlichkeit!"

Horst Stöcker
for the organizing committee

This volume is dedicated to
Walter Greiner
at the occasion of his sixtieth birthday

KOMSORN
PHUKET DEC 96

Welcoming Word

F.J.W. Hahne

University of Stellenbosch, Private Bag X1, Matieland, 7602 South Africa

It gives me pleasure to welcome you all to one of the most beautiful parts of the world - the Garden Route - where we have come to celebrate Walter Greiner's sixtieth birthday in true academic tradition by holding a high-level conference - presenting results and discussing them.

The focus of the conference is on several topics which Walter Greiner has initiated and in which he has been a leading figure for many years. Walter was one of the young German physicists who re-established physics and particularly nuclear physics in Germany after the devastating war. Walter played an important role in establishing the GSI and has been associated with it through all the years. Walter has trained many physicists for academia and for industry. He has played a major role in internationalising German physics and putting it very high on the world map. His institute was a place where people from the East, West and South were meeting at times when this was not general practice.

Walter has the critical attitude of a physicist, but he believes in good friendships and that friends should stick together. Although a very busy man, he is very hospitable and he can truly enjoy life.

The Institute in Frankfurt which Walter manages is like a big business, yet Walter attends to a lot of the detail of the work that is done there, particularly to the most fascinating parts of it.

Since many of our guests have some connection to Frankfurt and since its university is named after Goethe, I proposed a little competition among the participants for the most apt Goethe quote. I kick off with:

"Wer immer strebend sich bemüht, den können wir erlösen"

Following the challenge, another Faust quote was presented by Professor Hartmuth Arenhövel, which, even if there had been more competitors, could have been a clear winner:

"War es ein Gott der die Zeichen schrieb,
Dir mir das innere Toben stillen,
Das arme Herz mit Freude füllen
Und mit geheimnisvollen Trieb
Die Kräfte der Natur rings um mich her enthüllen?"

Walter and most of us have been striving to unveil the forces of nature throughout our lives. This conference represents yet another.

It is a pleasure to organise this conference for him. The relentless work done by Horst Stöcker, Eckart Stein and my secretary, Sanel de Swardt, is acknowledged. The Karos Wilderness Hotel was very accommodating during the organisation. Financial support by the African Foundation for Research Development is gratefully acknowledged.

Opening Remarks

Friedel Sellschop

Schonland Research Centre for Nuclear Sciences, University of the Witwatersrand, Johannesburg, South Africa

Delegates to this conference, one and all. I am very happy indeed to welcome you ...

to welcome you to this remarkable event

to welcome you to South Africa, the new South Africa, many of you for the first time to this country, with all its hopes and eager expectations for the future

to welcome you to this beautiful corner of South Africa, the Wilderness, with the confining parameters of the sea (the Indian Ocean), the vleis (to translate for your benefit "lagoons") and the mountains ...just out of the vineyards of the Western Cape, but fortunately still close enough to derive the most immediate benefits thereof).

We are delighted to have you with us and the more so because of the nature of this occasion ...

let me return to this occasion, or as I termed it earlier this remarkable event:

our conference has a challenging, indeed let us admit it, a flamboyant title

"Nuclear Physics at the Turn of the Millennium"

Let us turn to that font of all knowledge, the Oxford English dictionary ... to discover obviously that the millennium is a period of or anniversary of 1000 years, but an earlier interpretation (ca. 1820) referred to it as "a period of happiness and benign government" ,
and we find from Tennyson, the English poet (1809 - 1892, poet laureate from 1850) **"let thy feet, Millenniums hence, be set in midst of knowledge"** ...and I find that singularly appropriate to the person whose 6 score years we are recognising today

But the subtitle reads

"Structure of vacuum and elementary matter"

...tantalizing words, I am sure you will agree, and even a cursory glance at the titles of the papers to be presented this week speaks of a vast array of work at the cutting edge of our science, indeed of all science.

But what is the common denominator that links the various themes that make up our programme of this week ? ...almost without exception there is the input, usually the seminal, the pioneering precursor, the original and imaginative input of one man. I refer of course to Walter Greiner

Walter, we recognise this and we salute you.

For this conference has assembled from some 10 different countries of the world, in this far corner of Africa, to celebrate the first 60 years of Walter's life. his remarkably productive life, the life of a brilliant scientist, yes but more than that ...

for who of us has not had the opportunity to give witness to Walter the man, Walter the humanist, Walter the lover of music, Walter the courageous missionary for scientific research, specifically for research in the field of nuclear physics, to Walter the proud husband, father and grandfather, to Walter the friend ...to him friendship is a very precious thing. a most precious thing, and I am always reminded in my measure of Walter's friendship of the words of Shakespeare in King Richard II when he utters

"I count myself in nothing else so happy, as in a soul rememb'ring my good friends"

But who is this man, Walter, whose virtues I so extol, who has impacted on all our lives ? :

Walter Greiner was born in October 1935 in Thüringen, received his PhD from the University of Freiburg in Breisgau in 1961, was Assistant Professor at the University of Maryland, and then took up the Chair of Physics and the Directorship of the Institute for Theoretical Physics at the Johann Wolfgang von Goethe University at Frankfurt am Main in 1964, a post he holds to this day.

Of his manifold research works I can do but scant justice, but nevertheless let me try ...

one thinks of his pioneering work in **nuclear structure** (including topics such as the rotation-vibration model, the dynamic collective model of giant resonances, the general collective model, proton- neutron deformations and vibrations, superheavy nuclei, the 2-center shell model, and nuclear molecules)

his equally pioneering work in **nuclear reactions** (eigenchannel theory, and dispersion effects in electron scattering)

his imaginative original work in the **quantum electrodynamics of strong fields** (supercritical systems, spontaneous decay of the e^+e^- vacuum in overcritical electric fields, superheavy quasimolecules, delta electrons and positron emission in heavy ion collisions),

the prediction of **nuclear shock waves** and collective flow in high energy heavy ion collisions

cluster radioactivity and cold fission (fragmentation theory)

the study of **hot and dense nuclear matter** and its possible phase transitions

transport theories for describing complex high energy heavy ion reactions

Even if I have not managed to do adequate justice to his fertile mind and boundless energies, the list I have offered is nothing short of spectacular.

This has not surprisingly been recognised by others as well as by those gathered here today, and this is reflected in the award of the

Max Born prize in 1974

and the

Otto Hahn prize in 1982.

and many honorary doctorates from Wits in 1982, Tel Aviv in 1990, Strasbourg in 1991, Bucharest in 1992
he was elected an honorary member of the Roland Eotvos Society of Hungary in 1989, and of the Romanian Academy in 1993.

If I may take advantage of the words that Shakespeare places in the mouth of
Hamlet
"...What a piece of work is man!"
by which I mean of course, *this* man,
and Shakespeare picks up the same theme even more eloquently in his Julius
Caesar

> "His life was gentle, and the elements
> So mix'd in him that Nature might stand up
> And say to all the world **"This was a man!"**

Fortunately in the case of Walter Greiner, we say not "this was a man man"
but "this is a man".

Now Walter has always attracted at any one time quite a large number of
students and through their eyes one gets another perspective of the man. As a
visitor at the Institute, I soon found that one is wise to listen to the students.
Now as we all know, indeed we have all been the beneficiaries, Walter is un-
selfish when it comes to travel , —- he is the equivalent of the missionary of
yesteryear ——- he likes to travel. And I found during his absences, that his
students worked less hard, in fact they often partied, made jokes, drank beer,
recited joggerel and I liked in particular one quasi-poem that they liked to
intone whenever Walter was a-traveling ...now my mastery of German is not
good, so I will give you the poem in rather free translation, I will title it (with
acknowledgment to "Science with a Smile" Robert L Weber, IOP Publishing
Ltd 1992)

"Walter Greiner's research students lament"
and it goes something like this

> "Twinkle, twinkle learned Prof.
> How I wonder where you're off,
> Up above the world so high,
> in a Jumbo in the sky -
>
> To some conference afar,
> Where they see you as a star,
> Showing, with your mastermind,
> Work by suckers left behind

May you be a shining light!
Life's worth while when things go right,
May the questions never go
Way beyond the stuff you know!

So, good fortune when you fly
in your Jumbo, up so high,
Though we don't know where you're off,
Twinkle, twinkle, learned Prof! "

And so let us quite unashamedly give praise to Walter Greiner's great achievements in his first 60 years ... as you can read in the Bible in Ecclesiasticus

" Let us now praise famous men ... "

and that is in essence what this conference is about, a paean of praise, both a retrospective and a prospective overview of the very best of nuclear physics of our times, an occasion for celebration and appreciation, of joy and the stark pleasure of gaining new insights from one another. Let us enjoy ourselves in celebrating this notable event!

In saying that, I welcome you all both in my individual capacity and as the current President of the South African Institute of Physics.

Friedel Sellschop

1. Hillhouse, G. C.
2. Cindro, N.
3. Csernai, L. P.
4. Hamilton, J. H.
5. Mosel, U.
6. Greiner, W.
7. Stöcker, H.
8. Gallmann, A.
9. Sellschop, J. P. F.
10. Kienle, P.
11. Hess, P.
12. Simonoff, G.
13. Oganessian, Yu. Ts.
14. Sandulescu, A.
15. Lovas, I.
16. Kaun, K. H.
17. Richter, W. A.
18. Bass, S. A.
19. Vaagen, J. S.
20. Schukraft, J.
21. Sailer, K.
22. Roepke, G.
23. Kämpfer, B.
24. Braun-Munzinger, P.
25. Stachel, J.
26. Grosse, E.
27. Faessler, A.
28. Coffin, J. -P.
29. Jundt, F.
30. Gutbrod, H. H.
31. Gorenstein, M.
32. McLerran, L.
33. Glendenning N.K.
34. Weigel, M.
35. Sharpey-Schafer, J.
36. Symons, J.
37. Soff, S.
38. Satarov, L. M.
39. Wyngaardt, S.
40. Oeschler, H.
41. Schempp, A.
42. Fricke, B.
43. Scheid, W.
44. Dreizler, R.
45. Ramayya, A. V.
46. Müller, B.
47. Cleymans, J.
48. Herrmann, N.
49. Konopka, J.
50. Solov'yov, A. V.
51. Arenhövel, H.
52. Ludu, A.
53. Rischke, D.
54. Poenaru, D.
55. Soff, G.
56. Mishustin, I.
57. Eggers, H. C.
58. Greiner, C.
59. Aichelin, J.
60. Pelte, D.
61. Reinhardt, J.
62. Gyulassy, M.
63. Greiner, M.
64. Bleicher, M.
65. Bilpuch, E.
66. Spieles, Ch.
67. Lynen, U.
68. Eisenberg, J.
69. van der Ventel.B.I.S.
70. Lindsay. R.
71. Schramm, S.
72. Stein. E.
73. Britt, H. C.
74. Weber, F.
75. Nagle, J.

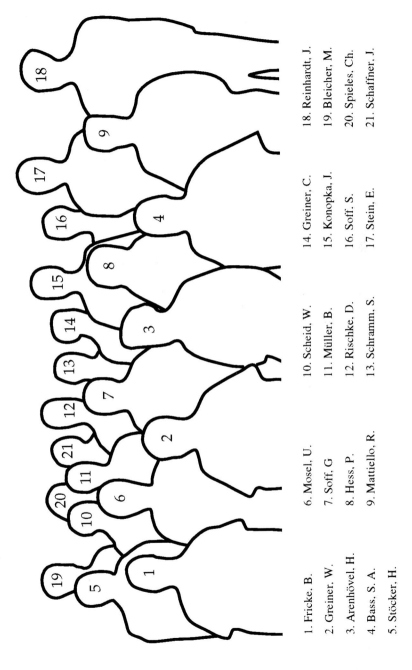

Walter Greiner and some of his former students

1. Fricke. B.	6. Mosel, U.	14. Greiner. C.	18. Reinhardt. J.
2. Greiner. W.	7. Soff. G	15. Konopka. J.	19. Bleicher, M.
3. Arenhövel. H.	8. Hess. P.	16. Soff. S.	20. Spieles. Ch.
4. Bass. S. A.	9. Mattiello, R.	17. Stein. E.	21. Schaffner. J.
5. Stöcker. H.	10. Scheid. W.		
	11. Müller. B.		
	12. Rischke. D.		
	13. Schramm. S.		

CONTENTS

1. Superheavy and Exotic Nuclei

2. Supercritical Fields and Atomic Physics

3. Superdense Matter and Particle Production

Epilogue

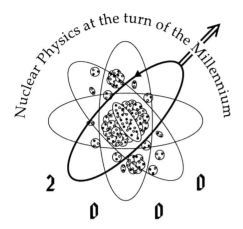

1. Superheavy and Exotic Nuclei

THE DISCOVERY OF ELEMENTS 110, 111 AND 112

S. HOFMANN

GSI, Max-Planck-Strasse 1, D-64291 Darmstadt, Germany

The new elements 110, 111 and 112 were synthesized and unambiguously identified in experiments at SHIP. Due to strong shell effects the dominant decay mode is not fission, but emission of alpha particles. Theoretical investigations predict that maximum shell effects should exist in nuclei near proton number 114 and neutron number 184. Our measurements give hope that isotopes of element 114 close to the island of spherical Superheavy Elements could be produced by fusion reactions using ^{208}Pb as target.

1   Introduction

Various theoretical calculations predicted the next double shell closures beyond ^{208}Pb at proton number Z=114 and neutron number N=184.[1,2,3] Experiments to search for superheavy nuclei near the predicted double magic ^{298}114 were negative so far.[4,5] In parallel, experimental methods were developed to produce and identify lighter new elements.[6,7] At SHIP these experiments were successful during the first ten years of UNILAC operation in identifying the elements Z=107-109.[8] An increased stability of nuclei approaching the neutron number N=162 and the proton number Z=108 was proved by the data. The cross-sections for production by fusion reactions with lead and bismuth targets decreased gradually down to values of 10 pb as measured for element 109. Still lower cross-sections could be expected for production of heavier elements. Reactions for synthesis of elements up to 114 using lead-isotopes as target were suggested, based on theoretical investigations.[9]

For further experimental investigation of new elements and small branching ratios, the set-up at SHIP was upgraded to reach a level of detection sensitivity of at least 1 pb.[10] Experiments with the new set-up were carried out starting in 1994. As a result, the three elements 110, 111 and 112 were discovered.[11,12,13,14] The reaction mechanism was investigated by excitation functions measured in small energy steps. The measured phenomena can be interpreted as fusion initiated by transfer (FIT).[15] New data on the binding energies of nuclei near the predicted magic number Z=114 were obtained by α-decay chains.

3

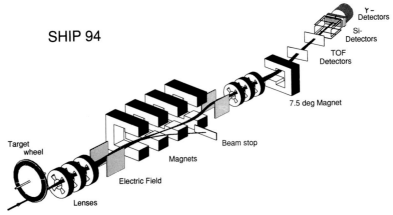

Figure 1: The modified separator SHIP and the new detection system.

2 Experimental Method

In parallel to the construction of SIS, the facilities at GSI for production of heavy elements at low beam energies were upgraded. At the UNILAC, a new high charge injector was built including an ECR-ion source. The separator SHIP was modified, and a new improved detection system was built (see Fig. 1).

The aims of the SHIP modifications were: 1. To increase the solid angle and thus increase the transmission for fusion-reaction products; 2. To reduce the background in order to accept higher beam currents and to increase simultaneously the significance of detected correlated events, and 3. to increase the overall stability of all components of the set-up in order to achieve high total experimental efficiencies. The improvements resulted in a sensitivity for detection of one heavy element decay-chain every 10 days at a cross-section of 1 pb.

3 Decay Properties of Heavy Elements

Experiments aiming at investigation of the reaction process and identification of elements Z=110 and 111 were carried out in 1994 in June/July (22 days) and October to December (77 days), respectively. In December 1995 the reaction ^{82}Se $+ ^{208}$Pb $\rightarrow ^{290}116^*$ was investigated in order to search for radiative capture process (32 days). Element 112 was identified in February 1996 (43 days). Beams of ^{50}Ti, ^{51}V, ^{58}Fe, ^{62}Ni, ^{64}Ni, ^{70}Zn and ^{82}Se were used with currents up

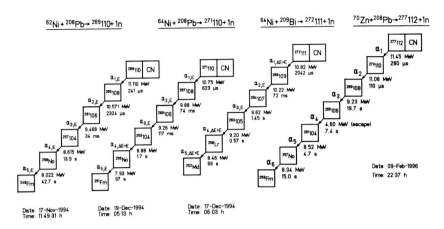

Figure 2: Four representative decay chains of the new elements 110, 111 and 112 discovered at SHIP during experiments in November/December 1994 and February 1996.

to $\approx$0.5 pμA. Targets of lead isotopes and ^{209}Bi were irradiated. The isotopes ^{269}110, ^{271}110, ^{272}111 and ^{277}112 were identified by 4, 9 ,3 and 2 decay chains, respectively. Four representative decay chains are shown in Fig. 2.

Recently, the stability of heavy and superheavy nuclei was investigated theoretically[16,17,18,19] using refined models based on the Nilsson-Strutinsky approach. The ground-state microscopic shell-corrections show two minima: One of about 8 MeV centered around $^{272}_{162}$110 with maximum deformation β_2=0.22, the other of about 9 MeV around $^{292}_{178}$114 forming an island of spherical superheavy nuclei (Fig. 3, top). The binding energies, β-, α-, and fission half-lives were calculated. The measured data, including the decays of ^{267}110 and ^{273}110 obtained in experiments in Berkeley[20] and Dubna[21], respectively, could be reproduced with good accuracy. A rough sketch of the half-lives obtained is shown in Fig. 3, bottom. As a result, we can deduce that most nuclei, that can be produced with stable projectiles and the available targets are predicted to be α emitters with half-lives between 1 μs and 1 s.

The nucleus $^{278}_{164}$114 is predicted to be deformed in its ground state and to decay by α emission with a half-life below 1 μs (see Fig. 3). Increased stability may occur also for spherical shapes due to the shell closure of the major shells $\nu j_{15/2}$ and $\nu i_{11/2}$ at N=164, (see Fig. 4 in Ref.[22]) in combination with the spherical shell closure for the protons at Z=114. The interplay with the increased stability of deformed shapes at neutron number N=162 may lead to existence of shape isomerism. Experimentally, the nuclei near ^{278}114 could be investigated by reactions of ^{70}Ge or ^{72}Ge beams with ^{208}Pb targets.

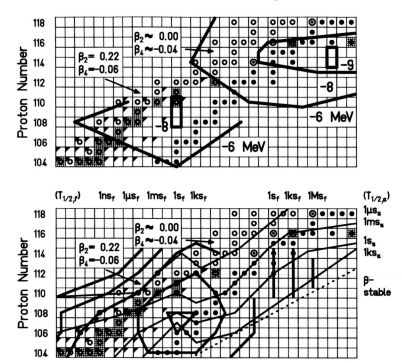

Figure 3: Rough sketch of calculated microscopic shell correction energies (top) and half-lives (bottom) of even-even nuclei; bold lines: partial fission half-lives; fine lines: partial α half-lives; broken line: β-stability line. The data are from Ref.[16,17,18,19] The fission half-lives of odd and odd-odd nuclei may be longer as a result of additional hindrance factors. The arrows point to the region of strongly deformed nuclei centered around ^{272}110 and to the region of spherical superheavies at around ^{292}114; increased stability of spherically shaped nuclei might be expected also near Z=114 and N=164, see text; open circles: compound nuclei of reactions (Ti to Kr) + ^{208}Pb →(Z=104 to 118); dots: compound nuclei of reactions (O to Ti) + ^{248}Cm →(Z=104 to 118); full triangles: known nuclei; shaded: compound systems investigated in experiments at SHIP.

The closest approach to the island of spherical superheavy nuclei can be achieved with stable projectiles and targets using ^{76}Ge and ^{208}Pb. The nuclei produced by one neutron emission or possibly by radiative capture could have almost spherical shapes, and they are predicted to decay by α emission with half-lives of ≈ 1 ms. The α decay would result in important information on the shell strength at $Z=114$ and, in addition, the decay chain down to nobelium would provide decay properties of five new heavy isotopes of elements from 112 down to 104. Synthesis of ^{283}114 or ^{284}114 is certainly the most interesting aim of the next series of experiments.

4 Reaction Properties

4.1 Excitation Functions

Information on optimum beam energies for production of heavy elements was obtained by excitation functions. The measured excitation functions for the elements 104 and 108 are shown in Fig. 4. The data points were narrow enough to allow for a safe determination of the positions for the cross-sections maxima. The beam energy for production of element 110 by the reaction ^{62}Ni+^{208}Pb $\rightarrow ^{269}$110+1n was obtained by linear extrapolation.

A systematics of measured cross-sections is shown in Fig. 5. An ordering as function of the projectiles isospin is obvious and allows for predictions by extrapolation. An estimate for production of ^{283}114 results in $\approx$1pb. The recently measured value of 1 pb for production of ^{277}112 is smaller than the extrapolated value, but still in agreement within the error bars. A smaller value may also be the result of a beam energy chosen not accurately enough to cover the maximum production cross-section. In case of element 112 isotopes, a cross-section "inversion" as function of isospin may occur, if the shell-correction energies of the fusion products influence the production probabilities. The isotope ^{275}112, which may be produced using a ^{68}Zn beam, is predicted to be stronger bound than ^{277}112, produced with a ^{70}Zn beam. A more complete excitation-function systematics for production of element 112 isotopes would be highly desirable.

4.2 Fusion Initiated by Transfer (FIT)

Fig. 4 shows that the largest cross-sections were measured "below the barrier". The energy relations determining the barrier are drawn in Fig. 6 in case of the reaction ^{64}Ni+^{208}Pb and a barrier according to the fusion model by Bass.[24]

A tunneling process through this barrier cannot explain the measured cross-sections. A semiclassical WKB approximation results in a tunneling

8

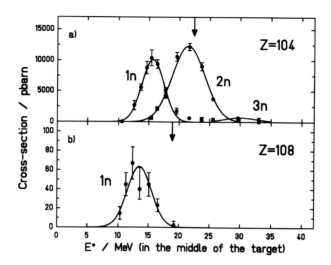

Figure 4: Measured excitation functions for production of element 104 and 108 by reactions of ^{50}Ti and ^{58}Fe projectiles with ^{208}Pb targets. The cross-sections are plotted as a function of the dissipated energy E*, calculated from the center-of-mass beam-energies in the middle of the target thicknesses and the Q-values using the mass tables of Ref.[23]. The lines are fits of gaussian curves through the data points. The arrows mark the interaction barriers of the reactions according to the fusion model by Bass.[24]

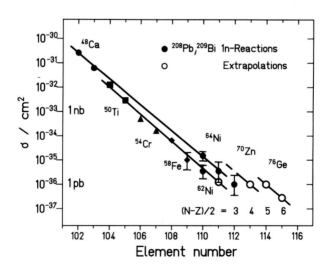

Figure 5: Systematics of measured cross-section maxima and extrapolations.

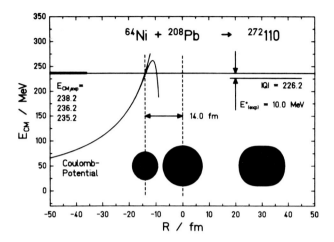

Figure 6: Energy relations resulting in maximum cross-sections.

probability of 2×10^{-21}, which is much too low to contribute to the measured cross-section. The conclusion is that additional effects must allow for fusion.

Compared to more sophisticated investigations of fusion,[25] we can make some simplifications in our case in order to explain the observations: 1. The reactions are limited to spherical target nuclei and nearly spherical projectiles. 2. The reactions are head-on collisions with angular momenta close to zero and, therefore, high rotational symmetry. 3. The reactions proceed at extremely low dissipative energies as a result of nearly compensating Q-value and center-of-mass energy.

Fig. 6 shows that the reaction partners came to rest before the fusion process starts. At a distance of 14.0 fm between the centers of the nuclei the initially given kinetic energy of 236.2 MeV is exhausted by the Coulomb potential. At that distance only nucleons at the outer surface are just in contact. Nevertheless, at that energy this heavy system has highest probability to fuse and to survive.

We remember that the kinetic energy of orbiting nucleons is low at the surface. Therefore, at the touching point of two nuclei in a central collision the probability is highest that nucleons or pairs of nucleons leave the orbit of one nucleus and move into a free orbit of the reaction partner. An adequate theoretical description could be obtained by use of the two-center shell model.[26]

Because of pairing energies and high orbital angular momenta involved, the transfer of pairs is more likely than that of single nucleons. The described process is a frictionless pair transfer happening at the contact point in a central collision at longitudinal momenta close to zero in irradiations of ^{208}Pb targets.

Already after the transfer of 2 protons from ^{64}Ni to ^{208}Pb the Coulomb barrier is decreased by 14 MeV allowing to keep the reaction partners in close contact and to continue fusion initiated by transfer. Important factors, which determine the cross-section at the very beginning of the fusion process, are: 1. The probability for a head-on collision and 2. The probability of proton transfer in competition to reseparation. These first steps of the fusion process at low energies can be investigated experimentally by measurement of transfer products in forward direction.

References

1. W.D. Myers and W.J. Swiatecki, *Nucl. Phys.* **81**, 1 (1966).
2. H. Meldner, *Ark. Fys.* **36**, 593 (1966).
3. U. Mosel and W. Greiner, *Z. Phys.* **A222**, 261 (1969).
4. P. Armbruster et al., *Phys. Rev. Lett.* **54**, 406 (1985).
5. Yu.Ts. Oganessian, et al., *Nucl. Phys.* **A294**, 213 (1978).
6. G. Münzenberg et al., *Nucl. Instr. and Meth.* **161**, 65 (1979).
7. S. Hofmann et al., *Nucl. Instr. and Meth.* **223**, 312 (1984).
8. G. Münzenberg, *Rep. Prog. Phys.* **51**, 57 (1988).
9. R.K. Gupta et al., *Z. Phys.* **A283**, 217 (1977).
10. S. Hofmann, *Journal of Alloys and Compounds* **213/214**, 74 (1994).
11. S. Hofmann et al., *Z. Phys.* **A350**, 277 (1995).
12. S. Hofmann et al., *Z. Phys.* **A350**, 281 (1995).
13. S. Hofmann et al., *GSI Nachrichten* **02-95**, 4 (1995).
14. S. Hofmann et al., *Z. Phys.* **A354**, 229 (1996).
15. S. Hofmann, *XV. Nucl. Phys. Conf.* **St. Petersburg**, 305 (1995).
16. A. Sobiczewski, *Physics of Particles and Nuclei* **25**, 295 (1994).
17. R. Smolanczuk et al., *Phys. Rev.* **C52**, 1871 (1995).
18. P. Möller and J.R. Nix, *J. Phys. G: Nucl. Part. Phys.* **20**, 1681 (1994).
19. S. Cwiok et al., *Nucl. Phys.* **A573**, 356 (1994).
20. A. Ghiorso et al., *Phys. Rev.* **C51**, R2293 (1995).
21. Yu.A. Lazarev et al., *Phys. Rev.* **C**, to be published (1996).
22. J.R. Nix and P. Möller, *Conf.* ENAM-95, Arles (1995).
23. G. Audi and A.H. Wapstra, *Nucl Phys.* **A565**, 1 (1993).
24. R. Bass, *Nucl. Phys.* **A231**, 45 (1974).
25. W. Reisdorf, *J. Phys. G: Nucl. Part. Phys.* **20**, 1297 (1994).
26. D. Scharnweber et al., *Nucl Phys.* **A164**, 257 (1971).

Dedicated to Prof. Walter Greiner on his 60-th birthday

SYNTHESIS AND RADIOACTIVE PROPERTIES
OF THE HEAVIEST NUCLEI

Yu.Ts. OGANESSIAN

Flerov Laboratory of Nuclear Reactions, Joint Institute for Nuclear Research,
141980 Dubna Moscow region, Russian Federation

Introduction

The problem of synthesizing new elements has a long history.

Based on the few atoms produced in nuclear reactions it was demonstrated that the radioactive properties of heavy nuclei confirm the main prediction of the macro-microscopic theory regarding a huge effect of nuclear shells on the spontaneous fission probability. As a result of the fission barrier emergence, determined by the nuclear structure, partial s.f. half-lives of heavy nuclei turn to be by 12–20 orders of magnitude large than the values predicted by the classical liquid drop model of nuclei.

As a result of high stability of spontaneous fission, isotopes of heaviest elements undergo α-decay with a half life $10^{-3} \div 10^1$ s.

What was the progress of these investigations and what are the prospects?

1 Nuclear shells and stability of heavy nuclei

Macro-microscopic investigations of the potential energy surface of nuclei at large deformations established that microscopic corrections drastically change the potential energy surface associated with fission [1-5].

Theory explains in general a number of experimental facts, fission barrier heights, shape isomerism in actinoid nuclei, spontaneous fission half-lives $(T_{s.f.})$ of transactinoids, substantial variations in $T_{s.f.}$ in the region $N = 152$ etc., which have not found any explanation in the classical liquid drop models.

Similarly to any other theory it possesses a certain predictive power, in particular for prediction of masses and radioactive properties of yet unknown superheavy nuclei. Such predictions were made in a number of papers. We are presenting here recent data from papers by Patyk, Smolanczuk and Sobiczewski [6-8] where there have been calculated masses and fission barriers as well as partial half-lives T_α and $T_{s.f.}$ of even-even nuclei with $Z = 100$–114 and $N = 140$–190.

Figure 1 presents a contour map of shell corrections to energy as a function of proton and neutron numbers. Significant changes in $T_{s.f.}$ of nuclei far from the $N = 152$ shell are determined to a great extent by another shell with $N = 162$. It should be noted that both neutron shells are referred to deformed nuclei in contrast to double magic nuclei such as ^{208}Pb ($Z = 82$, $N = 126$) possessing a spherical shape in the ground state. The maximum stabilization against spontaneous fission is expected for the nucleus $^{270}108$ ($Z = 108$, $N = 162$) for which the predicted $T_{s.f.}$ may reach $10^4 - 10^9$ s.

On the other hand, in the region of heavier nuclei, in the vicinity of the above mentioned deformed shells, one finds closed spherical shells $Z \approx 114$ and $N = 180-184$. A larger amplitude of shell corrections for spherical superheavy nuclei leads to as large (if not even larger) restrictions for the spontaneous fission.

Upon the whole, such a nontrivial situation may lead to interesting consequences.

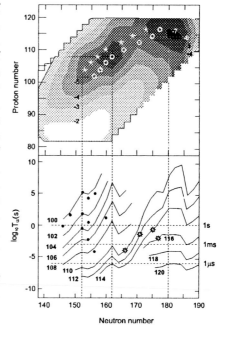

Figure 1. a) Contour map of the shell corrections to energy. Stars and cycles denotes the heaviest nuclides produced in cold and hot fusion reactions correspondingly.

b) Calculated α-decay half-lives (T_α). Black points — experimental data. Stars — T_α predicted for heaviest even-even isotopes of the elements 110, 112, 114 and 116 produced in hot fusion reactions with ^{48}Ca projectiles.

If one excludes spontaneous fission, than nuclei near the closed shells will undergo alpha and beta decays. The probabilities of these decays and, consequently, the halflife of superheavy nuclei will be determined by the masses of nuclei in their ground state. The latter may be calculated with the nuclear mass formula with the accuracy prevailing now at the description and extrapolation of nuclei masses on the basis of spectroscopic data.

It follows from calculations [8] that for a nucleus of $^{268}106$ ($N = 162$) T_α is equal to several hours (according to estimations by P.Moeller — several days) and for a nucleus of $^{298}114$ this value grows to several hundred years, possibly, thousands years! Really, we are talking here about very stable and very heavy nuclei. In case this is true, the nuclear structure expands significantly the limits of the Periodic

Table of Elements. This opens unique opportunities in neighboring sciences - atomic physics, inorganic chemistry which have a large experimental and theoretical basis.

Coming back to the issue of the spontaneous fission of superheavy nuclei, it is necessary to note the following circumstance.

The calculation of spontaneous fission half-life $T_{s.f.}$ in the dynamical way consists in the search for one-dimensional fission trajectory in a multi-dimentional deformation space, which minimizes the action integral corresponding to the penetration of the fission barrier. Although the calculated static barrier heights are about equal, differences in half-life estimates can be attributed to varying assumptions regarding the dynamical path through the fission and the consequent inertial mass.

In other words the complex structure of the potential surface and the variation of the inertial mass in the process of deformation may lead to different fission modes which are significantly different in time and, consequently, in their fission probability. Moreover, one and the same nucleus may have simultaneously two fission modes which was observed experimentally for heavy isotopes of actinide elements[9].

This situation becomes especially critical for the region of nuclei near deformed shells $Z = 108$ and $N = 162$. For example, Moeller et al.[10], taking ^{258}Fm as model for heavier nuclei. Assume that the path after the first barrier is short with the emerging fragments being nearly spherical and close to the doubly magic ^{132}Sn. On the other hand, Patyk et al.[6-7] calculate dynamical barriers that show a different path, higher inertial mass, and consequently much longer SF half-lives. This competition between static and dynamical features of the SF process which leads to so large differences in stability makes experiments that explore ground-state decay properties of nuclei around $N = 162$ and $Z = 108$ one of the most important tasks in heavy element research.

2 Reactions of synthesis

It is known, that the heaviest elements were synthesized in cold fusion reactions $^{208}Pb(HI,n)$. It has been experimentally observed that heavy ions with $A_I > 40$ undergo fusion with magic nuclei of ^{208}Pb deep in the subbarrier region, which lead to the formation of a compound nucleus with an excitation energy of 10–15 MeV. At such a small excitation the nuclear shell effects disappear, although not completely, which gives a certain stability to the system with respect to fission. The transition into the ground state occurs by emission of just one or two neutrons and γ rays[11,12].

14

The mechanism of such a process is not yet fully clear. This is evident, for example, in figure 2.

The process of ^{208}Pb nuclei fusion with ions of ^{16}O [13,14] and much heavier ions of ^{50}Ti, ^{58}Fe [15] or ^{64}Ni [16] falls under the general regularity of nuclei interaction at large distances. Despite a substantial growth of the Coulomb forces (from ^{16}O to ^{64}Ni the Coulomb energy grows nearly threefold) the threshold of the fusion reaction remains unchanged. This contradicts with numerous theoretical models of "extra-push" or "extra-extra-push" type in which the dynamic restrictions increase substantially the energy threshold of the fusion reaction [17-19].

At the same time, the decay of ^{224}Th can be satisfactorily described in the frame of statistical models but this can not be achieved for heaviest compound nuclei.

It is possible that in the cold-fusion reactions (HI,n) the emission of a neutron takes place at the stage of nuclear fusion and final compound nucleus decays further-on according to the laws of statistics.

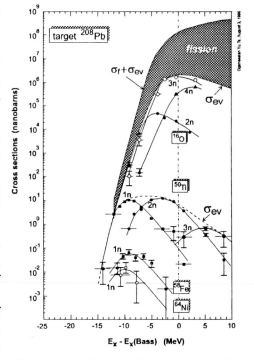

Figure 2. Cross sections σ_{xn} and $\sigma_{xn} + \sigma_f$ in the reaction $^{208}Pb + ^{16}O$. In the bottom part of the figure one can see the cross sections σ_{xn} in the reactions $^{208}Pb + ^{50}Ti$, ^{58}Fe and ^{64}Ni. The energy scal is presented as a difference $E_x - E_x$ (B_{Bass}). Solid lines are drawn through experimental points.

We hope that the coming joint experiment FLNR-GSI studying the reaction $^{86}Kr + ^{136}Xe \rightarrow ^{222}Th$ together with our earlier data on the measurements of $\sigma_{EVR}(E_x)$ and $\sigma_f(E_x)$ in the reaction $^{208}Pb + ^{16}O$ as well as the data on the mass and energy distribution of fission fragments $^{222-226}Th$ [20] will make the picture clearer.

The cross section of the cold fusion reaction for the heaviest elements is equal to only several pb which raises serious requirements to the luminosity of the experiment. The use of the ^{208}Pb target imposes a restriction on the value $(N-Z) \leq 48$ for $Z \leq 108$ which is somewhat away from the top of the predicted island of stability of the deformed shell $Z = 158$, $N = 162$.

In principle a significant growth in the number of neutrons in evaporation residues $(EVRs)$ up to $N-Z = 54$ can be obtained in fusion reactions between heavy actinide nuclei of the ^{244}Pu, ^{248}Cm type and projectiles such as ^{18}O, ^{22}Ne, ^{26}Mg. But in these reactions the excitation energy of the compound nucleus even at the Coulomb barrier is about 40 MeV (hot fusion).

Structural effects practically disappear at such a high excitation energy; their fission barrier is determined only by the macroscopic (liquid drop) component of the nucleus deformation energy $B_f \approx B_f(LD)$. It is well known that for transactinide nuclei $B_f(LD)$ is practically equal to zero. In the absence of a fission barrier the excited nucleus becomes totally unstable to fission which should lead to a strong decrease in the probability of its transition to the ground state via cascade evaporation of neutrons ($x \geq 4$). Under these conditions the survival of *EVRs* totally depends on the dynamic properties of the excited compound nucleus.

Investigation of excited nuclei fission dynamics by measuring the characteristics of pre-fission emission of gamma-quanta, neutrons and light charged particles was performed in numerous papers (see for example the overviews by Newton [21], Hilcher and Rossner [22].

We are most interested here in the region of heavy nuclei: $B_f(LD) \to 0$ with $E_x \approx$ 40–50 MeV and we are presenting here the experimental data obtained in collaboration with HMI (Berlin) for excited nuclei of Cf–Fm ($B_{LD} \approx$ 1.5-MeV) [23,24].

As is seen in figure 3a the contribution of pre-fission neutrons increases with the increase in excitation energy. The probability of pre-fission neutron emission can be calculated for the whole time interval of nucleus existence up to the moment of its splitting into two fragments. This time can be chosen as a parameter to obtain the best agreement with the experimental dependence $v_{pre}(E_x)$. For Cf nuclei $\tau_f \sim 3.5 \times 10^{-20}$ s.

Such experiments are used to measure the total number of neutrons emitted prior to reaching the scission point (pre-scission neutrons). A part of them had been emitted before the moment the nucleus reached the saddle point (pre-saddle neutrons). The number of pre-saddle neutrons can be defined from the excitation functions of the reaction $\sigma_{xn}(E_x)$ which determine the ratio of the widths Γ_n/Γ_f on each stage of compound nucleus deexcitation.

The values $\Gamma_n/\Gamma_{tot}(E_x)$ for nuclei with $Z = 98$ presented in figure 3b and cited from the data of

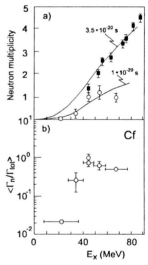

Figure 3. a) The number of pre-fission neutrons vs the excitation energy of ^{250}Cf nuclei. Black points: data of direct measurements preceeding the scission point (pre-scission neutrons). Open circles – pre-saddle neutrons obtained from the excitation functions of the *xn*-reactions;
b) Γ_n/Γ_{tot} vs E_x for ^{250}Cf.

Sikkeland et al. [25] testify to the fact that at $E_x \geq 40$ MeV the fission and neutron evaporation probabilities are comparable.

Note that for ^{250}Cf $(E_x = 80$ MeV) approximately 1/3 of the neutrons is emitted before reaching the saddle point.

At such a slow progress of deformation in the fission channel (viscous regime) even the heaviest nuclei with $B_{LD} \approx 0$ will have a finite probability of transition to the ground state through evaporation of neutrons. Quantitative data can be obtained only in direct experimental measurements of the evaporation residues formation cross sections for the heaviest excited nuclei.

The cross sections of the evaporation products formation in hot fusion reactions $(HI,5n)$ for $Z = 102$–105 obtained in studies by Andreev et al., [26,27], new results obtained by Lazarev et al. [28,31] for nuclei with $Z \geq 104$ are presented in figure 4. The same figure presents as well the results obtained by Hoffman et al. [15,16,32] for nuclei with $Z = 104$–112 in cold fusion reactions $(E_x = 10$÷15 MeV). As is evident from figure 4 the cross sections of $\sigma(HI,n)$ and $\sigma(HI,5n)$ for compound nuclei with $E_x \sim 10$ MeV (cold fusion) and $E_x = 50$ MeV (hot fusion) for the heaviest nuclei differ by approximately one order of magnitude in favour of the cold fusion reaction. This circumstance seems to be of importance in the problem of synthesizing heavier nuclei near the spherical shell $Z = 114$, $N = 180$–184.

The synthesis of nuclei with $Z = 114$–116 in cold fusion reactions using a ^{208}Pb target necessitates the increase of the ion mass to ^{76}Ge or ^{82}Se. The final products of the reaction (HI,n) will be isotopes of $^{283}114$ $(N = 169)$ and $^{289}116$ $(N = 173)$ located between the deformed and spherical shells. Note, that for the reaction $^{208}Pb(^{82}Se,n)^{289}116$ in the GSI experiments there was obtained the upper limit of the $\sigma_n \leq 5$ pb cross section [33].

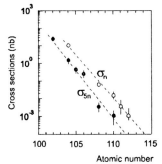

Figure 4. Cross sections of the reactions σ_n $(E_x \sim 10$–15 MeV) – cold fusion and σ_{5n} $(E_x = 50$ MeV) – hot fusion depending on Z of the compound nucleus.

One can assume that at the synthesis of neutron-rich superheavy nuclei certain advantages can be attained in hot fusion reactions of the type ^{244}Pu, $^{248}Cm + {}^{48}Ca$.

Because of a large excess of neutrons in the nucleus of ^{48}Ca the excitation energy of the compound nucleus at the Coulomb barrier is equal to $E_x^m = B + Q \approx 35$ MeV. The most probable channels of the reaction $(^{48}Ca,xn)$ corresponds to $x = 3$–4, which leads to the production of EVR with $Z = 114$, 116, $N = 174, 175$ and $N = 176, 177$ respectively. The cross section of the reactions even at $E_i - B_c > 0$ can be larger than the one observed at the synthesis of isotopes of elements *108* and *110* in the reactions ^{238}U or $^{244}Pu(^{34}S,5n)$ at $E_x = 50$ MeV.

The cross sections of the production of nuclei with $Z = 114$ calculated by Y.Abe et al. [34] on stochastic approaches of nuclear dynamics, produce the maximum value of $\sigma_{xn} \sim 10$ pb at $E_x \sim 30$ MeV. Other preliminary calculations by B.Pustylnik [35] based on the statistical model and describing the experimental results on the production of *EVR* in hot fusion reactions up to $Z = 110$, point to a strong dependence of the cross section of *xn* channels on the value of the shell correction energy dependence. They also point to a substantial growth of σ_{xn} in the region of $E_x \sim 30$ MeV (figure 5).

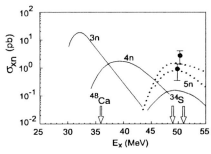

Figure 5. Calculated cross sections $\sigma_{xn}(E_x)$ in the reaction $^{244}Pu + ^{48}Ca$. Broken curves calculated cross sections $\sigma_{5n}(E_x)$ in the reactions $^{238}U + ^{34}S$ and $^{244}Pu + ^{34}S$. Points – experimental values. Arrows – excitation energy at the Coulomb barrier.

3 Observations of enhanced stability near closed deformed shells

Essentially, this was the underlying idea of a joint JINR (Dubna) – LLNL (Livermore) experiment on synthesis of *106* element heavy isotopes [29].

The ground-state decay properties of $^{266}106$ should be a quite sensitive probe of the theoretical predictions shown in figure 6a. If there is increased stability near $N = 162$ and $Z = 108$, the isotope $^{266}106$ should have a *SF*- or α-decay half-life of tens of seconds. Otherwise, $^{266}106$ should decay by *SF* with a half-life of ~100 μs, a $T_{s.f.}$ difference of ~10^5 or more. Thus a distinct signature for enhanced nuclear stability near $N = 162$ and $Z = 108$ would be the observation of the α decay of $^{266}106$ followed by the *SF* decay of the daughter nucleus $^{262}104$. A signature for the odd-A isotope $^{265}106$ would be the observation of its decay followed by decays of the known nuclides $^{261}104$ and $^{257}102$.

To produce $^{265}106$ and $^{266}106$ we used the complete fusion reaction $^{248}Cm + ^{22}Ne$ at bombarding energies which are expected to provide maximum cross sections for the *4n* and *5n* evaporation channels.

In a 360-hour irradiation of the ^{248}Cm target with a ^{22}Ne total ion beam dose of $1.6 \cdot 10^{19}$ produced on the U-400 accelerator (FLNR) by means of the Gas Filled Recoil Separator (GFRS) there have been synthesized two new most neutron-rich isotopes of element *106* with masses 265 and 266.

Both the isotopes $^{265}106$ (N = 159) and $^{266}106$ (N = 160) undergo mostly the α-decay with energies E_α = 8.71 $\div$ 8.91 and 8.63 $\pm$ 0.05 MeV correspondingly. The energy of α-decay of the even-even nucleus $^{266}106$ (Q_α = 8.76 MeV) determines its half-life T_α = 10–30 s.

Based on the six registered (α,sf) correlations to the α-decay of $^{266}106$ nucleus there was also determined the partial spontaneous fission half-life of the daughter nucleus $^{262}104$ (N = 158) $T_{s.f.}$ = $1.2^{+1.0}_{-0.5}$ s.

Radioactive properties of even-even isotopes of $^{262}104$ and $^{266}106$ give an indication of a substantial growth of heavy nuclei stability to spontaneous fission when approaching the closed shells Z = 108 and N = 162 (figure 6a,b).

The nuclei obtained in this experiment are in the process of an abrupt increase of stability to spontaneous fission like it has been predicted by macro-microscopic calculations by Patyk et al. [6,7]. Another spontaneous fission mode, characterized by a short way of tunneling through the fission barrier [10] and leading to a sharp decrease of $T_{s.f.}$ for $^{266}106$ is prohibited by more than 10^4 times.

The expected significant growth of T_α with the growing number of neutrons makes possible experiments studying the chemical properties of element 106 – $EkaW$ [36,37].

Among all possible target-ion combinations leading to the production of a $^{270}108$ nucleus with closed shells Z = 108 and N = 160, the reaction $^{238}U(^{36}S,4n)^{270}108$ seems to be the most promising one.

Investigations of fusion reactions $^{206-208}Pb(^{34,36}S,2-4n)$ in the course of which there were synthesized new neutron-deficient isotopes of Cf demonstrated high sensitivity and selectivity of the kinematic separator to detection of evaporation residues [38]. Along with that, because of great expenditure for a rare isotope ^{36}S (the natural abundance – 0.015%) by a PIG-type ion source it would be most difficult to carry out such an experiment. That is why in March–April 1994 experiments in

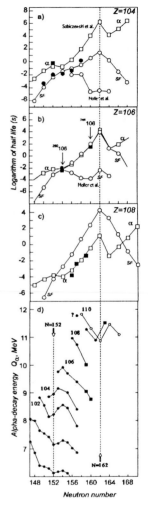

Figure 6. a,b,c) partial half-lives T_α and $T_{s.f.}$ for even-even isotopes with Z = 104, 106 and 108. Solid lines and open points – calculations. Black points – experiment. d) $Q_\alpha(N,Z)$. Black points – experiment values, open points – calculations for isotopes wih Z = 110.

Dubna were using a beam of a more abundant isotope ^{34}S enriched up to 90%.

At the irradiation of a ^{238}U target with a total ^{34}S-beam dose of $1.7 \cdot 10^{19}$, the position sensitive strip detectors of recoils registered $4(\alpha\text{-}\alpha)$ correlation events clearly pointing to the production of a new isotope of element *108* with a mass of 267 ($E_\alpha = 9.74$–9.88 MeV, $T_\alpha = 19^{+29}_{-7}$ s) [30].

And finally, in September-December of 1995 there were carried out experiments on the synthesis of the heaviest isotope of element *110*. At the irradiation of ^{244}Pu target with ions of ^{34}S with a total doze of $2.5 \cdot 10^{19}$ there were discovered few events pointing to an α-decay of an odd isotope of $^{269}110$, produced in the reaction $(^{34}S, 5n)$ [31].

The calculated and experimental values of partial periods T_α and $T_{s.f.}$ of isotopes of actinide elements $Z = 104$, 106 and 108 are represented in figure 6 (a,b,c) respectively. Upon the whole, experiments confirm not only qualitatively but also quantitatively the theoretical predictions of the stability of heavy nuclei.

The energy of α decays of heavy isotopes with $Z = 104$–110 produced in hot fusion reactions and that of lighter isotopes with $Z = 104$–112 in cold fusion reactions together with the earlier known data on nuclei with $Z \leq 104$ are presented in figure 6d. At passing the level of a deformed shell $Z = 102$, $N = 152$, like it has been expected, one can observe a leap in the value of Q_α. Quantitatively, at passing the shell $N = 152$ the change in the decay energy of the two isotopes of element *102* with $N = 151$ and 153 is equal to $\Delta Q \sim 0.12$ MeV. Note, that this small value plays a large role in the stability of deformed nuclei of transuranium elements.

Analogous effect is observed at passing the shell $N = 162$ for $Z = 110$. Here the value $\Delta Q \sim 0.6$ MeV. This is a direct proof of the existence of shell $N = 162$ predicted by the theory. The shell correction turns to be even larger than the one predicted in calculations [6-8] for this region of nuclei.

Out of the data presented above one can make a number of conclusions.

The masses of the heaviest nuclides, their decay energies and the time of life are in a good agreement with the predictions of the macro-microscopic theory, pointing to the significant role of the nuclear structure and first and foremost of nuclear shells for deformed superheavy nuclei.

For the known isotopes with $Z = 106$–112 the partial half-lives $T_{s.f.} > T_\alpha$. This circumstance is a direct indication to a decisive role of nuclear shells at the formation of the fission barrier and, consequently, in the stability of superheavy nuclei in the conditions when the liquid drop fission barrier $B^f_{LD} = 0$.

Neutron rich isotopes are most illustrative in this respect, since T_α grows sharply with the growth of the neutron number. Note, that these nuclei have been synthesized in hot fusion reactions.

4   Problems of synthesizing superheavy nuclei near closed spherical shells

The main provisions of the theory and the formalism of the calculation of specific properties of nuclei near the deformed shells can be applied to the region of heavier nuclei where a new growth of stability is expected which is due to the effect of spherical shells $Z = 114$, $N = 180$–184 [8].

In which way is it be possible to obtain experimental proves of the existence of these superheavy and superstable nuclides?

Unfortunately, no combination of stable isotopes chosen as a target and an ion can not form a compound nucleus with $Z = 110$–114 and $N = 180$–184.

That is why the essence of the problem is in the way to approach as close as possible the top of stability, i.e. how to produce heavy nuclei with $Z \sim 114$ with a maximum number of neutrons.

It is not difficult to understand that this can be achieved at the maximum excess of neutrons in the fusing nuclei with a minimum loss of neutrons in the process of compound nucleus deexcitation.

Figure 1b demonstrates that out of all possible reactions with extremely neutron rich ions of ^{48}Ca the maximum effect is achieved for nuclei with $Z = 114$ and $N = 174$, 175 produced in the fusion reaction $^{244}Pu(^{48}Ca,3$–$4n)^{289,288}114$.

How can one synthesize and identify these nuclides?

The fusion reactions kinematics of ^{244}Pu and ^{48}Ca is very little different from the one observed earlier in other reactions of the type $^{244}Pu + {}^{34}S$ or $^{238}U + {}^{40}Ar$. That is why the method of reaction products separation and registration used these last several years at GSI and FLNR can be applied in this case practically in full as well. In principle this approach ensures the luminosity of the experiment corresponding to the cross section ~ 1 pb or even less.

Identification of new nuclei is a more complicated issue. In all the earlier experiments after the α–decay of a new nuclide there was observed a chain of sequential α decays of already known isotopes, correlation with which determined Z and A of the nucleus synthesized. Now not only the initial nucleus is unknown but all the daughter decay products as well. That is why any event composed of sequential α decays with definite characteristics Q_α and T_α bears no direct information on the mass and charge of the initial nucleus. This means that one can not limit himself with just one experiment. At the same time the use of a very rare isotope of ^{48}Ca makes every experiment very expensive.

Further below a possible program of experiments on the synthesis of superheavy nuclei is suggested.

1. The experiment $^{232}Th(^{48}Ca,xn)^{280-x}110$ can be informative enough to determine the cross section of *EVR* formation in the reactions $(^{48}Ca,xn)$. At $x = 4$ in a short chain of α decays of nuclei $^{276}110 \xrightarrow{\alpha} {}^{272}108 \xrightarrow{\alpha} {}^{268}106$ the condition $T_{s.f.} \gg T_\alpha$ is satisfied. The final even-even nucleus $^{268}106$ ($N = 162$) according to calculations [8] will have $Q_\alpha = 8.0$ MeV and $T_\alpha \sim$ several hours. The isotope $^{264}104$ ($N = 160$) undergoing spontaneous fission with $T_{s.f.} \sim 10$ s will be its decay product. The $^{268}106$ nuclei can be extracted from the target by radiochemical methods: their consecutive $(\alpha - s.f.)$ decay can be registered with a high sensitivity. Taking into account the high intensity of the internal beam of the U-400 accelerator and the possibility to use a "thick target", one can achieve here a high sensitivity of the experiment ($\sigma_{min} \leq 1$ pb).

2. The reaction $^{244}Pu(^{48}Ca,xn)^{292-x}114$ is most effective for the progress in the direction of a maximum excess of neutrons in the nucleus $Z = 114$. The largest cross section is expected for channels $x = 3,4$ (figure 5). Since the calculations of nuclear properties have been done for even-even isotopes it is interesting to consider the case $x = 4$.

In the chain $^{288}114 \xrightarrow{\alpha} {}^{284}112 \xrightarrow{\alpha} {}^{280}110$ the ratio $T_{s.f.} > T_\alpha$ is fulfilled for the initial nucleus. For the isotope $^{284}112$ it is already $T_{s.f.} \sim T_\alpha$ and for $^{280}110$ $T_{s.f.} < T_\alpha$ For a short chain $(\alpha - s.f.)$ or, in a better case of $(\alpha - \alpha - s.f.)$ a conclusion that a decay of a super heavy nucleus takes place here, is determined to a large extent by fission characteristics of $^{280}110$. The fission of such an exotic nucleus may have unusual properties (high kinematic energy of fragments, symmetrical mass distribution, manifestation of neutron shells $N = 82$ and etc.). But at the implantation of the recoil nucleus into the front detector at a large depth the spectroscopy of fission fragments becomes rather problematic.

3. At the same time, if one uses as a target a lighter isotope ^{242}Pu leading to the production of the nucleus $^{286}114$ ($N = 172$), than the chain of consecutive α–decays gets longer until $Z = 104$. Note, that all the isotopes of the chain of daughter nuclei $112 \xrightarrow{\alpha} 110 \xrightarrow{\alpha} 108 \xrightarrow{\alpha} 106 \xrightarrow{\alpha} 104$ may be obtained in the reaction $^{238}U(^{48}Ca,xn)^{286-x}112$.

Thus, the first cycle of experiments on the synthesis of new superheavy nuclei includes the irradiation of ^{232}Th targets with a beam of ^{48}Ca (radiochemical extraction of the *EkaW* fraction and of-line detection of the decay of a $^{268}106$ nucleus) as well as the irradiation of ^{238}U and ^{244}Pu targets (in the on-line mode on kinematic separators).

Both the facilities – VASSILISSA (analog of SHIP at GSI) and the Dubna GFRS will be used in these experiments.

The Production of a ^{48}Ca ion beam

This is probably the key point of the problem of synthesizing new nuclei. The goal is to achieve the maximum intensity of the ^{48}Ca ion beam at the minimum expenditure of this rare and expensive isotope.

On the U-400 heavy ion cyclotron with an internal plasma source (PIG) there was obtained an ion beam of $^{48}Ca^{6+}$ with an intensity of about 0.1 pμA at the expenditure of the initial matter of ~3mg/h.

This result is unsatisfactory for long-term irradiations to achieve a beam dose of $\geq 10^{19}$.

To increase the intensity 5–10 times and decrease the expenditure of ^{48}Ca with a subsequent recuperation of the matter it is necessary to change radically the principle of production and acceleration of high charge ions.

In 1995–96 there was created an external ion source of ECR type and a channel of ion beam injection into the center of the U-400 accelerator. We assume that the new source ECR-4M will enable us to achieve an extracted beam intensity of ~0.5 pμA at the ^{48}Ca expenditure of ~0.5 mg/h (over 50% of the matter could than be extracted from the source chamber).

Supposedly, by the end of 1996 the technical part of this work will be completed and by mid 1997 we shall be able to start first experiments on the synthesis of superheavy elements.

Conclusion

Experimental investigations on the synthesis and study of properties of faraway transactinide elements confirm the predictions of macro-microscopic theory on the existence of closed shells in the region of heavy deformed nuclei. It has been demonstrated experimentally that nuclear structure plays a decisive role in the stability of superheavy nuclides.

Based on the experimental confirmation of the main provisions of the theory and after the introduction of a necessary correction into the calculation there have been predicted the properties of heavier nuclides in the region of spherical shells $Z = 114$ and $N = 180–184$. Here a substantial increase in the stability of nuclei is also expected.

All the nuclei synthesized by now, were obtained in fusion reactions with a formation of a compound nucleus, the transition of which to the ground state takes place with the emission of neutrons and gamma-rays.

Both the reactions of cold and hoot fusion of nuclei can be used for the synthesis of new nuclei. Nevertheless, new experimental data on the fusion mechanism are required, since a number of theoretical descriptions of the fusion dynamics of complex nuclear systems need a substantial reviewing. One can assume that the reactions of the type ^{244}Pu, ^{248}Cm + ^{48}Ca are still within the current potential of the accelerators and experimental technique. These potential, nevertheless, is still to be implemented.

Acknowledgments

In the process of preparing this report I had interesting and fruitful discussions with my colleagues Yu.A.Lazarev, V.G.Utyonkov, A.Eremin, B.Pustylnik, M.Itkis, A.Popeko as well as with A.Sobiczevsky, Z.Hoffman, G.Muenzenberg, P.Moeller and Y.Abe. I would like to express my deepest gratitude to them.

I also think it to be my duty to thank the Alexander Humbold Fund which did not only allowed me to ave intense contacts with my colleagues in Germany but stimulated also the cooperation between our Institutes.

I am grateful to E.Cherepanov and E.Schukina for their assistance in the preparation of this report and to V.Merzljakov for the translation of the Russian variant into English.

References

1. W.D. Myers and W.J. Swiatecki, *Nucl. Phys.* **81**, 1 (1966).
2. W.D. Myers and W.J. Swiatecki, *Ark. Fys.* **36**, 343 (1967).
3. M. Brack et al, *Rev. Mod. Phys.* **44**, 320 (1972).
4. V.V. Pashkevich, *Nucl. Phys.* A **169**, 275 (1971).
5. J.R. Nix, *Annu. Rev. Sci.* **22**, 65 (1972).
6. Z. Patyk et al, *Nucl. Phys.* A **533**, 132 (1991).
7. Z. Patyk and A. Sobiczewski, *Phys. Lett.* B **256**, 307 (1991).
8. R. Smolanczuk and A. Sobiczewski in *Proc. on XV Nucl. Phys. Conf. LEND-95*, St.Petersburg, April 18-22, 1995, World Sci., p.313.
9. E.K. Hulet et al, *Phys. Rev.* C **40**, 770 (1989).
10. P. Moeller, J.R. Nix and W.J. Swiatecki, *Nucl. Phys.* A **469**, 1 (1987); *Nucl. Phys.* A **492**, 349 (1989); P. Moeller and J.R. Nix, *Nucl. Phys.* A **549**, 84 (1992).
11. Yu.Ts. Oganessian in *Classical and Quantum Mechanical Aspects in Heavy Ion Collision, Lecture Notes in Physics* (Springer, Heidelberg, v.33, 1975).

12. G. Münzenberg, *Rep. Progrl. Phys.* 51, 57 (1988).
13. Yu.Ts. Oganessian et al, *JINR Rapid Communications* No 1[75]-96, p.123.
14. To be published.
15. S. Hofmann et al, *Z. Phys.* A **350**, 277 (1995).
16. S. Hofmann et al, *GSI-Nachricten* 02-95 (Darmstadt, 1995), p.4.
17. J.P. Blocki et al, *Nucl. Phys.* A **459**, 145 (1986).
18. P. Fröbrich, *Phys. Lett.* B **215**, 36 (1988).
19. D. Berdichevsky et al, *Nucl. Phys.* A **499**, 609 (1989).
20. M. Itkis et al, in *Proc. on XV Nucl. Phys. Conf. LEND-95*, St.Petersburg, April 18-22, 1995, World Sci., p.177.
21. J.O. Newton, *Sov. J. Part. Nucl.* 21, 321 (1990).
22. D. Hilscher and H. Rossner, *Preprint HMI-92-P23* Hil.-1, Berlin, 1992.
23. D.J. Hinde et al, *Nucl. Phys.* A **472**, 318 (1987).
24. E.M. Kozulin et al, *JINR FLNR Sci. Rep. 1991-92*, JINR Rep. E7-93-57, Dubna, 1993, p.70.
25. T. Sikkeland, J. Maly and D.F. Lebeck, *Phys. Rev.* **169**, 1000 (1968).
26. A.N. Andreyev et al, *Z. Phys.* A **345**, 389 (1993).
27. A.N. Andreyev et al, *Z. Phys.* A **344**, 225 (1992).
28. Yu.A. Lazarev et al, in *Proc. of Intern. School-Seminar on Heavy Ion Physics*, Dubna, 1993 (JINR E7-93-274, Dubna, 1993), vol.2, p.497.
29. Yu.A. Lazarev et al, *Phys. Rev. Lett.* **73**, 624 (1994).
30. Yu.A. Lazarev et al, *Phys. Rev. Lett.* **75**, 1903 (1995).
31. Yu.A. Lazarev et al, *JINR Preprint E7-95-552*, Dubna, 1995, submitted to *Phys. Rev.* C (in press).
32. S. Hofmann et al, *Z. Phys.* A **354**, 229 (1996).
33. S.Hofmann in *Proc. on XV Nucl. Phys. Conf. LEND-95*, St.Petersburg, April 18-22, 1995, World Sci., p.305.
34. Y. Abe et al, *Preprint Yukawa Institute Kyoto YITR-95-18*, December 1995.
35. B.I.Pustylnik et al, *Nucl. Phys.* NP A **553,** 735 (1993).
36. S.N. Timokhin et al, *JINR Rapid Communications* No.4[61]-93, Dubna, 1993, p.60; *J. Radioanal. Nucl. Chem. Letters* **212**, 31 (1996).
37. M. Schädel et al, *GSI Scientific Report 1995*, GSI 96-1, Darmstadt, 1996, p.10.

RELATIVISTIC BEAMS OF EXOTIC NUCLEI - A NEW ACCESS TO NUCLEAR STRUCTURE [a]

Gottfried Münzenberg

GSI Darmstadt, Planckstr. 1 64291 Darmstadt, Germany

Recent advances in experimental techniques for the investigation of exotic nuclei at the very limits of nuclear stability: the separation in-flight, 4π-spectroscopy, and the trapping of exotic nuclei are reviewed with respect to the new possibilities of physics with relativistic exotic nuclei. Perspectives and possible new developments in nuclear structure research leading into the next centennium will be discussed.

1 Introduction

Progress in physics is largely connected to progress in technical delvelopments and advanced experimental setups. They either permit the proof of theoretical predictions, or - what is even more exciting - give new insights in nature and are a challenge to improve and develop new theoretical models.

In recent years the new physics with relativistic beams of exotic nuclei already led to a number of interesting discoveries such as neutron halos and skins, revealing new phenomena observable in matter far off stability with an unbalanced neutron-to-proton ratio [1]. In this contribution a few examples of recent experimental developments in the field of nuclear structure research with relativistic heavy ions will be discussed.

2 Separation In-flight

The most successful method to acess the limits of nuclear stability is the separation in-flight [2]. It allows the detection of exotic nuclei produced with small rates as low as one atom per day and halflives down to microseconds. An example the separation of heavy-ion fusion products is the velocty filter SHIP [3]. sucessfully used for the discovery of the six heaviest elements known today: Ns, Hs, Mt, 110, 111, 112 [4,5,6] and the groundstate proton emission [7] (Fig. 1).

In-flight separation of relativistic exotic nuclei, produced by fragmentation of heavy ions in peripheral collisions, with the GSI projectile fragment separator [8,9] led to the discovery of the two doubly magic nuclei ^{100}Sn [10] and ^{78}Ni [11]. The ^{78}Ni was created by a new and powerful method to produce neutron rich isotopes, the fission of relativistic nuclei in flight, induced in peripheral

---

[a]For Walter Greiner to his 60th birthday

collisions. With this method more than hundred new isotopes were produced [12,13].

The clean separation and identification of nuclei in-flight allows to determine gross properties such as halflive, decay energy or even mass from the observation of few atoms. "Physics with single atoms", originally developed for the heaviest elements [14] is one of new tools to acess the very limits of groundstate nuclear matter. Fig. 2 shows the example of the discovery of ^{100}Sn on the basis of seven atoms produced by fragmentation of ^{129}Xe projectiles at 1095 A GeV together with the β-decay chain originating from a single tin atom. The halflife based on the decay of six atoms is $T_{1/2} = (0.94 \pm^{0.54}_{0.27})$ s, in good agreement with current predictions [15].

Nuclear reactions carried out with relativistic exotic nuclear beams of energies far above Coulomb barrier are a new tool to acess nuclear structure. In the first pioneering experiments nuclear matter radii were measured [16]. These experiments exhibited an abnormally large interaction cross-section for ^{11}Li, which could be interpreted in terms of a neutron halo. Recently also the existence of the proton halo could be proven for the nucleus ^{8}B [17].

The observation of nuclear halos and skins gave new impact on nuclear models. In the frame of a modern theory proton and neutron matter radii of the oxygen isotopes were calculated with the Skyrme Harteree-Fock and the relativistic mean field approaches, respectively [18]. Both models predict a significantly enhanced neutron skin towards the dripline. Experiments to verify these predictions are under way [19]. Up to now only the sodium isotopic chain has been studied [20], exhibiting a neutron skin for the neutron rich isotopes around A=30.

3 Nuclear Reactions and 4-π Spectroscopy

Secondary-beam experiments of the next generation require more complex setups to investigate nuclear reactions in complete kinematics. The in-flight separator serves as the source of relativistic isotopic beams fed into beamlines leading to various experimental setups. Such secondary-beam facilities are GANIL (Caen, France), MSU (East Lansing, USA), RIKEN (Tokio, Japan), and GSI (Darmstadt, Germany). The the GSI secondary beam facility is an example of a well equipped secondary beam course [21]: The fragment separator FRS with a dedicated area for decay spectroscopy, an injection line for secondary beams into the Experimental Storage Ring ESR, and the line to the experimental setups in the target hall: the ALADIN magnet with the LAND Large Area Neutron Detector, and the magnetic spectrometer KAOS.

The neutrons emitted from the breakup of relativistic halo nuclei reflect

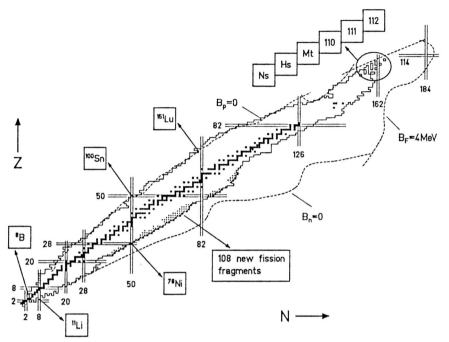

Figure 1: The present status of the chart of nuclei with the recent discoveries.

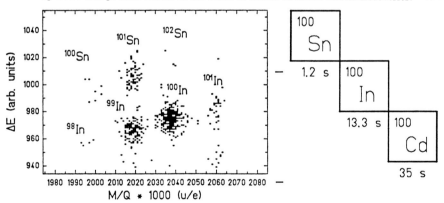

Figure 2: In-flight identification of ^{100}Sn and the β decay chain of a single ^{100}Sn atom .

the momentum distributions of the halo neutrons [22]. Nuclear breakup gives direct acess to the squared halo wavefunction. The neutron momenta proceed from the nuclear matter distributions by Fourier transformation. The density is related to the binding energy by the decay constant of the wavefunction.

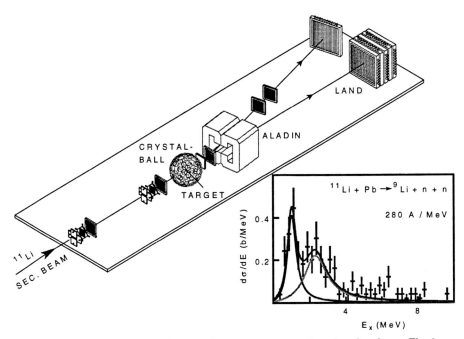

Figure 3: The GSI setup for the complete measurement of nuclear breakup. The inset diaplays the invariant mass pectrum of the ^{11}Li breakup at 280 AMeV.

Fig. 3 displays the GSI setup for complete 4π spectroscopy of breakup reactions [23]. The secondary beam hits the target located inside the Heidelberg Crystal Ball serving as a γ-calorimeter. The heavy fragments from the nuclear breakup are tracked by multiwire chambers and identified by the ALADIN magnet and scintillation detectors. The neutron momenta are measured using the LAND detector. The inset shows the measured sequential breakup of the ^{11}Li nucleus. Two unbound states at 1.1 MeV and 2.3 MeV, respectively are observed [24]. This type of setup allows the spectroscopy of nuclei at and beyond the driplines, e. g. halo nuclei ^{6,8}He, ^{11}Li, ^{11,14}Be, ^{17,19}B, unbound nuclei ^{5,7}H and ^{10}He, and nuclear clusters. It has been used for a first experiment of in-beam γ spectroscopy to investigate neutron rich isotopes in the sodium region [25].

4   Stored Relativistic Exotic Nuclei

The application of ion traps led to large progress in high-precision experiments. Storage rings are such traps of large acceptance for energetic particles. In the

GSI experimental storage ring instable nuclei in exotic ionic chargestates can be stored and cooled by electron cooling. First generation experiments are the direct mass measurements of exotic nuclei. The large acceptance of the heavy-ion storage ring allows to store and investigate a great number of nuclides at the same time, even in various ionic chargestates. Large mass regions can be scanned easily, important for a comparison to nuclear theories. In a recent experiment gold and bismuth fragments, separated by the FRS and injected into the ESR were investigated. About 100 new masses were measured. By combination with exisitng Q_α values parts of the proton dripline around radium could be fixed [26].

Fig. 4 shows an example of a section of a Schottky spectrum taken in the ESR to demonstrate the high resolution of this method. The groundstate of ^{52}Mn and its isomer, separated by only 378 KeV are well resolved. It was possible to measure the decay of the circulating beams to obtain the halflives from groundstate and isomer [27]. Experiments with exotic nulei, bare or as H-, He-, or Li-like systems to vary the electron density at the nuclear surface permit a new kind of of β decay experiments, relevant also for astrophysics, and will give new information on fundamental quantities such as the Fermi function. The high sensititivity of the Schottky noise analysis of cooled heavy-ion beams

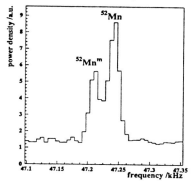

Figure 4: Direct mass measurement of the ^{52}Mn groundstate and isomeric state [27]

allowed to detect a single tungsten ion coasting in the ESR [28].

Experiments of the next generation will include high resolution nuclear reaction studies using the high phasespace density of the cooled beams using. As the internal gastarget of the ESR has a density of the order of 10^{13} atoms/cm^2 is crossed with the beam revolution rate of 10^6f pairing the effective density is 10^{19} atoms/s^{-1} cm^{-2}. The low target density allows the investigation of nuclear reactions such as (p,p'), (d,p), or Coulomb breakup at high precision, not disturbed by atomic collisions.

5 Secondary-Beam Physics at the Turn of the Millenium

Already the first generation of experiments with relativistic beams of exotic nuclei gave new access to the structure of nuclei far off stability. Reaction studies with instable beams complement the main source of information at present, the nuclear decay. The clean separation in-flight applicable to energetic nuclei permits to obtain basic nuclear properties from few atoms.

New experimental methods, such as the 4π-spectroscopy of γ-rays and particles in beam and off beam combined with in- flight identification of the heavy residues will give detailed nuclear structure information with highest possible sensitivity. Stored and cooled exotic nuclei are the key to high precision experiments such as direct mass measurements and decay studies of exotic nuclei in exotic nucler states. Reaction studies of highest resolution with cooled stored beams are coming up.

Energetic beams of instable nuclei produced in peripheral nuclear collisions suffer from low intensity of presently order of 10^8 particles/s. To obtain stronger beams and higher energies the already exisiting facilities are upgraded. New schemes of high-intensity stable beam acceleration to produce strong, energetic fragment beams are under discussion [29]. To achieve even higher secondary beam intensities, new secondary beam factories with accelerated secondary beams are coming up such as the SPIRAL at GANIL, the REX-ISOLDE at CERN, the Oakridge facility OREB, and the EXCYT at Catania [1]. The already operating ARENAS facility at Louvain-la-Neuve has demonstrated the feasibility of this approach. The GSI facility including the SIS and ESR will not only deliver the highest energies but also cover the largest enery range of all existing and planned facilities from 2 AGeV down to the Coulomb barrier.

The research with exotic nuclei aims to push the systematic studies of gross nuclear properties towards the very limits of stability. Detailed investigations of specific nuclei in shell- or deformend regions and of N=Z nuclei would yield new information on the hitherto not well predictable limits of the existence of nuclear matter determined by pairing and shell quenching of weakly bound systems far-off stabilty [30], the nuclear forces in unbalanced nuclear matter, and the proton-neutron residual interaction. The study of asymmetric, low-density matter in the nuclear skins and halos allows detailed studies of in-medium effects Specific for the weakly bound nuclei is the continuum-state interaction. For heavy nuclei the filling of large shells with large angular momenta is under investigation. Not yet discussed is the existence of cluster-regions or even - islands beyond stability and the possibility to implant strangeness into exotic nuclei to create of exotic hypernuclei [31] (Fig. 5)

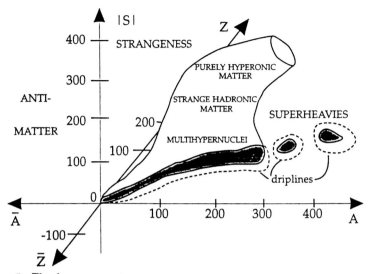

Figure 5: The future extension of the nuclear chart into the direction of antimatter and adding the new dimension of strangeness[31].

As we already have learned from the first nuclear absorption experiments, the nuclear mass, stability, and deformation does not yield sufficient information about the gross properties of the nucleus. The prediction of density distributions of neutrons and porotons is a stringent test for the nuclar models. Charge radii from laser-measured isotope shifts give relative values. The measurement of charge distributions of exotic nuclei by electron scattering with a small 100 MeV heavy-ion - electron collider seems feasible[32,33]. The luminosities achievable seem sufficient for such experiments[34]. Such a small electron ring could alternatively be filled with low energetic fully stripped instable nuclei and in alternatively serve as a trap-target to investigate efficiently nuclear reactions between exotic nuclei.

With the discovery of element 112 the magic proton number 114 has already been closely approached[6]. The magic neutron number N=184 seems to be unaccessible with the experimetnal methods used at present. Certainly this techniqe would open up new perspectives to approach the center of the spherical superheavy nuclei.

References

1. H. Geissel et al., Ann. Rev. Nucl. Part. Sci. 45(1995)163
2. G. Münzenberg in: Handbook of Nuclear Decay Modes, Eds. D. N. Poenaru and W. Greiner, De Gruyter, in press
3. G. Münzenberg et al., Nucl. Instr. and Meth 161(1979)65

4. P. Armbruster, Ann. Rev. Nucl. Part. Sci. 35(1985)135
5. G. Münzenberg, Radiochimica Acta 70/71(1995)193
6. S. Hofmann, this conference
7. S. Hofmann et al., Z. Phys. A305(1982)111
8. G. Münzenberg et al., Proc. First International Conference on Radioactive Nuclear Beams, Berkeley Cal. 1989, World Scientific 1990, 91
9. H. Geissel et al., Nucl. Instr. Meth B70(1992)286
10. R. Schneider et al., Z. Phys A348(1994)241
11. Ch. Engelmann et al., Z. Phys. A352(1995)351
12. M. Bernas et al., Phys. Lett. B331(1994)19
13. S. Czajkowski et al., Proc. Int. Conf. on Exotic Nuclei and Atomic Masses, ENAM95, Arles 1995, Editions Frontieres Gif-sur-Yvette, france, in press
14. G. Münzenberg et al., Z. Phys. A315(1984)145
15. R. Schneider et al., Phys. Scr. T56(1995)67
16. I. Tanihata et al., Phys. Lett. 160B(1985)380
17. W. Schwab et al., Z. Phys. A350(1995)284
18. C. Rutz, private communication and thesis Univ. Frankfurt
19. I. Tanihata et al., proposal for an experiment at the GSI projectile fragment separator
20. T. Suzuki et al., Phys. Rev. Lett 75(1995)3241
21. G. Münzenberg et al., Nucl. Instr. Meth B70(1991)265
22. P. G. Hansen et al. Ann. Rev. Nucl. Part. Sci. 45(1995)591
23. H. Emling, Priv. Comm. 1996
24. D. Aleksandrov et al., GSI Ann. Rep. 1995
25. P. Reiter et al., GSI Ann. Rep. 1995
26. Th. Radon et al., GSI annual report 1995 and in preparation
27. H. Irnich et al., Phys. Rev. Lett. 75(1995)4182
28. H. Schlitt et al., GSI annual report 1995 and in
29. N. Angert, private communication te
30. W. Nazarewicz et al., priv. comm. 1995
31. W. Greiner, Int. Journ. of Mod. Phys. E 5(1996)1
32. The K4-K10 project, a Dubna project study, Scientific Report 1989-1990, Dubna 1991, 217
33. The RIKEN upgrade, proposal, RIKEN 1995
34. I. Meshkov, private communication and G. Schrieder, private communication

NEW INSIGHTS INTO THE FISSION PROCESS: NEUTRON MULTIPLICITES, HYPERDEFORMATION, CLUSTERING, AND NUCLEAR STRUCTURE

J. H. HAMILTON[1], G. M. TER-AKOPIAN[1,2], A. V. RAMAYYA[1],
YU. TS. OGANESSIAN[2], A. V. DANIEL[1,2], J. KORMICKI[1],
B.R.S. BABU[1], S. J. ZHU[1,3,4], M. G. WANG[3], T. GINTER[1],
J. K. DENG[1,3], W. C. MA[5], G. S. POPEKO[2], Q. H. LU[1],
J. O. RASMUSSEN[6], S. ASZTALOS[6], I. Y. LEE[6], S. Y. CHU[6],
K. E. GREGORICH[6], A. O. MACCHIAVELLI[6], M. F. MOHAR[6],
S. PRUSSIN[7], M. A. STOYER[8], R. W. LOUGHEED[8], K. J. MOODY[8],
K. J. MOODY[8], J. F. WILD[8], J. D. COLE[9], R. ARYAEINEJAD[9],
Y.X. DARDENNE[9],M. W. DRIGERT[9], R. DONANGELO[10],
A. SANDULESCU[1,2]

[1]Vanderbilt University, Physics Department, Nashville, TN 37235
[2]Joint Institute for Nuclear Research, Dubna 141980, Russia
[3]Tsinghua University, Physics Department, Beijing, P.R. China
[4]Joint Institute for Heavy Ion Research, Oak Ridge, TN 37831
[5]Mississippi State University, Physics Department,
Mississippi State, MS 39762
[6]Lawrence Berkeley National Laboratory, Berkeley, CA 94720
[7]Univeristy of California/Berkeley, Berkeley, CA 94720
[8]Lawrence Livermore National Laboratory, Livermore, CA 94550
[9]Idaho National Engineering Laboratory, Idaho Falls, ID 83415
[10]Universidade Federal Do Rio de Janeiro, Brazil
[11]Institute für Theoretische Physik, der Universitat Frankfurt/Main,
D-60054 Germany
[12]Institute of Atomic Physics, Bucharest, P.O. Box MG-6, Romania

Abstract

Prompt $\gamma - \gamma - \gamma$ coincidence studies following spontaneous fission of ^{252}Cf were carried out at Gammasphere. Yields and neutron multiplicities were measured directly for Sr-Nd, Zr-Ce, Mo-Ba, Ru-Xe, and Pd-Te correlated pairs. Strong enhancements of the 7-10 neutron emission channels were seen only in Mo-Ba data. A new fission mode with enhanced 7-10 neutron yields and much lower total kinetic energy was discovered only in ^{108}Mo- ^{144}Ba, ^{107}Mo-^{145}Ba, and/or ^{106}Mo-^{146}Ba pairs. Analysis indicates one or more of ^{144,145,146}Ba are hyperdeformed at scission with 3:1 axis ratio. This new mode may arise from a cluster effect, for example, ^{106}Mo-$(^{12,14}$C + ^{132}Sn$)^{146}$Ba at scission or a neutron-ball cluster between the Mo-Ba fragments. States up to 20^+ were observed in ^{148}Ce and 19^- in ^{144}Ba. New level structures include a variety of new, identical bands and other structures.

Neutron Multiplicities, Hyperdeformation, and Clustering

Recently, we reported first direct measurements of the yields and neutron multiplicities of correlated fragment pairs of Zr-Ce and Mo-Ba in the spontaneous fission (SF) of ^{252}Cf (ref. 1). Yields for zero to ten neutron emission channels were determined for Mo-Ba pairs. The 8-10 neutron multiplicities for Mo-Ba pairs are considerably enhanced compared to those for gross (total) neutron multiplicities[2]. The enhancement of yields at high neutron multiplicity could be an indication of a second fission mode. Two or more fission modes were predicted theoretically[3-6] for the SF of the same nucleus and such are now known for a number of heavy nuclei[7,8]. To gain a better understanding of these enhancements for the Mo-Ba pairs, we have carried out further detailed analyses of the experimental yields. From our earlier data[9] and from our Gammasphere data[10] the neutron multiplicities for the odd-A and even-A Zr-Ce (40/58) and Mo-Ba (42/56) correlated pairs and for the next three strongest pairs of Sr-Nd (38/60), Ru-Xe (44/54), and Pd-Te (46/52) were extracted. As seen in Fig. 1, only the Mo-Ba pairs have strongly enhanced yields for 7-10 neutron emission. From both sets of data[9,10] we find that there are two fission modes in the breakup into Mo-Ba: first, the normal mode with large total kinetic energy (TKE) and broad mass distribution of primary fragments, and, second, a mode with much lower TKE and mass distribution limited to one or more of three ^{108}Mo-^{144}Ba, ^{107}Mo-^{145}Ba, ^{106}Mo-^{146}Ba pairs where at scission at least one or all of ^{144,145,146}Ba are hyperdeformed[9,10].

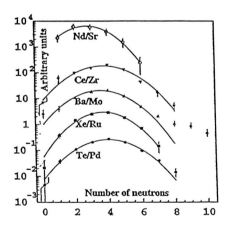

Fig. 1. Relative yields of the multiplicity distribution of prompt neutrons for ^{252}Cf.

The first experimental arrangement and data analysis procedures were described in our earlier paper[1]. In our second experiment, a ^{252}Cf source was placed at the center of Gammasphere with 36 detectors. The neutron multiplicity distributions extracted for all five pairs are shown in Fig. 1. The curves in Fig.1 are Gaussian fits to the distributions. A single Gaussian curve fits the full range of all the distributions except for the Mo- Ba correlated pairs. Note that only the Mo-Ba pairs show a strong enhancement for the 7-10 neutron emission yields. We emphasize again that these data are the first direct measurements of such yields and neutron multiplicities. Yields of the Mo-Ba fragment pairs obtained from our analysis of $\gamma - \gamma$ and $\gamma - \gamma - \gamma$ coincidence data are given in Table 1. To illustrate the quality of the data in Table 1, Fig. 2 gives examples of double gates on the $251.3((7/2^-) \longrightarrow (5/2^-))$-602.3 $(2 \longrightarrow 0)$ keV transitions in ^{103}Mo-^{140}Ba (the 9 neutron channel) and a background gate on 270-602.3 keV. Spectrum (a) clearly shows the lower and higher energy transitions in ^{103}Mo-^{140}Ba to definitively establish the 9 neutron channel, and (b) shows these are not from background coincidences. Similar spectra were analyzed for the other high neutron multiplicity channels.

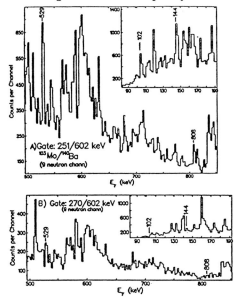

Fig. 2. Double gates on (a) ^{103}Mo/^{140}Ba, and (b) background/ ^{140}Ba. The ^{103}Mo/^{140}Ba peaks both disappear or are greatly reduced in double gates with one on a background to either side of either gate transitions in (b).

The yields extracted from the γ-rays emitted by the secondary fragments following neutron evaporation (Table 1) carry information about the mass and excitation energy distributions of the primary fission fragments of corresponding fixed charge splits. The unfolding of the extensive Mo-Ba data allowed us to extract the distribution $Y(A_L, E_L^*, A_H, E_H^* \mid Z_L, Z_H)$. A least squares fit of the calculated yields $Y_i^{calc}(A_L', A_H')$ to the pattern of the experimental yields after neutron evaporation $Y_i(A_L', A_L')$ was searched by assuming Gaussian forms for the mass and excitation-energy-distributions of primary fission fragments. If one assumes that only a single Gaussian fits the primary fragment mass distribution and the excitation energy distribution, no satisfactory fit can be obtained for the Mo-Ba data. However, a good fit to the data is obtained when we assume that two distinct fission modes contribute to the formation of the primary Mo-Ba fission fragments. Assuming a Gaussian distribution for the experimental yields after neutron evaporation $Y_i(A_L', A_H')$, a least squares fit was carried out between the experimental data and the calculated yields. Assuming that only one Mo-Ba primary fragment pair contributes to the second fission mode, good fits to the data in Table 1 were obtained. Very reasonable fits were obtained by assuming that the single primary fragment pair responsible for the second mode is ^{108}Mo-^{144}Ba or ^{107}Mo- ^{145}Ba, or ^{106}Mo-^{146}Ba. For each of these primary fragment pairs, essentially the same value of <TKE> = 153 ± 3 was found for the second fission mode as a result of the unfolding procedure. For the first fission mode <TKE> = 189 ± 1 MeV was obtained independent of which one of the above three pairs contributes to the second mode. Other parameters of the first mode also did not depend on the choice of the pair contributing to the second mode.

Table 1. High neutron multiplicity Yields of correlated Mo-Ba pairs in SF of ^{252}Cf from Gammasphere data. The yields are per 100 SF events normalized to the peak yields of Wall[11].

	^{138}Ba	^{140}Ba	^{141}Ba	^{142}Ba	^{143}Ba	^{144}Ba
^{102}Mo		0.042(2)	0.05(1)	0.05(1)	0.02(1)	0.05(1)
^{103}Mo		0.09(2)	0.15(3)	0.10(2)	0.20(7)	0.64(7)
^{104}Mo	0.07(3)	0.11(1)	0.19(2)	0.32(2)	0.49(5)	0.94(2)
^{105}Mo		0.11(2)	0.10(2)	0.71(7)	1.34(14)	1.21(12)
^{106}Mo		0.12(1)	0.39(4)	1.02(3)	1.13(11)	0.57(6)
^{107}Mo		0.11(2)	0.14(3)	0.23(2)	0.78(9)	1.18(2)
^{108}Mo		0.13(2)	0.14(2)	0.12(2)	0.18(2)	<0.01
ΣY	0.07(3)	0.71(5)	1.16(7)	2.55(9)	4.14(22)	3.59(15)

The ^{106}Mo-^{146}Ba primary pair gives the best fit to the 7-10 neutron multiplicities (see Fig. 3). However, the experimental data do not establish unambiguously which of the above three fragment pairs or their combinations provide for the existence of the second mode. However, two distinct fission modes in the case of the Mo-Ba split of ^{252}Cf are established unambiguously.

Mode 1 looks like the familiar fission mode of ^{252}Cf in that its principal characteristics ($<$TKE$>$, σ_{TKE}, A_H, σA_H) and excitation energies of the primary fragments are close to those that were known before [11]. The Mode 2 appears quite different because of its low $<$TKE$>$ and low Coulomb barrier. The low $<$TKE$>$ suggests an enormous elongation of the ^{252}Cf nucleus at its scission with a very high deformation of the fragments emerging after scission.

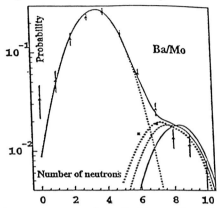

Fig. 3. Fit for Mode 2: 146/106 (left), 145/107 (center), and 144/108 (right). The X's are the difference between the data and the Gaussian fit to Mode 1.

From their large excitation energies, the barium fragments are those that are highly deformed in Mode 2. From their excitation energies, one can estimate at scission the ratios of the axes as: a/b = 2.8, 3.0 and 3.2 for ^{144}Ba, ^{145}Ba and ^{146}Ba, respectively. So, it is very probable that at scission, in Mode 2, the barium fragment or fragments are hyperdeformed with axis ratio of most likely 3:1! In the case when the pair ^{106}Mo- ^{146}Ba emerges at Mode 2, ^{106}Mo has about the same excitation energy as is the case for the first mode, and thus a "normal" deformation. The excitation energy of the Mo fragment increases considerably for the ^{107}Mo- ^{145}Ba pair and becomes comparable with that of the Ba fragment if the ^{108}Mo-^{144}Ba pair occurs in Mode 2.

It is possible that some reasonable modifications of the hypotheses underlying our unfolding procedure could alter somewhat the numerical results. However, we believe that the yields and multiplicity data of Mo-Ba necessarily lead to the conclusion that there are two distinct modes occurring in the Mo-Ba fission of the ^{252}Cf nuclei, and that apparently Mode 2 is not present in the other fragment pairs as seen in Fig. 1. Quite independent of the details, one can make the following general observations about the fragments: (1) in Mode 2 the Ba fragments are left in an unusually highly excited state from which 5-8 neutrons are evaporated, (2) the fragments have unusually low $<$TKE$>$ and, hence, unusually low Coulomb energy at scission which means the charge centers are much further apart, features consistent with hyperdeformation in the Ba fragment(s).

Recently Ćwiok et al.[12] looked at hyperdeformation and clustering in actinide nuclei. Systematic calculations of potential energy surfaces of the even-even Rn, Ra, Th and U isotopes were performed by using the shell- correction approach with the axially-deformed average Woods-Saxon potential. Their calculations yielded third minima in the PES, characterized by large elongations, $\beta_2 \sim 0.9$ and significant reflection asymmetry, $0.35 < \beta_3 \leq 0.65$, in addition to the ground state ($\beta_2 \sim 0.25$) and fission isomer ($\beta_2 \sim 0.6$) minima in several actinide nuclei. Rutz et al.[13] investigated ^{240}Pu, ^{232}Th and ^{226}Ra in a relativistic mean-field model and likewise found a third minimum for ^{232}Th at a hyperdeformed shape. Ćwiok et al.[12] suggested that the mass distribution of fission fragments should be greatly influenced by the structure of the HD minimum and the third saddle point. They also note that the observation of HD states in the actinides constitute an important confirmation of the shell structure of the nuclei. The unusual stability of these HD states is attributed to strong shell effects that are present in the average nuclear potential at shapes with 3:1 axis ratio.

In the same way[12] calculations were carried out for ^{252}Cf (ref. 9). The PES for ^{252}Cf is shown in Fig. 4 where β_2 extends between the second minimum and the outer barrier.

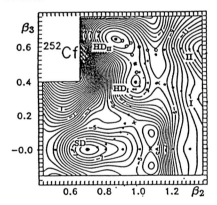

Fig.4. Potential energy surface for ^{252}Cf

There is a well-developed third minimum at $\beta_2 \sim 0.9$ and $\beta_3 \sim 0.7$ with the paths to scission shown. The static path for HD$_{II}$ is shown in Fig. 5 along with the strongly reflection asymmetric shape of ^{252}Cf in the third minimum. In the upper right-hand corner is the calculated nuclear shape just beyond point B where R is ^{146}Ba. It would be fascinating to see directly the HD minimum in ^{252}Cf.

It is possible that the HD, extracted for one or more ^{144,145,146}Ba fragments at scission, comes through the decay of this minimum.

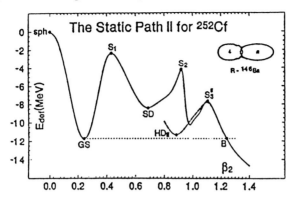

Fig. 5. Potential energy curve for ^{252}Cf as a function of β_2 along the static fission path II. The calculated shapes of ^{252}Cf in the minimum $HD_{II}(\beta_2 \sim 0.9, \beta_3 \sim 0.65)$ and at $\beta_2 = 1.4$ are shown together with the corresponding shapes of the left(L) and right(R) fragments.

Recently Donangelo et al.[13] repeated the analysis of the yields, and reached the same conclusion; that there is a new Mode 2 and that at scission one or more of ^{144,145,146}Ba is hyperdeformed with most likely ^{146}Ba. They further carried out analysis in which it is argued that the structure of the system at its scission point is that of a three cluster system, ^{106}Mo-^{14}C-^{132}Sn or $(^{12}$C+2n) [Fig. 6]

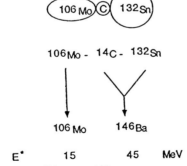

Fig. 6. Illustration of ^{106}Mo-C-^{132}Sn cluster configuration at scission and the reabsorption of ^{14}C to yield ^{106}Mo-^{146}Ba with excitation energies, E*, as shown.

At scission the carbon is reabsorbed to give the high excitation of the resulting ^{146}Ba fragment in the Mo-Ba split. In turn, the highly excited hyperdeformed

^{146}Ba leads to the large number of neutrons observed. Another possible origin of the anomalous high neutron multiplicities and low total kinetic energy of Mode 2 has been made by Greiner and Sandulescu. Their idea is that at scission the Coloumb barrier is low because of the formation of a neutron ball; for example, with a magic spherical shell model number of 8 neutrons between the Mo-Ba pairs. The ball then breaks up to yield the high neutron multiplicities. To test the ^{14}C-^{132}Sn idea, we have searched for coincidences between γ-rays emitted by ^{106}Mo and by ^{132}Sn which would be an alternate new ternary fission mode. We have clearly established α-ternary fission[14]. Our $\gamma - \gamma - \gamma$ data provide tentative evidence for ^{106}Mo-^{132}Sn coincidences. We expect to repeat our experiment with improved statistics at the full Gammasphere when it is available. We also will seek to study the high neutron multiplicities angular distribution to try to test the neutron ball hypothesis.

In conclusion, the observed coexistence of two fission modes in the spontaneous fission of ^{252}Cf involves a new type of bi-modal fission. For the Mo-Ba division of ^{252}Cf, the same fragments, ^{144}Ba, ^{145}Ba and/or ^{146}Ba, appear either with a "standard" or enormously high excitation energy in the first and second fission modes, respectively. Thus, ^{144}Ba, ^{145}Ba and/or ^{146}Ba are found in two states which are remarkable for their very different deformations at scission. The "normal" fission mode has features typical of the bulk of fission events of ^{252}Cf, whereas the second abnormal mode revealed here for the first time provides evidence for a hyperdeformed state or states in ^{144,145,146}Ba with $\simeq 3{:}1$ axis ratio. A third deep HD minimum in the ^{252}Cf PES could be the origin of the HD Ba-Mo split.

Identical Ground State Bands in $^{98,100}Sr$, $^{108,110}Ru$, $^{144,146}Ba$, $^{156-160}Sm$ *and Octupole States in* $^{144,146}Ba$

New, higher spin states in ^{98,100}Sr, ^{100,102}Zr are observed from SF. These have been used to extract values for the downward shifts in the energies of the superdeformed 0^+ ground states in ^{98}Sr, ^{100}Zr as a result of the interaction with the nearby 0_2^+ near-spherical states as observed in ^{74,76}Kr (ref.f 15). From plots of $\Delta E_\gamma = E_{\gamma 2} - E_{\gamma 1}$, the 0_1^+ energies are seen to be shifted down by 11(2) and 21(3) keV in ^{98}Sr and ^{100}Zr, respectively[16] with little, if any, shifts observed in ^{100}Sr and ^{102}Zr. These results are less than half the energy shifts of 23.3 and 46.9 keV derived from lifetime data in these nuclei[17] to call into question the energy shifts extracted in that work.

Lhersonneau et al.[18] noted the $4^+ \longrightarrow 2^+$ and $6^+ \longrightarrow 4^+$ transitions in ^{98,100}Sr are essentially identical. The levels of ^{98,100}Sr recently were extended to 10^+ (ref. 16). The $10^+ \longrightarrow 8^+$ to $2^+ \longrightarrow 0^+$ transition energies in ^{98,100}Sr are 688.6, 565.1, 433.0, 289.0, and 144.3 and 690.8, 567.2, 435.0, 287.2, and 130.2 keV, respectively. The new $8 \longrightarrow 6$ and $10 \longrightarrow 8$ transition energies are likewise identical. Moreover, if you subtract the 11 keV shift in the 0_1^+ energy, the $2^+_1 \longrightarrow 0^+_1$ energy in ^{98}Sr is 133(2) keV which is essentially identical with the same transition in ^{100}Sr. These are the first known identical bands seen in even-even nuclei in this low mass region. The identical nature of the bands in these two nuclei may be associated with the saturation of the collective motion immediately following the sudden onset of collectivity[18]. However, since the N=60,62 ^{100,102}Zr nuclei with similar large deformation do not have similar identical energies (The $10^+ \longrightarrow 8^+$ to $2^+ \longrightarrow 0^+$ transition energies in ^{100,102}Zr are 739.1, 625.0, 497.0, 352.0 and 212.6; and 756.6, 630.1, 486.1, 326.2, and 151.8 keV respectively), it may be that it is associated with the strength of the reinforcement of the proton and neutron shell gaps at the same deformation[19]. In some way, this reinforcement at Z=38, N=60 and 62 to make superdeformed double magic ^{98,100}Sr is different than that at Z=40, N=60,62. The ^{108,110}Ru nuclei (Fig. 7) provide another example of identical ground bands from 2^+ to 10^+ (ref. 20). Here N=66 is at the midshell where saturation of collectivity is expected. However, N=68 ^{112}Ru has larger J_1 moments of inertia (and presumably deformation). Its E_γ and J_1 are significantly different from those of ^{108,110}Ru.

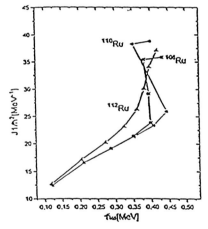

Fig. 7. J_1 vs. ω (ref. 16).

The neutron-rich Ba (Z=56), Ce (Z=58), Nd (Z=60) and Sm (Z=62) with A $\geq$ 142 provide important insights into the role of prolate quadrupole deformation and the role of stable octupole deformation in ^{148,150}Ce as N increases from the 82 closed shell. The energy levels for neutron-rich Ba, Ce, Nd were extended as high as 20^+ (ref. 21). Their moments of inertia (J_1) are shown in Fig. 9. One sees striking differences in how J_1 and J_2 change with increasing spin and N for these nuclei. For $^{142-148}$Ba, there are sizable jumps in J_1 between N = 86-88 and 90-92, but N = 88-90 have nearly identical J_1 and J_2 values that oscillate around each other up to spin 8^+. A strong upbend in J_1

occurs only in ^{146}Ba above 10^+ with only gradual upbends in ^{144,148}Ba. These crossings are likely associated with the alignment of $\nu i_{13/2}$ or $\pi h_{11/2}$ pairs. Perhaps more surprising, in N = 88,90 ^{144,146}Ba where the ground bands are essentially identical, the negative-parity bands that are intertwined with the positive-parity bands as a result of stable octupole deformation likewise are essentially identical with $\Delta E_\gamma /\ E_\gamma \leq 2\%$, $\Delta J_1 \mid J_1 \leq 2.4$ over a larger spin range from 3^- to 15^- and $\Delta J_2/J_2 \leq 2.9\%$ to 11^-, while J_1 and J_2 are much smaller for N=86 ^{142}Ba as seen in Fig. 8. This is the first report of essentially identical negative- parity bands with stable octupole deformation. However, for $^{144-150}$Ce and the new isotope, ^{152}Ce (ref. 21), there are no sharp changes in their J_1 values and no identical bands for N = 86-94 (Fig. 8). The $J_1 for^{148}$Ce is different and increases very rapidly with spin to cross J_1 for ^{150}Ce.

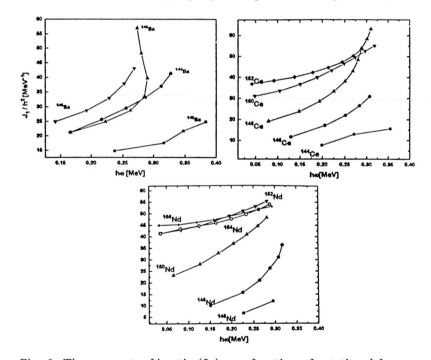

Fig. 8. The moments of inertia (J_2) as a function of rotational frequency $(\hbar\omega)$ for Ba, Ce and Nd nuclei.

The levels of $^{156,158}_{62}$Sm$_{94,96)}$ and recently discovered $^{160}_{62}$Sm$_{98}$ (ref. 22) are all characteristic of well-deformed shapes with a small, smooth decrease in the energies of the transitions at spins 2^+ to 14^+ as A increases (ref. 16). It is surprising to find three successive nuclei with such very similar E_γ, J_1 and J_2 moments of inertia (eg., constant differences, 2-4% for 156-158 and 3-4bands. Saturation of collectivity at midshell has been invoked to explain many identical bands. Clearly, the identical ground state bands and identical negative-parity bands for N = 88,90 $^{144,146}_{56}$Ba, and the absence of such identical bands in the similar N cerium nuclei and the new type identical bands in $^{156-160}$Sm, all of which are far from midshell in either neutrons or protons, present an interesting challenge for a theory of identical bands. Likewise, the quite different ways in which J_1 and J_2 change with with spin and with N from 86-94 in the Ba, Ce, Nd and Sm nuclei present another challenge.

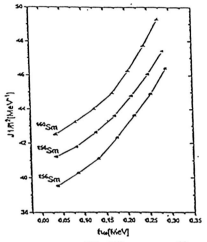

Fig. 9. J_1 for $^{156-160}$Sm nuclei[22].

Acknowledgement

G.M. T.-A. and A.V. D. express appreciation for the hospitality and financial support from Vanderbilt. Work is supported at Vanderbilt, INEL, LBNL and LLNL by the U.S. Dept. of Energy under grant and contracts DE-FG05-88ER40407, DE-AC07-76ID01570, DE-AC03-76SF00098, and W-7405-ENG48.respectively, and at the Joint Institute for Nuclear Research by grant 94-02-05584-a of the Russian Federal Foundation of Basic Sciences. The Joint Institute for Heavy Ion Research is supported by its members, U. of Tennessee, Vanderbilt U. and the U.S. DOE through contract DE-FG05-87ER40361 with the U. of Tennessee. Work at Tsinghua U. was partially supported by the National Natural Science Foundation of China.

References:

1. G.M. Ter-Akopian, et al., Phys. Rev. Lett. **73** (1994) 1477.
2. J.F. Wild et al., Phys. Rev. **C41** (1990) 640.
3. V.V. Pashkevich Nucl.Phys. **A161** (1971) 275; **A477** (1988) 1.

44

4. U. Brosa, S. Grossman, and A. Müller, Z. Nat **41a** (1986) 1341.

5. P. Möller, J.R. Nix, and W.J. Swaitecki, Nucl. Phys. **A492** (1989) 349.

6. U. Brosa, S. Grossman, and A. Müller, Phys. Rep. **197** (1990) 167.

7. M.G. Itkis et al., Z. Phys. **A320** (1985) 433.

8. E.K. Hulet et al., Phys. Rev. Lett **56** (1986) 313.

9. G.M. Ter-Akopian et al., Phys. Rev. Lett. (June 1996 to be published).

10. J.H. Hamilton et al., *private communication from Gammasphere collaboration, to be published.*

11. A.C. Wahl, At. Data Nucl. Data Tables **39** (1988) 1.

12. S. Cwiok et al., Phys. Lett **B322** (1994) 304.

13. R. Donangelo, J.O. Rasmussen, M.A. Stoyer and J.H. Hamilton *to be published.*

14. A.V. Ramayya et al., *elesewhere in these proceedings.*

15. R.B. Piercey et al., Phys. Rev. Lett. **47** (1981) 1514.

16. J.H. Hamilton et al., Prog. In. Part. Nucl. Phys. **35** (1995) 635.

17. H. Mach et al., Phys. Lett. **B230** (1989) 21.

18. G. Lhersonneau et al., in Spect Heavy Nuclei, IOP Conf No. **105** (1990).

19. J.H. Hamilton *in Treatise on Heavy Ion Science* Vol. **8**, Allen Bromley, ed., New York: Plenum Press (1989), pp. 2-98.

20. Q.H. Lu et al., Phys. Rev. **C52** (1995) 1348.

21. S.J. Zhu et al., J. Phys. G Nucl. Part. Lett. **21** (1995) L75.

22. S.J. Zhu et al., J. Phys. G Nucl. Part. Lett. **21** (1995) L57.

HIGHER DEFORMATIONS OF SUPERHEAVY NUCLEI

GUNAR HERING AND P. PAUL

Department of Physics, State University of New York at Stony Brook,
Stony Brook, NY 11794, USA

We present a phenomenological procedure using the $N_p N_n$ scheme that allows determination of β_4 and β_6 deformations of transactinide nuclei from the strength of the $2_1^+ \rightarrow 0_1^+$ transition alone. A sample experiment is outlined, and the $N_p N_n$ systematics is applied to the existing theoretical values to demonstrate that the shell closure at Z=126 and N=184 determines the nuclear deformations of transactinide nuclei.

1 Introduction

Since the late 1960's [1] it has been predicted that the island of superheavy nuclei would occur near a proton number Z=114. This prediction has survived all refinements until very recently, and the most recent shell correction calculations of Moeller and Nix [2] still predict an island of high stability at Z=114 and N=184, with a ridge of semi-stable nuclei connecting to the actinides. The programmatic search [3] for transactinides with long lifetimes, at GSI and elsewhere, has shown these predictions to be qualitatively correct by producing elements in the msec lifetime range up to [4] element 112. However, as they are approaching the nucleon numbers of the predicted shell closure, at least at the proton side, the observed lifetimes fail to show the predicted large increase in lifetime. It is in this context that one is tempted to search for an additional check on the predictive power of the shell correction calculations in transactinide nuclei.

The ridge of stability connecting actinides and the superheavy peak is thought to arise from the shell effect of (relatively) large higher-order deformations, β_4 and β_6. Experimental observation of such higher-order deformations would thus be an independent test of the shell correction calculations. In this paper we outline the possibilities of such measurements and the difficulties that have to be overcome. In considering possible experiments we have also established systematic trends that cast doubt on the validity of the Z=114 shell gap, except perhaps for a very narrow range of neutron numbers.

2 Phenomenological method

Higher-order deformations of ground states in deformed nuclei are usual obtained from inelastic scattering (see, e.g. [5]. This requires the use of stable or

45

long-lived targets. Beyond Z=100 such targets are not available and in fact even for Cf (Z=98) no measurement has been made. The heavier elements can only be produced in fusion-evaporation reactions, with very small cross sections, and therefore, another procedure must be used to measure, or infer, the exit higher-order deformations. We have recently shown[6] that the BE(2) value measurement of the $2^+_1 \rightarrow 0^+_1$ transition in the rotational ground-state band alone is sufficient to deduce the absolute value and the sign of β_4 if certain systematics are accepted. This is based on the definition of $B(E2)$

$$B(E2; 2^+ \rightarrow 0^+) = \frac{e^2}{16\pi} \cdot Q^2_{20} \tag{1}$$

in terms of the ground-state quadrupole moment Q_{20}

$$Q_{20} = \frac{3ZR^2_0}{\sqrt{5\pi}} \cdot \beta_2 \big(1 + 0.3604\beta_2 + 0.9672\beta_4 + 0.3277\frac{\beta^2_4}{\beta_2} + 0.321\frac{\beta^2_6}{\beta_2} +$$
$$0.954\frac{\beta_4\beta_6}{\beta_2} + 0.346\beta_2\beta_6 \tag{2}$$

(in the following we drop the identifier of the $2^+_1 \rightarrow 0^+_1$ transition.) This equation contains the effect of volume conservation and differs from the exact solution by at most 3%. It is clear that the β_4 and β_6 deformations influence Q_{20} with large coefficients. We can remove this influence by defining a *quadrupole* $B(E2)_Q$ (assuming for this discussion $\beta_6 = 0$, without loss of generality)

$$B(E2)_Q \equiv B(E2) \cdot \big(\frac{Q(\beta_4 = 0)}{Q(\beta_4 \neq 0)}\big)^2 \tag{3}$$

where $Q(\beta_4 \neq 0)$ is given by equation 2, and $Q(\beta_4 = 0)$ is the effective quadrupole moment with $\beta_4 = 0$. Equation 3 gives the $B(E2)$-value pertaining to a nucleus with the *same* quadrupole deformation, but no hexadecupole deformation.

It had been shown earlier[7] that the $B(E2)$-values of ground state bands over a large range of nuclei follow a compact, but non-linear, trajectory when the experimental values were emperically normalized by the nuclear mass number and plotted against the product $N_p N_n$, i.e. the number of valence neutrons and protons (or holes) counted from the respective nearest complete shell closure. For a comparison of $B(E2)$ values over a wide range of nuclei we express the transition strength in Weisskopf units:

$$B(E2)(W.u.) = \frac{25}{9} \frac{4\pi}{(r_0 \cdot A^{\frac{1}{3}})^4} B(E2)(e^2 fm^2) \tag{4}$$

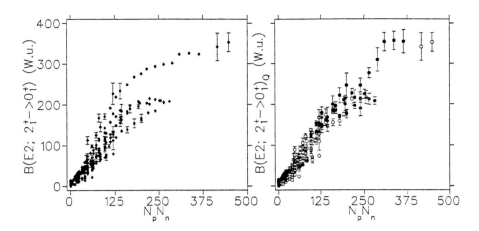

Figure 1: Left side: Experimental $B(E2)$ values plotted against $N_p N_n$ for all nuclei with $Z > 30$; The saturation of the rare-earth nuclei and the Actinides is clearly evident; right side: The same plot but now for the *quadrupole* $B(E2)_Q$.

The left side of Fig. 1 shows *all* available $B(E2)$-values for nuclei from Z=30 to 98 plotted against $N_p N_n$. The plot displays considerable scatter. The right side of the same figure then shows the same plot for the *quadrupole* $B(E2)_Q$ defined in equation 3. The result is remarkable: all the data collapse into a narrow almost linear band. (We now believe that the non-linearity seen in[7] was due to the effect of higher-order deformations). The most striking feature is that the $B(E2)_Q$ values below mid-shell fall exactly on the same curve as those above mid-shell for all major shells. This amazing feature is understood in the schematic model of Bertsch[8] in which orbitals get filled according to Harada's principle[9], i.e. according to their closeness to the body-fixed symmetry axis, with equatorial orbits filled first. Although this schematic model assumes that neutrons and protons fill the *same* major shell it can be extended to the case where neutron and proton orbitals are filled independently. By using the $N_p N_n$ scheme this is naturally incorporated. The two bends seen near $N_p N_n \approx 250$ and 375 occur when the major shells are filled to the middle (in the $N_p N_n$ scheme one counts particles or holes, depending on the nearest shell closure) and those orbitals have only a small influence on the deformation. The success of the $N_p N_n$ scheme is due to the fact that a *strong* n-p residual interaction drives the nuclear deformation.

We now concentrate on the actinide and transactinide nuclei ($Z > 82$) which are plotted in Fig. 2. For these nuclei the $N_p N_n$ scheme works especially well and in fact the *quadrupole* $B(E2)_Q$ shows no saturation, at least

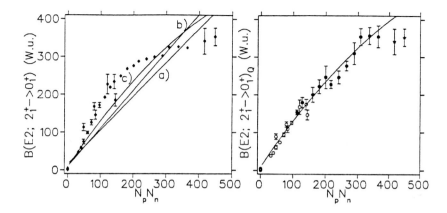

Figure 2: Left side: $N_p N_n$ plot of all known $B(E2)$ values for $Z \geq 82$ nuclei; right side: the same for the *quadrupole* $B(E2)_Q$. The superimposed lines are fits with parameters deduced from the rare-earth region (see text).

up to $N_p N_n \approx 400$. (Shell closures of Z=114, N=184 produce $N_p N_n = 464$ at mid-shell) The last two data points belong to ^{250,252}Cf and have *not* been corrected for a β_4 deformation since no data are available. At this point the plot takes on predictive power and allows us to extract information about the size and sign of β_4 if we assume identical deformations for neutron and proton distributions. One simply has to find that value for β_4 which brings the points onto the systematic line. For points that lie below the line before correction, such as is the case for Cf, β_4 will come out negative. The procedure can be tested against the cases where β_4 is known from experiment. Fig. 3 shows β_4-values extracted from experimental $B(E2)$'s in the actinide region and a systematic line obtained from fits to the rare-earth region. Values obtained with three different fitting procedures (see figure caption) are compared to the experimental β_4-values. The most sophisticated fit which includes the saturation effects agrees very well with the measured β_4-deformations and supports the predictive power of the procedure. In particular they show that β_4 changes its sign near $N_p N_n = 260$. The evident sign of the Cf deformations would extend that trend further. The quantitative extraction of β_4 for ^{250,252}Cf is the subject of an upcoming publication.

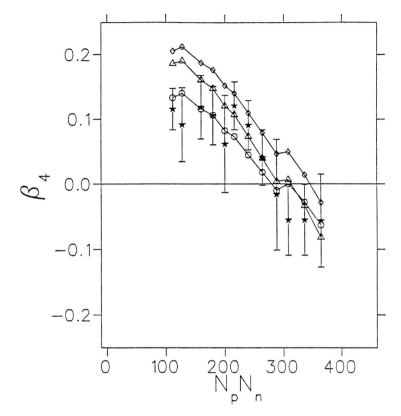

Figure 3: β_4-values for actinide nuclei obtained from three different fits to the $B(E2)_Q$ systematics and compared to the experimental values ($\star$). $\triangle = \beta_4$-values extracted from a linear fit including all $B(E2)_Q$'s in the rare-earth shell; $\diamond = \beta_4$-values assuming a linear fit to the $B(E2)_Q$'s excluding those showing saturation at mid-shell; $\circ = \beta_4$-values extracted by using parameters from a quadratic fit to all rare-earth $B(E2)_Q$-values (including saturation region).

3   Conceptual experiments

Large negative hexadecupole deformations have been predicted by recent shell correction calculations [10][12] for nuclei from Z=102 to Z=114. The procedure outlined in the previous section allows to measure these deformations. from the lifetime of the $2_1^+ \rightarrow 0_1^+$ transition alone. However, even this is not an easy experiment, because the energies of the first excited states typically have energies near 40 keV and, because of the high Z, are very highly converted.

The mean life of a rotational 2^+-state in theses nuclei is given by

$$\tau = 40.82 \times 10^{13} \cdot \frac{1}{(E/keV)^5} \cdot \frac{1}{(B(E2)/e^2b^2)} \cdot \frac{1}{(\alpha + 1)} (ps) \qquad (5)$$

To extract the β_4-value the $B(E2)$-value has to be measured with less than 10% error.

We use the ^{250}Fm nucleus as an example for typical numbers. This isotope has an $N_p N_n$-value of 432 assuming that the relevant proton shell closure occurs at Z=126. The N_n-value is below the neutron mid-shell and does not depend on the particular choice of next magic neutron number (i.e. N=178 or N=184). Using the scaled quadratic fit for the *quadrupole* $B(E2)_Q$ value in the actinide region gives an estimate $B(E2)_Q \approx 369$ W.u. for ^{250}Fm corresponding to a deformation $\beta_2 = 0.278$. Next, the β_4-value can be obtained from the systematics using the quadratic rare-earth fit (see Fig. 3). One finds $\beta_4 \approx -0.09$. These estimates lead to a $B(E2)=313$ W.U.$=29.3$ e^2b^2. This value is about 10% smaller than what one would expect from the an extrapolation of the experimental $B(E2)$ systematics (see Fig. 2 because of the large negative β_4-value. The energy of the 2^+-state (40 - 45 keV) allows L and M shell electron conversion. From conversion tables [13] one obtains $\alpha_L = 860 - 680$, and $\alpha_M = 536 - 185$, and a total conversion coefficient $\alpha_{tot} = 1400 - 850$. Thus the lifetime for the first E2 transition in ^{250}Fm is $\tau \approx 100$ ps. This makes such a lifetime measurement difficult, especially if the conversion electrons are observed. These have energies of about 15 keV (L-shell) and 37 keV (M-shell). ^{250}Fm can be produced in a reaction such as ^{16}O $+$ ^{238}U $\rightarrow$ ^{254}Fm $\rightarrow$ ^{250}Fm $+$ 4n with a maximal cross section [12] of about 1.5 μb. Although quite difficult, such experiments are feasible.

4   Determination of the next shell closures from the $N_p N_n$-scheme.

In principle, the $N_p N_n$ scheme systematics allows determination of the subsequent shell closures since one counts particles and holes from the nearest shell closure. If the number is derived from a wrong closure, the $B(E2)_Q$ systematics will be destroyed. So far we based the proton shell closure on the next proton magic number Z $= 126$. However, if there are substantial subshell gaps in the single particle energies the counting of N_P and N_N can be ambiguous. Such subshell gaps can be produced as a consequence of a strong n-p interaction, especially when the involved neutron and proton orbitals are spin-orbit partners with high spatial overlap. It has been shown from the E_{2+} and E_{4+}/E_{2+} systematics that recounting N_p and N_n from known (i.e. correct) subshell gaps leads to a restoration of the systematics, e.g. in the case of the

Z=64 subshell for neutron numbers N< 90. A 2.4-MeV subshell gap at Z=114 is a common feature [2] of all calculations of superheavy elements. We can now apply the idea of the previous section, namely, that in the $N_p N_n$ scheme the $B(E2; 2_1^+ \to 0_1^+)$ is symmetric around mid-shell, to the theoretical values in order to see whether a Z=114 subshell gap satisfies the systematics. For this purpose we chose among the many calculations those of Patyk and Sobiczweski [10]. The regular B(E2) values are obtained from their tabulated equilibrium deformations $(\beta_2, \beta_4, \beta_6)$ by use of equation which are then converted into the $B(E2)_Q$ values by using equation and assuming $\beta_4 = \beta_6 = 0$.

In constructing plots for the theoretical B(E2) values two scenarios, assuming proton shell closures at Z = 114 and Z = 126 for the counting of N_p numbers are shown in the left and right side of Figure 4, respectively. In each case various plausible neutron shell closures at N=162, 1164, 178 and 184 were assumed.

One observes that for the assumed Z=114 shell closure the $B(E2)_Q$-values follow a complex pattern with the values above mid-shell falling on a different branch from those below mid-shell, for all assumed neutron shell closures. In contrast the assumption of a closed proton shell at (or near) Z = 126 produces a tight curve for $B(E2)_Q$ for $N_n = 184$. Thus we conclude that the nuclear structure of the superheavy elements is determined by the shell gaps at Z = 126, N = 184, and that any Z = 114 gap is of a special nature, perhaps existing only for a small range of neutron numbers. Recent calculations of superheavy ground-state properties using a Hartree-Fock method [14] question the existence of a Z=114 shell gap altogether.

5 Summary

We have shown that the $B(E2)$ vs. $N_p N_n$ systematics of the $2_1^+ \to 0_1^+$ - tranmission in transactinide nuclei could be useful for first nuclear structure experiments that test the shell correction calculations in detail In this respect we note that the lifetimes are determined by α-decay which is not a very sensitive test of the calculated nuclear structure. The systematics also shows that the presumed Z=114 subshell gap is of peculiar character, perhaps similar to the Z=64 subshell gap which exists only for certain neutron numbers. The mere fact that the $N_p N_n$ scheme works so well demonstrates that the n-p interaction is strong which produces the dependency of the proton shell gap on neutron numcbers. If the Z=114 gap exists only for a small range of neutron numbers, the experimental search for the magic ultra-long lived superheavy nucleus might be unsuccessful in the end, at least in regular fusion-evaporation reactions.

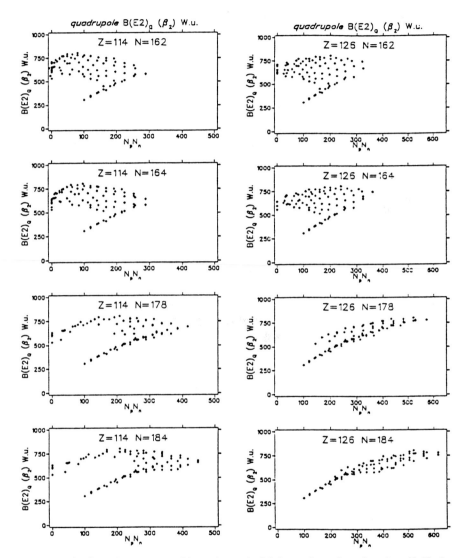

Figure 4: $B(E2)_Q$ values computed from theoretical deformations plotted against N_pN_n for nuclei with $92 \leq Z \leq 114$. The N_p values at the left side were computed from a spherical shell closure at Z= 114, while those at the right side assumed shell closure at Z=126. In each case four possible neutron shell closures at N = 162, 164, 178 and 184 are considered.

Acknowledgments

We dedicate this paper to the 60th birthday of Professor Walter Greiner.
Extensive discussions with N. V. Zamfir and R. F. Casten on this subject are gratefully acknowledged.This work was supported by the U.S. National Science Foundation.

References

1. U. Mosel and W. Greiner, Z. Phys. **222**,261 (1969)
2. P. Möller, J. R. Nix, W. D. Meyers, and W. J. Swiatecki, Atom. Data and Nucl. Data Tables **95**, 185 (1995)
3. G. Münzenberg, Rep. Prog. Phys. **51**, 57 (1988)
4. S. Hofmann et al., Z. Phys. A**350**, 277 and 281 (1995)
5. A. Guterman, D.L. Hendrie, P.H. Debenham, et al., Phys. Rev. **C39**, 440 (1985)
6. N.V. Zamfir, G. Hering, R. F. Casten, P. Paul, Phys. Lett. **B357**, 515 (1995)
7. R. F. Casten and N. V. Zamfir, Phys. Rev. Lett. **70**, 402 (1993)
8. G. F. Bertsch, Phys. Lett.**26B**, 130 (1968)
9. K. Harada, Phys. Letters **26B**, 81 (1968)
10. Z. Patyk, and A. Sobiczewski, Nucl. Phys. **A533**, 132 (1991)
11. S. Cwiok, S. Hofmann and W. Nazarewicz, Nucl. Phys. **A573**, 356 (1994)
12. N. Shinohara, S. Usuda, et al., Phys. Rev. **C34**, 34 (1986)
13. R. S. Hager and E. C. Seltzer, Nucl. Data **A4**, 1 (1968)
14. S. Cwiok, J. Dobaczewski, P.-H. Heenen, P. Magierski, W. Nazarewicz, JIHIR Report 96-02 (1996)

Dedicated to Professor Walter GREINER's 60th Anniversary

CLUSTER DECAY MODES
AND THE BEST COLD FISSIONING NUCLEUS

D.N. POENARU, R.A. GHERGHESCU

Institute of Atomic Physics,

P O Box MG-6, RO-76900 Bucharest, Romania

^{14}C, ^{20}O, ^{23}F, $^{24-26}$Ne, ^{28,30}Mg and ^{32,34}Si cluster radioactivities have been experimentally observed in Oxford, Orsay, Moskow, Berkeley, Geneva, Dubna, Argonne, Vienna, Milano, and Darmstadt in good agreement with our predictions. ^{12}C radioactivity of the proton-rich ^{114}Ba has been also detected. The neutron-rich ^{264}Fm should be the best cold-fissioning nucleus, owing to the strong shell effect of the doubly magic fragments ^{132}Sn. The optimum cold fission path in the plane of two independent shape coordinates, the separation distance R and of the radius of the light fragment R_2, is determined. The partial half-life for cold fission is estimated to be several orders of magnitude shorter than that of α-decay.

1 Introduction

In 1996 we celebrate the 100th anniversary of the discovery of radioactivity by H. Becquerel. Three different natural decay modes (α, β, and γ) have been known until 1940, when the spontaneous fission has been observed for the first time. A rich variety of other nuclear decay modes have been discovered and studied during the last two decades.[1,2] *Cluster radioactivities* are intermediate phenomena between fission and alpha decay, in which a parent nucleus $^A Z$ breaks apart into two fragments: the emitted cluster $^{A_e} Z_e$ and the daughter $^{A_d} Z_d$. The spontaneously emitted light fragment is a small nucleus (like ^{14}C, ^{24}Ne, etc) heavier than α particle, but lighter than the lightest fission fragment. A growing interest for these processes, has been manifested since 1984, when the first experiment on ^{14}C decay of ^{223}Ra has been reported,[3] confirming the predictions[4] of 1980. Our earlier works unifying the theory of cold fission, cluster radioactivities, and α-decay, as well as other theoretical models and experimental results, have been recently reviewed.[1,5]

Like for spontaneous fission or alpha decay, the basic explanation relies on the quantum mechanical tunneling through a potential barrier, which initially has been developed by Gamow (and independently by Condon and Gurney) in 1928. Theoretically any nucleus with $Z > 40$, for which the released energy is a positive quantity, $Q > 0$, can be a cluster emitter, but practically there is a severe selection imposed by the available techniques, requesting a short-enough halflife and a large enough branching ratio. The main experimental difficulty comes from necessity to select few rare events from an enormous background of alpha particles. Solid state track detectors (which are not sensitive to alpha particles), or magnetic spectrometers (in which alpha particles are deflected by a strong magnetic field), have been used to overcome this difficulty. The superconducting spectrometer SOLENO, at I. P. N. Orsay has been employed since 1984 to detect and identify the ^{14}C radioactivity. Its good energy resolution has been exploited in 1989 to discover[6] a *"fine structure"* in the kinetic *energy spectrum of* ^{14}C *emitted by* ^{223}Ra. Cluster emission leading to excited states of the final fragments have been considered for the first time in 1986 by Martin Greiner and Werner Scheid.[7] When the fine structure of ^{14}C radioactivity of ^{223}Ra has been discovered it was shown that the transition toward the first excited state of the daughter nucleus is stronger than that to the ground state. The most accurate experiment has been performed by Hourany *et al.*[8] with SOLENO using high quality ^{223}Ra sources implanted at ISOLDE CERN. The interpretation given by Sheline and Ragnarsson, according to which the main spherical component of the deformed parent wave function has a $i_{11/2}$ character, has been confirmed.

Within analytical superasymmetric fission (ASAF) model, we got a large hindrance factor, $H = 40$, for ^{24}Ne decay of ^{233}U (gs $5/2^+$) to the gs $(9/2^+)$ of ^{209}Pb. Other hindered transitions could be: ^{23}F decay of ^{231}Pa (gs $3/2^-$) to the 0^+ gs of ^{208}Pb $(H = 12)$; ^{14}C decay of ^{221}Fr (gs $5/2^-$) to the $1/2^+$ gs of ^{207}Tl $(H = 8.5)$; ^{14}C decay of ^{221}Ra to the $1/2^-$ gs of ^{207}Pb $(H = 9)$, and ^{24}Ne decay of ^{231}Pa to the gs of ^{207}Tl. The ^{14}C transition from ^{225}Ac (gs $(3/2^-)$) to the $(9/2^-)$ gs of ^{211}Bi seems to be not hindered.

The decay constant $\lambda = \ln 2/T$, characterising the well known exponetial law $e^{-\lambda t}$ of variation in time, of the number of parent nuclei, can be expressed as a product of three (model dependent) quantities $\lambda = \nu S P$, where ν is the frequency of assaults on the barrier per second, S is the preformation probability of the cluster at the nuclear surface, strongly dependent on the nuclear structure, and P is the quantum penetrability of the external potential barrier. According to our quasiclassical method, the preformation probability can be calculated within a fission model as a penetrabilty of the internal part of the barrier. Very often a diagram of $\log T$ versus $Q^{-1/2}$ for α emission or cluster radioactivity is called Geiger-Nuttal plot. In this kind of systematics the experimental or calculated points are considerably scattered. One can get one line for a given emitted cluster in a *universal curve* representation[9] $\log T = f(\log P)$. The up to now even-even half-life measurements are well reproduced (within a ratio 3.86, or rms=0.587 orders of magnitude).

2 Predictions and experimental confirmations

Any theory of cluster radioactivities with predictive power should give an answer to the following questions. Are these phenomena physically allowed? Could they be measured? In which region of parent nuclei can they be found? Which are the most probable emitted clusters? What is the order of magnitude of the emission rate? Four theoretical models answering at least some of these questions, have been reviewed by us in 1980, namely: fragmentation theory; penetrability calculations like in traditional theory of α-decay; numerical (NuSAF)- and analytical (ASAF) superasymmetric fission models. Earlier attempts since 1924, arrived to misleading conclusions.

A new superasymmetric peak, due to ^{38}S radioactivity, has been obtained in the ^{252}No fission fragment mass distribution calculation, based on the fragmentation theory and the two center shell model developed by the Frankfurt school. Eight decay modes by cluster emission, ^{14}C, ^{24}Ne, ^{28}Mg, ^{32,34}Si, ^{46}Ar, and ^{48,50}Ca, from sixteen even-even parents, ^{222,224}Ra, ^{230,232}Th, ^{236,238}U, ^{244,246}Pu, ^{248,250}Cm, ^{250,252}Cf, ^{252,254}Fm, and ^{252,254}No, have been predicted in 1980 by calculating the penetrability. Three variants of the numerical superasymmetric fission (NuSAF) models were developed since 1979 by adding to the macroscopic deformation energy of binary systems with different charge densities a phenomenological shell correction term, and by performing numerical calculations within Wentzel-Kramers-Brillouin (WKB) approximation.

A very large number of combinations parent – emitted cluster has to be considered in a systematic search for new decay modes. In order to check the metastability of more than 2000 nuclides with measured masses tabulated by Wapstra and Audi, against about 200 isotopes of the elements with $Z_e = 2$–28, this number is of the order of 10^5. The numerical calculation of three-fold integrals involved in the models mentioned above are too time-consuming. The large amount of computations can be performed in a reasonable time by using an analytical relationship for the halflife. Since 1980, we developed our ASAF model to fulfil this requirement. We started with Myers-Swiatecki liquid drop model adjusted with a phenomenological correction accounting for the known overestimation of the barrier height and for the shell and pairing effects in the spirit of Strutinsky method. Before any other model was published, we have estimated the half-lives for more than 150 decay modes, including all cases experimentally confirmed until now. A comprehensive table was produced by performing calculations within that model. Subsequently, the numerical predictions of ASAF model have been improved by taking better account of the pairing effect in the correction energy, deduced from systematics in four groups of parent nuclei (even-even, odd-even, even-odd and odd-odd). In a new table, published in 1986, cold fission as cluster emission has been included. The systematics was extended in the region of heavier emitted clusters (mass numbers $A_e > 24$), and of parent nuclei far from stability and superheavies. Since 1984, the ASAF model results have been used to guide the experiments. Also other theoretical groups have applied their own models to the cases we found to be interesting.

The main quantities experimentally determined are the partial halflife, T, and the kinetic energy of the emitted cluster $E_k = QA_d/A$. This equation is a direct consequence of the "cold" character of this decay mode – the total kinetic energy of the two fragments practically exhausts the released energy Q, which is shared between the two fragments. Spontaneous emission of ^{12,14}C, ^{20}O, ^{23}F, $^{24-26}$Ne, ^{28,30}Mg and ^{32,34}Si, from nuclei have been experimentally observed in Oxford, Orsay, Moskow, Berkeley, Geneva, Dubna, Argonne, Vienna, Milano, and Darmstadt.[10,11] Good agreement with our predictions have been obtained in an island of trans-francium parent nuclei, where the daughter nucleus is the doubly magic ^{208}Pb or some of its neighbours.

The island of proton-rich cluster emitters with $Z = 56-64$ and $N = 58-72$ with daughters around ^{100}Sn, we predicted since 1984, has been recently confirmed in Dubna and Darmstadt, where the ^{12}C radioactivity of the proton-rich ^{114}Ba has been observed.[12,13] A deeper discussion of the physics we can learn from the experimental results would be possible when the masses, the decay energies and the partial and total half-lives will be determined with a high accuracy. Until then we investigate the influence of some nuclear properties (masses, radii and interaction potentials) on the emission probability.[14]

Alternative approaches, extending either a fission theory to larger asymmetry or the traditional many-body explanation of α-decay to heavier emitted clusters, have been developed by Swiatecki, Fliessbach, Broglia, Bertsch, Buck, Gupta, Liotta, Silisteanu, Florescu, Delion, etc (see ref. 1 and the references therein).

3 Cold fission as cluster radioactivity

A fraction of the released energy (20–40 MeV from about 250 MeV) is spent to deform and excite the fragments in the usual fission phenomenon. They reach final ground states by neutron evaporation and γ-ray emission. Only the remaining part of the Q-value gives the total kinetic energy (TKE) of the fragments. Another process, called *cold fission*, in which the TKE practically exhausts the Q value (no excitation energy and compact shapes at the scission point), has been experimentally observed in two regions of nuclei: U, Np, Pu isotopes, as well as Fm, Md, No and other trans-fermium nuclei. While the new mechanism is very rare in the first group of nuclei, it is rather strong into the second one, giving rise to the *bimodal* character of the fission phenomena for some trans-fermium nuclei.

The unified approach of the three groups of decay modes (cold fission, cluster radioactivities and α-decay) within ASAF model is best illustrated by the example of ^{234}U nucleus, for which all these processes have been measured. Also we have studied many other cold- and bimodal fission processes viewed as cluster decays. For instance since 1986 we have studied the cold fission of ^{264}Fm. According to our calculations, performed within the analytical su- perasymmetric fission (ASAF) model, the cold fission mechanism could be the main decay mode of this neutron-rich nucleus, which has not been produced until now. It also should give the most pronounced symmetrical distribution of fission fragments, owing to the doubly magic character of ^{132}Sn fragments. The half-life of 5.7 μs for a "new fission path" with compact shapes calculated by Möller and Nix in 1994 is not very far from our $log_{10} T(s) = -5.6$.

4 Fission dynamics

Our extensive study of one-, two-, and three-dimensional fission dynamics in a wide range of mass asymmetry allowed us to find the nuclear shapes during the deformation process and to see that some confusions and errors have been made in this field, the most important one as a consequence of ignoring the center of mass motion in the calculation of nuclear inertia within Werner- Wheeler approximation. We found (see for example[15]) that the smoothed-neck influence is stronger for lower mass asymmetry.

In the parametrization of two intersected spheres assumed by ASAF model, the radius of the emitted fragment, R_2, was kept constant. Recently, a two- dimensional dynamics of the cold fission process was studied with both the separation distance R and the radius R_2 as independent variables.[16] Analyt- ical relationships, derived for the components of the nuclear inertia tensor have been obtained. In these calculations, we have used the same Yukawa- plus-exponential potential extended by us in 1979 to fragments with different charge densities. The cold fission character of the fission process under in- vestigation justify a nondissipative approach. In an intermediate stage of the deformation from one nucleus up to two fragments at the touching point, the mass asymmetry can be expressed as $\eta = (V_1 - V_2)/V$, where V_1, V_2, and V are the volumes of the two fragments and of the parent, respectively.

The measurable partial half-life

$$T = \frac{h \ln 2}{2 E_v S P_s} \tag{1}$$

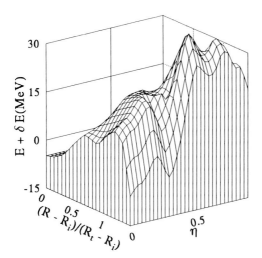

Figure 1: ^{264}Fm potential energy surface function of R and η with shell and pairing corrections included.

is determined by a product of three (model-dependent) quantities: a zero-point vibration energy, $E_v = h\nu/2$ (ν is the frequency of assaults on the barrier, and h is the Planck constant), a preformation probability of the emitted cluster into the parent nucleus, S, and the external barrier penetrability, P_s, given by

$$S = \exp(-K_{ov}) ; \quad P = \exp(-K_s) \tag{2}$$

They are calculated within the Wentzel-Kramers-Brillouin (WKB) approximation, where the action integrals, K_{ov} (for overlapping fragments) and K_s (for separated fragments), are expresed as

$$K_{ov,s} = \frac{2}{\hbar} \int\limits_{R_i,R_t}^{R_t,R_b} [2B(R)E(R)]^{(1/2)}dR \tag{3}$$

The least action trajectory in the plane R, R_2 is a curve $R_2 = R_2(R)$ along which the minimum value of K is obtained. By chosing a trajectory in the plane of the two independent shape coordinates, given in a parametric form: $R_2 = R_2(s)$; $R = R(s)$, and by taking $s = R$, an inertia scalar, $B(R)$, used to calculate the tunneling penetrability along this path is derived

$$B(R) = B_{RR}(R, R_2) + 2B_{RR_2}(R, R_2)\left(\frac{dR_2}{dR}\right)$$

$$+ B_{R_2R_2}(R, R_2)\left(\frac{dR_2}{dR}\right)^2 \tag{4}$$

from the components of the inertia tensor, and the two-dimensional problem is reduced to a single dimensional one. At the touching point and for separated fragments, $B(R)$ is equal to the reduced mass $\mu = mA_eA_d/A$, where m is the nucleon mass. $E(R)$ is the interaction energy of the two fragments from which the Q-value has been subtracted. R_i and R_b are the turning points of the WKB integral, $E(R_i) = E(R_b) = 0$.

Let us consider the region of the plane in which the geometry of two intersected spheres allows physically acceptable pairs of independent variables R, R_2. It is reasonable to assume that a physical solution of R_2 should be found in the range (R_e, R_0), while the corresponding R_1 can take values from R_d to R_0. Then, from the volume conservation and matching conditions, for every R, a certain range of R_2 values is permitted.

The maximum value of $B_{RR}(R, R_2)$ is reached at the touching point configuration where the reduced mass is obtained, $B_{RR}(R_t, R_e) = \mu$. For $R = $constant, B_{RR} decreases when the value of R_2 is increased up to the allowed limit.

It is convenient to choose different laws of variation $R_2 = R_2(R)$ describing a possible sequence of shapes during the fission process. From several tested variation laws $R_2 = R_2(R)$ the following four have been choosen to illustrate the least action principle: second-; third-; fourth order polynomials, and exponential function. These choices lead to a minimum value of the action integral. The best result ($K_{ov}^{min} = K_0$) was found with the second order polynomial law; the three others are less convenient, yielding higher K_{ov} values by 3.1% (third order polynom), 3.6% (exponential function), and 5.7% (fourth order polynom). By adding a correction energy, allowing to reproduce the Q-value estimated from a mass table, the fission barrier is reduced. We get $K = K_{ov} + K_s = 34.597$ and $E_v = 0.141$ MeV, hence $log_{10} T(s) \simeq -5$, very close to the preceding values mentioned above. The potential energy surface (see Fig. 1) with shell and pairing corrections included, $E + \delta E$, shows the typical two minima (ground state and shape isomeric state) at symmetrical split, where the lowest potential barriers lead to the highest yield at symmetry in this case.

For alpha-decay, we estimate the following half-lives: $log_{10} T(s) = 10.77$, 11.00, 12.49, 12.67, 12.86, 13.75, and 14.32, which correspond to $Q_\alpha = 5.905$,

5.865, 5.615, 5.585, 5.555, 5.415, and 5.330 MeV, obtained by taking the masses of the parent ^{264}Fm and of the daughter ^{260}Cf, from different calculated mass tables. One can see that the cold fission process should be several orders of magnitude stronger than the α-decay.

Nuclear physics of near future will shad more light on the exciting aspects of cluster radioactivities and of other nuclear molecular states – offering a unifying point of view for many, thus far disconnected phenomena.

References

1. D.N. Poenaru and W. Greiner, in *Nuclear Decay Modes*, ed. D. N. Poenaru (IOP Publishing, Bristol, 1996), p. 275.
2. W. Greiner, M. Ivaşcu, D.N. Poenaru, and A. Săndulescu, in *Treatise on Heavy Ion Science, Vol. 8*, ed. D.A. Bromley (Plenum Press, New York, 1989), p. 641.
3. H.J. Rose and G.A. Jones, *Nature* **307**, 245 (1984).
4. A. Săndulescu, D.N. Poenaru and W. Greiner, *Sov. J. Part. Nucl.* **11**, 528 (1980).
5. D.N. Poenaru and W. Greiner, in *Handbook of Nuclear Properties*, eds. D.N. Poenaru and W. Greiner (Oxford University Press, Oxford, 1996), p. 131.
6. L. Brillard, A.G. Elayi, E. Hourany, M. Hussonnois, J.F. Le Du, L.H. Rosier and L. Stab, *C. R. Acad. Sci. Paris* **309**, 1105 (1989).
7. M. Greiner and W. Scheid, *J. Phys. G.* **12**, L229 (1986).
8. E. Hourany et al, *Phys. Rev.* C **52**, 267 (1995).
9. D.N. Poenaru and W. Greiner, *Physica Scripta* **44**, 427 (1991).
10. E. Hourany, in *Nuclear Decay Modes*, ed. D.N. Poenaru, (IOP Publishing, Bristol, 1996), p. 350.
11. R. Bonetti and A. Guglielmetti, in *Nuclear Decay Modes*, ed. D.N. Poenaru, (IOP Publishing, Bristol, 1996), p. 370.
12. Yu.Ts. Oganesyan, Yu.A. Lazarev, V.L. Mikheev, Yu.A. Muzychka, I.V. Shirokovsky, S.P. Tretyakova and V.K. Utyonkov, *Z. Phys.* A **349**, 341 (1994).
13. A. Guglielmetti et al, *Nucl. Phys.* A **583**, 867 (1995).
14. D.N. Poenaru, W. Greiner, and E. Hourany, *Phys. Rev.* C **51**, 594 (1995).
15. M. Mirea, D.N. Poenaru and W. Greiner, *Z. Phys.* A **349**, 39 (1994).
16. R.A. Gherghescu, W. Greiner and D.N. Poenaru, *Phys. Rev.* C **52**, 2636 (1995).

COLD FISSION AND ALPHA TERNARY FISSION STUDIES FOLLOWING THE SPONTANEOUS FISSION OF ^{252}CF

A. V. RAMAYYA[1], J. H. HAMILTON[1], B.R.S.BABU[1], S.J. ZHU[1,2,3],
L.K. PEKER[1,4],T. N. GINTER[1], J. KORMICKI[1,5], W. C. MA[6],
J. D. COLE[7], R. ARYAEINEJAD[7], K. BUTLER-MOORE[7],
Y.X. DARDENNE[7]M.W. DRIGERT[7], G. M. TER-AKOPIAN[8],
YU. TS.OGANESSIAN[8], J.O. RASMUSSEN[9], S. ASZTALOS[9], I. Y. LEE[9],
A.O. MACCHIAVELLI[9],S.Y.CHU[9], K.E. GREGORICH[9], M.F. MOHAR[9],
S. PRUSSIN[10], M.A. STOYER[11], R.W. LOUGHEED[11], K.J. MOODY[11],
J.F. WILD[11], A. SANDULESCU[12], A. FLORESCU[12]

[1]Department of Physics, Vanderbilt University, Nashville, TN 37235
[2]Physics Department, Tsinghua University, Beijing, P.R. China
[3]Joint Institute For Heavy Ion Research, Oak Ridge, TN 37831
[4]Brookhaven National Laboratory, Upton(LI), NY 11973
[5]ORISE, Oak Ridge, TN 37831
[6]Physics Department, Mississippi State University, Mississippi State,
MS39762
[7]Idaho National Engineering Laboratory, Idaho Falls, ID 83415
[8]Joint Institute for Nuclear Research, Dubna, 141980, Russia
[9]Lawrence Berkeley National Laboratory, Berkeley, CA 94720
[10]Nuclear Eng. Department, University of California, Berkely, CA 94720
[11]Lawrence Livermore National Laboratory, Livermore, CA 9455
[12] Institute of Atomic Physics, Bucharest, P.O. Box MG-6, Romania
[13] Institute für Theoretische Physik der J.W. Goethe Universität, D-60054,
Frankfurt am Main, Germany

Abstract

The isotopic yields for the cold fission follwing the spontaneous fission of ^{252}Cf have been extracted from the triple γ-coincidence data collected using the Gammasphere at Lawrence Berkeley National Laboratory. Cold fission yields were extracted by analyzing all the events corresponding to a pair and normalizing them to the calculated total yields. The experimental results are compared with the theoretical values calculated using a double folding potential and M3Y nucleon-nucleon forces.

1. Introduction:

The experimental discoveries of new spontaneous decays with emission of clusters like ^{14}C, ^{20}O, ^{23}F, ^{24}Ne, ^{28}Mg and $^{32-34}$S (cluster radioactivity) have confirmed the theoretical predictions based on the idea that there are cold rearrangements of large number of nucleons from one ground state of the A-nucleon system to two ground states of the A_1 and A_2 $(A=A_1+A_2)$ nucleon systems [1]. These new super asymmetric decay modes, intermediate between

α-decay and symmetric fission, can be interpreted as "Pb decays": The spherical double magic nature of the $^{208}_{82}\text{Pb}_{126}$ nucleus(or the neighboring nuclei with $Z \approx 82$ and $N \approx 126$) causes a deep valley in the mass and charge asymmetry potential, such that the barrier penetrabilities for channels involving ^{208}Pb become comparable with the penetrabilities in α-decay. The new decays are difficult to observe because of the large background of α-particles.

Cold fragmentation in spontaneous fission or in thermal-neutron induced fission, i.e., where the fragments with maximum kinetic energies close to the corresponding Q values or Q plus the binding energy of the neutron is difficult to observe because of the large background of the hot fission fragments.

In this paper, we present the new experimental techniques to extract the cold fission yields from the spontaneous fission of ^{252}Cf. We compare our results with the calculations of Sandulescu et al. [2]. Furthermore, we also present tentative evidence for α-ternary fission and double fine-structure.

2. Experimental & Data Analysis Techniques:

A ^{252}Cf source of strength $\sim 6 \times 10^4$ fissions/s was placed at the center of the early implementation of Gammasphere with 36 Ge detectors and Ge LEPS at Lawrence Berkeley National Laboratory. A total of 9.8 $\times 10^9$ triple or higher fold coincidence events were recorded. A three dimensional "cube" was built using RADWARE. In addition, we have also recorded $\gamma - \gamma$, $x - \gamma$ coincidences and two dimensional matrices were built from these events.

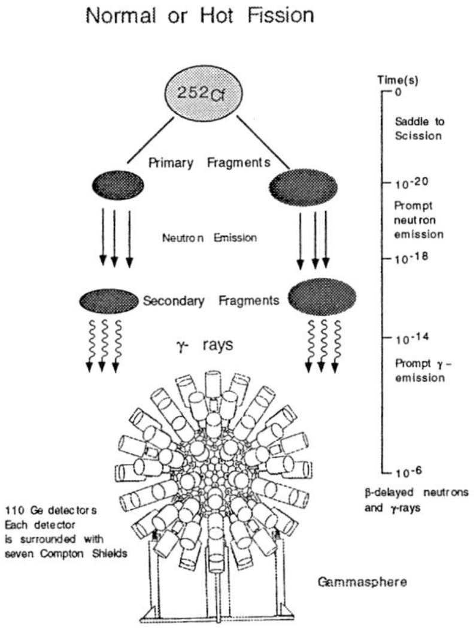

Fig.1. Spontaneous fission processes and Gammasphere

In the fission of ^{252}Cf, about 100 nuclei are produced. After the fission of ^{252}Cf into two primary fragments, these primary fragments emit several neutrons until the excitation energy of the primary fragments is below the neutron binding energy. The resulting secondary fragments emit several γ-rays. These processes are shown in Fig. 1. The primary and secondary fragment energy distributions [3] are shown in Fig. 2. Apart from small shifts towards lower energies, neutron evaporations does not drastically alter the characteristics of single-fragment energy distributions. Hence, observing the γ-rays emitted by the secondary fragments, one can extract the information on the yields of the primary fragments. Furthermore, the total intensities of the $2^+ \longrightarrow 0^+$ ground state band transitions observed in the de-excitation of the even-even fission products correspond to a high degree of accuracy (5%) to the total independent yields of these isotopes. Based on this observation, the independent yields and neutron multiplicities of a number of individual correlated pairs of fission fragments of ^{252}Cf were determined 1994 [4].

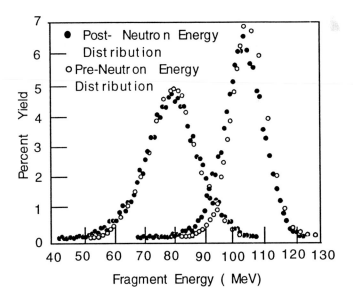

Fig. 2. Post-and Pre-neutron energy distributions in spontaneous fission of ^{252}Cf.

Since the α-particle is inert and cannot be excited, we have only fine structure in α-decay where a ground state of a parent decays to the excited states of the daughter as shown in Fig. 3.

Cold Fission (Double Fine Structure) α – decay (Fine Structure)

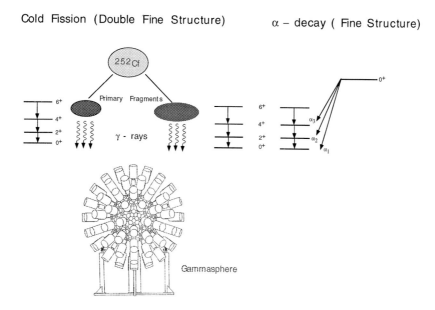

Fig.3. Cold fission processes and double fine-structure

Since both the primary fragments can be excited in cold fission, we have double fine structure.

In order to search for events corresponding to zero neutron channels the $\gamma - \gamma - \gamma$ coincidence events produced in the spontaneous fission of ^{252}Cf were anlyzed by setting double gates on the transitions in a particular (e.g. ^{106}Mo; Fig. 4.)

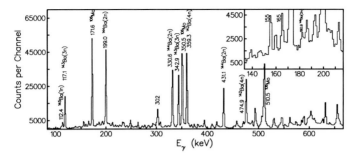

Fig.4. Sum coincidence spectrum obtained by double gating on all the possible transitions in ^{106}Mo. The inset shows the $2^+ \longrightarrow 0^+$ 180.9 kev transition in ^{146}Ba.

even-even nucleus and the corresponding γ-rays of its correlated partners (all Ba isotopes) were analyzed. As a cross check, the coincidence spectrum obtained by setting a double gate on 180.9 and 332.0 keV γ-rays in ^{146}Ba is shown in Fig. 5.

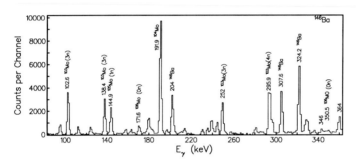

Fig. 5. Coincidence spectrum obtained by double gating on 180.9 and 332.0 keV γ-rays in ^{146}Ba.

The $\gamma - \gamma$, and x-γ data were used to identify low energy transitions in a given nucleus. In majority of cases, only the $2^+ \longrightarrow 0^+$ transition intensity of the partner was used to extract the zero neutron channel yields. After correcting for the efficiency and the internal conversion, the resulting areas for all the neutron channels were summed and the areas for the individual channels were then divided by the resulting sum to extract the relative fraction of the events for each neutron channel. Furthermore to obtain the zero neutron channel yields per 100 fission events, our above yields were then normalized to the integrated yields for the partner ^{106}Mo computed by Wahl [5]. The results including the 2n, 4n, 6n etc., are fitted with gaussian to check for internal consistency.

3. Results and Discussion:

Table 1. gives the zero neutron yields obtained in the present investigation along with the theoretical calculations by Sandulescu et al. [2] using folded M3Y nucleon-nucleon potential. The yields for ^{108}Mo- ^{144}Ba and ^{110}Ru-^{142}Xe have large uncertainities because of the very nearly identical energies of the 2^+ states in ^{108}Mo, ^{104}Mo and ^{110}Ru and ^{108}Ru nuclei respectively.

Table 1: Experimental and theoretical cold fission yields

^AX_Z- ^AY_Z	Y_{exp}	Y_{theory}	^AX_Z-^AY_Z	Y_{exp}	Y_{theory}
$^{96}Sr_{38}$-$^{156}Nd_{60}$	0.02(1)	0.022	$^{104}Mo_{42}$-$^{148}Ba_{56}$	0.02(1)	0.0202
$^{98}Sr_{38}$-$^{154}Nd_{60}$	0.08(2)	0.04	$^{106}Mo_{42}$-$^{146}Ba_{56}$	0.05(2)	0.0143
$^{100}Sr_{38}$-$^{152}Nd_{60}$	0.05(2)	0.024	$^{108}Mo_{42}$-$^{144}Ba_{56}$	0.15(6)*	0.0324
$^{100}Zr_{40}$-$^{152}Ce_{58}$	0.12(4)**	0.062	$^{110}Ru_{44}$-$^{142}Xe_{54}$	0.10(5)	0.036
$^{102}Zr_{40}$-$^{150}Ce_{58}$	0.02(1)	0.072	$^{112}Ru_{44}$-$^{140}Xe_{54}$	0.04(2)	0.038
$^{104}Zr_{38}$-$^{148}Ce_{58}$	0.02(1)	0.135	$^{116}Pd_{46}$-$^{136}Te_{52}$	0.05(2)	0.003
$^{99}Sr_{38}$-$^{153}Nd_{60}$	0.30(1)†	0.009	$^{107}Mo_{42}$-$^{145}Ba_{56}$	0.24(6)	0.072
$^{103}Zr_{40}$-$^{149}Ce_{58}$	0.10(1)	0.026	$^{111}Ru_{44}$-$^{141}Xe_{54}$	0.14(2)	0.008
$^{105}Mo_{42}$-$^{147}Ba_{56}$	0.20(1)	0.063			

*The Peaks in ^{108,104}Mo are very difficult to strip.

**In this case we have used the 182.5 $(4^+ \longrightarrow 2^+)$ keV transition in ^{152}Ce.

†Here we used the 271.9 kev transition from the second excited state in ^{153}Nd.

We would like to mention that only the pairs for which the spectra are clean could be analyzed. One can see, from table 1, that the zero neutron yields are $\approx 10^{-2}$ per 100 fission events. In a few cases we also present in Table 1 our results for odd-odd fragmentations which are a factor of 3-5 larger than the yields for the even-even cases. The details of the calculations are presented by Sandulescu et al [2]. The values of the theoretical relative yields and the experimental yields should not be compared directly since the former values are integrated yields (over the internal excitation energies of the final fragments upto the binding energy of the last neutron). The experimental yields corrrespond only to transitions leading to ground state in the final fragments. Nevertheless, the trends for both quantities should be very similar.

Using the triple γ-coincidence technique, the double fine-structure in the neutronless fission of ^{252}Cf could be detected. Fig. 6 shows the level populations relative to the $2^+ \longrightarrow 0^+$ transition which is normalized to 100 for the two even-even Mo-Ba pairs including the relative transition probabilities to the first 2^+, 4^+, and 6^+ states of rotational band built on the 0^+ ground state.

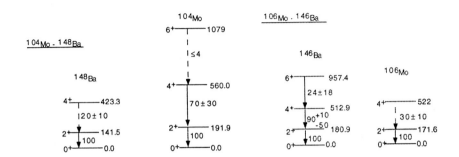

Fig. 6. Relative level populations of the lowest excited states in ^{104}Mo- ^{148}Ba and ^{106}Mo-^{146}Ba.

For these nuclei, in ^{252}Cf data, we have observed the ground state band upto $\sim 16^+$. It is evident that in the cold even-even splittings, only a few lowest levels of the ground state rotational bands are populated. This fact supports the dynamical β- stretching model for cold fission [6].

It is possible for the fissioning nucleus to undergo cold fission by splitting into three fragments one of them being an α-particle or some other heavy fragment. This process is known as ternary fission. When ^{252}Cf undergoes α-ternary fission, for example, we get the ^{142}Xe and ^{106}Mo fragments accompanied by an α-particle. In the present investigation, we also tried to get an estimate of the yields for the α-ternary fragmentation. Fig. 7 shows the coincidence spectrum obtained by double gating on 287.1 and 403.5 keV γ-rays in ^{142}Xe.

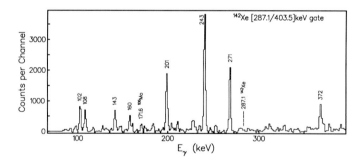

Fig.7. Coincidence spectrum obtained by double gating on 287.1 and 403.5 keV γ-rays in ^{141}Xe.

We can see a peak at 171.6 keV corresponding to the $2^+ \longrightarrow 0^+$ in ^{106}Mo fragment. However, one should consider the evidence as preliminary because of the lack of several cross-checks needed.

Acknowledgements

The work at Vanderbilt University is supported in part by the U.S. Department of Energy under grant No. DE FG05 88ER40407. The work at Tsinghua University is partially supported by the National Natural Science Foundation and the Nuclear Industrial science Foundation of China. The Joint Institute for Heavy Ion Research is supported by its members, University of Tennessee, Vanderbilt University and the US Department of Energy through contract NO. DE-FG05- 87ER40361 with University of Tennessee. The work at Idaho National Engineering Laboratory is supported in Part by the U.S. Department of Energy under contract No.DE-AC07-76ID01570. The work at Lawrence Berkeley National Laboratory and Lawrence Livermore National Laboratory is supported in part by the U.S. Department of Energy under grant Nos. DE-AC03-76SF00098 and W-7405-ENG48 respectively.

References

1. A. Sandulescu and W. Greiner   Rept. Prog. Phys. **55**, (1992) 1423

2. A. Sandulescu, A. Florescu, F. Carstoiu, W. Greiner, J.H. Hamilton, A.V. Ramayya and B.R.S. Babu   (To be published in Phys. Rev. C July 1996 and also contribution to this conference)

3. H.W. Schmitt, W.E. Kiker, and C.W. Williams   Phys. Rev. **B137**, (1965) 837

4. G.M. Ter-Akopian, J.H. Hamilton, Yu. Ts. Oganessian, J. Kormicki, G.S. Popeko, A.V. Daniel, A.V. Ramayya, Q.Lu, K. Butler-Moore, W.-C. Ma, J.K. Deng, D.Shi, J. Kliman, V. Polhorsky, M. Morhac, W. Greiner, A. Sandulescu, J.D. Cole, R. Aryaeinejad, N.R. Johnson, I.Y. Lee, and F.K. McGowan   Phys. Rev. Lett. **73** (1994) 1477

5. A.C. Wahl, Atomic Data & Nuclear Data Tables **39**, 1, (1988)

6. A. Florescu, A. Sandulescu, C. Cioaca, and W. Greiner   J. Phys. G: Nucl.Part. Phys., **19** (1993) 669

ISOTOPIC YIELDS FOR THE BINARY AND ALPHA TERNARY COLD FISSION OF ^{252}Cf, ^{248}Cm AND ^{242}Pu

A.Sandulescu$^{a,b,c,d)}$, A.Florescu$^{a,c,d)}$, F.Carstoiu$^{a)}$, J.H.Hamilton$^{c)}$, A.V.Ramayya$^{c)}$ and B.R.S.Babu$^{c)}$

$^{a)}$ Institute for Atomic Physics, Bucharest, P.O.Box MG-6, Romania

$^{b)}$ Institut für Theoretische Physik der J.W.Goethe Universität, D-60054, Frankfurt am Main, Germany

$^{c)}$ Physics Department, Vanderbilt University, Nashville, TN 37235, USA

$^{d)}$ Joint Institute for Heavy Ion Research, Oak Ridge, TN 37831, USA

The cluster model for cold rearrangements of large groups of nucleons in binary and ternary fragmentations is applied to neutronless fission of heavy nuclei. The corresponding isotopic yields for the spontaneous binary and alpha ternary cold fission of ^{252}Cf, ^{248}Cm and ^{242}Pu are predicted. We use the WKB penetrabilities through a double folding potential barrier. Realistic fragment ground-state deformations are used for the calculations.The most probable fragmentations predicted are supported by the preliminary neutronless yields extracted from γ - γ - γ coincidence studies in the spontaneous fission of ^{252}Cf recorded with the early implementation Gammasphere. Also the double fine structure in such binary and ternary cold fragmentations has been experimentally observed for the first time. A new phenomenon was discovered: the alpha ternary cold (neutronless) fission.

1 Introduction

In recent times, many new experimental data concerning the spontaneous cold fragmentations of nuclei have been obtained. These include exotic decays with emission of heavy clusters having masses from $A_L = 12$ to 34 [1, 2] and the cold fission of many actinide nuclei with fragment masses from ≈ 70 to ≈ 166 atomic mass units [3 − 7]. They all confirm the theoretical predictions based on the idea of the cold rearrangements of large groups of nucleons from the ground state of the initial nucleus to the ground states of the two final fragments [1, 8].

The existence of the so-called fusion valleys or "cold" valleys on the potential energy surfaces of the fission-prone heavy nuclei [8, 9, 10] proved to be a key ingredient for the prediction [8] and later for the interpretation of heavy cluster decays (the ^{208}Pb or ^{100}Sn valleys), or for the understanding of cold fission of the actinides (especially the ^{132}Sn valley). In all these situations the final fragments have compact shapes at the scission point and almost zero excitation energy. It has been shown that the transitions from the fission valley to the fusion valley along the fission path can qualitatively explain the cold fragmentation of the actinides [11]. Also the cluster decay model where the ground state deformations of the final fragments are very important [12, 13] is able to explain quantitatively the mass and charge yields in thermal neutron-induced cold fission [14].

An extreme case is that of the bimodal fission observed for the Fm and Md isotopes [15] where the predominant fragmentations are close to two double magic ^{132}Sn nuclei. Here, two clearly distinct fission channels are observed, one with very high total kinetic energy (TKE) which practically exhausts the disintegration energy (Q-value), corresponding to the fusion valley, and the second one at much lower TKE's which proceeds through the usual fission valley with elongated shapes. The Cf, Cm and Pu isotopes studied in the present paper constitute a transition region between the lighter actinides such as Th and U, and the bimodal fission region of Fm and Md nuclei.

Recently, in the first direct observation of spontaneous cold (neutronless) fission of ^{252}Cf, for a few pairs of fragments the ground band gamma cascades were accurately detected in both light and heavy partners with the early implementation Gammasphere using the triple-gamma coincidence technique [5, 6, 16]. It may be pointed out that the yields measured in the above mentioned experiments are integrated yields. Up to now the yields in cold fission were measured as a function of TKE of the fragments, or as a function of their total excitation energy (TXE= Q -TKE). It is known that these yields increase strongly with the decrease of TKE's or equivalently with the increase of the final fragment TXE's. In the present experiments, based on triple-gamma coincidence of the lowest transitions in both fragments, most of the fragmentations are usually leading to higher excitations of the final nuclei which later on are decaying to the lowest states by gamma cascades. Evidently there are also splittings which leave the fragments in their ground or first excited states, but with lower probabilities. We are calling these experimentally determined yields as integrated yields due to the fact that they collect the contributions of all (neutronless) transitions over a whole range of fragment TXE's from zero up to at least the neutron binding energy, from where the evaporation of a first neutron becomes possible.

In these experiments [5, 16] the double fine structure in the transitions leading to the final fragments was observed directly for the first time in the cold fission decays, similar to the well-known fine structure revealed in the alpha decays and even in heavier cluster decays of odd-mass nuclei. In alpha decay of even-even and odd-A heavy nuclei the transitions are observed not only to the ground state of the final nuclei, but also to some of their low excited states [17]. These could be states belonging to the rotational ground, beta or gamma bands in even-even nuclei, or to some rotational bands built on intrinsic quasi-particle states in odd-A final nuclei. It was early noted [18, 19, 20] that the alpha transitions in odd-A nuclei exhibited reduced widths ranging from values similar to those of neighbouring even-even nuclei down to much smaller values. When the transition proceeds to a final state with no change in the single particle configuration of the last odd nucleon, a 'favoured' transition is observed, with an alpha reduced width comparable to those in neighbouring even-even nuclei. Other 'unfavoured' transitions will be nevertheless observed in some cases in the same nucleus, if they lead to the ground state or other low states of the daughter nucleus and if their decay energy is high enough. Consequently many alpha-emitting nuclei are presenting a fine structure in their alpha transitions. In the last years the fine structure was also discovered for some heavy cluster-emitting nuclei, e.g. for the ^{14}C-transitions in ^{223}Ra [21]. In a similar way, the cold (neutronless) fission of heavy nuclei could leave both final fragments on some low excited states and a double fine structure of the transitions is observed for those cases.

In this paper we present a first estimation of the isotopic yields for different cold binary and alpha ternary fragmentations of the nuclei ^{252}Cf, ^{248}Cm and ^{242}Pu in spontaneous cold fission using only the barrier penetrabilities, and new experimental data of such fragmentations. A new phenomenon was discovered: the alpha ternary cold (neutronless) fission. In the present estimations no preformation factors are considered. For the evaluation of the potential barriers between the final fragments we used the M3Y nucleon-nucleon forces. We found that the yields are very sensitive to the fragment's quadrupole deformation. No octupole or higher deformations were taken into account.

2 The Binary Cold Fragmentation Yields

In the present paper we evaluated the nuclear plus Coulomb interaction between two coaxial deformed fragments with the help of the double folding M3Y potential defined [23] as

$$V_{M3Y}(R) = \int d\mathbf{r}_1 d\mathbf{r}_2 \; \rho_1(\mathbf{r}_1) \; \rho_2(\mathbf{r}_2) \; v(\mathbf{r}_{12}) \tag{1}$$

which contains the corresponding nucleon-nucleon interaction [24]

$$v(\mathbf{r}_{12}) = v_{00}(r_{12}) + \hat{J}_{00}\delta(\mathbf{r}_{12}) + v_{01}(r_{12}) \; \tau_1 \cdot \tau_2 + \frac{e^2}{r_{12}} \tag{2}$$

where $\mathbf{r}_{12} = \mathbf{R} + \mathbf{r}_2 - \mathbf{r}_1$.

The two final nuclei are viewed as coaxial spheroids ("nose-to-nose" configuration) with nuclear density

$$\rho(\mathbf{r}) = \rho_0 \left[1 + \exp \frac{1}{a} \left(r - \frac{R_0}{c} \left(1 + \beta_2 Y_2^0(\cos\theta) \right) \right) \right]^{-1} \tag{3}$$

with the constant ρ_0 fixed by normalizing the proton and neutron density to the Z proton and N neutron numbers, respectively, the diffusivity $a = 0.5$ fm and $R_0 = r_0 A^{1/3}$ with $r_0 = 1.12$ fm. Here β_2 is the quadrupole deformation and c is the usual constant which ensures the volume conservation condition.

We computed the double folded deformed potential barrier, eq.(1), by making a general multipole expansion of the potential [25, 26].

For the cold (neutronless) fragmentation of a heavy nucleus, we considered that the two final nuclei are in their ground states or in the first one or two excited states, usually belonging to the ground state rotational band or other rotational bands built on intrinsic quasiparticle states. Due to the inherent ambiguities of different final channels (e.g. fragment deformations, level densities and channel radii) we assumed that the cluster preformation probabilities are similar for all possible splittings and consequently we used the same frequency factor ν for the collisions with the fission barrier for all fragmentations. Henceforth in our calculations we are neglecting the preformation factors for different channels. Our relative isotopic yields computed below correspond to those cold (neutronless) fragmentations with both final nuclei emitted in their ground state. Consequently the theoretical isotopic yields are not integrated over a domain of TXE values, like the experimental yields mentioned above, but are resulting for zero total excitation energy.

We estimated the penetrability through the double folded potential barrier in the framework of the WKB approximation. The accurate evaluation of the Q-values is very important since the WKB penetrabilities are very sensitive to them. We obtained the Q-values from recent experimental mass tables [27].

The deformation parameters which we used were taken from the tables of Ref. [28], and for some isotopes we employed the deformation values deduced from cold fission data [12, 14] which are similar or slightly larger in some cases (e.g. 5% to 8% larger for the Sr, Zr and Mo isotopes) than those from [28]. The calculated values of the penetrabilities are very sensitive to the assumed deformations of the final fragments, since a 10% increase in the β_2 values leads up to an order of magnitude increase of the penetrabilities. Consequently our calculated penetrabilities should contain an uncertainty factor of about ten. Nevertheless, the relative yields should not change significantly. Of course, the higher multipoles like the octupole can play an important role bringing about additional uncertainties.

The relative isotopic yields $Y(A_L, Z_L)$ are defined as

$$Y(A_L, Z_L) = P(A_L, Z_L)/ \sum_{A_L Z_L} P(A_L, Z_L) \qquad . \qquad (4)$$

The calculated isotopic yields for the binary cold (neutronless) fragmentations of ^{252}Cf, ^{248}Cm and ^{242}Pu are given in Fig. 1. As we see, for ^{252}Cf the highest yields are obtained for fragment masses $98 \leq A_L \leq 110$ and $142 \leq A_H \leq 154$, and for fragment charges $38 \leq Z_L \leq 44$ and $54 \leq Z_H \leq 60$ respectively. For ^{248}Cm, the highest yields are obtained for masses $101 \leq A_L \leq 116$ and $132 \leq A_H \leq 147$, and for charges $40 \leq Z_L \leq 46$ and $50 \leq Z_H \leq 56$ respectively. For ^{242}Pu the highest yields are obtained for masses $98 \leq A_L \leq 110$ and $132 \leq A_H \leq 144$, and for charges $38 \leq Z_L \leq 44$ and $50 \leq Z_H \leq 56$ respectively.

The theoretical penetrabilities and isotopic yields predicted by the present cluster model are not simply correlated with the highest Q-values for the corresponding cold fragmentations. In fact, from the twenty-five mass splits shown in Fig. 1 for ^{252}Cf, in eleven cases the highest yield is attained by those charge splits which have not the highest Q-values but larger deformation of one or both fragments.

The recent experiment of cold fragmentations in the spontaneous fission of ^{252}Cf [5, 6, 16] permitted for the first time the direct observation of neutronless fission channels for the light fragments $^{96-100}$Sr, $^{100-104}$Zr, $^{104-108}$Mo, $^{110-112}$Ru and ^{116}Pd in coincidence with the corresponding heavy fragments, for which the isotopic yields were determined. In Table 1 we present the integrated isotopic yields for the above cold neutronless fragmentations. In order to search for events corresponding to zero neutron channels the γ - γ - γ coincidences events produced in the spontaneous fission of ^{252}Cf were recorded with early implementation Gammasphere. A γ - γ - γ "cube" was built using the RADWARE program. The "cube" data were analyzed by setting two gates, one on the $2^+ \rightarrow 0^+$ transition and the other on the $4^+ \rightarrow 2^+$ transition in a particular even-even nucleus and the corresponding γ-rays of its correlated partner nuclei were analyzed. The peak areas of the γ rays were then corrected for efficiency and internal conversion. In the majority of the cases only the $2^+ \rightarrow 0^+$ transition intensity of the partner was used to extract the zero neutron channel yield. After the efficiency correction, the resulting areas for all the neutron channels were summed and the areas for the individual channels were divided by the resulting sum to get the relative fraction of the events for each neutron channel. Furthermore to obtain the zero neutron channel yields per 100 fission events, our above yields were then normalized to the integrated yield for the partner computed by Wahl [29]. The results including 2n, 4n, 6n etc. channels are fitted with a Gaussian to check for internal consistency. The yields for ^{108}Mo -^{144}Ba and ^{110}Ru -^{142}Xe have large uncertainties because of the very nearly identical energies of the 2^+ states in ^{108}Mo, ^{104}Mo and ^{110}Ru, ^{108}Ru nuclei, respectively. We should like to mention that only the pairs for which the spectra are known could be considered. In a few cases, we also present in Table 1 our results for odd-odd fragmentations which are a factor of 3-5 larger than the yields for the even-even cases. For odd-Z fragmentations, the spectra are too complicated to extract useful information from the present data. Higher yields for the odd-odd fragmentations of ^{252}Cf are also observed by Hambsch et al. [3] for low TXE values (TXE $\leq$ 9 MeV). One can see from Table 1 that the zero neutron yields are of the order of 5×10^{-2} per 100 Cf fission events.

The values of the theoretical relative yields (Fig. 1) and experimental yields (Table 1) for ^{252}Cf should not be compared directly since the latter values are integrated yields (over the internal excitation energies of the final fragments, at least up to the binding energy of the last neutron) while the former ones are yields corresponding only to transitions to the ground states of the final fragments. Nevertheless the trends for both quantities should be very similar, i.e. larger theoretical yields will indicate that also larger experimental yields could be observed for the same cold (neutronless) fragmentation.

Using the triple-gamma coincidence technique the double fine structure in the neutronless fission of ^{252}Cf was detected. In Fig. 2-a are shown the level populations, relative to the $2^+ \rightarrow 0^+$ transition which is normalized to 100, for the fission fragments from ^{252}Cf obtained for two even-even Mo -Ba pairs, including the relative transition probabilities to the first $2^+, 4^+$ and 6^+ excited states of the rotational band built upon the 0^+ ground state. For comparison in Fig. 2-b are shown the complete ground state rotational bands for the same fragments, observed for all fission channels.

3   The Alpha Ternary Cold Fragmentation Yields for ^{252}Cf

We made also a theoretical estimation of the isotopic yields for the spontaneous cold alpha ternary fission of ^{252}Cf. We evaluated the potential barrier between the three final nuclei by summing up the inter-fragment potential along their trajectories as function of time. We used the Coulomb and the M3Y nuclear forces exerted between the deformed fragments. The alpha particle was considered as spherical.

The double folded potential between two of the final fragments was computed starting from eq.(1). The trajectories of the three final fragments were calculated as a function of time, starting from the scission point configuration, by using pure Coulomb forces

$$F_{ij} = \frac{e^2 Z_i Z_j}{r_{ij}^2} \quad , \qquad r_{ij}^2 = (x_j - x_i)^2 + (y_j - y_i)^2 + (z_j - z_i)^2 \qquad (5)$$

where the z coordinates are always zero since the three bodies are moving in a constant plane. The trajectories were calculated in the laboratory frame of reference with the x axis taken as the initial fissioning axis of the two heavier fragments, by integrating numerically the equations of motion

$$v_{ki}(t) = v_{ki}(t') + a_{ki}(t') \, dt \qquad (6a)$$

$$k_i(t) = k_i(t') + v_{ki}(t') \, dt + \frac{1}{2} \, a_{ki}(t') \, dt^2 \qquad (6b)$$

along the $k = x, y$ coordinates, with $t = t' + dt$ and v_{ki} the velocity components of the fragments. The acceleration components were calculated as

$$a_{ki} = \sum_j \frac{F_{ij}}{m_i} \cdot \frac{k_j - k_i}{r_{ij}} \qquad . \qquad (7)$$

At the scission configuration the two heavier fragments were assumed to be coaxial spheroids in contact through their tips. The same scission configuration was considered for all the ternary fragmentations, with the alpha particle initially situated in a position at $y_o = 1.0$ fm above the contact point between the two heavier fragments. The alpha particle initial velocity was directed along the y axis. The initial velocities of the two heavier fragments were along the x axis and in opposite directions. For the initial kinetic energy of the alpha particle we assumed a value of 5 MeV, and initial kinetic energies of 5 MeV for each of the heavier fragments.

We calculated the penetrabilities through the corresponding potential barrier using the WKB approximation. In Fig. 3 we present the disintegration energies Q and the relative isotopic yields, calculated from eq.(4), for the alpha cold ternary spontaneous fission of ^{252}Cf. We present in Table 2 the integrated isotopic yields which were measured for some even-even Kr, Sr, Zr, Mo, Ru and Pd isotopes. The method employed for extracting these yields from the experimental spectra is similar with the procedure used for the binary fragmentations with the difference that ternary yields are added to the binary ones and later on normalized to Wahl tables [29].

4 Conclusions

Using a cluster model we calculated the relative isotopic yields for the cold (neutronless) fission of ^{252}Cf, ^{248}Cm and ^{242}Pu. We have shown that this simple cluster model is able to predict almost correctly the most important cold binary and alpha ternary fragmentations observed in the spontaneous cold fission of the nucleus ^{252}Cf.

It was experimentally shown that in the cold (neutronless) even-even splittings only the lowest levels of the ground state rotational bands are populated (see Fig. 2-a). This fact supports the dynamical beta-stretching model for cold fission [14] which assumes only a coaxial motion of the fragments at the scission configuration. Since other vibrational states corresponding to the kinking and bending modes between fragments at the scission point were not observed in the present experiment, it is possible that other 0^+ beta-phonon states could be observed in the future. In the odd and odd-odd fragments, with higher isotopic yields, the beta-phonon states are coupled to the one- or many-quasiparticle states leading to a more complicated energy level landscape and to higher level densities. Such transitions can be easily detected by using two back to back ionization chambers for measuring the TKE in coincidence with the gamma-ray cascades in both fragments.

We should mention here that a striking feature of the cold fission yields close to the highest TKE values permitted by the Q-values [3, 4, 6] is the fact that many odd-odd splittings have values larger than the neighbouring even-even fragmentations. This feature of cold fragmentations suggests that either the cold fission yields are influenced by the level density of the fragments [7, 22] or that the deformations of odd fragments are larger than for the corresponding even ones, since in this case the barrier height is lowered and within our model we obtain increased penetrabilities and higher yields.

It was already shown that for cold binary fragmentations, the ground state deformations are a key ingredient for the correct prediction of the most favoured splittings and of the isotopic yields [13, 14, 22]. The simple cluster model which we used in this paper for calculating the isotopic yields associated to the alpha-accompanied cold fission, also predicts a large number of favoured ternary splittings in which one or both heavier fragments are well deformed in their ground states. Some of the favoured even-even alpha-accompanied cold fragmentations were discovered with an important yield for the first time, as a new phenomenon: alpha ternary cold (neutronless) fission.

The determination of the scission point configurations in the fission of heavy nuclei starting from the experimental kinetic energy and angular distributions of the light charged particles emitted in ternary fission, has been a great hope for many years. Unfortunately too many unknown parameters are associated with the initial scission point configurations in the case of the usual "hot" ternary fission. But for the new phenomenon: alpha ternary cold fission, the initial scission configurations could be much easier determined. Experimental studies of the kinetic energy and angular distributions of the light clusters emitted in alpha ternary cold fission will provide new insights on the fragmentation processes of the heavy nuclei.

In order to determine more exactly the double fine structure in cold fission one of the new implementation Gammasphere with 90 detectors has to be used. In the future the theory can be further centered on studying new cases of neutronless cold fragmentations, especially the odd-odd ones which should have yields larger than the even-even splittings. Thus further experimental and theoretical studies of the double fine structure in cold fission decays can open up new insights into cold fragmentation phenomena.

References

[1] A.Sandulescu and W.Greiner, Rep. on Progr. in Phys., 55 (1992) 1423

[2] P.B.Price, Nucl. Phys. A502 (1989) 41c

[3] F.-J.Hambsch, H.-H.Knitter and C.Budtz-Jorgensen, Nucl. Phys. A554 (1993) 209

[4] A.Benoufella, G.Barreau, M.Asghar, P.Audouard, F.Brisard, T.P.Doan, M.Hussonnois, B.Leroux, J.Trochon and M.S.Moore, Nucl. Phys. A565 (1993) 563

[5] J.H.Hamilton, A.V.Ramayya, J.Kormicki, W.C.Ma, Q.Lu, D.Shi, J.K.Deng, S.J.Zhu, A.Sandulescu, W.Greiner, G.M.Ter-Akopian, Yu.Ts.Oganessian, G.S.Popeko, A.V.Daniel, J.Kliman, V.Polhorsky, M.Morhac, J.D.Cole, R.Aryaeinejad, I.Y.Lee, N.R.Johnson and F.K.McGowan, J. of Phys. G: Nucl. Part. Phys., 20 (1994) L85

[6] G.M.Ter-Akopian, J.H.Hamilton, Yu.Ts.Oganessian, J.Kormicki, G.S.Popeko, A.V.Daniel, A.V.Ramayya, Q.Lu, K.Butler-Moore, W.-C.Ma, J.K.Deng, D.Shi, J.Kliman, V.Polhorsky, M.Morhac, W.Greiner, A.Sandulescu, J.D.Cole, R.Aryaeinejad, N.R.Johnson, I.Y.Lee and F.K.McGowan, Phys. Rev. Lett., 73 (1994) 1477

[7] W.Schwab, H.-G.Clerc, M.Mutterer, J.P.Theobald and H.Faust, Nucl. Phys. A577 (1994) 674

[8] A.Sandulescu and W.Greiner, J.of Phys. G : Nucl. Phys., 3 (1977) L189

[9] P.Moller, J.R.Nix and W.J.Swiatecki, Nucl.Phys. A469 (1987) 1

[10] S.Cwiok, P.Rozmej, A.Sobiczewski and Z.Patyk, Nucl. Phys.
A491 (1989) 281

[11] J.F.Berger, M.Girod and D.Gogny, Nucl. Phys. A428 (1984) 23c

[12] A.Sandulescu, A.Florescu and W.Greiner, J. of Phys. G: Nucl.
Part. Phys., 15 (1989) 1815

[13] F.Gönnenwein and B.Borsig, Nucl. Phys., A530 (1991) 27

[14] A.Florescu,A.Sandulescu,C.Cioaca and W.Greiner, J. of
Phys. G : Nucl. Part. Phys., 19 (1993) 669

[15] E.R.Hulet, J.F.Wild, R.J.Dougan, R.W.Lougheed,
J.H.Landrum, A.D.Dougan, M.Schadel, R.L.Hahn, P.A.Baisden,
C.M.Henderson, R.J.Dypzyk, K.Summerer and G.R.Bethune,
Phys. Rev. Lett., 56 (1986) 313

[16] J.H.Hamilton, A.V.Ramayya, B.R.S.Babu, W.C.Ma, S.J.Zhu,
J.Kormicki, T.N.Ginter, G.M.Ter-Akopian, Yu.Ts.Oganessian,
J.O.Rasmussen, M.A.Stoyer, I.Y.Lee, S.Y.Chu, K.E.Gregorich,
M.Mohar, S.G.Prussin, J.D.Cole, R.Aryaeinejad, Y.X.Dardenne
and M.W.Drigert (to be published)

[17] L.Rosenblum, C.R. Acad. Sci., Paris, 188 (1929) 1401

[18] A.Bohr, P.O.Fröman and B.R.Mottelson, Dan. Mat. Fys. Medd.,
29 no.10 (1955)

[19] A.Sandulescu, Nucl. Phys., 37 (1962) 332

[20] J.O.Rasmussen, in Alpha, Beta and Gamma-Ray Spectroscopy,
Ed. K.Siegbahn, North Holland, Vol. 1 (1965) 701

[21] E.Hourani, L.Rosier, G.Berrier-Ronsin, A.Elayi,
A.C.Mueller, G.Rappenecker, G.Rotbard, G.Renou, A.Liebe,
L.Stab and H.L.Ravn, Phys. Rev. C44 (1991) 1424

[22] V.Avrigeanu, A.Florescu, A.Sandulescu and W.Greiner,
Phys. Rev. C52 (1995) R1755

[23] G.R.Satchler and W.G.Love, Phys. Rep. 55 (1979) 183

[24] G.Bertsch, J.Borysowicz, H.McManus and W.G.Love, Nucl. Phys.
A284 (1977) 399

[25] F.Carstoiu and R.J.Lombard, Ann. of Phys. (N. Y.)
217 (1992) 279

[26] A.Sandulescu, R.K.Gupta, W.Greiner, F.Carstoiu and M.Horoi,
Int. J. of Mod. Phys., E 1 (1992) 379

[27] A.H.Wapstra, G.Audi and R.Hoeckstra, At. Data and Nucl. Data
Tables, 39 (1988) 281

[28] P.Möller, J.R.Nix, W.D.Myers and W.J.Swiatecki, At. Data
and Nucl. Data Tables, 59 (1995) 185

[29] A.C.Wahl, At. Data and Nucl. Data Tables, 39 (1988) 1

Table1. The experimental isotopic yields Y_{exp} for the binary cold (neutronless) fragmentation of ^{252}Cf, per 100 fission events.

A_L/Z_L	A_H/Z_H	Y_{exp}
96/38	156/60	0.02 ± 0.01
98/38	154/60	0.08 ± 0.02
100/38	152/60	0.05 ± 0.02
100/40	152/58	0.12 ± 0.04
102/40	150/58	0.02 ± 0.01
104/40	148/58	0.02 ± 0.01
104/42	148/56	0.02 ± 0.01
106/42	146/56	0.08 ± 0.05
108/42	144/56	$0.15 \pm 0.06^*$
110/44	142/54	0.10 ± 0.05
112/44	140/54	0.04 ± 0.02
116/46	136/52	0.05 ± 0.02
99/38	153/60	0.30 ± 0.10
103/40	149/58	0.10 ± 0.01
105/42	147/56	0.20 ± 0.07
107/42	145/56	0.24 ± 0.06
111/44	141/54	$0.14 \pm 0.06^{**}$

$^{*)}$The ^{108,104}Mo peaks are very difficult to strip in this case
$^{**)}$The 150 keV transition in ^{111}Ru was used to obtain this result

Table2. The experimental alpha ternary isotopic yields for the cold (neutronless) fragmentation of ^{252}Cf, per 100 fission events.

A_L/Z_L	A_H/Z_H	Y_{exp}
92/36	156/60	0.06 ± 0.01
96/38	152/58	0.03 ± 0.02
98/38	150/58	0.28 ± 0.16
100/38	148/58	0.08 ± 0.01
100/40	148/56	0.25 ± 0.05
102/40	146/56	0.13 ± 0.01
104/40	144/56	0.07 ± 0.02
106/42	142/54	0.05 ± 0.03
108/42	140/54	0.03 ± 0.01
112/44	136/52	0.55 ± 0.25
116/46	132/50	0.11 ± 0.05

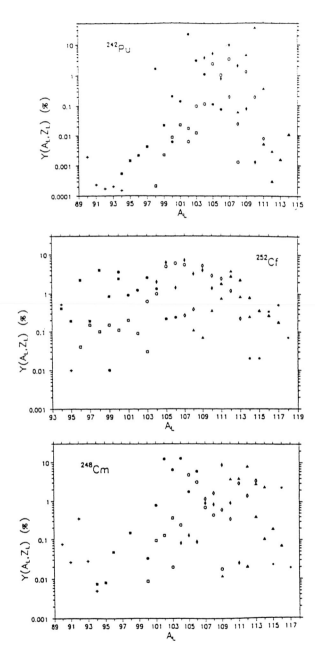

Fig.1. The calculated relative isotopic yields for the cold binary fragmentations of ^{252}Cf, ^{248}Cm and ^{242}Pu. The shown isotopes are Kr ($+$), Sr ($\blacksquare$), Y ($\square$), Zr ($\bullet$), Nb ($\circ$), Mo ($\blacklozenge$), Tc ($\lozenge$), Ru ($\blacktriangle$), Rh ($\triangle$), Pd ($*$).

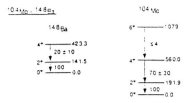

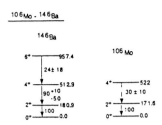

Fig.2 – a. Relative level populations of the lowest excited states of fission fragments of ^{252}Cf obtained in two zero-neutron even-even Mo / Ba splittings. The transition $2^+ \rightarrow 0^+$ is normalized to 100.

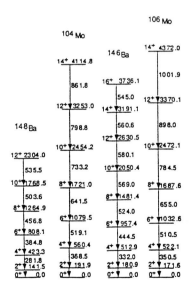

Fig.2 – b. The ground state bands observed for the ^{104}Mo -^{148}Ba and ^{106}Mo -^{146}Ba pairs as separated fragments.

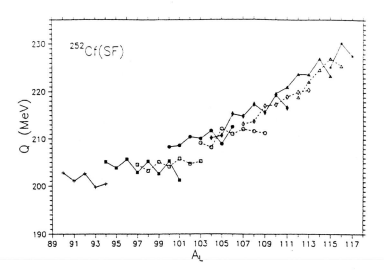

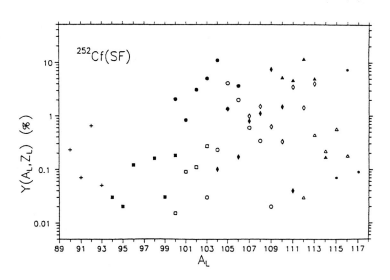

Fig.3. The Q-values (in MeV) and the calculated relative isotopic yields for the alpha ternary cold fragmentation of ^{252}Cf. The shown isotopes are Kr (+), Sr (■), Y (□), Zr (●), Nb (○), Mo (◆), Tc (◊), Ru (▲), Rh (△), Pd (∗).

SURPRISING DINUCLEAR-LIKE CONFIGURATIONS IN THE SCATTERING OF Ni+Ni

L. VANNUCCI, M. BETTIOLO

*Istituto Nazionale di Fisica Nucleare, Laboratori Nazionali
di Legnaro, 35020 Legnaro, Italy*

U. ABBONDANNO, G. VANNINI

*Dipartimento di Fisica dell'Università di Trieste,
34127 Trieste, Italy, and Istituto Nazionale di Fisica Nucleare,
Sezione di Trieste, 34127 Trieste, Italy*

M. BRUNO, M. D'AGOSTINO, P.M. MILAZZO

*Dipartimento di Fisica dell'Università di Bologna,
40126 Bologna, Italy, and Istituto Nazionale di Fisica Nucleare,
Sezione di Bologna, 40126 Bologna, Italy*

N. CINDRO [a]

Rudjer Bošković Institute, 10001 Zagreb, Croatia

R.A. RICCI

*Dipartimento di Fisica dell'Università di Padova,
35131 Padova, Italy and Istituto Nazionale di Fisica Nucleare,
Laboratori Nazionali di Legnaro, 35020 Legnaro, Italy*

T. RITZ, W. SCHEID

*Institut für Theoretische Physik der
Justus-Liebig-Universität, 35392 Giessen, Germany*

Excitation functions and angular distributions of $^{58}Ni+^{58}Ni$ and $^{58}Ni+^{62}Ni$ scattering at energies just above the Coulomb barrier have been measured around $\theta_{cm} = 90°$ in energy steps $\Delta E_{cm} = 0.25$ MeV from $E_{cm} \simeq 110$ MeV to $E_{cm} \simeq 120$ MeV for $^{58}Ni+^{58}Ni$ and from $E_{cm} \simeq 110$ MeV to $E_{cm} \simeq 118$ MeV for $^{58}Ni+^{62}Ni$. Evidence for structure of non-statistical character has been found in the angle-summed excitation functions; this evidence is corroborated by the analysis of the angular distributions. This is the first time that non-statistical structure in elastic and inelastic scattering is reported with high confidence level for this mass and excitation energy ranges. Attempts are presented to understand the nature of this structure, including the presence of intermediate dinuclear states and virtual states in a potential well.

---

[a] Invited speaker

1 Introduction and motivation

While resonances in collisions of light nuclei have been widely reported [1], the literature does not show data on resonant behavior for systems heavier than $^{28}Si+^{28}Si$ [2]. Interesting in this respect is the prediction of the orbiting-cluster model (OCM, ref. [3]) of the possibility of resonance observation in collisions of medium-mass nuclei leading to composite nuclei with closed neutron or proton $g_{7/2}$ shells (e.g. $^{28}Si+^{66}Zn$ leading to the composite ^{94}Ru or $^{46}Ti+^{58}Ni$ leading to ^{104}Sn). According to the same calculation, $^{58}Ni+^{58}Ni$ and, to a lesser degree, $^{58}Ni+^{62}Ni$ are also favorable cases for the observation of resonant behavior. In the bombarding energy region of the Coulomb barrier the model predicted spin values of $50\hbar$ to $60\hbar$. Earlier experiments of elastic and inelastic scattering of ^{46}Ti on ^{58}Ni [4] gave encouraging results, producing evidence for non-statistical structure in the energy dependence of the angle-summed elastic and inelastic cross-sections at $E_{cm} = 99$ and 100.6 MeV.

Based on this encouraging knowledge, we started a search for non-statistical configurations in medium-mass nuclei (composite masses ~ 100) that could be in some way related to the so-called quasi-molecular resonances observed in lighter composite systems ($A \sim 50$). Our first step was to look for non-statistical structure in various types of excitation and/or correlation functions; our second step was to try to relate this structure to a particular type of configurations. This step consisted in comparing the observation with calculations based on intermediate di-nuclear state model (IDS, ref. [5]) and the potential-well model (ref. [6]).

2 Experimental results

The described measurement was performed at the LN Legnaro tandem accelerator facility. Beams of ^{58}Ni were accelerated and directed to targets of ^{58}Ni and ^{62}Ni. The elastic and various inelastic scatterings (Q = -1.45 and -2.90 MeV for $^{58}Ni+^{58}Ni$ and Q = -1.17, -1.45 and -2.62 MeV for $^{58}Ni+^{62}Ni$) have been measured in angular ranges around $90°_{cm}$. The excitation functions of the angle-summed ($76° \leq \theta_{cm} \leq 104°$) elastic and inelastic (Q = -1.45 MeV first excited level of ^{56}Ni) scattering differential cross sections of $^{58}Ni+^{58}Ni$ are shown in Fig. 1, top to bottom. Correlated peaks appear at 5 energies.

The angular distributions of elastically scattered particles for $^{58}Ni+^{58}Ni$ showed marked periodicity, which could be attributed to the quantum effect of identical-particle scattering; for $^{58}Ni+^{62}Ni$, some periodicity was observed only at the energies of the correlated peaks in the excitation functions.

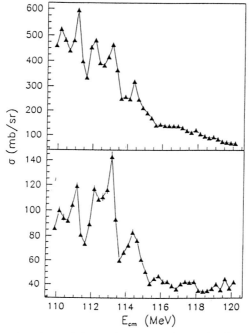

Figure 1: Excitation functions of the angle-summed elastic (top) and inelastic scattering (bottom) differential cross-sections of $^{58}Ni + ^{58}Ni$.

3 Statistical analysis

In this section we present the results obtained from the $^{58}Ni + ^{58}Ni$ and $^{58}Ni + ^{62}Ni$ data in the frame of a statistical treatment of the excitation functions (Subsection 3.1) and the angular distributions (Subsection 3.2). In Subsection 3.1 we use the method adopted by refs. [7] and [8]; in Subsection 3.2 we apply the method developed in refs [9] and [10].

3.1 Statistical analysis: excitation functions

To single out the non-statistical structure in the excitation functions, we use the data from two or more reaction channels of each of the collision systems and calculate

1. the sum of experimental angle-summed excitation functions

Table 1: Structures of non-statistical origin ($\sim$ 99% confidence level) observed in $^{58}Ni+^{58}Ni$ and $^{58}Ni+^{62}Ni$ elastic and inelastic scattering (see also subsection 3.1).

System	$E_{lab}(MeV)$	$E_{cm}(MeV)$
$^{58}Ni+^{58}Ni$	222.5	111.25
	224.5	112.25
	226.5	113.25
	229.0	114.50
$^{58}Ni+^{62}Ni$	224.0	115.70
	225.5	116.50

$$\sigma_{tot}(E) = \sum_{m=1}^{M} (\sigma_m(E)), \tag{1}$$

2. the summed deviation function (ref. [11])

$$D(E) = \frac{1}{M} \sum_{m=1}^{M} \left(\frac{\sigma_m(E)}{<\sigma_m(E)>} - 1 \right) \tag{2}$$

3. the energy-dependent cross-correlation function (ref. [11])

$$C(E) = \frac{2}{M(M-1)} \sum_{m'>m=1}^{M} \left(\frac{\sigma_{m'}(E)}{<\sigma_{m'}(E)>} - 1 \right)$$
$$\left(\frac{\sigma_m(E)}{<\sigma_m(E)>} - 1 \right) [R_{m'}(0)R_m(0)]^{-\frac{1}{2}}, \tag{3}$$

where M is the total number of correlated excitation functions, $\sigma_m(E)$ is the differential cross section of the mth excitation function at the bombarding energy E and $<>$ denotes the corresponding running average over an interval ΔE_{cm}. The $R_m(0)$ and $R'_m(0)$ are autocorrelation functions defined as

$$R_m(\epsilon) = \frac{1}{N} \sum_{n=1}^{N} \left(\frac{\sigma_m(E_n + \epsilon)\sigma_m(E_n)}{<\sigma_m(E_n + \epsilon)><\sigma_m(E_n)>} - 1 \right), \tag{4}$$

calculated for $\epsilon = 0$ and called variances of the excitation functions with N measured points. The value selected to average out the statistical fluctuations

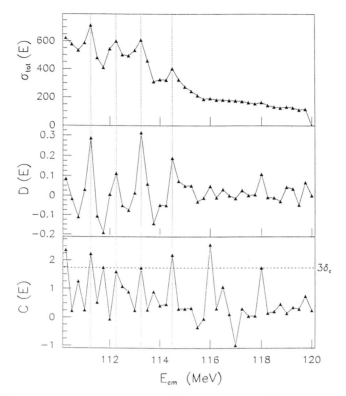

Figure 2: Top to bottom: the angle-summed (elastic and inelastic) excitation function, the deviation function and the cross-correlation function for $^{58}Ni+^{58}Ni$.

(the fine structure) was chosen as $\Delta E_{cm} = 1.0$ MeV, equal to four energy steps in the measurement of the excitation functions.

The basic idea underlying such an analysis is as follows: if structures of non-statistical nature are present, correlated peaks must appear in the three functions (1), (2) and (3). Figs. 2 and 3 show, top to bottom, $\sigma_{tot}(E)$, $D(E)$ and $C(E)$ for, respectively, $^{58}Ni+^{58}Ni$ and $^{58}Ni+^{62}Ni$. For the former system, $\sigma_{tot}(E)$ is the sum of the elastic and the inelastic (Q = -1.45 MeV) cross sections; for the latter system, $\sigma_{tot}(E)$ is the sum of all four channels measured in this reaction.

Fig. 2 $\left(^{58}Ni+^{58}Ni\right)$ shows correlated maxima at $E_{cm} = 111.25$, 112.25, 113.25 and 114.5 MeV. The analysis eliminated the peak at $E_{cm} = 110.25$ MeV that was present in the excitation functions of Fig. 1. This is most likely due to our averaging procedure (the peak lies too close to the limit of the measuring energy interval). Fig.3 ($^{58}Ni+^{62}Ni$) shows correlated maxima at $E_{cm} = 115.7$ and 116.5 MeV that originate from the peaks in the elastic excitation function.

To establish confidence limits of the non-statistical nature of the observed structures, we have drawn the limit of 3 standard deviations on the cross-correlation functions in Figs. 2 and 3 (horizontal dashed lines, bottom). Owing to the finite energy interval, the standard deviation δ_c of $C(E)$ is given by [8]

$$\delta_c = \left[\frac{2}{M(M-1)(K-1)} \right]^{\frac{1}{2}} \tag{5}$$

with M the number of correlated excitation functions and K the number of data points in the averaging energy interval. All the observed correlated peaks are outside the $3\delta_c$ limit. This fact confirms with high confidence level ($\sim$ 99%) that the underlying structure is of non-statistical character. To our knowledge, this is the first time that non-statistical structures in the elastic and inelastic scattering of two medium-mass nuclei have been observed at such high excitation energies with such a high confidence level.

The energies of the non-statistical structures are listed in Table 1.

3.2 Statistical analysis: angular distributions

The aim of the analysis presented in this subsection is to reveal periodicities in the angular distributions, and, if possible, to relate them to the values L of the angular momenta associated with the observed structures. We used the method described in refs. [9] and [10]. As it minimizes the effect of uncorrelated statistical fluctuations, this method is particularly suited to single out small intermediate-width structures as well as to reveal small-amplitude periodicities in the experimental data. The procedure is described in detail in ref. [12]; here, we only briefly state the results.

First, periodicities in the angular distributions, $\Delta\theta_{per}$, were obtained. Then, the obtained periodicities were used to deduce an estimate of the range of angular momenta (partial waves) effective in the scattering process, by postulating $L_{ded} = 180°/\Delta\theta_{per}$. The values of $\Delta\theta_{per}$ and the corresponding values of L_{ded} are shown in Table 2. For comparison, the angular momenta L_{OCM} predicted by the orbiting cluster model [3] for this range of masses and energies are also shown in the table. It is gratifying that the values of L_{ded} for the two systems, $^{58}Ni+^{58}Ni$ and $^{58}Ni+^{62}Ni$, are similar and that the L_{ded} and L_{OCM} values are also in reasonable concordance. This consistency, as well as the fact that in the latter system periodicity could be observed only at

Table 2: Angular distribution analysis of $^{58}Ni+^{58}Ni$ and $^{58}Ni+^{62}Ni$ at the energies of the peaks in the excitation functions.

System	E_{lab} (MeV)	E_{cm} (MeV)	$\delta\theta_{per}$ (deg)	$L_{ded} =$ $180°/\delta\theta_{per}$	L_{OCM} ($\hbar$)
$^{58}Ni+^{58}Ni$	220.5	110.25	3.0^a	60 ± 2	55
	222.5	111.25	3.1^a	58 ± 2	57
	224.5	112.25	3.1^a	58 ± 2	59
	226.5	113.25	3.2^a	56 ± 2	60
	229.0	114.50	3.1^a	58 ± 2	62
$^{58}Ni+^{62}Ni$	224.0	115.70	3.0^b	60 ± 1	67
	225.5	116.50	3.4^b	53 ± 1	68

^{a} deduced from a Legendre-polynomial analysis;

^{b} obtained from the statistical analysis of angular distributions (Subsect. 3.2)

the energies of the correlated peaks, speak in favor of an underlying dinuclear rotating configuration responsible for the observed non-statistical structures.

4 Comparison with model approaches

In this section we compare the experimental results with the model approach developed in ref.[5] and with a model calculation based on constructing virtual states in a potential well (refs[6] and[12]).

In two recent papers Kun[5] proposed a model for extracting periodic non-statistical structure from the excitation functions. In the model, the formation of an intermediate state is ascribed to the preferential excitation of dinuclear states rotating with a stable angular velocity. Two parameters are let free in the calculation: the angular velocity ω and the total width Γ. The best-fit values of these parameters are obtained by comparing the experimental auto correlation functions with the model expression

$$R(\epsilon) \simeq exp\left(-\frac{2\pi\Gamma}{\hbar\omega}\right) cos\left(\frac{2\pi\epsilon}{\hbar\omega}\right) \qquad (6)$$

Typical fits are shown in Fig. 4; best-fit values of ω and Γ are shown in Table 3. We stress that the best fit values of Γ are close to the observed widths of the structures in the excitation functions ($\Gamma \sim 1$ MeV). Table 3 also reports the life-times τ of the studied systems, calculated as $\hbar/\Gamma$, and, tentatively, the

Table 3: Best-fit parameters and related quantities from the autocorrelation analysis of $^{58}Ni+^{58}Ni$ and $^{58}Ni+^{62}Ni$, using Eq. 6.

System	Best-fit values $\Gamma(MeV)$	Best-fit values $\hbar\omega(MeV)$	Calculated $\mathcal{I}_{rel} = \mu R^2$ $(10^{-41}MeVs^2)$	$L_{stat} = \mathcal{I}_{rel}\omega(\hbar)$	$\tau = \hbar/\Gamma$ $(10^{-22}s)$
$^{58}Ni +^{58}Ni$	0.73	1.14	2.80	73	9.0
$^{58}Ni^* +^{58}Ni$	0.72	1.12	2.80	72	9.1
$^{58}Ni +^{62}Ni$	0.59	0.82	2.97	56	11.2
$^{58}Ni^* +^{62}Ni^*$	0.68	0.98	2.97	67	9.7
$^{58}Ni +^{62}Ni^*$	0.54	0.96	2.97	66	12.2
$^{58}Ni^* +^{62}Ni$	0.53	0.94	2.97	64	12.4

average angular momenta $L_{stat} = \mathcal{I}_{rel}\omega/\hbar$ ($\mathcal{I}_{rel}$ is assumed to be the moment of inertia of two osculating rigid nuclei, with $r_0 = 1.25$ fm).

It appears from Table 3 that the decay of the two colliding systems, $^{58}Ni+^{58}Ni$ and $^{58}Ni+^{62}Ni$, is rather rapid (order of 10^{-21} s), testifying of the short lifetime of the supposed rotational dinuclear system. Even so, the high values of the corresponding angular momenta lead to reasonable rotation angles (1.5 rad), a fact that sheds a different light on the whole approach.

Another approach is based on the search for pockets in the interaction potential of the two nuclei. Such pockets would generate quasi-bound states and virtual resonances responsible for the structure observed in the various excitation functions. Virtual states near the Coulomb barrier were predicted several years ago in ref. [6]. Recently, we calculated the folding potentials of $^{58}Ni+^{58}Ni$ for values of the angular momenta from $L = 56\hbar$ to $64\hbar$ in the sudden approximation using a double-folded Yukawa integral; details are given in ref. [12]. The resulting excitation functions of the elastic and inelastic scattering do show structures, albeit less pronounced than the experimental data.

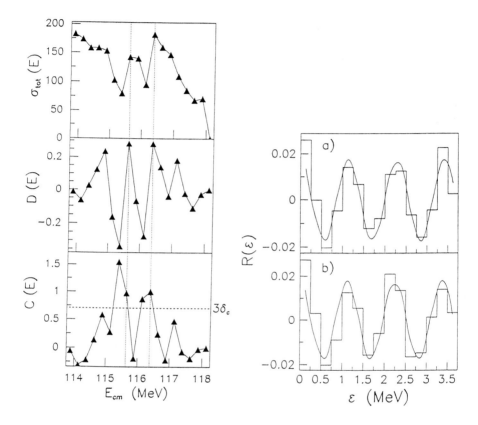

Figure 3: Same as Fig. 2, for $^{58}Ni+^{62}Ni$.

Figure 4: Histograms: the experimental energy autocorrelation functions calculated for (a) $^{58}Ni+^{62}Ni$ elastic, (b) the double ($Q = -2.62MeV$) excitation, (c) the single excited level ($Q = -1.45MeV$) in ^{58}Ni and (d) the single excited level ($Q = -1.17MeV$) in ^{62}Ni. The sinusoidal curves are the corresponding best fits to the histograms obtained with the expression (6).

5 A short summary

In this contribution we have presented the data on the $^{58}Ni+^{58}Ni$ and $^{58}Ni+^{62}Ni$ elastic and inelastic scattering, taken in the bombarding energy range of the Coulomb barrier. These data have been used in a search for intermediate configurations in the composite system, the presence of which is suggested by phenomenological models. The analysis of the scattering process reveals the presence of intermediate non-statistical structures, with a high level of confidence. In view of the explored mass and energy ranges, this by itself is a surprising finding. Moreover, some features of the scattering process, stressed by the appropriate statistical analysis, speak for the idea that the observed structures are related to weak dissipation of energy in the collision process; the slow rate of energy dissipation may favor the formation of dinuclear configurations in the early stages of the nucleus-nucleus collision.

References

1. For a recent review of the subject see, e.g., U. Abbondanno and N. Cindro, *Int. Jour. Mod. Phys.* **E2**, 1 (1993).
2. R.R. Betts, S.B. DiCenzo and J.F. Petersen, *Phys. Lett.* B **100**, 117 (1981).
3. N. Cindro and M. Božin, *Ann. Phys. (N.Y.)* **192**, 307 (1989).
4. U. Abbondanno et al., *Int. Jour. Mod. Phys.* **E3**, 919 (1994).
5. S. Yu. Kun, *Phys. Lett.* B **257**, 247 (1991); *Europhys. Lett.* **26**, 505 (1994).
6. R. Könnecke, W. Greiner and W. Scheid, in Nuclear Molecular Phenomena, ed. by N. Cindro, North Holland Publ. Co., Amsterdam 1978, pg. 109
7. D. Počanić et al., *Nucl. Phys.* A **444**, 303 (1985).
8. A. Sarma, *Z. Phys.* A **337**, 23 (1990).
9. G. Pappalardo, *Phys. Lett.* **13**, 320 (1964).
10. E. Gadioli et al., *Nuovo Cimento* **38**, 1105 (1965).
11. L.C. Dennis, S.T. Thornton and K.R. Cordell, *Phys. Rev.* C **19**, 777 (1979).
12. L. Vannucci et al., *Z. Phys. A*, in press.

STRUCTURE OF FUNDAMENTAL MATTER AT THE DRIPLINES

JAN S. VAAGEN

SENTEF, Department of Physics, University of Bergen, Norway

with RNBT collaboration

T. ROGDE, B.V. DANILIN, S.N. ERSHOV,
M.V. ZHUKOV and I.J. THOMPSON

Recent developments of radioactive nuclear beam techniques have allowed us to reach the neutron drip line for light nuclei. Although we are only at "the end of the beginning", hiking along the rugged coastlines of fundamental matter has been highly rewarding. Striking halo clusterizations of matter into a normal core and a loosely bound low density veil of neutrons seem to be more a rule than an exception. Thus we may study matter with very low density. We discuss the development of a theoretical understanding of dripline matter and present results of recent investigations of what specific features the continuum of a system with a halo ground state may have.

1   The rugged coastlines of fundamental matter. Halos.

By adding neutrons to light nuclei, surprising dripline structures have been discovered. The Helium chain (see fig. 1) is an example; adding a neutron to an alpha particle does not produce a bound system, adding one neutron more does; ^{6}He a loosely bound halo-like system. The same happens if we continue. While ^{7}He and ^{9}He are unbound, ^{8}He is bound and ^{10}He nearly bound; 2 protons and 8 neutrons! The neutron dripline for light nuclei resembles the Norwegian coast, being filled with skerries and underwater rocks.

Substantial evidence has by now been accumulated for halo like phenomena in nuclear physics: Light nuclei at the neutron dripline which exhibit a clear separation between a "normal" core nucleus and a loosely bound low-density veil of neutrons. This offers new and very interesting possibilities for studies of neutron matter in a low-density background and, vice versa, the response of the core nucleus in a cloud of neutrons. If we try to support our thinking by using shell model pictures with a simple local field, the core and halo neutrons have no simple relations to a common field. For heavier dripline nuclei we should not take for granted that the shell structure from the stability line is necessarily preserved.

The birth and early development of dripline physics must be credited to visionary experimentalists who, supporting their own work by practitioners' theory models, discovered that old rules are no longer applicable [1]. That ^{11}Li has a r.m.s. size which old scaling rules would ascribe to a nucleus with more than twice that number of nucleons, was truely a remarkable discovery. If distributed within the large r.m.s. radius, the average density would be less than half that of ordinary nuclei. The unravelling of the mystery as being due to a two-neutron halo is an international experimental achievement. The main tool for investigating halo features has been breakup reactions induced on fixed targets, with the weakly bound nuclei as secondary radioactive beams.

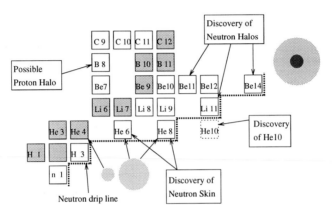

Figure 1: The neutron halo and neutron skins were discovered from measurements of matter radii.

The physics of the halo phenomena seems to be the homefield for few-body cluster models. To understand why these new clustering features emerge from the nuclear many-body problem is a challenge for the future. In the mean-time few-body approaches, where the interactions between the constituents are tuned to available experimental data, provide the practitioners with tools for learning how cluster phenomena are interrelated. This approach is a natural continuation of the way it all started, but theory now adds better underpinning and substantially greater depth.

The emphasis in this paper is on weakly bound light neutron rich nuclei, mainly of borromean type [2], and where pronounced halo features are present. The most exciting nuclei found so far *are* borromean; while none of the binary channels have bound states, the full system is still bound, but weakly and often with only one bound state. Typical examples are the two-neutron halo nuclei ^{6}He and ^{11}Li, each having just one bound state, with two-neutron separation energies 1.0 MeV and 0.3 MeV respectively. The only one-neutron halo nucleus known so far is ^{11}Be which has two bound states, both of halo type, with one-neutron separation energies 0.5 MeV and 0.2 MeV. A new and possibly even more extended candidate is ^{19}C, which is being investigated for the moment. Fig. 1 gives a relevant part of the chart of isotopes.

The physics of dripline nuclei is that of threshold phenomena, emphasizing the increased importance of the continuum [3]. The quest for unambiguous pictures of the underlying new structures calls for adequate tools. Such a probe is the $s_{1/2}$ orbital motion: it is the main component of the $1/2^+$ ground state of ^{11}Be, apparently a virtual state in ^{10}Li, and probably a large $(s)^2$ component in the halo of ^{11}Li. Orbits with low angular momentum, hence low centrifugal barriers, are crucial halo inducers.

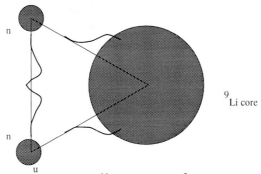

Figure 2: The geometrical structure of ^{11}Li and its core ^{9}Li with r.m.s radii 3.2 and 2.3 fm, exhibiting borromean "residence in forbidden regions". The ranges of the binary interactions are indicated.

Fig. 2 gives an impression of the average (r.m.s.) geometrical separations in ^{11}Li, indicating also the ranges of the interactions between the constituents. The situation is truely one of "residence in forbidden regions" [4]. The figure corresponds to three-body calculations [2], the gross structure is however insensitive to finer details of the calculations.

The abnormal size of the halo systems is obviously related to the unusually small separation energies. Since the halo nucleons reside in the nuclear periphery, the deficiency of oscillator expansions becomes acute, and even more so than for form factors for particle transfer reactions, a well known problem from past years. Recall that the maximal classical turning point for an oscillator orbit is roughly $r_{max}^2 \sim b^2(N + \sqrt{N^2 - \ell^2})$ where $b = (\hbar/M\omega_0)^{1/2}$ is the oscillator length, N the oscillator number and ℓ the orbital angular momentum. For s-motion we get $r_{max} = b\sqrt{2N}$.

It was not clear from the first experiments if ^{11}Li and ^{6}He had point-like dineutron structures or if the two neutrons were significantly spatially separated on the average. Three-body theory [2] predicted the latter situation, as shown in fig. 2. These predictions have been substantiated by experiment.

Another question, in particular relevant for ^{11}Li and ^{11}Be, is to what degree the core nucleus is polarized or excited as part of the halo system. Although this question still requires further investigations, core excitations is important in ^{11}Be, apparently less so in ^{11}Li; but what we have learned so far will not cloud the basic picture of a core and halo neutrons, which on the average are separated so much that they reside in the tails of the cluster interactions. Thus experiments show that static electromagnetic moments are nearly the same for ^{11}Li and ^{9}Li [1]. Few-body halo models have in fact been very successful [2].

The most natural physical building blocks for describing such systems are, just like for transfer reactions, appropriate overlap functions $\mathcal{O}_C^A(x) = \langle \Psi_C(C) , \Psi_A(C, x) \rangle$ (where x are the translation invariant halo coordinates) between a given state A for the halo nucleus and a possible state C for the core, regarded as a separate nucleus. In a few-body model with the core in its ground state, the overlap coincides with the few-body wave function.

The asymptotic behavior of a one-neutron overlap between bound states is simply $\exp(-\kappa_C^A r_{1C})$ where r_{1C} is the relative coordinate between the core and the valence neutron while the decay constant $\kappa_C^A = \sqrt{2\mu_{1C} S_C^A}/\hbar$ is governed by the separation energy between the two states being connected.

The asymptotic behavior becomes essentially more complicated already for two halo neutrons [5]. For borromean nuclei [2], however, it simplifies to $\exp(-\kappa_C^A \rho)$ where ρ is the hyperradius, expressed in n-n and (nn)-C relative distances as

$$\rho^2 = \frac{1}{2} r_{nn}^2 + \frac{2C}{C+2} r_{(nn)C}^2 \tag{1}$$

while the decay constant now is $\kappa_C^A = \sqrt{2M S_C^A}/\hbar$ and S_C^A the two-neutron separation energy while M is the nucleon mass. If an initial expansion of the three-body wave function on hyperspherical harmonics (HH) is used, the three-body Schrödinger equation becomes a set of coupled equations for radial amplitude functions $\chi_{[K]}(\rho)$ in the hyperradius ρ, K being the hypermoment quantum number associated with the "angular" conjugate of the hyperradius.

For borromean systems there is no binary binding. So why is the system bound? Part of the explanation is related to the three-body HH centrifugal barrier, which is proportional to $(K + 3/2)(K + 5/2)/\rho^2$ where $K = 2n + \ell_{(nn)C} + \ell_{nn}$ with (n=0, 1, 2 ...) is the hypermoment. Even if $K = 0$, implying that the total orbital angular momentum is zero, the three-body centrifugal barrier does not vanish contrary to the two-body case. For sufficient binding between the bodies, this may provide a barrier inside which bound motion may be established. As Efimov [6] first pointed out, the attraction in ρ contains components of asymptotic behaviour ρ^{-3} for the diagonal terms.

2   Halo analogue states - the persistence of halo structure.

May we use isospin transitions to detect the halo, not only its size but also its inner orbital structure? An example is the halo-neutron, core-neutron overlap Gamow-Teller (GT) transition ^{11}Li(gs; $3/2^-$) $\rightarrow$ ^{11}Be($1/2^-$; 0.32 MeV) between a two-neutron and a one-neutron halo state, recently discussed by T. Suzuki and T. Otsuka [7]. To obtain agreement with data a substantial ($\sim$50 %) s^2 component seems required in the ^{11}Li halo.

For halo nuclei with a ^{4}He core ($T=T_3=0$) the total isospin coincides with that of the halo. When we look at the energy differences between a two-neutron halo nucleus like ^{6}He and the other members of the isobaric supermultiplet, we notice significant energy shifts. We may however ask if this implies drastic changes in the wave function. Our answer is no; the internal structure of the triplet wave functions is largely preserved if we change halo neutrons into protons. The symmetry operates at the level of wave functions, not energies. This also applies to the iso-singlet ^{6}Li ground state, leading to maximum B(GT) values. For other halo systems like ^{11}Li, also the core carries isospin. Still we may ask if not only a spatial separation but also a separation of the system into nearly decoupled core and halo isospins is a useful physical concept. We have suggested the term *halo analogue (HA) states*. Thus, we have recently [8] studied HA states in the ^{11}Be spectrum within the framework of a three-body (^{9}Li+n+p) approach. Very recent experiments seem to support the validity of the concept. Three-body bound and scattering HA states, with structure similar to the two-neutron halo state of ^{11}Li are obtained in ^{11}Be at excitation energies above 18 MeV and comparable with experimental spectra. The influence of the "deuteron" like HA state with T=3/2 on beta-delayed deuteron emission from ^{11}Li was also studied and we found that the shape of the deuteron spectrum and branching ratio to the ^{9}Li+d channel are sensitive to the position of this HA state in ^{11}Be.

3   The borromean continuum; halo excitation

The nucleus ^{6}He, for which we have good information about the binary cluster interactions, serves as the "benchmark" nucleus that a model with self-respect should reproduce. The structure of the 0^+ (0.973 MeV) ground state, shown in [2], has in fact been reproduced by a number of few-body procedures. The excited 2^+ state, a narrow resonance at about 0.8 MeV above the two-neutron threshold, has also been consistently reproduced, see ref. [9] and [10]. The known spectrum of ^{6}He contained up to quite recently only these two states and then a desert in the three-body $\alpha+n+n$ continuum up to the ^{3}H + ^{3}H threshold at about 13 MeV.

Using DWIA theory appropriate for dilute halo matter we have recently probed the structure of the low-lying ^{6}He continuum (below two-body threshold) via calculations of charge-exchange (Fig. 3) and inelastic scattering. For

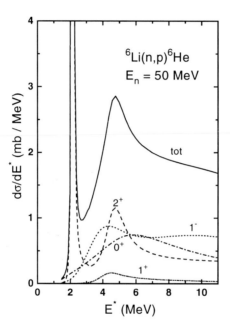

Figure 3: Predicted charge-exchange cross section for the ^{6}He continuum. The total magnitude and multipole decomposition are given.

a Borromean system the HH treatment involving the hyperradius ρ can be extended to the continuum. The mathematical scheme bears resemblance with that of a particle in a strongly deformed field $V_{K\gamma,K'\gamma'}(\rho)$. Here we will only comment on the new 2_2^+ resonance and the so-called "soft dipole mode" suggested in [11]. To this end, as in binary systems, we look for pockets in the diagonal potentials $V_{K\gamma,K\gamma}(\rho)$, and we study the energy dependence of three-body phase shifts (extracted from diagonal elements of the S-matrix) and of eigenphases. To discriminate between genuine 3-body resonant behaviour and response in particular binary partitions, we study how energy correlation plots vary with choice of Jacobi coordinates. Like for the much narrower 2_1^+, the 2_2^+ behaviour shown in Fig. 4 for "di-neutron" $\{r_{nn}, r_{\alpha(nn)}\}$ coordinates stays the same when we change to the $\{r_{\alpha n}, r_{n(\alpha n)}\}$ system. Potential pockets and phase shifts also support a classification of these responses as due to a genuine 3-body resonance.

This is not the case for the dipole response peak which shows preference for the di-neutron partition. It is easier to say what it is not than what it is: it is not a simple di-neutron oscillation either. Nature does not seem to quite favor an idea that created a lot of enthusiasm, at least not in ^{6}He.

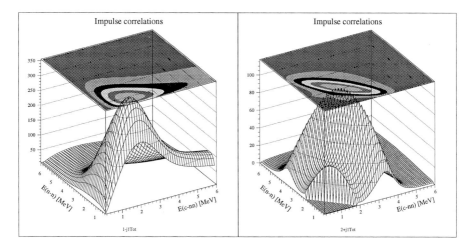

Figure 4: Correlated response for 1^- "soft dipole"(left), and 2_2^+ resonance (right).

The question is then: May soft-dipole resonances be realized in other halo systems? The natural candidate would be the halo nucleus par excellence, ^{11}Li. A problem for theory has been the rudimentary information on the binary n-^{9}Li channel, with implications for the question about the relative proportion of p^2- and s^2-motion in the ground state of ^{11}Li.

The only calculations that agrees with the available data (MSU and RIKEN) for the electric dipole response in ^{11}Li, have a large component of s^2-motion. They produce again a strength accumulation at low energies, but no genuine dipole resonance. The exploration of the ^{11}Li continuum is not a closed chapter.

4 Outlook

> *There is nothing more difficult to take in hand, more perilous to conduct, or more uncertain in its success than to take the lead in the introduction of a new order of things,*
>
> *because the innovator has for enemies all those who have done well under the old condition, and lukewarm defenders in those who may do well under the new.* (Machiavelli)

The evidence for something new is however becoming increasingly convincing. New experimental possibilities allow us to study fundamental matter along new dimensions, and we are presently only at "the end of the beginning". Also at AGS energies are halo nuclei ^{6}He and ^{8}He being produced in heavy-ion collisions. A main objective of some of these current studies is to search for strangelets, i.e. to explore the strangeness degree of freedom. A calculation of the $^6_{\Lambda\Lambda}$He hypernucleus within the three-body hyperharmonic procedure was already carried out by one of us (Danilin) a few years ago. Aspects of the strangeness degree of freedom will certainly be addressed in future investigations.

The fact that the halo particles are nearly free provides a new and very interesting starting point for the continuum multi-cluster problem. The potential role of halo nuclei in nucleosynthesis in supernovae, by combining alpha particles and neutrons in the outskirts of the exploding iron core, is a field of current investigation.

References

[1] See T. Tanihata, J. Phys. G: Nucl. Part. Phys. **22** (1996) 157 and references therein; P.G. Hansen, A.S. Jensen and B.Jonson, Ann. Rev. Nucl. Part. Sci. **45** (1995) 591.

[2] M.V. Zhukov *et al.*, Phys. Rep. **231** (1993) 151, and references therein. The Borromean rings, included in the heraldic symbol of the Princes of Borromeo, are three rings interlocked in such a way that if any of them were removed, the other two would also fall apart.

[3] W. Nazarewicz, T.R. Werner and J. Dobaczewski, Phys. Rev. **C50** (1994) 2860 and references therein.

[4] K. Riisager, Rev. Mod. Phys. **66** (1994) 1105

[5] J.M. Bang *et al.*, Phys. Rep. **125** (1985) 253

[6] V.M. Efimov, Comments Nucl. Part. Phys. **19** (1990) 271

[7] T. Suzuki and T. Otsuka, Phys. Rev. **C50** (1994) R555-R558

[8] M.V. Zhukov et al., Phys. Rev. **C52** (1995) 2461

[9] B.V. Danilin *et al.*, Phys. Lett. **B302** (1993) 129

[10] A. Csótó, Phys. Rev. **C48** (1993) 165

[11] P.G. Hansen and B. Jonson, Europhys. Lett. 4 (1987) 409; K. Ikeda, INS report JHP-7 (1988) [in Japanese]; Y. Suzuki, Nucl. Phys. **A528** (1991) 395

THE "HOLE" IN HELIUM

L. Wilets, M. A. Alberg*
*Department of Physics, Box 351560, University of Washington, Seattle, WA
98195-1560, USA*

J. Carlson
*Theoretical Division, Los Alamos National Laboratory, Los Alamos, New Mexico
87545, USA*

W. Koepf
*School of Physics and Astronomy
Raymond and Beverly Sackler Faculty of Exact Sciences
Tel Aviv University, 69978 Tel Aviv, Israel*

S. Pepin and Fl. Stancu
*Université de Liège
Institut de Physique B.5
Sart-Tilman, B-4000 Liège 1, Belgium*

The measurement and analysis of electron scattering from ^{3}He
and ^{4}He by Sick and collaborators reported 20 years ago remains a
matter of current interest. By unfolding the measured free-proton
charge distribution, they deduced a depression in the central point
nucleon density, which is not found in few-body calculations based
on realistic potentials. We find that using wave functions from such
calculations we can obtain good fits to the He charge distributions
under the assumption that the proton charge size expands toward
the center of the nucleus. The relationship to 6-quark Chromo-
Dielectric Model calculations, is discussed. The expansion is larger
than than the predictions of mean field bag calculations by others
or our CDM calculations in the independent pair approximation.
There is interest here in the search for a "smoking gun" signal of
quark substructure.

*permanent address: Department of Physics, Seattle University, Seattle WA 98122,
USA

103

I. INTRODUCTION

The charge distribution of nuclei has been the subject of experimental studies for more than forty years. Electron scattering and muonic atoms now provide detailed descriptions of the full range of stable, and many unstable nuclides. Unique among the nuclides are the isotopes ^{3}He and ^{4}He because they exhibit a central density about twice that of any other nuclide. There is a long-standing apparent discrepancy between the experimentally extracted charge distributions and detailed theoretical structure calculations which include only nucleon degrees of freedom.

McCarthy, Sick and Whitney [1,2] performed electron scattering experiments on these isotopes up to momentum transfers of 4.5 fm^{-1} yielding a spatial resolution of 0.3 fm. They extract a "model independent" charge distribution, which means that their analysis of the data is not based upon any assumed functional form for the charge distributions. Their charge distributions are shown in Fig. 1. Taken alone, they do not appear to be extraordinary. However, using a finite proton form factor, which fits the experimentally measured rms radius of about 0.83 fm, they unfolded the proton structure from the charge distributions to obtain the proton point distributions. For both isotopes there is a significant central depression of about 30% extending to about 0.8 fm. Sick [2] also presented results where relativistic and meson effects are included. These are shown for ^{4}He in Fig. 2. One note of caution here is that it is not possible to subtract these effects from the experimental data in a completely model-independent way.

One might assume that such a central depression is to be expected because of the short-range repulsion of the nucleon-nucleon interaction. So far, this is not borne out by numerous detailed theoretical calculations, none of which finds a *significant* central depression, certainly not of the above magnitude. Relatively smaller central depressions are found in Green's function Monte Carlo (GFMC) calculations of the alpha particle for realistic models of the two- and three-nucleon interaction. (see Fig. 4 below.)

The status of theoretical structure calculations through mass number 4 is very satisfactory at present. Given any assumed interaction, the few body problem can be solved to within tenths of an MeV in energy and the wave function can be calculated to a precision better than that required for the present discussion.

In using a nuclear wave function to construct a charge distribution, one must assume a nucleon charge density and the possibility of meson exchange contributions. While the meson exchange contributions in the transverse channel are well-constrained (at least at moderate momentum transfer) by current conservation, no such constraint is available in the longitudinal channel. Indeed, meson exchange current contributions are of relativistic order and hence one must be careful when interpreting them with non-relativistic wave functions.

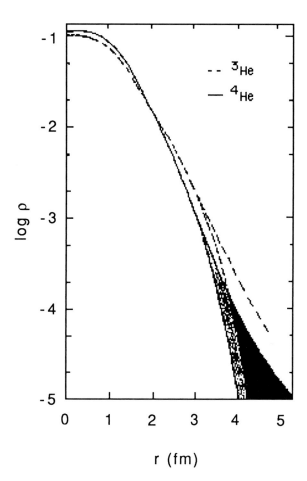

FIG. 1. Model-independent charge distributions extracted from experiment. Reproduced from McCarthy *et al.*[1].

Given these caveats, it is possible to reproduce reasonably well the longitudi-

106

nal form factors of 3- and 4-body nuclei within a nucleons plus meson-exchange model. [3,4] The current and charge operators are constructed from the N-N interaction and required to satisfy current conservation at non-relativistic order. The resulting meson-nucleon form factors are quite hard, essentially point-like. [3] This raises the possibility of explaining the form factors in quark or soliton based models, which would describe the short-range two-body structure of the nucleons in a more direct way than is available through meson exchange current models. For example, see the discussion of a model by Kisslinger *et al.* [6] below.

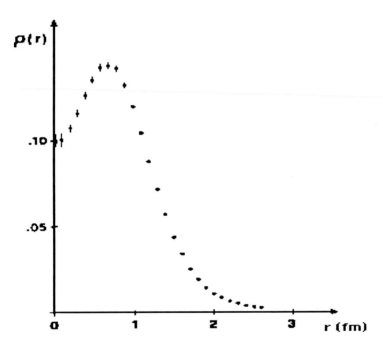

FIG. 2. Point-proton density distributions obtained by unfolding the finite free proton form factor, allowing for meson exchange corrections and relativistic effects. Reproduced from Sick[2].

We present here a possible explanation of the electric form factors which does not involve a hole in the point-proton density distribution, but rather is consistent with theoretical few body calculations. It involves the variation of the proton charge size as a function of density, or as a function of nucleon-nucleon separation. This is not depicted as an average 'swelling' of the nucleon, but as a result of short-range dynamics in the nucleon-nucleon system. The results presented here are preliminary but encouraging.

II. QUARK SUBSTRUCTURE OF NUCLEI AND NUCLEONS

Within the context of soliton models, there have been numerous calculations of the nucleon size in nuclear media. Most of these involve immersion of solitons in a uniform (mean) field generated by other nucleons. Another approach has been the 3-quark/6-quark/9-quark bag models, which has been applied to various nuclear properties, including the EMC effect. It has been applied by Kisslinger et al. [6] to the He electric form factors with some success.

In a series of papers, Koepf, Pepin, Stancu and Wilets [5,7,8] have studied the 6-quark substructure of the two-nucleon problem, and in particular obtain the variation of the quark wave functions with inter-nucleon separation. Contrary to previous expectations, the united 6-quark cluster does not exhibit a significant decrease in the quark momentum distribution function in spite of an increase in the volume available to the individual quarks. [7] This is due to configuration mixing of higher quark states. Such a momentum decrease was proffered as an explanation of the EMC effect. However, the united cluster does have approximately twice the volume of confinement of each 3-quark cluster, and the quarks extend to a volume nearly three times that of the 3-quark clusters, again enhanced by configuration mixing of excited states.

In Fig. 3 we exhibit the proton rms charge radius r_p extracted as follows from the calculations of Pepin et $al.$ [8]: the abscissa gives the effective nucleon-nucleon separation r_{NN} obtained by the Fujiwara transformation; the soliton-quark structure is a 6-quark deformed composite. The proton rms radius is defined to be

$$r_p = \sqrt{<r^2> - r_{NN}^2/4} \tag{1}$$

where the quark density used in calculating $<r^2> = \int \rho_q r^2 d^3r$ is the six-quark density normalized to unity.

For well separated solitons, the r_{NN} is just the separation of the soliton centers and $r_p = 0.83$ fm as indicated by the horizontal line. Large defor-

108

mations (near separation) are difficult to calculate so that the figure does not reproduce well the separation region. Shown also in the figure is a gaussian approximation fitted to $r_{NN} = 0$, $r_{NN} = 1$ fm, and the asymptotic region, see Eq. (5) below, with $A = 0.45$, $s = 0.92$ fm. Then the charge distribution due to two-body correlations is

$$f_p(\boldsymbol{r}, \boldsymbol{r}; r) = \Big\{ \delta_{ip} \, \exp\big[- |\boldsymbol{r} - \boldsymbol{r}|^2 / b^2(r_{ij})\big]$$
$$+ \, \delta_{jp} \, \exp\big[- |\boldsymbol{r} - \boldsymbol{r}|^2 / b^2(r_{ij})\big] \Big\} / 2 \, \pi^{3/2} b^3(r_j) \qquad (2)$$

where we assumed and indicate explicitly that b is a function of the distance r_{ij} between the nucleons i and j, as we expand upon later, and that the proton and neutron functions are the same. Here "p" stands for "proton" and the Kronecker deltas pick out protons among i and j.

Proton Form Factor

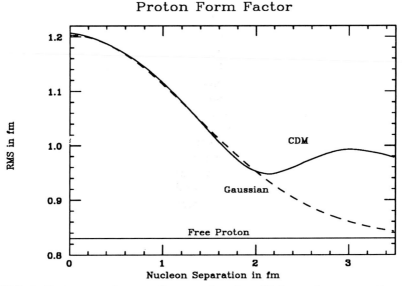

FIG. 3. Proton rms charge radius as a function of inter-nucleon separation. The dashed line is a gaussian approximation, normalized to the free value.

Using the independent pair approximation (IPA) and Eq. (2) we find the charge distribution by employing a two-body correlation function $\rho_2(\boldsymbol{r}, \boldsymbol{r})$,

$$\rho_{ch}(r) = \sum_{i<j} \int d^3 r_i \int d^3 r_j \, \rho_2(r \,, r \,) f_p(r \,, r \,; r)/3 \,. \qquad (3)$$

There are six pairs (i, j). Each proton appears three times; Hence the factor $1/3$.

III. PHENOMENOLOGICAL DENSITY-DEPENDENT PROTON FORM FACTOR

To obtain some qualitative feeling for the expansion of the proton charge size with nucleon density, we assume a proton form factor f_p (differing from Eq. (2)) with a size that depends simply on the local density and hence on the distance from the center of the nucleus (r'). Then

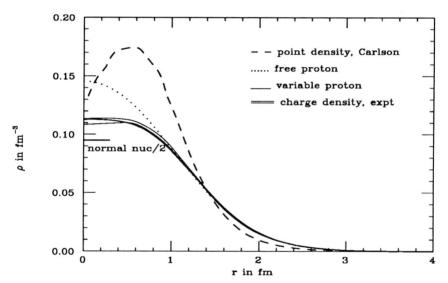

FIG. 4. ^{4}He density distributions including a fit with a variable proton charge size. The dashed curve labelled "Carlson" is based on a Green's function Monte Carlo calculation [4]

$$\rho_{ch}(r) = \int d^3r' \rho(r') f_p(\mathbf{r}, \mathbf{r}) \,, \tag{4}$$

where ρ is the (theoretical) point proton density

$$f_p(\mathbf{r}, \mathbf{r}') = \exp\left[-|\mathbf{r} - \mathbf{r}'|^2/b^2(r')\right] / \pi^{3/2} b^3(r') \tag{5}$$

and we choose

$$b(r') = b_0[1 + Ae^{-r'^2/s^2}] \,, \tag{6}$$

with $b_0 = \sqrt{2/3}\,0.83$ fm, the free proton value. A fairly good fit to the data was found with $A = 0.45$ and $s = 0.65$ fm corresponding to $b(0)$ equal to the central value given in Fig. 3. The best fit, with only slightly better χ-squared, was obtained with $A = 2.10$ and $s = 0.13$ fm, which does not seem to be reasonable, in that the A is too large and the s too small.

IV. THE INDEPENDENT PAIR APPROXIMATION

In the spirit of the independent pair approximation, the charge distribution was calculated using Eq. (3) with ρ_2 the two-particle correlation function [4]. The proton size parameter $b(r_{12})$ was first taken from the gaussian fit to the calculations of Pepin et $al.$ [8]. The improvement over the free constant proton size, as shown in Fig. 5 (dot-dash) was small.

A phenomenological fit to the data was made with a parameterized b given by

$$b(r_{12}) = b_0[1 + Ae^{-(r_{12}/s)^n}] \,, \tag{7}$$

where $n = 2$ is of the gaussian form. $n > 2$ yields a sharper transition. Indeed, $n \to \infty$ yields a step function. Recall that the model of Kisslinger et $al.$ corresponds to a step function.

We examined $n = 2$, 4 and 6. Although the A and s were different in each case, the quality of fits were very similar. The corresponding best fit values of (A, s, n) for the three n's were (2.185, 0.883, 2), (0.976, 1.245, 4), (0.774, 1.34, 6). In Fig. 5 we show the results for $n=2$ since the others are indistinguishable to the eye.

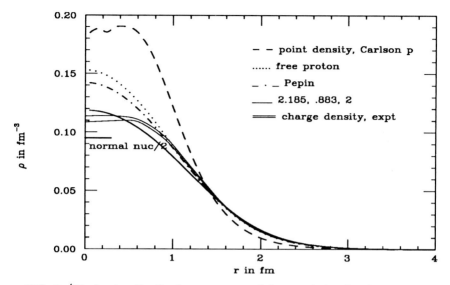

FIG. 5. ^{4}He density distributions constructed from variational point densities and two-body correlation functions in a parameterized variational calculation by Carlson *et al.*[4]. The curve labelled "Pepin" uses the gaussian fit of Fig. 3, based on the calculations of Pepin *et al.*[8]. The solid curve is a phenomenological fit as described in the text

V. CONCLUSIONS

We have obtained a phenomenological fit to the ^{4}He charge distribution by assuming a proton size which increases with increasing density. More specifically, we minimize the charge distribution χ-squared using a two-parameter gaussian function of r, the distance from the center-of mass.

We would like to identify the variable proton size with the structure function of Fig. 3 derived from 6-quark N-N studies in the spirit of the independent pair approximation. Fairly good agreement with experiment was obtained with a phenomenological parameterization of the proton size function.

The inadequacy of the previous calculation might be due to

• A constant confinement volume was assumed for the six-quark structure as a function of deformation. It may be that the intermediate volume (between separated and united clusters) is larger.

• The independent pair approximation may be invalid at the high densities of the central region.

• Meson effects should be recalculated using the quark structure functions given (say) by the six-quark IPA model.

Items 1 and 3 are topics for further investigation. In addition, one must study the predictions of such models for quasi-elastic scattering. In the quasi-free regime, nucleon models produce a good description of the data as long as realistic nucleon interactions, including charge exchange, are incorporated in the final-state interactions. [9] Unlike the charge form factor, two-body charge operators are expected to play a much smaller role here [4]. The combination of the two regimes provides a critical test for models of structure and dynamics in light nuclei.

ACKNOWLEDGMENTS

This contribution is dedicated to Prof. Walter Greiner on the occasion of his sixtieth birthday.

We wish to thank C. Horowitz for valuable discussions. This work is supported in part by the U. S. Department of Energy.

---

[1] J. S. McCarthy, I. Sick and R. R. Whitney, Phys. Rev. C **15**, 1396 (1977).

[2] I. Sick, *Lecture Notes in Physics*, **87**, 236 (Springer, Berlin, 1978).

[3] R. Schiavilla, V. R. Pandharipande, and D. O. Riska, PRC 40, 2294 (1989) .

[4] R. Schiavilla and D. O. Riska, Phys. Lett. **244B**, 373 (1990); R. B. Wiringa, Phys. Rev. C **43**, 1585 (1991); J. Carlson, Nucl. Phys. **A522**, 185c (1991).

[5] W. Koepf, L. Wilets, S. Pepin and Fl. Stancu, Phys. Rev. C **50**, 614 (1994).

[6] L. S. Kisslinger, W.-H. Ma and P. Hoodbhoy, Nuc. Phys. **A459**, 645 (1986); W.-H. Ma and L. S. Kisslinger, Nuc. Phys. **A531**, 493 (1991).

[7] W. Koepf and L. Wilets, Phys. Rev.C **51**, 3445 (1995).

[8] S. Pepin, Fl. Stancu, W. Koepf and L. Wilets, Phys. Rev. C **53**, 1368 (1996).

[9] J. Carlson and R. Schiavilla, Phys. Rev. Lett. **68**, 3682 (1992); Phys. Rev. C**49**, R2880 (1994).

Coffee break and gathering for the conference picture

1. Francis Jundt and Andre Gallmann, Dan Strottman and Igor Mishustin in the second row, Judah and Nelly Eisenberg further back.

2. Jens Konopka and Tanja Bittner, Oda Stöcker, Werner Scheid, Miklos Gyulassy, Laszlo P. Csernai, and Istvan Lovas.

3. Eric Uggerhøj and Mrs. Uggerhøj, Andrey Solov'yov, Leonid Satatrov and Mark Gorenstein.

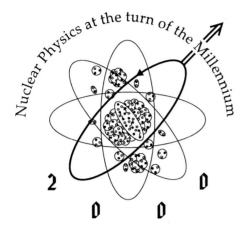

Nuclear Physics at the turn of the Millennium

2. Supercritical Fields and Atomic Physics

ELECTRON-POSITRON PAIR PRODUCTION IN ATOMIC RELATIVISTIC HEAVY ION COLLISIONS [a, b]

N. GRÜN, C. MÜLLER, W. SCHEID, T. STEIH, and R. TENZER

Institut für Theoretische Physik der Justus-Liebig-Universität Giessen

1 Introduction

Electron-positron pairs can be produced by the electromagnetic fields of highly charged heavy ions in atomic relativistic heavy ion collisions. With atomic collisions we mean collisions with impact parameters larger than the sum of the nuclear radii, i.e. the nuclear forces have no influence on the collision. An overview over this field of physics has recently been given in the monography of Eichler and Meyerhof [1]. From the theoretical side various methods were developed to estimate the cross section for electron-positron production with the capture of the electron into the K-shell, since this cross section limits the lifetime of heavy ion beams in relativistic heavy ion colliders [2]. Experiments with relativistic heavy ions are rare. Measurements of the electron-positron pair production in peripheral collisions of S^{16+} (200 GeV/nucleon) on fixed targets of Al, Pd and Au were carried out by Vane et al. [3] at CERN. The electron capture from electron-positron pair production was observed by Belkacem et al. [4] with a 0.96 GeV/nucleon U^{92+} beam on a Au target. Very recently an experiment with a Pb^{82+} (160 GeV/nucleon) beam on a Au target was performed by the Oak Ridge-Stockholm group, but has not yet been completely analyzed[5].

In this paper we present some newer developments in our investigations of the pair production process, namely the treatment of the time-dependent Dirac equation by the finite element method, the calculation of correlations between the electron and positron emission and the analysis of the S^{16+} (200 GeV/nuleon) + Au^{79+} experiment with distorted wave approximations. An overview over the older work of the Giessen group has been given in Refs. [6,7].

---

[a]Dedicated to Professor Walter Greiner on the occasion of his 60th birthday.

[b]Work supported by BMBF (06 GI 740), GSI (Darmstadt) and DFG.

2 Solution of the time-dependent Dirac equation by the finite element method

In our semiclassical approach we treat the motion of the nuclei and the electromagnetic fields classically and the electron-positron field by quantum field theory. For simplification we neglect the interaction between the leptons (electrons, positrons) and let the nuclei move on straight lines with constant velocities. Under these assumptions one has to solve the time-dependent Dirac equation for the wave functions of the electrons (positrons). In the target rest frame this equation is given by

$$\frac{i\partial\psi(\vec{r},t)}{\partial t} = \left[\vec{\alpha}\vec{p} + \beta - \frac{Z_T e^2}{r} - \frac{\gamma Z_P e^2(1 - \alpha_z v_P)}{((x-b)^2 + y^2 + \gamma^2(z - v_P t)^2)^{1/2}}\right]\psi(\vec{r},t).$$

(1)

Here, v_P is the projectile velocity in z-direction, $\gamma = (1 - v_P^2)^{-1/2}$ the Lorentz factor, and Z_T and Z_P the charge numbers of the target and projectile. Since for small impact parameters ($b < \hbar/(mc) = 386 fm$) and high charge numbers the solution of Eq.(1) can only be calculated with nonperturbative methods, we started to apply the finite element method to this equation. In order to reduce the computing time, we consider nearly central collisions in the first studies by setting b=0. Then the problem is cylindrically symmetric around the internuclear axis which is in z-direction. For the calculation of the electron-positron pair production with capture of the electron into the K-shell one may solve Eq.(1) with the initial condition of $\psi(\vec{r}, t \to -\infty) = \varphi_{1s_{1/2}}(\vec{r}, t)$ (time reversed problem) and project $\psi(\vec{r}, t \to \infty)$ on the negative continuum states. Therefore, we study the time-development of the $1s_{1/2}$-state under the influence of the electromagnetic field of the projectile charge within the finite element method.

Assuming a zero impact parameter and using cylindrical coordinates (z, ρ, φ) we can write the wave function as

$$\psi_\mu(\vec{r}, t) = \rho^{-1/2} \begin{pmatrix} U_1^\mu(\rho, z, t)e^{i(\mu-1/2)\varphi} \\ U_2^\mu(\rho, z, t)e^{i(\mu+1/2)\varphi} \\ U_3^\mu(\rho, z, t)e^{i(\mu-1/2)\varphi} \\ U_4^\mu(\rho, z, t)e^{i(\mu+1/2)\varphi} \end{pmatrix},$$

(2)

where $\mu = \pm 1/2$ is the eigenvalue of the z-component of the angular momentum operator. If this ansatz is inserted into Eq.(1), we get the following coupled differential equation for the spinor $U^\mu = (U_1^\mu, U_2^\mu, U_3^\mu, U_4^\mu)^T$

$$i\frac{\partial U^\mu}{\partial t} = (\alpha_x p_\rho + \alpha_y \frac{\mu}{\rho} + \alpha_z(p_z + eA_z) + \beta - e\varphi)U^\mu,$$

(3)

where A_z and φ are the electromagnetic potentials given in Eq.(1). The spinor U^μ is expanded in finite element functions $\nu_i(\rho)$ and $w_j(z)$:

$$U^\mu(\rho, z, t) = \sum_{i=1}^{N_\rho} \sum_{j=1}^{N_z} \begin{pmatrix} \xi_{ij}^{(1)}(t) \\ \xi_{ij}^{(2)}(t) \\ \xi_{ij}^{(3)}(t) \\ \xi_{ij}^{(4)}(t) \end{pmatrix} \nu_i(\rho) w_j(z). \tag{4}$$

In this case U^μ depends on $4 \times N_\rho \times N_z$ time-dependent coefficients $\xi_{ij}^{(k)}$. Projecting Eq.(3) with the finite element functions one obtains a system of differential equations of first order in time for the coefficients $\xi_{ij}^{(k)}$ which is solved by the alternating direction Crank-Nicolson approach [8]. In this method the two dimensional problem is reduced to successive one-dimensional problems, which decreases the computational time considerably.

Up to now we have tested the method by studying the time development of a stationary $1s_{1/2}$-wave function bound at an Uranium nucleus and of a $1s_{1/2}$-wave function moving with a velocity v=c/2 along the z-axis. In both cases the exact analytical wave functions are known and can be used to control the numerical results. For the stationary case we applied linear form functions for the finite elements with grids of 400 and 800 elements in the ρ- and z-directions, respectively. On a grid which was exponentially condensed towards the nucleus, we found an excellent constancy of the expectation value of energy with time. The $1s_{1/2}$-wave function moving with a velocity c/2 kept a realistic shape up to the usual times of 10^{-20} s needed for the calculation of the electron-positron pair creation in relativistic heavy ion collisions. After these pre-investigations we conclude that the finite element method can advantageously be used for the nonperturbative treatment of the reactions of interest.

3 Correlations between electrons and positrons

In order to learn more about the dynamics of the electron-positron production we considered correlations between these particles in the energy and angle of emission. The pair creation is described in a coordinate system where the nuclei have equal constant, but opposite velocities v_o moving on straight lines parallel to the z-axis. Then the potential is given by

$$V = -\gamma e^2 \left(\frac{Z_T}{r_+}(1 + \alpha_z v_o) + \frac{Z_P}{r_-}(1 - \alpha_z v_o) \right) \tag{5}$$

with

$$r_\pm = \left((x \pm \frac{b}{2})^2 + y^2 + \gamma^2 (z \pm v_o t)^2 \right)^{1/2}.$$

The time-dependent Dirac equation, $i\partial\psi/dt = (H_o + V)\psi$, is solved within a basis consisting of wave packets formed with the eigenfunctions of the free Dirac Hamiltonian H_o:

$$\varphi_\alpha(\vec{r}, t) = \frac{1}{\sqrt{\Delta_\alpha}} \int_{\Delta_\alpha} d^3p \, \varphi(\vec{r}, \vec{p}, s_\alpha) e^{-iE(p)t}, \qquad (6)$$

$$\varphi(\vec{r}, \vec{p}, s) = \frac{1}{(2\pi)^{3/2} E^{1/2}} u(\vec{p}, s) e^{i\vec{p}\vec{r}}, \qquad (7)$$

where $\vec{p}$ is the momentum, Δ_α a small volume in momentum space around a mean momentum $\vec{p}_\alpha$, and s_α the quantum number for spin projection: $s_\alpha = \pm 1/2$. Expanding the solutions of the Dirac equation with the wave functions (6)

$$\psi_\alpha = \sum_\beta a_{\beta\alpha}(t) \varphi_\beta(\vec{r}, t), \qquad (8)$$

we obtain the following system of coupled equations for the expansion coefficients

$$i\frac{\partial a_{\alpha\beta}}{\partial t} = \sum_\gamma < \varphi_\alpha \mid V \mid \varphi_\gamma > a_{\gamma\beta}. \qquad (9)$$

This system has to be solved with the initial conditions $a_{\alpha\beta}(t \to -\infty) = \delta_{\alpha\beta}$ for $E_\beta < 0$ and $a_{\alpha\beta}(t \to -\infty) = 0$ for $E_\beta > 0$. Thiel et al. applied Eq.(9) for the calculation of the number of electron-positron pairs in Pb^{82+} (10 GeV/nucleon) + Pb^{82+} collisions as a function of the impact parameter [9]. In the same publication we extended the fermionic formalism to a bosonic one, interpreting an electron-positron pair as a boson, and calculated the many-pair production in Pb^{82+} (200 GeV/nucleon) +Pb^{82+} collisions.

The occupation number for the electron in the state α after the collision is given by

$$N_\alpha^{e^-} = \sum_{\beta < F} \mid a_{\alpha\beta}(t \to \infty) \mid^2, \qquad (10)$$

where the sum runs over the states of the negative continuum. A correlation between the electrons and positrons can be defined as

$$C_{\alpha\beta}^{e^-e^+} = N_{\alpha\beta}^{e^-e^+} - N_\alpha^{e^-} \cdot N_\beta^{e^+}. \qquad (11)$$

Here, $N_{\alpha\beta}^{e^-e^+}$ is the occupation number for a pair with the electron in the state α and the positron in the state β, $N_\alpha^{e^-}$ and $N_\beta^{e^+}$ are the occupation numbers of the electron and positron in the states α and β, respectively. The coefficients $C_{\alpha\beta}^{e^-e^+}$ measure the correlation between the leptons, since they vanish if the particles are uncorrelated. They are obtained in terms of the expansion coefficients $a_{\alpha\beta}(\alpha > F, \beta < F)$:

$$C_{\alpha\beta}^{e^- e^+} = \sum_{\gamma < F} \mid a_{\alpha\gamma}^*(t \to \infty) a_{\beta\gamma}(t \to \infty) \mid^2 .\qquad(12)$$

As an example we present a calculation for the collision system Pb^{82+} (10 GeV/nucleon)$+Pb^{82+}$ at b=386 fm. The momentum space was divided in cubes with a grid length of 1 MeV/c up to a maximal momentum of 3MeV/c. This crude division leads already to 136×2(spin)$\times 2$ (positive and negative continuum)=544 channels. Fig.1 shows a correlation diagram in the center of

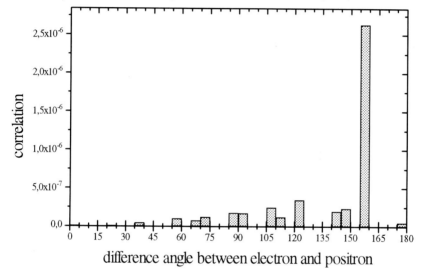

Figure 1: Correlation in the difference angle between the directions of the positron and electron with the electron emitted under 101° with respect to the beam direction in the center of momentum system for the scattering system Pb^{82+} (10 GeV/nucleon)$+Pb^{82+}$ at b=386 fm. The electron and positron have the same energy of 2.6 MeV.

momentum system as a function of the difference angle between the directions of the electron and positron which both have about the same energy of 2.6 MeV. The electron is emitted perpendicularly to the beam (z-) axis. We find the largest correlation for the case that the positron is also perpendicularly emitted, but backwards with respect to the direction of the electron. If the electron is emitted in the direction of the beam axis, the correlation of the emission angle of the positron depends on the ion charges. The emission of the positron is partly in the beam direction, but also in the 60 to 70° direction with respect to the beam axis. Further calculations are necessary with a finer discretization of the momentum space in order to understand the dynamics of the electron-positron emission.

4   Free pair production in perturbation theory

For collisions with smaller charge numbers or larger impact parameters one can apply the time-dependent perturbation theory to the calculation of probabilities or cross sections for pair production. Choosing the target rest frame as coordinate system one can calculate the transition amplitude in first order perturbation theory as follows ($\vec{r}' = (x - b_x, y - b_y, \gamma(z - v_P t))$):

$$a_{ep}^{(1)}(\vec{b}) = -i \int_{-\infty}^{\infty} dt < \psi_e \mid H - i\frac{\partial}{\partial t} \mid \psi_p >, \tag{13}$$

where the Hamiltonian can be separated as

$$H = H_o - \gamma Z_P e^2 (1 - v_P \alpha_z)/r', \tag{14}$$

$$H_o = \vec{\alpha}\vec{p} + \beta - Z_T e^2/r. \tag{15}$$

The simplest choice for ψ_e and ψ_p are the Coulomb-Dirac solutions φ_e and φ_p of $H_o\varphi_{e,p} = i\partial\varphi_{e,p}/\partial t$. This approximation is denoted as Relativistic Born Approximation (RBA). Because of the far reaching Coulomb potential of the projectile ion, one introduces distortion factors in front of the Coulomb-Dirac solutions of the target ion [10]:

$$\psi_{e,p} = L_{e,p}(\vec{r}')\varphi_{e,p}(\vec{r}, t). \tag{16}$$

The Relativistic Symmetric Eikonal approximation (RSE) makes use of distortion factors of the form:

$$L_{e,p}(\vec{r}') = exp(i\xi'_{-,+} ln(p'_{-,+} r' + \vec{p}'_{-,+} \vec{r}')). \tag{17}$$

Here, $\xi'_+ = Z_P/v'_+$ and $\xi'_- = Z_P/v'_-$ are the Sommerfeld parameters of the positron and electron, respectively, with the velocities v'_+ and v'_- or the momenta $\vec{p}'_+$ and $\vec{p}'_-$ in the projectile system. The advantage of the eikonal phase factors is that the wave functions (16) remain normalized. In the Relativistic Symmetric Continuum Distorted Wave approximation (RSCDW)[10] one chooses a Coulomb factor as distortion (upper sign for the electron, lower sign for positron):

$$L_{e,p}(\vec{r}') = e^{\pm\pi\xi'_{-,+}/2}\Gamma(1 + i\xi'_{-,+})_1 F_1(-i\xi'_{-,+}, 1, \mp i(p'_{-,+} r' + \vec{p}'_{-,+} \vec{r}')). \tag{18}$$

The distortion factors have the same asymptotic behaviour as the eikonal factors (17),

$$exp(i\xi ln z) = lim_{z \to \infty} e^{\pi \xi/2} \Gamma(1 + i\xi)_1 F_1(-i\xi, 1, -iz) \qquad (19)$$

with $z = \gamma v r' + \gamma \vec{v} \vec{r}'$. There exist also further theories with other distortion factors, e.g. asymmetric distorted wave theories, where the initial and final states of the leptons are multiplied with different distortion factors. The differential cross section for free pair production is obtained by integrating the differential transition probability over the impact parameter:

$$\frac{\partial^6 \sigma}{\partial \epsilon_+ \partial \epsilon_- \partial \Omega_+ \partial \Omega_-} = \epsilon_+ \epsilon_- p_+ p_- /c^4 \int \sum_{\mu_+ \mu_-} | a_{ep}^{(1)}(\vec{b}) |^2 d^2 b. \qquad (20)$$

We applied the discussed theories to the collision system S^{16+} (200 GeV/nucleon) $+Au^{79+}$, where data were measured by Vane et al.[3] at CERN. In the calculation of the transition matrix elements (13) we used Sommerfeld-Maue functions as a relativistic approximation of the Coulomb-Dirac functions.

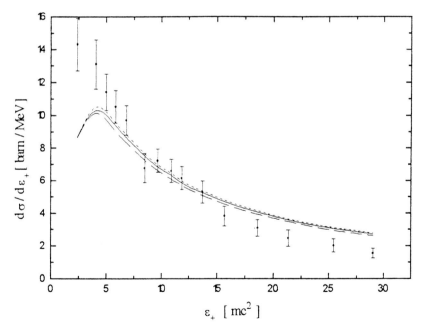

Figure 2: Single-differential cross section as a function of the positron energy for the system S^{16+} (200 GeV/nucleon)$+Au^{79+}$. Experimental data of Vane et al. [3]; solid curve: RSCDW; dotted curve: RSE; dashed curve: RBA.

Fig. 2 shows the calculated single-differential cross sections for the production of positrons as a function of their energy in comparison with the experimental data. The three different theories, RBA, RSE and RSCDW, yield nearly the same differential cross sections indicating that the distortion factors are not so important for this system. The experimental single differential cross section was integrated over the interval $\epsilon_+ = 1$ to 17 MeV, and then an experimental cross section for free pair production of $\sigma = 85 \pm 22$ barn was found for the system $S^{16}(200 \text{ GeV/nucleon})+Au^{79+}$. If the theoretical single differential cross sections shown in Fig. 2 are added up in the same interval, we get 76, 79 and 78 barn for the RBA, RSE and RSCDW, respectively. Finally, when we compare calculated cross sections for free pair production in the system $S^{16+} + Au^{79+}$ as a function of the S^{16+} energy per nucleon, we note that for high incident energies the distortion factors do not need to be taken into account.

References

1. J. Eichler and W.E. Meyerhof, *Relativistic Atomic Collisions* (Academic Press, New York) 1995.
2. H. Gould,*Atomic Physics Aspects of a Relativistic Nuclear Collider*, Lawrence Berkeley Laboratory, Rep. No. LBL 18593 UC-28(1984).
3. C.R. Vane, S. Datz, H.F. Krause, C. Bottcher, M. Strayer, R. Schuch, H. Gao and R. Hutton, *Phys.Rev.Lett.* <u>69</u>, 1911 (1992).
4. A. Belkacem, H. Gould, B. Feinberg, R. Bossingham, W.E. Meyerhof, *Phys.Rev.Lett.* <u>71</u>, 1514 (1993).
5. R. Schuch, *Workshop on Dynamical QED-Effects in Relativistic Heavy Ion Atom Collisions*, GSI (Darmstadt), April 1996.
6. J. Thiel, J. Hoffstadt, N. Grün and W. Scheid, in *Frontier Topics in Nuclear Physics*, ed. by W. Scheid and A. Sandulescu (Plenum Press, New York) 1994.
7. J. Thiel, N. Grün and W. Scheid, in *Hot and Dense Matter*, ed. by W. Greiner, H. Stöcker and A. Gallmann (Plenum Press, New York) 1994.
8. G.F. Carey and J.T. Oden, *Finite Elements: Computational Aspects*, (Prentice-Hall, Englewood Cliffs, N.J.) 1984.
9. J. Thiel, J. Hoffstadt, N. Grün and W. Scheid, *Z.Phys.D* <u>34</u>, 21 (1995).
10. G. Deco and N. Grün, *Phys.Lett.* A <u>143</u>, 8 (1990).

TRANSFER AND EXCITATION IN VERY HEAVY ION-ATOM COLLISIONS

B. FRICKE

Fachbereich Physik, Universität Kassel, D-34109 Kassel

Besides QED and the many-electron aspect the third main problem in Atomic Physics is the description of the time-dependence of collision systems. We present here a method which solves the time-dependent Dirac-equation for such a system in the Dirac-Fock-Slater approximation including the many-electron effects in the inclusive probability description. Various collision systems are discussed. As a result we present the possibility to even describe a complicated many-electron collision like Pb^{82+} on a solid state Pb target with the possibility to deduce the level structure of the innermost levels up to very small internuclear distances.

1 General Problems in Modern Atomic Physics

From a very general point of view the problems in modern theoretical atomic physics can be defined by the following 3 statements:

1. The electro-magnetic interaction which governs the interaction is NOT known exactly. Of course the leading $1/r$ term in the interaction is dominant, but the rest is highly non-local and very complicated. The field which is defined by this first statement is the QED. This field will be discussed in this Conference by the talk of Gerhard Soff.

2. Already in classical mechanics a three body system is non-trivial. Many electron-systems in quantum mechanics are even more non-trivial. The usual theory which nowadays can be applied in general for such systems are based on the level of the Hartree- or Dirac-Fock method. Everything beyond this level is called "Correlation effect" (by definition). Generations of physicists and quantum chemists are devoted to this problem. A lot of results have been achieved during recent years, but there is still a lot to do; especially for open shell atoms with more than two electrons.

3. Time dependent effects in collision systems. In connection with this conference it is of great interest to discuss the possibility to get information of the innermost levels in the quasi-molecules for very heavy systems, because (up to now) this is probably

the only chance to get experimental information of such a system for very large fields, where the structure of the Vacuum may play a dominant role.

In general all three problems of interest have to be taken into account and discussed simultaneously! Of course everyone knows that this is not possible.

Under certain circumstances one can e.g. avoid problem no. 2 by using a very heavy collision system with only one electron like the system Pb^{82+} on Pb^{81+}. Again this is not yet possible but this may be a type of experiment coming up in the next Millenium.

What can be done in this context is:
- neglect in the first place QED or add just one-electron QED contributions in perturbation theory
- use the many-electron contributions in Hartree- or Dirac-Fock approximation, which then allows an interpretation of an experiment which can be done already nowadays like Pb^{82+} on Pb as a solid state target.

2 Description of Time-dependent Ion-Atom Collisions

2.1 The general problem

Any ion-atom collision can be described by the simple equation

$$A^{n+} + B \quad \rightarrow \quad A^{*p+} + B^{*m+} + q\,e^{-}$$

The experimental variables in such a system are the energy of the colliding ion and the initial ionization. If the energy is large enough that the motion of the nucleus can be described by a classical trajectory then a third variable is the impact parameter b. During and after such a collision x-rays (light) and electrons (Auger- and δ-electrons) can be emitted. If the emission occurs in the interaction region of the two systems d   u   r   i   n   g the collision then the X-rays and Auger-electrons are called MO X-rays and MO Auger-electrons (MO = Molecular Orbit).

2.2 Description with a correlation diagram

If the collision energy is large enough to describe the nuclear motion by a classical trajectory and the collision energy is small enough that the collision energy still allows a molecular quantum mechanical description of the electrons then correlation diagrams are the most physical way to describe the collision. Such correlation diagrams describe the total energy of the quantum mechanical system as function of internuclear distance as function of the quantum numbers for the full electronic system.

In case of the very heavy ion-atom collisions with very many electrons we instead use the simpler form of the correlation diagram where the single particle levels are given as function of the internuclear distance. This diagram eliminates the nucleus-nucleus interaction and is, at least for the inner electronic levels which is of interest in this connection, a relatively good approximation. Fig. 1 shows such a schematic correlation diagram as an example.

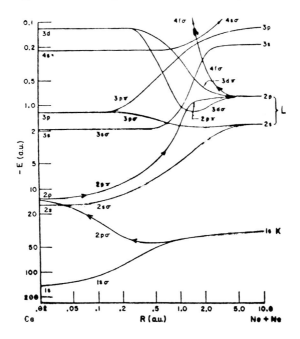

Fig.1 Schematic correlation diagram of the system Ne - Ne. The arrows present the one-particle way to understand excitation and transfer as well as ionization in such a picture.

This diagram also shows the two main processes which occur in such a system. The first is the inner electron transfer and excitation from the K-shell of one of the partners via the rotational dynamic coupling (which is described below) to the L-shell of the other collission partner. The other possibility is the transfer of a L-electron to a higher shell via a molecular level which is connected to levels in the vicinity of the continuum and thus ionize.

2.3 Theoretical description

To describe such a complicated many electron-ion atom collision properly one has to take into account
- all the electrons in the collision with good energy eigenvalues and a good potential function to describe the motion of the nucleus
- a relativistic description, because for high Z systems any non-relativistic description is wrong from the very beginning
- the time-dependence of the collision process.
Thus one has to solve the time-dependent Dirac-equation

$$H \Psi = i h \, \partial/\partial t \, \Psi \,. \tag{1}$$

For the many-particle wavefunction Ψ we choose an antisymmetric Slaterdeterminant which is built from single-particle time-dependent wavefunctions $\psi_i(t)$. These wavefunctions are calculated as solution of the Time-Dependent Dirac-Fock-Slater equations (TDDFS) which one gets if the Slaterdeterminant Ψ is introduced in equ. 1

$$h^{TDDFS} \, \psi_i(t) = i h \, \partial/\partial t \, \psi_i(t) \qquad i = 1,N \tag{2}$$

where N is the number of electrons involved. The $\psi_i(t)$ itself are expanded as linear combinations of solutions ϕ_j of the static molecular DFS equation for a fixed internuclear distance R:

$$\psi_i(r,t) = \Sigma \, a_{ij}(t) \, \phi_j(r;R(t)) \,. \tag{3}$$

If this ansatz is introduced in equ. 2 one gets the so called coupled channel equations which can be solved for the amplitudes a_{ij}.

$$\partial/\partial t \, a_{im}(t) = \Sigma - a_{ij}(t) \, \langle \phi_j | \partial/\partial t | \phi_m \rangle \, \exp(-i/h \int (\varepsilon_i - \varepsilon_j) dt) \qquad i = 1,N \tag{4}$$

he details of this procedure is given in Ref.1. The set of amplitudes$\{ a_{ij} \}$ allows to reconstruct the many particle wavefunction Ψ introduced in equ. 1 for every kind of many-particle electron configuration including the information if in a given experiment certain states are measured or not. This method of Inclusive Probability can be found in Ref. 2-4.

3   Results for Ion - Atom collisions with gas targets

3.1   Triple differential x-ray cross-sections

The most complicated example which we ever have calculated was the system S^{16+} on Ar which is a gas type experiment published by Schulz et.al. (5) . The result of the experiment and our calculation (6) for the triple differential cross section to observe at least two X-rays after the collision from either Ar (P_{AA}) or S (P_{SS}) or one X-ray each of both sorts (P_{SA}) is given in Fig. 2.

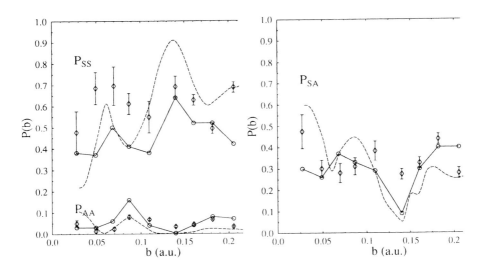

Fig. 2 Impact parameter dependent probability P to observe two simultaneous X-rays from the collision system 16 MeV S^{16+} on Ar. The two indices S (Sulphur) and A (Argon) denote the origin of the X-rays. The experimental values are from Schulz et al. ( ); the full curves denote our TDDFS results (6) and the broken curves our simpler SCF results (7).

3.2   Results for MO X-rays

The second possibility to get direct information from such a collision system is the direct observation of MO X-rays. The result of our full scale many-particle calculation (8) is given in Fig. 3 where the spectrum from the collision system 20 MeV Cl^{16+} on Ar is compared with the experiment (9).

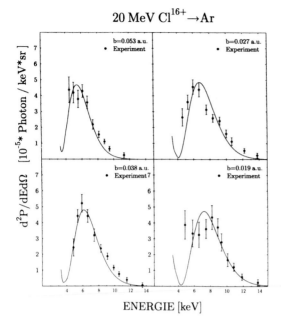

$$20 \text{ MeV Cl}^{16+} \rightarrow \text{Ar}$$

Fig.3 MO X-ray spectrum for the system 20 MeV Cl[16+] on Ar for four different impact parameters. The full curve is our calculation (8). The experiment is from Schuch et al. (9).

4 Results for Ion - Atom collision with Solid State targets

4.1 Difference between gas target and solid state target experiments

In a gas target experiment an ion collides only once with a target atom. This exactly is also the assumption in the theoretical procedure described above.

In a solid state target the incoming ion interacts during the passage with every atom in the solid which is nearer than the interaction distance of the ion which is about the diameter of its electron cloud. The largest distance such a ion can have in the solid to the next partner is half of the distance between two nearest neighbour atoms. So every ion undergoes a large number of collisions with atoms in the solid with large impact parameters. In order to simulate this we perform calculations for a number of large impact paramaters where we learn how the electron cloud of the ion excites, ionizes or transfers electrons. The time involved in these collisions is so short that this ion has nearly no chance to rearrange so that it is highly excited and differently charged after a few layers. If the ion then undegoes a collision with very small impact parameter (which leads to a measurable deflection) it will be measured in the coincidence experiment. But the relevant excitation and ionization is quite different because it is not anymore the originaly charged ion in its ground state.

4.2 One example: Cu on Ni

One such example with 50 MeV Cu on Ni is given in Fig. 4 where the effect of thick as well as thin targets can be seen. The experiment is done by   et al. (10) and the theory is published in ref. 11. The prediction of the $2p_\pi$ -$2p_\sigma$ theory Taulbjerg et al. (12) is included as well.

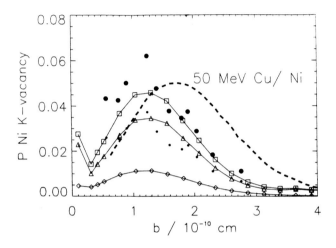

Fig. 4 Ni K-vacancy probability in a 50 MeV Cu on Ni collision as function of impact parameter. The two upper solid curves give the results of a thick and a thin target calculation. The dots are the experimental points (10). The dashed line is the Taulbjerg, Briggs theory (12).

The P(b)-maximum of the old theory (12) and the results of many experiments came out to be different positions for ion-solid state collisions whereaes for the ion-gas experiments the maximum seemed to be at the same position. In a paper by Kambara et al. (13) this effect was attributed to the solid itself as an unexplained "solid-target" effect. We have shown in several publications (14-16), and Fig. 4 is an example, that our theory always is at or very near the experimental maximum. In couse of these research we found the reason for this discrepancy. It is the fact that at very small internuclear distances not only the relativistic equivalent to the $2p_\pi$ -$2p_\sigma$ rotational coupling contributes but also the other two levels of the united L-shell contribute which in a non-relativistic treatment do not couple (14). This changes the structure of the impact parameter dependent transfer probability drastically to smaller b already for systems with a united Z around 50 and even below.

5 Further experiments

Having this in mind we propose that one now should start with experiments for very heavy ion-atom collisions in order to learn about the behaviour of the innermost levels of systems with a united Z in the superheavy region. The quality of the theoretical description is now so good that
- accurate correlations diagrams can be calculated,
- full coupling treatment including all rotational and radial couplings of the inner shells can be performed and
- a many-particle interpretation is possible which allows to take into account the electron excitation and transfer in a collision of a naked projectile in a solid state target.
- In addition MO X-ray calculations and experiments are under way which also should provide information about the inner level structure.

This then will hopefully allow to determine the level behaviour of the innermost level of the united system like the one given in Fig.5 for the system Pb on Cm which was calculated with 96 electrons.

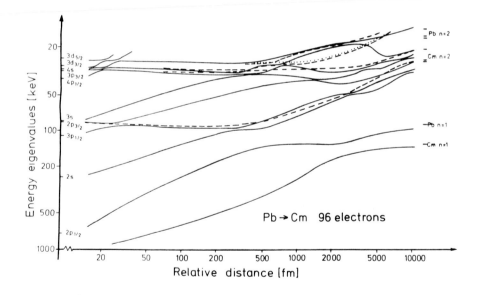

Fig. 5 Correlation diagram of the system Pb on Cm which takes into account 96 electrons. The interesting point is, if in a real experiment the structure of the innermost level, which drastically increases in binding for smaller internuclear distances, behaves like this prediction or is changed by other effects.

6 Conclusions

As long as one is not able to perform experiments on collision systems which contain only one electron, **many electron effects become strong** and have to be taken into account when one tries to compare with an experiment. They modify the one-particle energy levels in the correlation diagram and every theoretical prediction has to be modified by the Inclusive Probability interpretation.

Relativistic effects become strong. They produce not only large large spin-orbit splittings and changes of the wavefunctions and their radial and angular behaviour but also produce very different radial and rotational coupling behaviour between the relativistic levels. This last effect especially produces a radical change in the impact parameter dependent transfer probability for K-shell excitation. This allows to understand ion-solid target collisions in the way discussed above. Thus the so called "solid-target effect" (at least for inner shell transfer) does not exist!

As the consequence of this improved understanding one should try to start now experiments like Pb^{81+} or Pb^{82+} on solid Pb. Already in these type of experiment with the observation of transfer- and excitation-probabilties as well as the MO X-ray spectrum as function of impact parameter one should be able to deduce the inner shell level structure in the region of superheavy elements. This in turn would allow to get information on the QED corrections on the level structure or the other missing influences like the transient magnetic field.

References

1. P. Kürpick, W.-D. Sepp, B. Fricke Phys. Rev. **A51**, 3693 (1995)
2. J.F. Reading, Phys. Rev. **A8**, 3262 (1973)
3. J. Reinhardt, B. Müller, W. Greiner, B. Soff, Phys. Rev. Lett. **43**, 1307 (1979)
4. P. Kürpick, H.-J. Lüdde, Comp. Phys. Comm. **75**, 127 (1993)
5. M. Schulz, E. Justiniano, J. Konrad, R. Schuch, A. Salin, J. Phys. **B20**, 2057 (1987)
6. P. Kürpick, W.-D. Sepp, H.-J. Lüdde, B. Fricke, Nucl. Instr. Meth. **B94**, 183 (1994)
7. P. Kürpick, B. Thies, W.-D. Sepp, B. Fricke, J. Phys. **B24**, L139 (1991)
8. K. Schulze in preparation
9. C.H. Annett, B. Curnutte, C.L. Cocke, Phys. Rev. **A19**, 1038 (1979)
10. R. Schuch et al., Phys. Rev. **A37**, 3313 (1988)
11. P. Kürpick, B. Fricke, C.L. Cocke, Phys. Rev. **A53**, 1 (1996)
12. K. Taulbjerg, J.S. Briggs, J. Phys. **B8**, 1895 (1975)
13. K. Taulbjerg, J.S. Briggs, J. Vaaben J. Phys. **B9**, 1351 (1975)
14. P. Kürpick, T. Bastug, W.-D. Sepp, B. Fricke Phys. Rev. **A52**, 2132 (1995)
15. P. Kürpick et al., Phys. Lett. **A207**, 199 (1995)
16. P. Kürpick, B. Fricke, W.-D. Sepp, J. Phys. **B28**, L127 (1995)

RADIATIVE CORRECTIONS IN STRONG ELECTROMAGNETIC FIELDS *

G. SOFF, T. BEIER, G. PLUNIEN, M. GREINER, C. R. HOFMANN

Institut für Theoretische Physik, TU Dresden, D-01062 Dresden, Germany

We discuss various contributions to the Lamb shift and to the hyperfine splitting in hydrogen-like heavy ions. Furthermore the Lamb shift in lithium-like ions is considered. Radiative corrections in strong electromagnetic fields are evaluated. The results are compared with available experimental data.

1   Test of QED in strong fields

During the last few years the precision of Lamb-shift measurements for highly-charged ions has increased considerably[1,2,3]. An experimental precision of about 1 ppm for atomic transition energies is foreseeable[4]. Thus it becomes possible to test quantum electrodynamics (QED) in the strong electromagnetic fields[5] that are prevailing in heavy ions. As an example we display in Fig. 1 the expectation values of the electric field strength for the strongest bound electron states in hydrogen-like atoms for the entire range of nuclear charge numbers Z. The calculation for $\langle \phi_{nlj} | E(r) | \phi_{nlj} \rangle$ was accomplished utilizing the solutions of the Dirac equation

$$\left(c\alpha p + \beta m_e c^2 - e\,V(r) \right) \phi_{nlj} = E_{nlj}\,\phi_{nlj} \qquad (1)$$

for the bound states ϕ_{nlj}, where $V(r)$ denotes the potential due to the nuclear charge distribution. For this charge distribution we employed homogeneously charged spheres with radii $R = 1.2$ fm $\times\, A^{1/3}$ with A being the nuclear mass number. The electric field strength $E(r)$ was obtained by $E(r) = -\partial/\partial r\, V(r)$. Fig. 1 exhibits the increase of the expectation value of the electric field strength from 10^{10} V/cm for hydrogen to 1.8×10^{16} V/cm for uranium for the $1s_{1/2}$ state. In particular very heavy atoms are best suited to generate the strongest electric field strength being accessible to experimental investigations.

---

*This contribution is dedicated to Prof. Dr. Walter Greiner on the occasion of his 60th birthday.

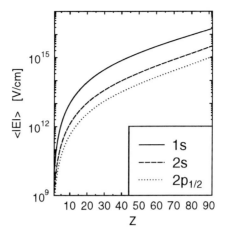

Figure 1: Expectation values $< |E| >$ of the electric field strength for the lowest-lying bound states for hydrogen-like atoms with nuclear charge numbers Z.

2 Contributions to the Lamb shift in uranium

The Feynman diagrams contributing to the first order radiative corrections of the electron binding energy in the Coulomb potential are plotted in Fig. 2. The energy shift of a bound state ψ caused by the first-order self energy correction can be written as[6]

$$
\begin{aligned}
\Delta E \;=\;& 4\pi i\alpha \int \mathrm{d}t\, \mathrm{d}^3x\, \mathrm{d}^3y\, \overline{\psi}(y)\gamma^\nu S_F(x,y)\gamma^\mu \psi(x)\, D(x,y) \\
& -\delta m \int \mathrm{d}^3x\, \overline{\psi}(x)\psi(x) \quad .
\end{aligned}
\tag{2}
$$

Here $S_F(x,y)$ denotes the Green function of the Dirac equation for an external Coulomb field. Numerical evaluation of the self-energy correction[7,8] for an $1s$ electron yields $\Delta E_{\mathrm{SE}} = 355.05$ eV in uranium. The vacuum polarization diagram results in an effective potential for the electron which reads[9]

$$
U(\vec{x}_2) = 4\pi i\alpha \int \mathrm{d}(t_2 - t_1) \int \mathrm{d}^3x_1\, D_F(x_2 - x_1)\, \mathrm{Tr}\{\gamma_0 S_F(x_1, x_1)\} \quad .
\tag{3}
$$

D_F is the photon propagator. The energy shift caused by the vacuum polarization can be computed according to

$$\Delta E = \int \mathrm{d}^3 x_2 \, \phi_n^\dagger(\vec{x}_2)\phi_n(\vec{x}_2) \, U(\vec{x}_2) \quad . \tag{4}$$

As result we obtained for the $1s$ electron in uranium $\Delta E_\mathrm{VP} = -88.60$ eV. On the 1 eV level of precision we need to calculate also the Feynman diagrams of order α^2. These diagrams are displayed in Fig. 3. The numerical results for the diagrams which are computed till now are comprised in the Tables 1 and 2 and will be discussed in the next section.

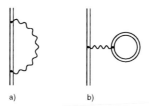

Figure 2: Feynman graphs corresponding to QED corrections of first order in α. The double solid line corresponds to the bound electron, the wavy line to the photon. (a) electron self energy, (b) vacuum polarization.

3   Comparison with experimental data for uranium

In Table 1 we give an overview of all corrections known up to now for the lowest lying states of uranium. The finite size correction is deduced assuming a Fermi distribution with $\langle r^2 \rangle^{1/2} = 5.860$ fm for the nuclear charge distribution. The resulting potential is included in the Dirac equation, and the difference between these energy eigenvalues and those of the point nucleus solution yields the finite size correction. The uncertainty of this value is merely the absolute difference to the binding energy obtained from a homogeneously charged sphere with the same rms-radius for the nuclear charge distribution[10].

The first-order QED corrections presented here were obtained utilizing a uniform sphere model for the nuclear charge distribution. Almost all QED corrections of the order α^2 have been evaluated during the last few years. The individual contributions are displayed in Fig 3. The corresponding values for the energy shift in uranium are listed in Table 1. The reducible part of the SESE a) diagram forms a gauge-invariant set together with the SESE b) and c) diagrams and is the only one which remains to be compiled.

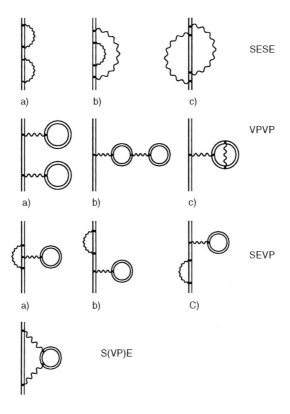

Figure 3: Feynman graphs corresponding to the second order QED corrections.

From Table 1 the total contribution of the α^2-diagrams is found to be of magnitude 1 eV and therefore of the same size as binding energy corrections due to nuclear properties, i.e. the finite mass of the nucleus and its polarizability[11].

The finite nuclear mass leads to a motion of the nucleus and thus causes a reduced mass correction and also a relativistic recoil correction. The sum of both was recently compiled by Artemyev *et al.* [12]. All in all the corrections for the uranium ground state amount to 464.7 eV. The Lamb shift for one single state includes all corrections except the ordinary reduced mass correction and hyperfine corrections due to the nuclear spin. Therefore we subtract the reduced mass correction from the total sum to obtain the proper Lamb shift.

The most recent experimental value[3] confirms the theoretical predictions. Measurements increased in precision to 429 ± 63 eV achieved by Stöhlker *et al.* [2] and to the value of 470 ± 16 eV achieved by Beyer *et al.* [3] recently. An

Table 1: Binding energies and first- and second-order QED corrections of $1s_{1/2}$-, $2s_{1/2}$- and $2p_{1/2}$-electrons in H-like uranium ($Z = 92$). All values are given in eV.

	$1s_{1/2}$	$2s_{1/2}$	$2p_{1/2}$	Ref.
E_B (point nucleus)	-132279.96	-34215.49	-34215.49	
finite size	198.82 (36)	37.77 (08)	4.42 (01)	10
VP	-88.60	-15.64	-2.70	
SE	355.05	65.42	9.55	7,8
SESEa (irr)	-0.97	-0.08	0.01	14
SESEa (red) + SESE b,c	remain to be compiled			
VPVPa	-0.22	-0.04	0.00	15
VPVPb,c	-0.72	-0.12	-0.02	18
SEVP a,b,c	1.14	0.21	0.02	17,15
S(VP)E	-0.13	-0.02	0.00	15
Recoil	0.51	0.13	0.09	12
Nucl. pol.	-0.18	-0.03	0.00	11
Sum of corr.	464.70 (36)	87.60 (08)	11.37 (01)	
Total E_B	-131815.26 (36)	-34127.89 (08)	-34204.12 (01)	
Red. mass	0.30	0.08	0.08	
L. S. (Th.)	464.40 (36)	87.52 (08)	11.29 (01)	
L. S. (Exp.)	470 (16)			3

experimental uncertainty of 1 eV or better seems to be reasonable in the near future[4].

For Li-like uranium, the actual experimental precision is even better. The shift between the $2s_{1/2}$ and the $2p_{1/2}$ levels was measured[1] to be 280.59 ± 0.09 eV corresponding to a precision of 3×10^{-4}. The theoretical contributions to this shift are displayed in Table 2. The first line in this Table corresponds to the Relativistic Many Body Perturbation Theory calculation. This calculation differs from the full QED result, which is still absent, by the following details: in RMBPT there are 1) no retardation 2) no virtual pairs (i.e. no negative energy intermediate states) 3) no cross-photon Feynman graphs. The rough estimate of these corrections leads to the possible error given in the first line of Table 2. The small two-photon reference state corrections, also absent in RMBPT, are given separately. The nuclear size corrections are included in RMBPT, as well as in SE, VP and corresponding screening corrections.

Table 2: $2p_{1/2}$–$2s_{1/2}$ shift in Li-like U

Correction	Numerical value	Reference
RMBPT	322.33(15) eV	Lindgren et al.[16,17]
SE	−55.87 eV	Lindgren et al.[16]
SE screening	1.55 eV	Lindgren et al.[16]
VP	12.94 eV	Persson et al.[15]
VP screening	−0.39 eV	Persson et al.[15]
SESEa (irr)	0.09 eV	Mitrushenkov et al.[14]
SEVP	−0.19 eV	Lindgren et al.[17]
S(VP)E	0.02 eV	Persson et al.[15]
VPVPa	0.03 eV	Persson et al.[15]
VPVPb,c	0.10 eV	Schneider, Greiner and Soff[8]
Two-photon (box)	0.04 eV	Lindgren et al.[17]
Two-photon (cross)	−0.02 eV	Labzowsky and Tokman[19]
Recoil	−0.08 eV	Artemyev et al.[12]
Nucl. pol.	0.03 eV	Plunien et al.[11]
Total theory	280.58(15) eV	
Experiment	280.59(9) eV	Schweppe et al.[1]

The transition probability of this transition was calculated to be $1.65 \cdot 10^{10}$ s^{-1} which corresponds to a lifetime of 61 ps of the $1s^2 2p\, 2^2P_{1/2}$ state. This number was obtained utilizing the MCDF computer code of[13] which also allows to include radiative corrections. For the $1s^2 2p\, 2^2P_{3/2}$ state the transition rate to the $1s^2 2s\, 2^2S_{1/2}$ state was found to be $8.5 \cdot 10^{13}$ s^{-1} corresponding to a lifetime of $1.2 \cdot 10^{-14}$ s.

The analysis based on Table 2 leads to the conclusions that the further refinement of the $2p_{1/2} - 2s_{1/2}$ energy shift calculation for hydrogenlike uranium requires first the full QED calculation of one- and two-photon exchange graphs

and secondly the reduction of the nuclear size uncertainty. We should emphasize that the nuclear polarization and recoil have to be taken into account on
this level of accuracy.

4   The hyperfine splitting

Recent measurements[20] of the hyperfine splitting in hydrogen-like ^{209}Bi^{82+}
allow for testing QED in the presence of a strong magnetic field. The hyperfine
splitting is in leading order strongly influenced by the vacuum polarization and
the self-energy correction.

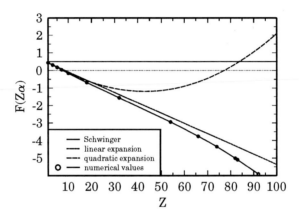

Figure 4: Graphical representation of the function $\tilde{F}$ (full line with circles) in comparison
with the earlier $Z\alpha$-expansion results. The horizontal full line displays the Schwinger result,
the linear (quadratic) expansion in $Z\alpha$ is represented by short-dashed (long-dashed) lines.

The first contribution was calculated by Schneider et al.[21] in an Uehling-
like approximation, while the latter was computed by Persson et al.[22] to all
orders in the external nuclear charge $Z\alpha$. The energy-splitting caused by the
self-energy correction can be cast into the form[22]

$$\Delta E^{\mathrm{QED}} = \frac{\alpha}{\pi} \Delta E^{\mathrm{1.ord.}} \tilde{F} \tag{5}$$

where $\Delta E^{\mathrm{1.ord.}}$ denotes the full relativistic first-order energy splitting. In
Fig. 4 the function $\tilde{F}$ is plotted for the exact calculation and for several
approximations[23].

5 Conclusions

We presented theoretical values for the Lamb shift in hydrogen-like uranium which are obtained by taking into account the radiative corrections of second order in the fine structure constant α. We discussed also the Lamb shift in lithium-like uranium. The theoretical results are in fair agreement with available experimental data. Precise measurements of the Lamb shift and of the hyperfine splitting in highly charged systems represent a critical test of QED in strong electromagnetic fields. However, these precision tests are bounded by nuclear properties, which cannot be deduced rigorously, e.g., nuclear charge distributions, virtual nuclear excitations, and nuclear magnetization distributions.

Acknowledgments

This work was supported by the DFG, by the BMBF, and by GSI (Darmstadt). We gratefully acknowledge very fruitful discussions with P. J. Mohr, W. R. Johnson, H. Persson, I. Lindgren, L. Labzowsky, S. Schneider, and W. Greiner.

References

1. J. Schweppe, A. Belkacem, L. Blumenfeld, N. Claytor, B. Feinberg, H. Gould, V. E. Costram, L. Levy, S. Misawa, J. R. Mowat, and M. H. Prior, *Phys. Rev. Lett.* **66**, 1434 (1991)
2. Th. Stöhlker, P. H. Mokler, K. Beckert, F. Bosch, H. Eickhoff, B. Franzke, M. Jung, T. Kandler, O. Klepper, C. Kozhuharov, R. Moshammer, F. Nolden, H. Reich, P. Rymuza, P. Spädtke, and M. Steck, *Phys. Rev. Lett.* **71**, 2184 (1993)
3. H. F. Beyer, G. Menzel, D. Liesen, A. Gallus, F. Bosch, R. Deslattes, P. Indelicato, Th. Stöhlker, O. Klepper, R. Moshammer, F. Nolden, H. Eickhoff, B. Franzke, and M. Steck, *Z. Phys.* D **35**, 169 (1995)
4. D. Liesen, H. F. Beyer, and G. Menzel, Comments *At. Mol. Phys.* **32**, 23 (1995)
5. W. Greiner, B. Müller, J. Rafelski, *Quantum Electrodynamics of Strong Fields* (Springer, Berlin, 1985)
6. S. S. Schweber, *An Introduction to Relativistic Quantum Field Theory* (Harper and Row, New York, 1961)
7. P. J. Mohr, *Phys. Rev.* A **46**, 4421 (1992)
8. P. J. Mohr, G. Soff, *Phys. Rev. Lett.* **70**, 158 (1993)

9. G. Soff and P. J. Mohr, *Phys. Rev.* A **38**, 5066 (1988)

10. T. Franosch and G. Soff, *Z. Phys.* D **18**, 219 (1991)

11. G. Plunien, G. Soff, *Phys. Rev.* A **51**, 1119 (1995)

12. A. N. Artemyev, V. M. Shabaev, V. A. Yerokhin, *Phys. Rev.* A **52**, 1884 (1995)

13. I. P. Grant, B. J. McKenzie, P. H. Norrington, D. F. Mayers, N. C. Pyper, *Comp. Phys. Comm.* **21**, 207 (1980)

14. A. Mitrushenkov, L. Labzowsky, I. Lindgren, H. Persson and S. Salomonson, *Phys. Lett.* A **200**, 51 (1995)

15. H. Persson, I. Lindgren, L. Labzowsky, G. Plunien, T. Beier, and G. Soff, submitted to *Phys. Rev.* A

16. H. Persson, I. Lindgren, and S. Salomonson, *Phys. Scr.* **T46**, 125 (1993); I. Lindgren, H. Persson, S. Salomonson and A. Ynnermann, *Phys. Rev.* A **47**, 4555 (1993)

17. I. Lindgren, H. Persson, S. Salomonson, V. Karasiev, L. Labzowsky, A. Mitrushenkov and M. Tokman, *J. Phys.* B **26**, L503 (1993)

18. S. M. Schneider, W. Greiner and G. Soff, *J. Phys.* B **26**, L529 (1993)

19. L. Labzowsky and M. Tokman, *J. Phys.* B **28**, 3717 (1995)

20. I. Klaft, S. Borneis, T. Engel, B. Fricke, R. Grieser, G. Huber, T. Kühl, D. Marx, R. Neumann, S. Schröder, P. Seelig, L. Völker, *Phys. Rev. Lett.* **73**, 2425 (1994)

21. S. M. Schneider, W. Greiner, G. Soff, *Phys. Rev.* A **50**, 118 (1994)

22. H. Persson, S. M. Schneider, W. Greiner, G. Soff, I. Lindgren, *Phys. Rev. Lett.* **76**, 1433 (1996)

23. N. M. Kroll, F. Pollock, *Phys. Rev.* **85**, 876 (1952); R. Karplus, A. Klein, *Phys. Rev.* **85**, 972 (1952); D. E. Zwanziger, *Phys. Rev.* **121**, 1128 (1961); S. J. Brodsky, G. W. Ericson, *Phys. Rev.* **148**, 26 (1966); J. R. Sapirstein, *Phys. Rev.* **51**, 985 (1983)

CHANNELING OF A CHARGED PARTICLE IN A BENT CRYSTAL

A.V. Solov'yov

*A.F.Ioffe Physical-Technical Institute, Russian Academy of Sciences ,
194021 St. Petersburg, Russia*

We consider the channeling process of charged particles in a bent crystal. Invoking simple assumptions we derive a criterion, which determines whether channeling occurs or not. The same criterion using the Dirac equation is obtained. We demostrate that the centrifugal force acting on the particle in the bent crystal significantly alters the effective transverse potential. The channeling probability and the de-channeling probability due to tunneling of the particle under the barrier in the effective transverse potential are estimated. These probabilities depend on the specific scaling parameter characterizing the process. We calculate the contribution to the radiation spectrum arising due to the curvature of the channel and demonstrate that this contribution is rather significant.

1   Introduction

Channeling of charged particles in a crystal is the well known phenomenon (for references [1,2,3]). Radiation phenomena during the interaction of positrons and electrons with crystals have been studied in numerous theoretical and experimental papers (for review see e.g.[4]).

Channeling was also successfully used to bent ion beams [5,6,7]. The basic effect is that charged particles traveling through a crystal nearly parallel to a crystal axis experience the collective electric field of the ions and are thus steered into the interatomic region (for positively charged particles) or into the vicinity of the atoms (for negatively charged particles). Due to this effect ion beams can be bent much more strongly than with external electric or magnetic fields. The efficiency is, however, typically smaller than 10 per cent. In Ref. [8] the question whether the channeling effect can also be used to focus beams, especially ion beams was discussed. To this end a crystal is needed in which the crystal axes are no longer parallel, but are slanted more and more the farther away they are from the axes of the beam.

In the present paper (see also [9]) we analyse the channeling process of electrons and ions in a bent crystal and derive the condition determinig the channeling process in a bent crystal. We estimate the probability of channeling and investigate its dependence on the characteristic parameters. We discuss the following specific features of the channeling process in a bent crystal: the alteration of the transverse potential due to the centrifugal force acting on the particle, the polarization of the states of the transverse motion and the tunneling of the particle through the barrier in the effective transverse potential. Within the framework of the quasiclassical method[3] we calculate the spectrum of radiation originating due to the bending of the particle's trajectory.

2   CHANNELING CONDITION IN A BENT CRYSTAL

2.1   Qualitative consideration

If a fast charged particle moves under a very small angle relative a crystal axes or a crystal plane, then the transverse motion of the particle is strongly influenced by the field of the crystal (see e.g. [2]). As it was demonstrated by Lindhard the condition for the particle to be trapped in the transverse directions by the field of the channel is as follows, $\Theta \leq \Theta_L = \sqrt{2U_0/m_0 c^2 \gamma}$. In this equation Θ is the angle between the direction of the incident beam and the crystal axes or the crystal plane, Θ_L is the critical Lindhard angle; U_0 is the depth of the transverse effective potential; m_0 is the mass of the particle; c is the velocity of light; γ is the relativistic Lorentz factor. This condition arises from the requirement the transverse kinetic energy of the particle in the channel has to be less then the depth of the transverse potential.

Now let us assume that the channel has a certain curvature radius equal to ρ. Moving along the channel a particle oscillates in the transverse plane. It is qualitatively clear, that the channeling effect in the bent crystal takes place, if the bending angle Θ_1 of the channel for one period of the transverse oscillation of the particle is much smaller then the Lindhard angle: $\Theta_1 \ll \Theta_L$. Using simple relations between the variables, characterizing the process:

$$L_1 = \rho \cdot \Theta_1, \quad L_1 = c \cdot T_1, \quad T_1 = \frac{2R}{v_\perp}, \quad v_\perp \sim \sqrt{\frac{2U_0}{m}}, \quad m = m_0 \gamma,$$

we derive from the following condition

$$C = \frac{\varepsilon_\| R}{U_0 \rho} \ll 1 \tag{1}$$

Here $\varepsilon_\| \approx m_0 c^2 \gamma$ is the kinetic energy of the longitudinal motion of the particle in the channel. We shall see that any physical quantity characterizing the channeling process in a bent crystal is a function of the parameter C, appearing in (1).

2.2 The equation of motion

As was demonstrated in [9], keeping only the potential term in the second order Dirac equation, growing with the energy $\varepsilon \gg |U(r)|$, and neglecting the spin-momentum coupling and squared potential terms, which do not depend on ε, leads to the following equation, describing the channeling process:

$$-\frac{\hbar^2}{2m}\left\{\frac{\partial^2}{\partial\rho'^2} + \frac{\partial^2}{\partial z^2}\right\}\Psi(r'_\perp;\rho') + \left\{U(r'_\perp) - \frac{\varepsilon_\parallel}{\rho_0}\rho'\right\}\Psi(r'_\perp;\rho') =$$

$$= (\varepsilon_\perp - \varepsilon_\parallel\frac{\rho'_0}{\rho_0})\Psi(r'_\perp;\rho') \qquad (2)$$

where ρ' and z are the two orthoganal coordiantes in the transversal plane of the particle motion. Axis ρ' is oriented in the direction of the curature center of the channel; ρ_0 is the curvature radius; ρ'_0 is the initial local coordinate of the particle into the channel. In our consideration, we assume that $\rho_0 \gg \rho', \rho'_0$.

This equation contains an additional centrifugal potential terms in both sides of the equation. Comparing these terms in (2) with the typical depth U_0 of the potential $U(r_\perp)$, we come to the conclusion that if the condition (1) is fulfilled then the centrifugal potential terms are small and can be neglected. In this case the equation (2) coincides with the one obtained for the linear channeling case. It proves that channeling in a bent crystal has to take place indeed if condition (1) is fulfilled.

The equation (2) determines the specific features of the channeling process in a bent crystal, which we briefly discuss in the next section.

3 SPECIFIC FEATURES OF CHANNELING IN A BENT CRYSTAL

3.1 The effective potential of the transverse motion

The motion of the particle in the transverse plane is determined by the effective potential acting on the particle, which appears in the equation of motion (2). The effective energy contains two parts - the transverse potential of the channel $U(r'_\perp)$ and the centrifugal potential contribution. The latter one increases with the energy of the particle into the channel and is inversely proportional to the channel's curvature. In the limit of the linear channel ($\rho_0 \longrightarrow \infty$) this contribution vanishes. The transverse energy of the particle in the channel depends on the initial coordinate of the particle into the channel ρ'_0 arising in the corresponding term in the right hand side of (2).

The potential $U(r'_\perp)$ in (2) has two parts

$$U(r'_\perp) = V(r'_\perp) + \frac{M^2_\perp}{2mr'^2_\perp}, \tag{3}$$

where $V(r'_\perp)$ is the transverse potential of the axial channel and $M_\perp$ is the angular momentum of the particle's rotation in the channel. Vibrations of atoms forming the channel influence the shape of the transverse potential $V(r'_\perp)$. It makes the shape of $V(r'_\perp)$ slightly different at different temperatures. The shape of $V(r'_\perp)$ is well known in many crystals. There are analytic approximations for the transverse potential $V(r'_\perp)$ (see e.g. [3]), which depend both on crystal parameters and temperature. The shape of the effective potential (2) depends on many variables: $M_\perp$, m, $\varepsilon_\parallel$, ρ_0 and the angular position φ of the particle in the channel ($\rho' = r_\perp \cdot \cos\varphi$).

3.2 The polarization of the transverse energy states

The equation (2) is analogous to the equations of two or one-dimensional atom placed in the external uniform electric field. This analogy lets us make some conclusions about the transverse motion of the particle. Moving along the channel particle populates one of the discrete levels in the transverse potential. In a bent crystal these energy levels become shifted and splitted compared the linear channel case like the energy levels of an atom in an external electric field.

This splitting can be calculated as

$$\Delta\varepsilon_{n;M_\perp} = -\frac{1}{2}\alpha_{n;M_\perp} \cdot \left(\frac{\varepsilon_\parallel}{\rho}\right)^2 \tag{4}$$

where $\alpha_{n;M_\perp}$ is the polarizability of the state $|n; M_\perp >$.

The states of the transverse motion become polarized due to the bending of the channel. The induced dipole moment of the state with the principal quantum number n and the angular momentum $M_\perp$ ca be estimated as

$$D_{n;M_\perp} = e \cdot \alpha_{n;M_\perp} \cdot \frac{\varepsilon_\parallel}{\rho} \sim eR \cdot \left(\frac{\varepsilon_\parallel R}{U_0\rho}\right) = eR \cdot C \ll eR \tag{5}$$

This condition shows that if (1) is fulfilled then the polarization of the transverse states is small. The induced dipole moment $D_{n,M_\perp}$ is much smaller then typical values of the transition dipole matrix elements between states of the transverse motion.

3.3 The tunneling effect

The outer wing of the effective transverse potential of a given bent channel has a barrier. Therefore a particle captured by the potential can leave the channel by tunneling under the potential barrier in the effective potential.

Let us estimate the probability of this process. It is obvious that the tunneling of the particle with the large relativistic mass can be considered as quasiclassical. Therefore the probability of de-channeling from a certain level in the transverse potential is equal to the probability of the particle's tunneling under the barrier, which can be calculated quasiclassically. The result of the calculation for the square-well potential is given by the expression [9]

$$W \sim \exp\left\{-\frac{4}{3}\frac{\varepsilon_\perp}{\varepsilon_\parallel}\frac{\rho\sqrt{2m|\varepsilon_\perp|}}{\hbar}\right\} \tag{6}$$

The shape of the potential determines the predexponential factor omitted in (6).

The quantity $\varepsilon_\perp$ in (6) is the binding energy of the particle in the transverse potential, which can be estimated as $\varepsilon_\perp \sim U_0$. The momentum appearing in the exponent is approximately equal to $\sqrt{2m|\varepsilon_\perp|}/\hbar \sim 1/R$. Then one obtains that the expression in the exponent is proportional to $1/C$, which is large if the condition (1) is fulfilled. This result shows that the probability of de-channeling due to the tunneling effect turns out to be exponentially small. It becomes large only then the channeling effect itself is practically absent.

3.4 The probability of the channeling process

Let us now estimate the probability of the particle to be captured by the field of a bent channel. If the channel is linear then the particle populates one of the discrete levels in the effective transverse potential. In case of a bent channel the continuous spectrum of the energy levels co-exists with the spectrum of the quasi-discrete levels. The channeling effect occurs only if the particle populates one of the levels of the quasi-discrete spectrum. Assuming that all the states in the discrete and the continuous spectra are being initially populated with equal probability, we come to the conclusion that the probability for channeling can be estimated as $W = N_D/(N_D + N_C)$ where N_D and N_C is the number of states in the discrete and continuous spectra respectively.

Esitmating N_D and N_C in the effective potential, one can obtain the following dependency of W on the parameters of the system [9]:

$$W = \frac{1}{1 + a \cdot C^{3/4}} \tag{7}$$

Here the parameter C is the main parameter of the problem defined in (1). The constant $a = (4U_0R/Z_{ef})^{1/4}$ depends on the potential characteristics and can be considered as an empirical constant, typically $a \sim 2$. The result (7) shows again that the probability of the channeling in a bent crystal is determined by C.

3.5 Radiation by a particle channeled in a bent channel.

The radiation emitted by a particle moving in a bent channel differs from that in the case of a linear channel. The additional contributions to the radiation spectrum arise due to the curvature of the channel and disappear in case of a linear channel [9].

In this paper we consider only the radiation generated by the additional acceleration in the field of a bent channel. This problem can be easily solved using the results of the theory of synchrotron radiation in a uniform magnetic field [3]. In this consideration one has to replace the magnetic field by the transverse force, $\varepsilon_{\parallel}/\rho$, acting on the particle in a bent channel.

The results of our calculations are plotted in figure 1. The intensity in figure 1 is defined as $\omega dN/d\omega/L$ and given in $1/cm$. The photon frequency is plotted as $\omega/\varepsilon_{\parallel}$. This figure shows that with increasing of energy of the electron the radiation intensity increases and the frequency maximum of the emitted photons is shifted to the higher values. With increasing curvature of the channel the intensity of the radiation decreases. These dependencies are qualitatively clear. The electron radiates more intensively, when its acceleration is larger.

The derived intensities of radiation are rather large compared to those obtained for linear channeling [3]. For example, the maximum intensity of the radiation of the $150GeV$ electron alined with $< 110 >$ in Ge at $T = 100K$ is approximately equal to $20cm^{-1}$. This intensity is acieved in a bent crystal for the $50GeV$, when $\rho = 10$ as it is clear from figure 1. In this case the main parameter $C \sim 1$. The main condition (1) becomes violated and the probability of channeling decreases rapidly, but it is still rather large, being on the level of 30 per cent, as follows from (7) and the estimates performed in the previous section. At smaller curvatures (see curves $\rho = 5cm$ and $\rho = 2cm$ in the upper part of figure 1) or higher electron energies (see curve $E = 250GeV$ in the lower part of figure 1) the intensity of radiation in a bent crystal exceeds substantially the intensity of radiation in the linear case. At the same time the probabilty of channeling decreases according to (7). As a result of these two opposite tendencies the effective intensity of the photon emission becomes comparable again with the intensity of the radiation of the electron in the linear

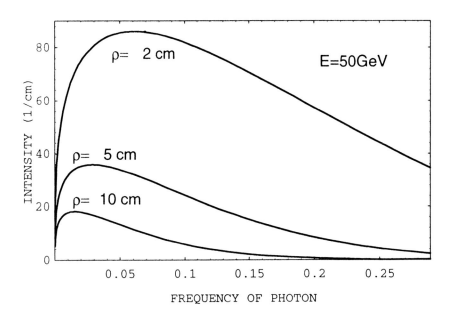

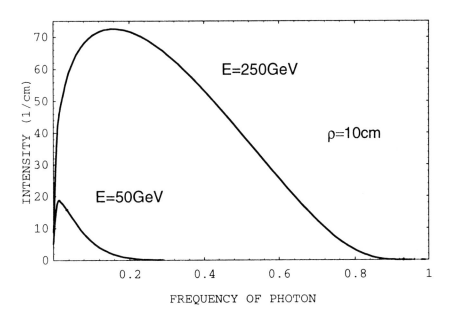

Figure 1: The radiative spectra of an electron channeling in a bent crystal calculated at different energies and curvatures of the channel. The intensity is given in $1/cm$.

channel, even if $C \sim 5$. The case $C \sim 5$ corresponds to the curve $\rho = 2$cm in the upper part of figure 1 and to the one with $E = 250$GeV in the lower part. In this case the probability W, (7), is approximately equal to 15 per cent.

The total radiative energy loss is determined by the parameter $\delta = \hbar\omega_0/\varepsilon_{\parallel}$, where $\omega_0 = c\gamma^4/\rho$ is the frequency corresponding to the maximum in the photon distribution. This parameter can be expressed in the form

$$\delta = \frac{\hbar\omega_0}{\varepsilon_{\parallel}} = \frac{\gamma^2}{m_0 c\rho} = C \frac{U_0}{m_0 c^2} \cdot \frac{\hbar}{m_0 cR} \gamma$$

If $\delta \sim 1$, then almost all energy of the particle is transferred to the emitted photons. The representation of δ in such a form shows that under the channeling conditions δ becomes comparable to unity at $\gamma \sim 10^7$.

4 ACKNOWLEDGMENTS

The authors are thankful to the Deutsche Forschungsgemeinschaft, the GSI and BMFT for support.

[1] Relativistic Channeling, edited by A.Carrigan and J.Ellison, Plenum (1987).

[2] B.L.Berman, S.Datz, Channeling Radiation-Experiments, In: Coherent Radiation Sources.- Berlin/Heidelberg/New York/Tokyo: Springer-Verlag, p.165 (1985).

[3] V.N.Baier, V.M.Katkov, V.M.Strakhovenko, Phys.Lett.A114, p.511 (1986); Phys.Lett. A117, p.251 (1986), Sov.Phys.JETP, v.65, p.686 (1987); High Energy Electromagnetic Processes in Oriented Monocrystals, Novosibirsk, Nauka (1989) (in Russian).

[4] J.Kimball and N.Cue, Phys.Rep., v. 125, p.69 (1985)

[5] S. P.Moller, E.Uggerhoj, H.W.Atherton, M.Clément, N.Doble, K.Elsener, L.Gatignon, P.Grafstrom, M.Hage-Ali and P.Siffert, Phys.Lett., v.256B, p.91 (1991).

[6] A.F.Elishev et al, Phys.Lett., v.88B, p.387 (1979)

[7] J.F.Bak, G.Melchart, E.Uggerhoj, I.S.Forster, P.R.Jensen, H. Mads Boll, S.P.Moller, H.Nielsen, G.Petersen, H.Schiott, J.J.Gross and P.Siffert, Phys.Lett., v.93B, p.505 (1980)

[8] A.Schäfer and W.Greiner, J.Phys.G:Nucl.Part.Phys., v.17, L217-221 (1991).

[9] A.V.Solov'yov, A.Schäfer and W.Greiner, Phys.Rev.E, v.53, p.1029 (1996).

QED-PROCESSES IN STRONG CRYSTALLINE FIELDS

E. UGGERHØJ

Institute for Storage Ring Facilities, University of Aarhus,
DK-8000 Aarhus C, Denmark

The paper discusses some recent results on the penetration of multi-GeV electrons/positrons penetrating strong fields in single crystals of diamond Si, Ge and W. Along axial directions radiation emission is enhanced up to two orders of magnitude with photon multiplicities up to ten or more. Pair production is enhanced up to one order of magnitude. This leads to a very fast shower formation with radiation lengths in axial directions (10–50) times shorter than the corresponding Bethe-Heitler values. The very strongly enhanced radiation emission leads to a reduction in transverse energies—cooling.

1 Introduction

Relativistic particles incident on crystalline targets in directions close to axial/planar directions will interact with the electromagnetic fields from the atomic rows/planes. The interaction length for GeV particles can be very large ∼mm. This means that the particles have a fair chance of scattering coherently on many atoms along their way. This coherence in scattering leads to strongly enhanced radiation emission — *coherent bremsstrahlung*.[1] *Channeling* is another coherent process leading to a strong steering effect, different for positive and negative particles.[2] Here the interaction with axes/planes can be described by a potential – the continuum potential – obtained by smearing the charges along the rows/planes of atoms.[2] The corresponding electromagnetic fields are $E \sim 10^{11}$ V/cm at a distance of 0.05–0.1 Å from the axis. In the GeV region it turns out that the continuum description can be used for incidence angles much larger than the channeling angles (cf. Ref.[3] for a review). When the energy of the particles gets into the multi-GeV region, the photon emission causes a significant recoil. On the other hand, the pure motion of the particle still remains classical. For a general description of penetration phenomena at such energies, a proper QED calculation has to be used; but this is very complicated. Fortunately, the Baier group[4] has shown that the so-called *constant field approximation* (CFA) is a good approximation for small incidence angles to crystal axes/planes. The critical angle θ_0 for this approximation is given by $\theta_0 \equiv U/mc^2$. Here, U is the depth of the axial/planar potential and m the rest mass of the electron. For incidence angles $\theta \gg \theta_0$ the Born approximation is justified and for $\theta \ll \theta_0$ the CFA is a good approximation. It should furthermore be pointed out that probabilities for QED processes in the electromagnetic fields from axes/planes are determined by the magnitude of these fields in the rest system for the incoming particle. Hence, the quantum parameter *chi* is given by $\chi = \gamma\, E/E_0$, where $E_0 = 1.3 \times 10^{16}$ V/cm and γ is $10^5 - 10^6$. This means that crystalline targes are unique for investigating QED processes in strong fields because here large χ-values (≥ 1)can be easily obtained.

Theoretical models for radiation emission and pair production was also developed by Kononets and Ryabov.[5-6] By introducing phenomenological constants to incorporate the influence of angular momenta on the transverse electron distribution they get good agreement with our CERN-results.[7-9] Very recently, J. Augustin *et al.*[10] used a single string model and solved the Dirac equation in cylindrical coordinates. The main objective being to check whether their results agree with the quasi-classical results.[4-6] A short description of the experimental technique is given in the following, followed by typical experimental results from the most recent investigations. In July 1995 a workshop on 'Channelling and other coherent crystal effects at relativistic energy' was held in Aarhus. Interested readers are referred to the conference proceedings appearing as a topical issue of NIM B.

2 Experimental

For investigation of the predicted strong field effects, the experimental setup used by NA-43[11] in the H_2 beam of the North Hall is unique. The two drift chambers 40 m apart on the incident side gives an angular resolution of 3–4 μrad. There are two positions for crystal mounting on high precision goniometers inside dedicated vacuum chambers. In Vac. Chamber I one probes the crystal with e^+/e^-, in Vac. Chamber II with photons. The photons may be tagged with the combination of DC3, Bend3 and DC4. Crystals may be cooled to liquid nitrogen temperature. DC1-DC2-DC3 allows to measure the scattering taking place in crystal I. With C (converter) and SSD is measured the average photon multiplicity. Finally, the pair spectrometer represented by Bend4, DC5 and DC6 is used to determine single photon energies from the created pairs.

The NA-43 detector is, as can be seen, a multi-purpose setup in which many aspects of strong QED effects may be studied. In the following the experimental crystalline yields are normalized to those from amorphous foils of the same thickness. Radiation spectra are given as power spectra.

3 Results

Fig. 2 depicts the dramatic radiation emission when 150 GeV electrons traverse a 1.5 mm thick diamond crystal along the <100> direction. The photon spectra are given as enhancements compared to the Bethe-Heitler yields. It is seen that close to axial directions the photon yields are enhanced up to two orders of magnitude. The Lindhard channeling angle is only 25 μrad, so the strong field effects are found for incidence angles much larger than the critical angle for channeling as predicted by Baier et al.[4] The photon multiplicities for the cases in Fig. 2 are shown in Fig. 3. The large photon multiplicities (5–10) along an axial direction correspond to a strongly reduced mean free path λ_f for photon emission, which for amorphous diamonds is about 4 cm. So the average mean free path for photon emission is reduced by (1–2) orders of magnitude.

Fig. 4 shows photon spectra from 150 GeV electrons incident on the 0.5 mm thick <100> diamond crystal. The spectra are shown for increasing incidence angles to the <100> axis. The beam divergence was ±30 μrad with the center of the beam core parallel to the {110} plane. For incidence angles around 0.3 mrad to the <100> axis a high energy photon peak appears at $\sim$ 0.7 times the electron energy E_0. The lower energy photons with maximal enhancement around $0.2 \cdot E_0$ is attributed to channeling radiation (Ch R) from the {110} plane.[9] The energy of the high-energy photon peak increases for increasing incidence angle Θ like normal coherent bremsstrahlung. In fact, this effect arises from coherence in radiation emission when the projectiles cross the planar rows of atoms at an angle of 0.3 μrad in the present case. This effect is found in other crystals also and is discussed by Baier et al.[12]

The strongly enhanced radiation emission along crystalline directions gives rise to a dramatic, increased energy loss compared to amorphous foils. In Fig. 5 is shown the measured radiative energy losses for 70, 150 and 240 GeV electrons traversing Si, Ge and W crystals. The energy losses along crystalline directions are enhanced about 25 times compared to amorphous targets.

Since radiation emission and pair production (PP) are inverse processes, a strong enhancement in PP along crystalline directions is also expected. Fig. 6 represents some recent results for high energy gamma rays penetrating a 3 mm thick W crystal in directions close to the <100> direction. The first results of this type were found for Ge crystals.[13] The enhancement for PP as a function of crystal type is discussed in detail by Baier et al.[4] In general, the maximal enhancement is about 10. The threshold photon energy for which the crystal PP equals the Bethe-Heitler value decreases for increasing atomic numbers of the target. As can be seen in Fig. 6, this threshold for W is about 10 GeV. For details see ref.[4]

The influence of strong crystalline fields on shower formation can be seen in Fig. 7, where pulse height spectra from a solid-state detector placed just behind a 25 mm Ge crystal is depicted. The pulse height spectra correspond to secondary particles from the shower formation when 40, 149 and 287 GeV electrons are incident along the <110> direction in Ge and in a random direction. Clearly, there is practically full separation between the random and aligned spectra in the case for 287 GeV electrons. The angular resolution of these effects are found to be about 100 μrad for the planar cases. The effect can be used as a directional sensitive calorimeter.

4 Conclusion

From the present experimental investigations of QED-processes, like radiation emission, pair production and shower formation, it is obvious that the strong electromagnetic fields along crystalline directions enhance very strongly QED-processes. These effects have a large variety of applications, such as production of high energy photons, directional sensitive calorimeters and particle production. The strong crystalline fields are also today used for bending GeV-TeV beams and extracting such beams from storage rings.

5   Figures

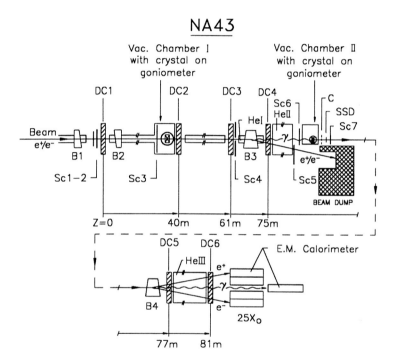

Figure 1: Experimental setup used in the NA-43 experiment at CERN. Scintillators are designated by Sc, drift chambers by DC and deflection magnets by B. HeI etc. are Helium tanks (introduced to reduce the amount of material along the beam line) while C is a calibrated convertor and SSD the solid state detector.

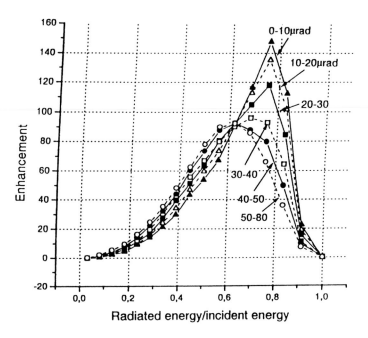

Figure 2: Radiation emission spectra from 150 GeV electrons incident on a 1.5 mm thick diamond crystal. The incidence directions to the <100> axis are given in the figure.

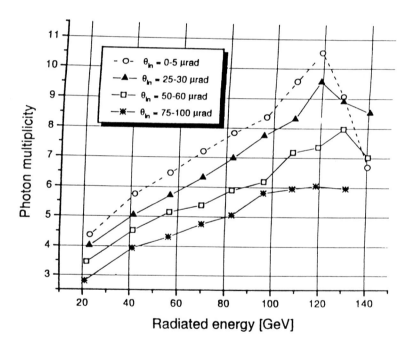

Figure 3: Average photon multiplicities for 150 GeV electrons incident on the 1.5 mm diamond crystal (Fig. 2). The multiplicities are given as a function of radiated energy for different incidence angle regions given in the figure.

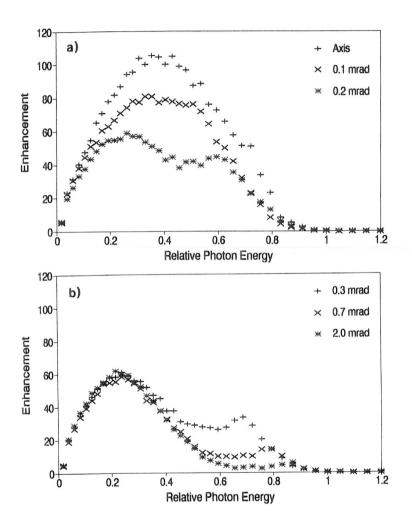

Figure 4: Radiation spectra from 150 GeV electrons incident on the 0.5 mm thick diamond crystal as a function of incidence angles to the <100> axis — angles given in the figures. The beam divergence was ±30μrad with the center of the beam core around the {110} planes.

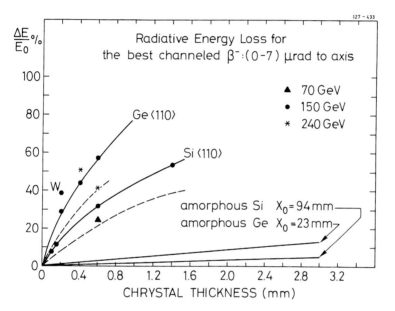

Figure 5: Radiative energy loss for electrons incident on <110> Si and Ge crystals of varying thickness given in mm. The incidence angle region was (0–7) μrad to the <110> axis for full-drawn curves. The dashed curves correspond to the largest incidence angle regions of present data (Figures 8 and 10). The particle energies are given on the figure — shown is also the value for a W crystal.

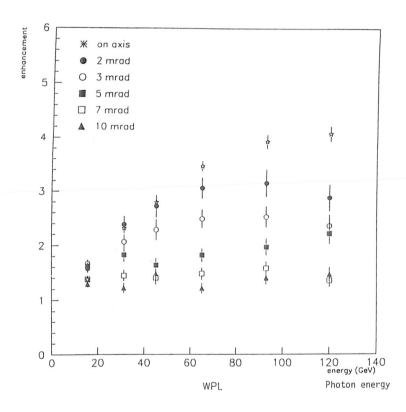

Figure 6: Pair production for high energy photons incident on a 3 mm thick <100> W crystal. The yields are normalized to the Bethe-Heitler yields and shown for different incidence angles to the <100> direction.

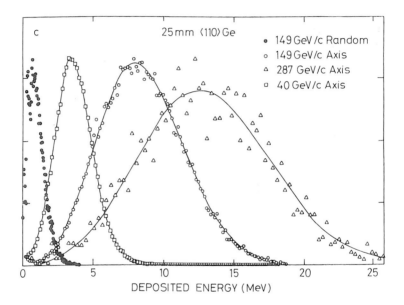

Figure 7: The pulse height spectrum as a function of deposited energy for the solid-state detector (SSD) behind the target. The target was 25 mm thick <110> Ge. The incidence beam energies are given in the figure.

References

1. G. Diambrini Palazzi, Rev. Mod. Phys. **40**, 611 (1968).
2. J. Lindhard, K. Dan. Vid. Selsk. Matt. Fys. Medd. **34**, 14 (1965).
3. A.H. Sørensen and E. Uggerhøj, Nucl. Sci. Appl. **3**, 147 (1989).
4. V.N. Baier, V.M. Katkow and V.M. Strakhovenko, Nucl. Instr. and Meth. B **69**, 258 (1992); and
   V.N. Baier, V.M. Katkov and V.M. Strakhovenko, Sov. Phys. Usp. **32**, 972 (1989).
5. Y. Kononets and Ryabov, Nucl. Instr. & Meth. B **48**, 269 (1990) and **48**, 274 (1990).
6. Y. Kononets, J. Moscow Phys. Soc. **2**, 71 (1992).
7. R. Medenwaldt *et al.*, Phys. Lett. B **242**, 517 (1990).
8. R. Medenwaldt *et al.*, Phys. Lett. B **260**, 235 (1991).
9. R. Medenwaldt *et al.*, Phys. Lett. B **281**, 153 (1992).
10. J. Augustin, A Schäfer and W. Greiner, Phys. Rev. A **51**, 1367 (1995).
11. K. Kirsebom, R. Medenwaldt, U. Mikkelsen, S.P. Møller, A.H. Sørensen, E. Uggerhøj, T. Worm, K. Elsener, S. Ballestrero, P. Sona, S.H. Connell, J.P.F. Sellschop, Z. Vilakazi, M. Hage-Ali, P. Siffert, J.-P. Stoquert, C. Biino, R.O. Avakian, K. Ispirian and S.P. Taroian, NA-43 collaboration.
12. V.N.Baier, V.M. Katkov and V.M. Strakhovenko, Nucl. Instr. & Meth. B **69**, 258–267 (1992).
13. J.F. Bak *et al.*, Phys. Lett. B **202**, 615 (1988), and Phys. Lett. B **213**, 242 (1988).

DYNAMICS OF FULLERENE COLLISIONS

R. SCHMIDT, O. KNOSPE

Technische Universität Dresden, Institut für Theoretische Physik,
D–01062 Dresden, Germany

Results from a systematic theoretical study of $C_{60}^+ + C_{60}$, $C_{70}^+ + C_{60}$ and $C_{70}^+ + C_{70}$ collisions using molecular dynamics combined with density–functional theory are presented. It is demonstrated that the reaction channels known from nuclear heavy–ion collisions – deep inelastic scattering, fusion, fragmentation and multi-fragmentation – are realized in atomic cluster–cluster collisions as well. In particular three–body collisions, two oppositely charged atomic clusters can be trapped in a bound state ("Rydberg clusters"). Basic properties of this new class of Rydberg systems are discussed.

1   Introduction

It is now well known that nuclear physics concepts can be applied to understand many properties of atomic clusters. Walter Greiner was one of the first who has stressed this point. [1] On the occasion of his 60th birthday it is a great pleasure for us to demonstrate how concepts developed by him and his co–workers for nuclear heavy–ion collisions (HIC) are very useful in order to understand the basic mechanism of atomic cluster–cluster collisions (CCC) as well. So, the most striking characteristic of CCC is the appearance of a number of well defined reaction channels. These are deep inelastic scattering, fusion, fragmentation and multifragmentation, which are expected to occur in this order with increasing collision energy. Exactly the same basic reaction channels were found in nuclear HIC. [2,3] Their atomic counterparts, meanwhile realized also experimentally in fullerene–fullerene collisions, [4,5] will be discussed in section 2. In "exotic" three–body collisions, two oppositely charged atomic clusters can be trapped in a bound state. This new class of Rydberg systems – Rydberg clusters – is presented in section 3.

2   Reaction channels in fullerene–fullerene collisions

2.1   Fusion and deep inelastic scattering

Our systematic investigation of fullerene–fullerene collisions using molecular dynamics combined with density–functional theory in the local density approximation and in a linear–combination–of–atomic- orbitals representation (hereafter referred to as quantum molecular dynamics – QMD) includes at

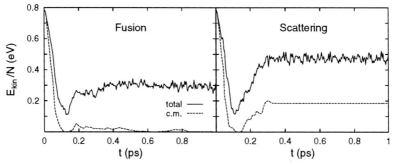

Figure 1: Calculated total kinetic energy per atom (full lines) and centre–of–mass kinetic energy per atom of the relative motion between the two colliding fullerenes (dashed lines) for QMD simulations of two central $C_{70}^+ + C_{70}$ collisions at a collision energy $E = 110$ eV, zero initial temperature and the same principal orientation. The different random orientations around the symmetry axes lead to different reaction channels – fusion (left part) and deep inelastic scattering (right part). [8]

present $C_{60}^+ + C_{60}$ [6,7,4], $C_{70}^+ + C_{60}$ and $C_{70}^+ + C_{70}$ collisions. [8] The ingredients of the QMD approach have been described previously. [9] In this chapter, the fusion–relevant range of collision energies $50 < E < 250$ eV (given in the centre–of–mass frame throughout this paper) is considered. A large number of collision events was simulated for zero cluster temperature as well as for a temperature of $T = 2000$ K in both clusters. In the case of $C_{60}^+ + C_{60}$, the whole relevant range of impact parameters ($b = 0 \ldots 14$ a.u.) has been covered.

For all three collisional systems considered, fusion and deep inelastic scattering were found to be competing reaction channels in the energy range above the fusion barrier (see below) before fragmentation becomes the dominating reaction channel at higher collision energies (see sect. 2.2).

The discriminating mechanism between fusion and scattering in these collisions can be recognized by investigating the kinetic energy as a function of time, [4] which is shown for two $C_{70}^+ + C_{70}$ events in fig. 1. The general increase of the centre–of–mass (and also the total) kinetic energy after the system has reached the distance of closest approach (corresponding to zero centre–of–mass energy in fig. 1) indicates that a part of the stored potential energy can be converted back into kinetic energy of relative motion. If this "bouncing–off" is suppressed (left part of fig. 1) due to the rearrangement of atoms in the contact zone a fusion event will be detected, whereas if the increase of kinetic energy proceeds to a large enough extent (right part of fig. 1) the effective repulsion of the (more or less) intact fullerene structure leads to a scattering event. [8]

The values obtained for the fusion barriers are compared with the experimental data [5] in table 1. The theoretical barriers for $T = 0$ are significantly

Table 1: Fusion barriers for $C_{60}^+ + C_{60}$, $C_{70}^+ + C_{60}$ and $C_{70}^+ + C_{70}$ collisions obtained from QMD simulations at zero temperature (second column) and at $T = 2000$ K for projectile and target[8] (third column) compared with the experimental data[5] (fourth column).

	QMD		Experiment
	$V_B(T=0)/eV$	$V_B(2000\text{ K})/eV$	$V_B/(\text{eV})$
$C_{60}^+ + C_{60}$	80	60	60 ± 1
$C_{70}^+ + C_{60}$	94	70	70 ± 6.5
$C_{70}^+ + C_{70}$	104	75	76 ± 4

larger than the experimental values (cf. second and fourth column in table 1), which has been understood to be an effect of the finite cluster temperature in the experiment. The available phase space is considerably enlarged due to the additional energy and the softening of the tight fullerene structure. Unfortunately, the actual cluster temperature in the experiment cannot be measured but only roughly estimated by indirect methods.[5] We have chosen for our simulations a temperature of $T \approx 2000$ K for projectile and target, which fits into the experimental estimation (1800 . . . 2000 K[5]). The resulting fusion barriers for $C_{60}^+ + C_{60}$, $C_{70}^+ + C_{60}$ and $C_{70}^+ + C_{70}$ agree perfectly with the experimental values (cf. third and fourth column in table 1).[8]

Besides the microscopic (and extremely computer–time consuming) QMD simulations, phenomenological fusion models based on transparent physical assumptions about the reaction mechanism can provide an intuitive picture of the complicated real situation. Such a fusion model for fullerene–fullerene collisions has been derived[4] in which the following simple expression for the fusion cross section as a function of the collision energy has been obtained:

$$\sigma_{CF}(E) = \begin{cases} \sigma_0 \dfrac{(E - V_B)^2}{E} & ; \quad V_B \leq E \leq E_{cr} \\ \dfrac{\sigma_1}{E} - \sigma_2 & ; \quad E > E_{cr} \end{cases} \tag{1}$$

with the model parameters σ_0, σ_2, V_B and E_{cr} ($\sigma_1 = \sigma_0(E_{cr} - V_B)^2 + \sigma_2 E_{cr}$ is fixed to guarantee continuity at $E = E_{cr}$). As a first test of the fusion model, eq. (1) is fitted to the experimental fusion cross section for the three considered collisional systems. The results as a function of $1/E$ are shown in fig. 2, where the experimental values for V_B were used, and the critical energy E_{cr} was located (approximately) at the maximum of the experimental cross section. The energy dependence of the fusion cross section is in very good agreement with the experimental data, which demonstrates that the assumptions concerning

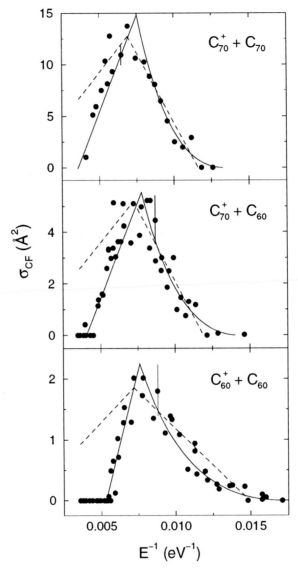

Figure 2: Absolute fusion cross section as a function of the inverse collision energy for C_{60}^+ + C_{60} (lower part), C_{70}^+ + C_{60} (middle) and C_{70}^+ + C_{70} (upper part). The full lines represent the results of the phenomenological fusion model according to eq. (1)[8] in comparison with the experimental data[5] (dots) and with a model with energy–independent fusion probability[5] (dashed lines). The model parameters in eq. (1) are adjusted to give the best overall agreement with the data (V_B is fixed at the experimental values from table 1).[8]

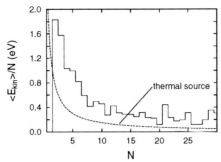

Figure 3: Mean translational kinetic energy per atom $\langle E_{kin}\rangle/N$ as a function of the cluster size N in near central $C_{60}^{+} + C_{60}$ ($E = 500$ eV) collisions (Histogram). The dashed curve corresponds to the expectations of a thermal source (see text).[6]

the fusion probability reflect the essential physical aspects properly.[8] In the case of $C_{70}^{+} + C_{70}$, there might be an indication of a plateau around E_{cr}, which could not be explained with the model (1). However, the accuracy of the experimental data will have to be improved before further details of the cross section can be discussed.

2.2 Multifragmentation and collective–flow effects

Spontaneous multifragmentation and collective–flow effects are macroscopic phenomena like shattering of glass, or shock wave propagation. Their existence in nuclear heavy–ion collisions with a large but *finite* number of degrees of freedom is nontrivial, and their specific nuclear (quantum) properties enlarged widely our knowledge about nuclear dynamics far from equilibrium (e.g., about the validity range of macroscopic concepts, like fluid dynamics[3]). From a more general point of view, this holds for *any* many–body system with a *finite* number of particles, like atomic clusters.

In this chapter, $C_{60}^{+} + C_{60}$ collsions at a bombarding energy of 500 eV are considered with an excitation energy of 30 eV for the projectile ion to obtain realistic experimental conditions. The whole range of (fragmentation-) relevant impact parameters ($b \leq 14$ a.u.) was investigated.[6,7]

In fig. 3, the mean translational kinetic energy per atom $\langle E_{kin}\rangle/N$ of all clusters with $N \leq 30$ is shown (histogram). It is compared to that of a thermal source which simply yields $\langle E_{kin}\rangle/N = 3kT/2N$ (dashed curve). The temperature $kT = 1$ eV has been estimated from the mean internal excitation energy per atom of the clusters.[7] The nearly N–independent shift of the calculated $\langle E_{kin}\rangle/N$ compared to that of a thermal source is a first clear signal of an additional mechanism that contributes to the final translational cluster velocities in terms of collective–flow effects. [6]

One of the universal features of multifragmentation concerns the power law in the mass distribution of the fragments, which is observed in very different fields of physics, e.g., in the mass distribution of the fragments in high–energy nuclear reactions, for the collision debris of macroscopic stones or in the size distribution of asteroids in the planetary system. The calculated mass distribution of the multifragmentation events for our collisional system exhibits a simple power–law distribution $\sim N^{-\tau}$ up to the largest clusters formed ($N \leq 94$) with an exponent between $\tau \approx 1$ and $\tau \approx 1.5$. [7]

3 Rydberg clusters

The interest in the fundamental question of the transition between quantum and classical physics, already discussed in Schrödingers pioneering work [10], has been renewed considerably (in particular, the Kepler dynamics of electronic wave packets formed by a superposition of Rydberg states). Despite of the intensive efforts in the last years, the ultimate goal of these investigations – the preparation of a Rydberg wave packet localized with respect to radial *and* angular coordinates moving along a Kepler orbit – was not yet reached experimentally. Apparently, it is difficult to excite a "classical" state (a spatially localized wave packet) starting from the "quantum–mechanical limit" (the electronic ground state of an atom). Alternatively, we have recently proposed to form Coulomb systems by objects, the sizes of which are in the transition region from microscopic to macroscopic dimensions, e.g. two oppositely charged atomic clusters strongly bound in states with arbitrarily high quantum numbers ("Rydberg clusters"). [11] In this case, mass and charge can be changed independently leading to an infinite variety of systems. The motion of such massive objects is expected to be in essence classical, i.e. characterized by spatially localized wave packets. In addition, qualitatively new phenomena are expected, because an atomic cluster represents a system with a large number of internal degrees of freedom, and their influence on the wave–packet dynamics opens an interesting new field of research. [12] One of the most challenging questions concerns the possible formation process of these states and, thus, their experimental verification. In contrast to electronic Rydberg states, Rydberg clusters can not (or hardly) be formed by electronic transitions. Instead, *preformed* and *unbound* positively and negatively charged cluster ions (as e.g. existing in the plasma of a C_{60}–laser desorption source) can be used as starting point. Assuming a large intercluster distance and a high orbital angular momentum between both cluster ions, their relative kinetic energy may dissipate during the collision with a third particle leading to a bound state, e.g. as

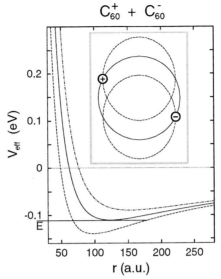

$C_{60}^+ + C_{60}^-$

Figure 4: Rydberg clusters $C_{60}^+ + C_{60}^-$: effective potential V_{eff} as a function of the intercluster separation r for three different angular momenta L = 8, 9 and $10 \cdot 10^3 \, \hbar$ (dashed, solid and dashed–dotted line, respectively) and the corresponding classical trajectories (insert) to a given total energy E = -0.1102 eV. The same length scale is used in both drawings.[11]

presented in fig. 4. The possibility of such formation processes in (effective three–body) $C_{60} + C_{60}^+ + C_{60}^-$ collisions has been demonstrated by molecular dynamics simulations.[11]

In contrast to electronic Rydberg systems, the large binding energy of Rydberg clusters[11] guarantees stability against external perturbations. Therefore, their lifetime is expected to be determined mainly by two processes, i.e. electromagnetic radiation and mutual neutralization. However, macroscopically large lifetimes were estimated provided that the states are formed with sufficiently large angular momenta (or, equivalently, at large intercluster separations).[11]

4 Summary

We have shown that phenomena like fusion, deep inelastic scattering and multifragmentation connected with collective–flow effects can be probed in atomic CCC and discussed in terms of nuclear HIC. Especially the existence of collective flow in CCC represents an interesting result for these systems with a large but still finite number of degrees of freedom. The simultaneous investigation in different fields of physics may reveal some universal features of these phenomena.

Second we have demonstrated that atomic physics concepts can also be probed with atomic clusters on a new ground. So, the experimental verification and investigation of the predicted "Rydberg clusters" may contribute to our understanding of the fundamental question of the transition between quantum and classical physics.

Acknowledgements

Financial support by the Deutsche Forschungsgemeinschaft through Schwerpunkt "Zeitabhängige Phänomene und Methoden in Quantensystemen der Physik und Chemie" and by the EU through HCM networks "Formation, stability and photophysics of fullerenes" and "Collision induced cluster dynamics" is gratefully acknowledged.

References

1. W. Greiner in *Nuclear Physics concepts in the study of atomic cluster physics*, Lectures notes in physics, Vol. **404**, eds. R. Schmidt, H.O. Lutz and R. Dreizler (Springer, Berlin 1992), p. 3

2. J. Maruhn, W. Greiner, W. Scheid in *Heavy Ion Collisions*, vol. 2, ed. R. Bock (North–Holland, Amsterdam 1980), p. 399

3. W. Scheid, H. Müller, W. Greiner, Phys. Rev. Lett. **32**, 741 (1974); H. Stöcker, W. Greiner, Phys. Rep. **137**, 277 (1986)

4. F. Rohmund, E.E.B. Campbell, O. Knospe, G. Seifert, R. Schmidt, Phys. Rev. Lett. **76**, 3289 (1996)

5. F. Rohmund, A.V. Glotov, K. Hansen, E.E.B. Campbell, J. Phys. B (1996), in print

6. R. Schmidt, J. Schulte, O. Knospe, G. Seifert, Phys. Lett. A **194**, 101 (1994)

7. J. Schulte, O. Knospe, G. Seifert, R. Schmidt, Phys. Lett. A **198**, 51 (1995)

8. O. Knospe, A.V. Glotov, G. Seifert, R. Schmidt, J. Phys. B (1996), in print

9. G. Seifert and R. Schmidt, J. of Modern Physics B **96**, 3845 (1992) and New J. Chem. **16**, 1145 (1992)

10. E. Schrödinger, Naturwiss. **14**, 137 (1926)

11. O. Knospe, R. Schmidt, Z. Phys. D **37**, 85 (1996)

12. O. Knospe, R. Schmidt, Phys. Rev. A (1996), in print

Recent applications of energetic particle beams with diamond targets[a]

Friedel Sellschop and Simon Connell
Schonland Research Centre for Nuclear Sciences, University of the Witwatersrand, Johannesburg, South Africa

Early nuclear based methods of interrogation of diamond, dealt predominantly with questions of composition, major, minor and at trace levels. Recent studies have addressed more dynamic properties and turned to unique applications. The nuclear contribution to the study of and with diamond, both natural and man-made, has produced information and insights not accessible by other methods of physics.

1 Earliest appreciation of the key physical properties of diamond

The earliest records addressing diamond are Biblical, and attest to its hardness. It did not take long however before this hardness became associated with (the need for) hardihood and diamond became a male affectation, being worn on clothing or affixed to the hilts of swords to ensure the desired outcome. The chemical inertness of diamond soon became recognised, inflating the value both intrinsic and mystical. The hydro-phobicity of the diamond surface is likely to have been identified at an early era, since this property is exploited in the early mythology of the recovery of diamonds from a rocky valley in the East. But evidently the imperative to work diamond by cleaving and polishing, revealed the recognition of yet another physical property, that of the high refractive index and consequently the unique beauty of the well-polished stone. This unleashed the demand from Kings to make diamond a central focus in the insignia of office such as crowns, orbs and sceptres. Necessarily Queens soon saw to it that their insignia of omnipotence flamboyantly displayed the same evidence....... and from the circle of Queens the demand spread to courtesans, to the wives of the rich and evermore the less rich, so that even some hundreds of years ago diamonds were exchanged between couples in the loving act of betrothal.

The start of the industrial application of diamond is lost in the mists of time, but the (simple) wire sawing with diamond powder of stones (particularly marble) for statues and funereal monuments became common a long time ago, and persists essentially unaltered in fact to this day. Mechanical abrasion, sawing and drilling applications with diamond become ever more sophisticated

---

[a] We dedicate this paper to Walter Greiner in honour of his 60'th birthday

and important, but the major objective of exploiting diamond, suitably doped, for unique and demanding electronic applications taking advantage (inter alia) of the wide band gap, the high thermal conductivity and the high electrical resistivity remain tantalisingly beyond our grasp despite progress in, for example the controlled doping of diamond.

In the diamond physics and diamond applications fields, attention must be drawn to the production of ever- bigger and - better diamonds produced by the high temperature, high pressure simulation by man of Nature's processes. No less dramatic has been the progress of true diamond growth by chemical vapour deposition in pressure-temperature regimes well outside the diamond-stable region as demarcated by the Berman-Simon line, leading to many new applications such as in providing (thin) diamond windows. It has become evident that in CVD diamond growth, the role of hydrogen is of seminal importance.

In the field of diamond physics, new research opportunities have been created through the growth of diamonds of excellent quality composed of isotopically pure carbon-13.

2 The first generation of nuclear interrogation of diamond

The need for analytical methods when studying precious materials, such as diamond, which preserve the specimen in an essentially undamaged state, is obvious, and this is in large measure satisfied by nuclear probes. The need for high sensitivity has become increasingly apparent .... and again, nuclear methods frequently have this distinction.

Diamond is composed of carbon, and even here in the ^{13}C : ^{12}C ratio lies important data on conditions of genesis. Following the pioneering work of Kaiser and Bond the major defect in the impurity spectrum characteristic of the dominant class (Type 1a) of natural diamond has been accepted to be due to nitrogen. That represented the bulk of information on the composition of diamond. Then a comprehensive study by instrumental neutron activation analysis revealed convincingly (1) that the picture was far more complex and that on the trace level, there was a large set of impurity elements, with a definitive chemistry, representative of the composition of the magma in the upper mantle of the earth where diamond genesis or crystallization had taken place. However this technique is effectively blind to the lightest of elemental impurities since neutron absorption produces either stable isotopes or very short-lived nuclear states. Fortunately charged particle (activation) analysis is particularly apt for these lighter nuclides so that they powerfully fill in the vital gap of the presence in diamonds of particularly the light volatiles, hydrogen, nitrogen and oxygen (2) ..... revealing inter alia that in regard to the work of Kaiser and Bond nitrogen must be considered a major impurity in diamond, but not the major impurity.

The outcome of these comprehensive nuclear analytical analyses is that all natural diamonds are defective in the sense that they carry inclusions of dimension circa 5 to 10 Ångstrom, apparently homogeneously distributed, submicroscopic it is true, but carrying in their chemical composition unique information on the nature of the upper mantle of the Earth ! These have been appropriately termed magma droplets, characterised by some 58 elements of which the three volatiles (hydrogen, nitrogen and oxygen) play a major role. The magmatic goo is both water-rich and gas-rich. Hydrogen in particular presents a definitive challenge !, which we will address In the second generation of nuclear research into diamond. The first generation of nuclear interrogation of diamond has been chronicled (3), (4), (5) . This embraces analysis of diamond, ion channeling and dechanneling studies in diamond, time differential perturbed angular correlation and distribution in demarcating preferred dopant sites in diamond, muon and muonium spin rotation in studying the dynamic behaviour of hydrogen in diamond.

3 The second generation of nuclear interrogation (study and application) of diamond.

While the earlier work on muon and muonium spin rotation in diamond revealed that the muonium atom occupied two distinctive sites, one at the tetrahedral interstitial position in the diamond lattice, with isotropic character, the other at the bond-centred site, which is anisotropic with symmetry about the $< 111 >$ axes. This is important information since it indicates the sites that the simplest dopant, hydrogen, will take up also. Two recent works addressed the sensitivity of the technique to defect structures and sought to measure for the first time the nuclear hyperfine component in diamond.

Natural diamonds are traditionally classified into four types (1a, 1b, 2a and 2b) with about 98% being in the type 1a class. However optical and electrical measurements have suggested quite convincingly that the higher nitrogen content associated with type 1a diamonds, is associated frequently with two specific defect geometries, coded as type 1aA and type 1aB, with the occasional occurrence of a mixed type 1aAB. It is both a reflection of how difficult the problem is and an indictment of the state of the field that the structures ascribed to these two dominant defects are as yet unproven. Current dogma is that the A defect is due to two nearest neighbour substitutional nitrogen atoms, viz. $[N - N]^0$, and the B defect is due to the N_4V structure, a cluster of 4 nitrogen atoms around a vacancy.

We are now taking the Muon / Muonium spin rotation technique to new levels of refinement in an attempt to interpret inter alia these primary defect structures. First experiments sought to see if the four types of natural diamond revealed any distinctive features in MuSR ...... not only were the precession frequencies measured, but also the relaxation rates of the precession due to depolarization. There are indeed distinctive features, such as for example the absence of the tetrahedral signal in some type 1a diamonds implying that muonium at this site could diffuse and be rapidly trapped at the defects characteristic of the type 1a diamonds in question, thereby being rapidly depolarized. The bond-centred muonium on the other hand is much more tightly bound and thus does not as readily migrate and become prey to traps. We have tested this also by deliberately creating vacancies (by electron irradiation) and substantiating the above conclusions. A simple two-stage trapping model appears to account quite satisfactorily for the observed effects, including the temperature dependence. There are thus real prospects of using muons to delimit to a greater degree than hitherto possible the structure of at least the most dominant defects in diamond.

The second of the recent advances in this field relates to the determination of the nuclear hyperfine interaction as between the unpaired electron that is a partner to the muonium ($\mu^+ - e^-$) atom in the bond-centred site, and the nearest neighbour nuclei. Unfortunately in diamond, the carbon host nuclei have zero nuclear spin, but this can be remedied by using a diamond composed solely of carbon-13 nuclei, which of course have one half integer spin. It was established in a seminal study (6) that the nuclear hyperfine structure of bond-centred muonium in pure ^{13}C diamond could indeed be resolved using time-differential transverse-field muon spin rotation. The measured nearest-neighbour ^{13}C hyperfine parameters can be compared to recent ab initio Hartree-Fock type cluster calculations. The ratios of the measured ^{13}C hyperfine parameters and free atom values indicate that 92% of the unpaired electron spin density resides on the two nearest-neighbour atoms, in marked contrast with the situation for the analogous muonium centre in Si where only 41% of the spin density is on the two nearest neighbours. The very high concentration of the unpaired electron spin density on the two nearest neighbour nuclei in the case of diamond, together with the observation that the next-nearest-neighbour hyperfine interaction is found to be too small to be measured, attests to the fact that the unpaired electron is highly localized. Furthermore, from the ratio of the s and p spin density on the nearest neighbour nuclei one can estimate that the bond containing the muonium is stretched (relaxed) by as much as 31%!

These muonium studies are in large measure inspired by the fact that muonium is to all relevant respects a light isotope of hydrogen, and the significance of hydrogen in relation to the properties of diamond cannot be underestimated. An isolated interstitial hydrogen atom is one of the simplest defects in a crystalline semiconductor. Consequently, the electronic structure, the associated lattice deformation, and the diffusivity and reactivity of hydrogen have long since been recognised by theorists and experimentalists alike as being of cardinal importance. With the discovery that hydrogen can passivate the electrical activity of shallow impurities in a wide variety of semiconductors, and given the fact that hydrogen is present in many steps during the processing of semiconductor devices, it is evident that it is important to understand the role of hydrogen for technological control also. While we have pursued much of our study of hydrogen in diamond through the behaviour of muonium, we have studied hydrogen itself directly also, using as nuclear probes the resonant heavy ion reactions

$$p(^{19}F, \alpha\gamma)^{16}O \qquad (1)$$
$$\text{and} \quad p(^{15}N, \alpha\gamma)^{12}C \qquad (2)$$

which can measure the depth profile of hydrogen in the bulk material under investigation, through sweeping the incident energy of the heavy ions from values that place the resonance on the surface of the sample to quantifiable depths. By this means, it has been discovered (2) that all diamonds contain hydrogen in their bulk together with a specifically surface coverage. The truly surface coverage of this layer of hydrogen is bonded to the carbon atoms, as revealed through the bending and stretching vibrations of the C - H bond which, in appropriate geometry, contribute a measurable Doppler spread to the width of the nuclear resonance (7). There is some evidence also of some tailing of the surface concentration of hydrogen into the near-surface of region of the sample, and of migration of hydrogen away from the region interrogated by the beam, diffusing back as soon as beam-activation is removed. Energy-stepped resonance studies are however time-consuming so that recently we have preferred to use the elastic recoil detection technique (ERDA : an inverse version of Rutherford scattering where the heavy ion is incident on the hydrogen in the sample, the proton then recoiling out from the sample —- a depth distribution of the hydrogen is obtained in a single measurement). Such nuclear probes have been illuminating in a variety of more dynamic studies : so, for example, when one deliberately implants hydrogen at a specific depth into diamond, a depth chosen to be greater that that of the ERDA probing heavy ion, no migration of the hydrogen is observed even up to 1200°C. This can only be interpreted as meaning that the hydrogen is trapped in the damage caused by its own implanting beam. This tight binding / trapping of the implanted hydrogen is confirmed by creating a specific heavily damaged layer nearby and finding no accumulation of trapped hydrogen through migration therein. In this regard diamond appears to be quite different to silicon in which hydrogen readily migrates at such temperatures. Also one must infer that the crystal environment in which hydrogen as implanted resides is quite different from that in which hydrogen occurs in natural diamond, where our earlier arguments attest to the fact that hydrogen would concentrate in water-rich and gas-rich phases in the ubiquitous magma droplets.

These are but examples of recent nuclear studies that probe in a unique and dynamic way the world of diamond on the atomic and indeed nuclear scale ...... in this second generation of nuclear studies in diamond.

Acknowledgments

It is a pleasure to acknowledge the contributions of many generations of students and colleagues to these many and diverse studies. Specifically we would like to thank our collaborators Tom Anthony for the skilled growing of pure ^{13}C diamonds, Achim Richter for his conception and creation of an electron accelerator with electron beam optical charecteristics of unsurpassed quality, Walter Greiner for his unwavering faith in our work in the field of Diamond Physics, and Krish Bharuth-Ram for his support and participation. We pay tribute to the unique skills of our colleague Mick Rebak in the preparation of diamond (targets) for nuclear interrogation. The support of Messrs de Beers Industrial Diamonds (Pty) Ltd and of the Foundation for Research Development is recorded with appreciation. An especial statement of appreciation is recorded by one of us (JPFS) for the award of the Max Planck Research Prize of the Max Planck Society and the Alexander von Humboldt Foundation which contributed substantially to these studies.

References

1. Fesq HW, Bibby DM, Erasmus CS, Kable EJD and Sellschop JPF
"A comparative trace element study of Diamonds from Premier, Finsch and Jagersfontein mines"
Physics and Chemistry of the Earth 9 (1975) 817 - 836

2. Sellschop JPF et al.
"The nature of the state of hydrogen on the surface and in the bulk of natural and synthetic diamond (using ion beam techniques)"

Vacuum 45 (1994) 397 - 402

3. Sellschop JPF
"Nuclear probes in physical and geochemical studies of natural diamond"
in
"The Properties of Diamond"
pp. 107 - 163
Ed. Field JE
Academic Press, 1979

4. Sellschop JPF
"Nuclear probes in the study of diamond"
in
"The properties of natural and synthetic diamond"
pp. 81 - 179
Ed. Field JE
Academic Press, 1992

5. Sellschop JPF
"Applied Nuclear Physics of our times : the example of Diamond Physics"
in
"Nuclear physics of our times"
pp. 301 - 311
Ed. Ramayya AV
World Scientific, 1993

6. Schneider, Kiefl, Chow, Johnston, Sonier, Estle, Hitti, Lichti, Connell, Sellschop, Smallman, Anthony and Banholzer
"Bond-Centred muonium in diamond: resolved nuclear hyperfine structure
Phys. Rev. Lett. 71 (1993) 557 - 560

7. Jans S, Kalbitzer S, Oberschachtsiek P and Sellschop JPF
"^{15}N Doppler spectroscopy of ^{1}H on diamond"
Nucl. Inst. and Meth. in Phys. Res. B85 (1994) 321 - 325

1. Harold and Donna Britt, Sybille Mersmann and Uli Lynen.
2. Yuri Oganessian and Bärbel Greiner in the middle.

1. A. Schempp, J. S. Vaagen and others on the boat.

2. M. K. Weigel showing U. Lynen where to go.

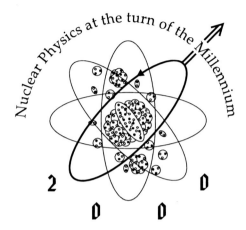

3. Superdense Matter and Particle Production

REALISTIC FORCES, MEDIUM DEPENDENCE AND HEAVY ION COLLISIONS[1]

Amand Faessler [2]

Institut für Theoretische Physik
Universität Tübingen
Auf der Morgenstelle 14
72076 Tübingen, Germany

Abstract

Heavy ion collisions at intermediate energies between 100 and 2 000 MeV per nucleon are described using quantum molecular dynamics (QMD) and relativistic quantum molecular dynamics with realistic forces. In detail we are discussing two points:

(i) We show how QMD can be extended to a relativistic treatment (RQMD) using a scalar and a vector potential. The directed transversal flow turns out to be sensitive to the different forces used and to the relativistic treatment.

(ii) In a second part we are looking how the mass of a nucleon depends on the density and the temperature. This dependence can be tested in antiproton production.

1 Introduction

The collision between two heavy ions goes via the nucleon-nucleon or in general the baryon-baryon interaction. At low relative momenta the interaction is attractive while at high relative momenta it is strongly repulsive.

Matrix elements of a potential which is at short range so strongly repulsive cannot be calculated in perturbation theory. The distortion of the wave functions at short relative distances must be taken into account with the help of the Brueckner theory where for the relative wave functions the Bethe Goldstone equation is solved. The resulting Brueckner reaction matrix elements do not only depend on the incoming and outgoing momenta of the two baryons, but due to the Pauli operator and the selfconsistent single particle energies in the denominatior they depend also on the medium: the matrix elements are dependent on the momentum and on the density distribution of the surrounding nucleons and on the starting energy [1], [2], [3].

---

[1]Supported by the BMBF under contract no. 06TU746(2) and GSI TÜFÄT
[2]e-mail: Amand.Faessler@uni-tuebingen.de

For the description of the heavy ion reactions we use non-relativistic and relativistic Quantum Molecular Dynamics (QMD).

Finally one is solving the Hamilton equations

$$\overset{\circ}{\mathbf{r}}_n(t) = \frac{\vartheta H(1,...,A;t)}{\vartheta \mathbf{P}_n(t)}$$
$$\overset{\circ}{P}_n(t) = -\frac{\vartheta H(1,...,A;t)}{\vartheta \mathbf{r}_n(t)} \tag{1}$$

and adds a collision term if two particles are coming close to each other. In the collision term one is applying Monte Carlo methods to determine (weighted with the corresponding cross sections) if one scatters the two interacting particles or if they react and produce for example a nucleon and a delta resonance or some other baryonic states or if they produce in addition to the two baryons also mesons, photons or even three baryons and an antibaryon. For this one needs energy dependent cross sections for scattering, reactions and particle productions. The difference between QMD and Vlasov-Uehling-Uhlenbeck (or Boltzmann-Uehling-Uhlenbeck) consists essentially in the fact that for QMD one is simulating completely each event before one is averaging over the outcome of several events (an ensemble). In VUU one is averaging in each time step of the simulation over the whole ensemble of events. Thus VUU (Vlasov-Uehling-Uhlenbeck) is essentially a one-body theory which cannot give correlations which influence for example a coincidence measurement.

If we want to probe the EOS, we have to reach thermal equilibrium. That means that in momentum space we have to come from a Fermi distribution which corresponds to two spheres to one sphere with a smeared out Fermi surface, indicating a temperature. In figure 1 we are showing the anisotropy ratio.

$$\langle R_a \rangle = \frac{\sqrt{\langle Px^2 \rangle} + \sqrt{\langle Py^2 \rangle}}{2\sqrt{\langle Pz^2 \rangle}} \tag{2}$$

For the collision ^{40}Ca on ^{40}Ca with 400 MeV/A bombarding energy in the laboratory system.

Figure 1 shows that even for central collisions of Ca on Ca at 400 MeV/A in the laboratory system one does not get thermal equilibrium corresponding to an anisotropy ratio $\langle R_a \rangle = 1$. Only in central collisions for very heavy systems one is coming close to thermal equilibrium.

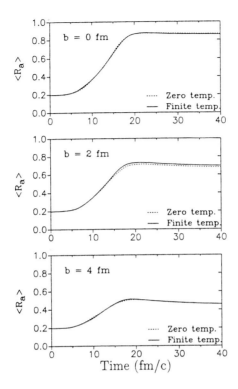

Fig. 1: *Anisotropy ratio* $\langle R_a \rangle$ *as defined in equation (2) as a function of the reaction time in units of fm/c for* ^{40}Ca *on* ^{40}Ca *with a laboratory bombarding energy* $E_{lab} = 400 MeV/A$. *The upper part shows a central collision with the impact parameter* $b = 0$ *fm, while the lower parts show results for impact parameters* $b = 2$ *and* 4 *fm. Even for a central collision the solid line which indicates the anisotropy ratio does not reach the thermal equilibrium value of* $\langle R_a \rangle = 1$.

2 Relativistic QMD

Already Dirac tried to extend classical many body dynamics to a relativistic treatment. But his theory was essentially interaction free [4]. Only Samuel [5] at the beginning of the eighties was able to give a formulation which can include interactions between the particles in a consistent way. Sorge, Stoecker and Greiner formulated in 1988 the theory in such a way, that it can be used in a fully covariant QMD (RQMD) [6]. We in Tuebingen, developed our computer code for RQMD along the lines of Samuel [5] and Sorge, Stoecker and Greiner [6]. Opposite to the code in Frankfurt, we included for the selfconsistant field and for a relative time contstraints all the nucleons and not only those which are nearby at the reference nucleon.

A quantity which turned out to be very sensitive to different types of forces and also to the relativistic treatment is the directed transversal momentum.

$$\langle P_x \rangle = \frac{1}{A} \sum_{i=1}^{A} sign(P_{zi}) P_{xi} \tag{3}$$

Here the beam axis is defined in the z direction and the reaction plane is the x-z plane. The transversal components of the momenta of the baryons are added up in the direction of the reaction plane in the center of mass system, counting the x components of the momenta with the positive sign in beam direction and with a negative sign opposite to the beam direction.

The covariant definition of the times for each particle attributes different times in each time step to each individual particle. These times vary according to the Lorentz transformation if one transforms from the laboratory to the center of mass system and from there into the centre of mass system of the two colliding hadrons. Such Lorentz transformations have to be done several million times if one is simulating an ensemble of events. Figure 2 shows [9] the average transversal directed flow as a function of time for the collision ^{40}Ca on ^{40}Ca with a laboratory energy of 1 GeV/A and an impact parameter b = 2 fm. Due to the length contraction RQMD yields a higher transversal flow then a QMD treatment. But one puts the times of all particles equal, what is done by several groups, one obtains also for $RQMD^*$ a transversal flow which is of the same order of magnitude as in QMD. The relativistic contraction is lost if the times are put equal and the transversal flow is reduced. For QMD^* we used the non-relativistic treatment, but included the length contraction in the initial condition. One obtains a too large transversal flow.

3   Dependence of the mass of a nucleon on density and temperature probed by antiproton production in nucleus-nucleus collisions

The selfconsistent potential in which the nucleons and in general the baryons are moving in a nucleus and also in a nucleus-nucleus collision can be described by a scalar potential mainly due to σ-meson exchange and by a four vector potential which is repulsive between two nucleons and is mainly due to ω exchange. The scalar potential has the same transformation property under Lorentz transformation as the mass of the nucleon. Since the scalar potential is density and temperature dependent, the effective mass of the nucleon which comprises the mass and the scalar potential can be described by a density and temperature dependent effective mass.

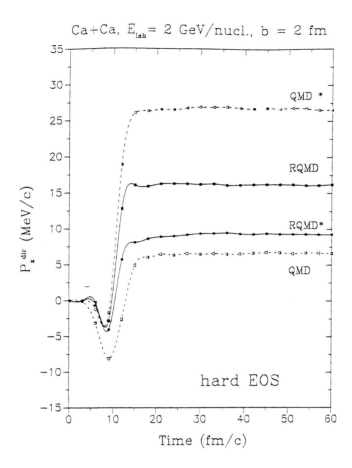

Fig. 2: Average directed transversal flow as a function of time for the reaction ^{40}Ca on ^{40}Ca with a laboratory bombarding energy of 1 GeV/A and an impact parameter of b = 2 fm. Figure 8 shows four results, all are calculated with realistic G matrix elements of the Bonn potential. QMD is the non-relativistic simulation while RQMD is the covariant relativistic formulation. For RQMD* all times are put equal. The increase in directed transversal flow is lost if the difference in times is not included. In QMD* we used a non-relativistic treatment but included the length contraction in the starting configuration.

The interaction of the quarks with the vacuum condensates breaks chiral symmetry and gives the quarks a finite mass of the order of 310 MeV, which one calls the constituent mass. These constituent quarks have to be pictured as a valence quark surrounded by a gluon cloud and by quark-antiquark pairs.

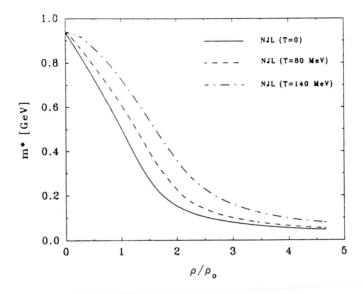

Fig. 3: Quantitative dependence of the nucleon or the quark mass on the density expressed by the saturation density ρ_0 in nuclear matter for three temperatures $T = 0$ MeV (solid line), for $T = 80$ MeV (dashed line) and for $T = 140$ MeV (dashed dotted line). The dependence is calculated [10] for quarks using the Nambu-Jona-Lasinio model in thermo field dynamics. The same result has been obtained by Weise and co-workers [14] by more conventional means.

For the nucleon mass we are assuming the same relative density and temperature dependence as for the constituent quarks. To probe this temperature and density dependence [10], [11] we look far below the nucleon-nucleon threshold (about 5.6 GeV) into antiproton production in heavy ion collisions.

$$B + B \to B + B + B + \bar{p} \qquad (4)$$

Since we have only two baryons in the initial state and four baryons in the final state (see eq. (4)), the threshold is lowered considerably by decreasing the effective mass of the nucleons and the antiproton due to the finite density of nuclear matter as shown in figure 3. After the production the antiproton is moving in the field of the other nucleons. The scalar field which is responsible for the change of the effective mass and its dependence on the density and the temperature is the same for the antiproton. But the vector potential is changing sign. In each time step the local density at the position of the antiproton is taken from the simulation of the heavy ion reaction in QMD and the corresponding effective mass is taken for the antiprotons. The temperature is not so well defined, because one does not obtain thermal equilibrium in the heavy ion collisions considered here. We therefore treat the temperature as a parameter which we adapt to the data. This gives then a value for the temperature at the point and time inside nuclear matter during the heavy ion collision at the point where the antiproton is produced. The results are shown in figure 4.

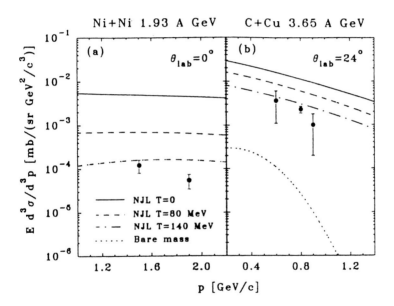

Fig. 4: Invariant cross section for antiproton production as a function of the momentum of the produced antiproton. The left-hand side shows data from GSI for antiproton production in the collision of Ni on Ni at 1.93 GeV/A for $\theta_{lab} = 0°$. The data [15] are compared with the theoretical curves calculated for $T = 0$, $T = 80$ and $T = 140$ MeV. The curve for the bare nucleon and antinucleon mass of 938 MeV is far below the cross section scale shown in the figure. To get agreement with the date, we need at the point and the time where the antiproton is produced a "temperature" of $T = 140$ MeV. On the right-hand side the invariant antiproton production cross section is shown again as a function of the antiproton momentum for the

raction C on Cu with 3.65 GeV/A at $\theta_{lab} = 24°$ measured at Dubna [16]. Here we are still below the threshold of about 5.6 GeV. But the fact that with 3.65 GeV/A we are closer to the threshold shows up by the fact that the different curves for different temperatures ($T = 0$ MeV solid line; $T = 80$ MeV dashed line and $T = 140$ MeV dashed dotted line) are closer together. The result for the bare nucleon and antinucleon masses ($M_N = 938MeV$) is here shown on the lower part of the figure (dotted line).

The Nambu-Jona-Lasinio model [13] seems to reproduce the density and temperature dependence roughly correct. We have to assume that locally at the time where the antiproton is produced we have an extremely high temperature of 140 MeV.

4 Summary

The description of heavy ion collisions at intermediate energies between 100 and 4 000 MeV per nucleon is a complicated nuclear many-body problem which should be described with the realistic nucleon-nucleon interaction. But the realistic interaction is highly momentum dependent and has a strong short range repulsive core which can only be treated with Brueckner theory. The Brueckner reaction matrix elements depend on the local momentum distribution of nucleons through the Pauli operator and the selfconsistent single particle energies. The situation is shown in fig. 1. In practically none of the cases we obtain full thermal equilibrium.

Due to the strong momentum dependence of the realistic interaction, the selfconsistent field which a nucleon is experiencing is attractive inside the target and inside a projectile but repulsive at the interphase where the target and the projectile nucleons meet with high relative velocities. This leads to a higher transversal directed momentum. The value of the directed transversal momentum is even increased in a relativistic treatment due to the relativistic contraction.

In a second point we probed the dependence of the mass of the nucleon on density and temperature, looking into the antiproton production in heavy ion collisions. The density and momentum dependence of the mass of the nucleons are calculated in the Nambu-Jona-Lasinio model.

At the end I would like to thank the collaborators, Prof. N. Ohtsuka, Dr. S. Huang, Dr. E. Lehmann, Dr. R. Puri, Dr. G. Batko, Dr. Ch. Fuchs, Dr. L. Sehn, Dr. T. Marujyama and Dr. Matin who worked with me on the problems I presented in this talk.

References

[1] R. Sator, A. Faessler, S. B. Khadkikar, S. Krewald, Nucl. Phys. **A359** (1981) 467 and N. Ohtsuka, R. Linden, A. Faessler, Nucl. Phys. **A490** 715

[2] T. Izumoto, S. Krewald, A. Faessler, Nucl. Phys. **A341** (1980) 319 and **A357** (1981) 471

[3] N. Ohtsuka, R. Linden, A. Faessler, Phys. Lett. **199** (1987) 325

[4] P. A. M. Dirac, Ref. Mod. Phys. **21** (1994) 392

[5] J. Samuel, Phys. Rev. **D26** (1982) 3475 and 3482

[6] H. Sorge, H. Stoecker, W. Greiner, Ann. Phys. **192** (1989) 266

[7] T. Maruyama, S. W. Huang, N. Ohtsuka, A. Faessler, Nucl. Phys. **A534** (1991) 720

[8] C. Fuchs, E. Lehmann, L. Sehn, F. Scholtz, T. Kubo, J. Zipprich, A. Faessler, to be published in Nucl. Phys. A

[9] E. Lehmann, R. K. Puri, A. Faessler, G. Batko, S. W. Huang, Phys. Rev. **C51** (1995) 2113

[10] T. Maruyama, K. Tsushima, A. Faessler, Nucl. Phys. **A535** (1991)**493** and **A537** (1992) **303**

[11] Ohtsuka, S. W. Huang, E. Lehmann, G. Batko, K. Tsushima, A. Faessler,

[12] G. Batko, A. Faessler, S. W. Huang, E. Lehmann, R. K. Puri, J. Phys. **G20** (1994) 461

[13] Y. Nambu, G. Jona-Lasinio, Phys. Rev. **122** (1961)345 and **124** (1961) 246

[14] U. Vogl, W. Weise, Progr. Part. Nucl. Phys. **27** (1991) 195

[15] A. Schroeter, E. Bedermann, H. Geissel, A. Gillitzer, J. Homolka, P. Kienle, W. Koenig, B. Povh, F. Schumacher, H. Stroeher, Nucl. Phys. **A553** (1993) 775c

[16] A. A. Baldin et al. Nucl. Phys. **A519** (1991) 407c

Results from the EOS TPC

T.J.M. Symons for the EOS collaboration

Nuclear Science Division, Lawrence Berkeley National Laboratory
CA 94720 USA

The EOS detector has been used at the LBL Bevalac to study central collisions of heavy nuclei. The detector is described and some results on flow of nuclear matter are presented. It is shown that a compact description can now be given for many of the phenomena observed in these reactions.

1 Introduction

The study of nuclear collisions at relativistic energies began in earnest in 1971, with the acceleration of nitrogen nuclei at the Bevatron in Berkeley and at the PPA in Princeton. So this conference should perhaps celebrate the silver jubilee of this field as well as the birthday of one of its most ardent devotees!

The field has certainly come a long way in those twenty five years, but perhaps not quite as far as had been hoped. Some reasons for this slow development are not hard to find. The first obvious one is that measuring and understanding these processes has proven to be much more difficult than anyone expected. Today, we realise that we require a full understanding of the collision dynamics before we can infer the existence of density isomers, pion condensates, new phases of matter, or even make the most unsophisticated observations on the properties of nuclear equation of state. We also know that we need sophisticated detectors to make the measurements. Twenty five years ago, this was not so obvious and it took a few years for the idea to settle in.

Another reason is that about ten years ago, just when both the experimentalists and theorists were beginning to get a grip on the problem, a significant fraction of the community shifted its attention from the region of dense nuclear matter to the new challenges of quark-gluon dynamics at Brookhaven and CERN. While this was a natural progression for the field, there was much unfinished business at the Bevalac. When looking forward to the next quarter century with the advent of RHIC and LHC, one hopes that workers in the field will be a little more disciplined.

With these cautionary remarks about the past, it is a pleasure to be able to say that the future once again looks very bright. One reason for this is that data of the quality which the theorists were demanding at the Bevalac ten years ago is now finally available from the FOPI detector at GSi and from EOS. The second is that there is an excellent opportunity at the AGS to trace

TOF wall

MUFFINS

MUSIC II

EOS TPC

HISS MAGNET
13 kG

BEAM

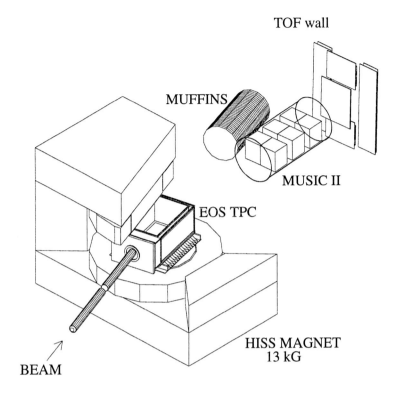

Figure 1: Experimental layout of the EOS spectrometer, as used at the Bevalac.

the evolution of the collective effects observed at lower energies to a point of maximum density. Finally, the CERN program which initially suffered from 'Bevalacitis' (too light beams and inadequate equipment) has invested in heavy beams and new detectors and is poised to fulfill its early promise.

EOS has a connection to all three of these areas of opportunity. The connection to SIS/Bevalac physics is obvious. It may be less well known that a new collaboration has been formed and the detector has been moved to the AGS where it is being used to study the energy dependence of flow[1]. Finally, the technology used in EOS is at the heart of the NA49 experiment at CERN which is being used to provide full tracking for Pb+Pb collisions at the highest available energies.

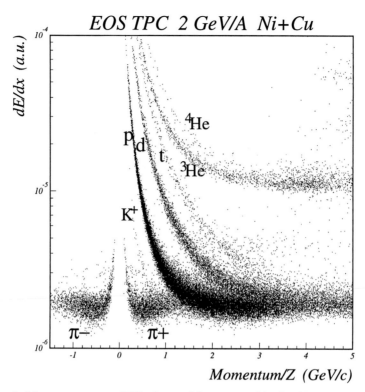

Figure 2: Momentum versus dE/dx for particles measured in the TPC, demonstrating isotopic separation.

2   Description of the Detector

The principal component of the EOS detector[2] is a time projection chamber (TPC) mounted inside a large superconducting dipole magnet. The TPC measures both the position and the energy loss of charged particles passing though the chamber from which, knowing the magnetic field, one may deduce their momenta, charge and velocity, and, in many cases, identify the particle type. In addition, for the runs at the Bevalac, a multiple-sampling ion chamber (MUSIC) was used to measure the charges and position of the forward going projectile fragments. There were also time of flight detectors, a zero-degree neutron detector and assorted detectors upstream of the spectrometer for measuring the position of the beam.

The layout of the spectrometer as used at the Bevalac is shown in figure 1. The dipole geometry has essentially complete acceptance for the forward hemisphere in the laboratory. This means that in the centre of mass for equal mass collisions at energies around 1 GeV/nucleon, the detector provides excellent acceptance for projectile fragments and particles produced at mid-rapidity, at the expense of very poor acceptance for target fragments. In this respect EOS has similar properties to the streamer chamber used at Berkeley in the late seventies and early eighties and very different ones from the plastic ball which was optimised for the target hemisphere. EOS is, however, far superior to the streamer chamber both in the number of tracks which can be recorded per collision and in particle identification. In figure 2, we show a graph of momentum versus energy loss for a typical run. All the light isotopes up to lithium are clearly identified and the chamber provides charge resolution up to carbon at which point the MUSIC detector takes over.

Although the concept of the TPC is now over twenty years old[8], the design of EOS was an important step in the development of these devices. EOS has a two dimensional pad-plane (15000 pads in an area of $2m^2$) which presents special difficulties since the physical density of the pads is so high. To solve this problem, custom-designed integrated circuits are mounted directly on the pad plane to process, digitize and multiplex the signals. This allows a considerable reduction in the size of the electronic system over more conventional schemes. Essentially the same solution, with incremental improvements, has now been applied to the NA49 TPC at CERN and will be used for the STAR detector at RHIC. These two detectors have very much larger numbers of pads but quite similar pad densities and the technology works well.

As stated above, the MUSIC detector is mounted down-stream of the detector and provides tracking and charge resolution for the projectile fragments. This has proven to be especially advantageous for the study of multi-fragmentation. The EOS geometry is similar to that of ALADDIN at GSI, but, because of the power of the TPC, EOS is essentially the only device which has been capable of measuring the momenta of all charged fragments from protons to beam particles in an asymmetric collision.

Finally, one may note that perhaps the most striking feature of EOS is how different it is in form and concept from the 4π detector at GSI (FOPI) which was designed and built at the same time. If EOS can be thought of as an electronic streamer chamber, then FOPI is the descendent of the plastic Ball. In the long run, this will prove to be a great advantage since the systematic errors of the two devices are quite different allowing much greater confidence if new effects can be observed in both detectors.

3 Results

EOS was used at the Bevalac for a relatively short time, just a few months before the accelerator was closed in early 1993. Under these circumstances, hard choices had to be made regarding the beams and energies to run and the targets to bombard.

It was decided to address two main lines of research. The first was the the study of nuclear multi-fragmentation, a topic which has received much attention because of its close relation to the dynamics of the nuclear liquid-gas phase transition. The idea is to prepare a nuclear system in which the excitation energy is comparable to its total binding energy. This is usually done by colliding equal mass nuclei at energies of a few tens of MeV/nucleon at accelerators such as GANIL or the NSCL at MSU. In our case, we studied very asymmetric collisions (Au + C at 1 A.GeV) to prepare a system of appropriate excitation energy but moving at high velocity. This has the great advantage that the fragments are kinematically focussed forward in the laboratory and that essentially all the charged particles can be captured in EOS. Using these data, we first determined the critical exponents for the decay of the composite system and showed that they are close to the values expected for a second order phase transition of a liquid-gas system[4,5]. This is clearly an important result, although some of the assumptions underlying the analysis have been questioned. However, we note that a recent, more thorough analysis of the data[6], paying close attention to the dynamics, still support the original conclusions.

The second line of research has been the study of hydrodynamic flow, a phenomenon first observed by the Plastic Ball[12]. To investigate this, high statistics samples were taken for Au+Au over the entire available energy range, and smaller samples of La + La, Ni + Ni and other systems. The data have been analysed using the formalism developed by Odyniec and Danielewicz[11]. The first point to note is that the EOS results are in good agreement with those of the Plastic Ball. There has been some discussion that the light fragments may be a more sensitive to flow phenomena than the protons. We have studied this carefully[9] and while the values of the transverse momenta are certainly larger, it should be noted that essentially all the flow data for isotopes up to ^4He can be reproduced assuming that the coalescence model is valid. This is an important finding because it implies that the flow observables are established before the fragments are formed and supports the view that fragment production takes place at freeze-out. It also means that flow observables calculated using models such as the BUU transport model, which do not explicitly contain fragment production can still be compared to the data. These remarks should be qualified by noting that this picture certainly breaks down for heavier isotopes, but it is still open to debate whether this tells us more about the nature of flow or the limits of the coalescence model.

The next aspect of these data which we have studied is the so-called radial flow of nuclear matter. This concept was first introduced as the blast-wave model by Siemens and Rasmussen[14] who had observed that one could reconcile temperature values deduced from different fragments measured in inclusive reactions if it was assumed that there was a uniform expansion velocity of the nuclear matter upon which the thermal motion was superimposed. This formalism has now been applied to the exclusive data measured in EOS[8] and we find that excellent fits to the particle spectra are obtained if one assumes a single temperature and redial flow velocity. At the highest energies studied, we find that the radial velocity β is 0.33 and a temperature approaches 90MeV. This implies an approximately equal partition of energy between collective and thermal modes.

Finally, we have studied the azimuthal distribution of spectra relative to the flow axis, where enhanced yields have been reported out of plane. This phenomenon, known as squeeze-out was predicted by hydrodynamic models[15] and first observed by the Plastic Ball[12] and Diogene detectors[13]. Attempts to parametrize the effect have not been particularly successful. Interestingly, we find that we can develop a remarkably simple parametrization if we take into account the radial flow motion just mentioned[10]. The cross sections are described best if we first transform to the directed flow frame and then perform the radial flow analysis allowing the radial velocity to vary with the azimuthal angle. Excellent fits to the triple differential cross sections are then obtained using a single temperature, a directed-flow angle, a radial flow velocity in flow coordinate system, and a sinusoidal modulation around the flow axis of that flow velocity. The quality of the fits can be seen in figure 3. This is an appealing physical picture. The fact that one describes such a complicated experimental situation with a single temperature gives us confidence in the use of hydrodynamic and thermal models, and the retardation of velocity in the reaction plane is presumably a geometric effect which can be predicted by these models.

To summarize, we can now parametrize all the observed phenomena relating to the triple-differential cross sections of light fragments produced in central collisions (except at the very lowest transverse momenta) using the coalescence model and just four other physically reasonable parameters. This is something of a change from earlier days when each new experiment seemed to find a new bump or wrinkle and the situation appeared to be growing increasingly complex. It is worth remembering that this field was started with

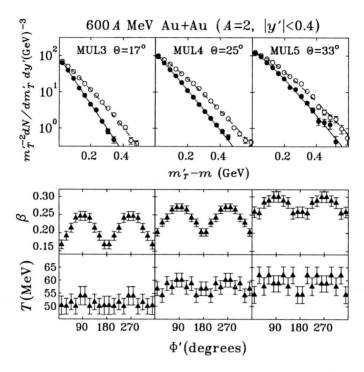

Figure 3: The upper panels show deuteron transverse mass spectra at $\Phi = 0^0$ (solid circles) and $\Phi = 90^0$ (open circles). The curves are fits using a spherical shell at temperature T, expanding at velocity, β. The lower panels present best-fit parameters as a function of Φ. Taken from Ref. 10.

the expectation one could use macroscopic concepts such as hydrodynamics and thermodynamics and that at least some of the microscopic complications would be integrated out. While there are still many open issues on the experimental side, particularly with respect to particle production, we have a solid understanding of the experimental situation (unless FOPI turns up a big surprise) which may be supportive of a return to these more intuitive models. Unfortunately, however, the bottom line is that we have yet to relate any of these phenomena convincingly and unambiguously to parameters of the nuclear equation of state, the real goal of our field. It is to be hoped that the quality and scope of these new results from EOS and FOPI will stimulate a new round of theoretical work in the field with this goal in mind.

Acknowledgments

In the introduction, we noted that there have been two rather distinct eras in the experimental study of nucleus-nucleus collisions at SIS/Bevalac energies, with different experimental techniques and, for the most part, different groups of people. The same is true on the theoretical side: new faces moved in as the others moved on to higher energies and the quark-gluon domain. We are fortunate, however, that one institution, the Theoretical Physics Institute at the University of Frankfurt, under the leadership of Professor Walter Greiner, has held to a steady course of interest in experiments at all energies and has provided the intellectual backbone for our field. I would like to thank Professor Greiner particularly for his direct and effective support of EOS during times when it was unclear that funding would be made available for the project. He was helpful in many ways in bringing EOS to completion.

I wish to thank G. Rai and H.G. Ritter for useful comments on this manuscript. This work is supported in part by the U.S. Department of Energy under Contracts/Grants Nos. DE-AC03-76SF00098, DE-FG02-89ER40531, DE-FG02-88ER40408, DE-FG02-88ER40412, DE-FG05-8840437, and by the Natioanl Science Foundation under grant PHY-9123301.

References

1. G.Rai, et al. BNL Experiment E895 proposal, 1994
2. G.Rai, et al., IEEE Trans. Nucl. Sci. 37, 56 (1990)
3. D. Nygren, LBL internal report, 1974
4. M.L. Gilkes et al., *Phys. Lett.* B **381**, 35 (1996)
5. J.P. Elliott et al., to be published in Phys. Lett.
6. J.A. Hauger et al., *Phys. Rev. Lett.* **77**, 235 (1996)
7. M.D. Partlan et al., *Phys. Rev. Lett.* **75**, 2100 (1995)
8. M.A. Lisa et al., *Phys. Rev. Lett.* **75**, 2662 (1995)
9. S. Wang et al., *Phys. Rev. Lett.* **74**, 2646 (1995)
10. S. Wang et al., *Phys. Rev. Lett.* **76**, 3911 (1996)
11. P. Danielewicz and G. Odyniec *Phys. Lett.* B **157**, 146 (1985)
12. H.H. Gutbrod, A.M. Poskanzer and H.G. Ritter Rep. Prog. Phys. **52**, 1267 (1989)
13. J. Gosset et al. in The Nuclear Equation of State, edited by W. Greiner and H. Stöcker, NATO ASI, Ser. A. Vol B216(Plenum, New York, 1989), p. 87.
14. P.J. Siemens and J.O. Rasmussen, *Phys. Rev. Lett.* **42**, 880 (1979)
15. H. Stöcker and W. Greiner, Phys. Rep. **137**, 277 (1986)

MULTIFRAGMENTATION AND THE SEARCH FOR THE LIQUID-GAS PHASE TRANSITION IN NUCLEAR MATTER

U. Lynen[1] for the ALADIN collaboration:

R. Bassini,[2] M. Begemann-Blaich,[1] Th. Blaich,[3] H. Emling,[1] A. Ferrero,[2] S. Fritz,[1] C. Groß,[1] G. Immé,[4] I. Iori,[2] U. Kleinevoß,[1] G. J. Kunde,[1†] W. D. Kunze,[5] V. Lindenstruth,[1‡] M. Mahi,[1] A. Moroni,[2] T. Möhlenkamp,[6] W. F. J. Müller,[1] B. Ocker,[1] T. Odeh,[1] J. Pochodzalla,[1§] G. Raciti,[4] Th. Rubehn,[1‡] H. Sann,[1] M. Schnittker,[1] A. Schüttauf,[5] C. Schwarz,[1] W. Seidel,[6] V. Serfling,[1] J. Stroth,[1] W. Trautmann,[1] A. Trzcinski,[7] G. Verde,[4] A. Wörner,[1] H. Xi,[1†] E. Zude,[1] B. Zwieglinski[7]

[1] *Gesellschaft für Schwerionenforschung, D-64291 Darmstadt, Germany*
[2] *Dipartimento di Fisica, Università di Milano and I.N.F.N., I-20133 Milano, Italy*
[3] *Institut für Kernchemie, Universität Mainz, D-55099 Mainz, Germany*
[4] *Dipartimento di Fisica dell' Università and I.N.F.N., I-95129 Catania, Italy*
[5] *Institut für Kernphysik, Universität Frankfurt, D-60486 Frankfurt, Germany*
[6] *Forschungszentrum Rossendorf, D-01314 Dresden, Germany*
[7] *Soltan Institute for Nuclear Studies, 00-681 Warsaw, Hoza 69, Poland*

The caloric curve of the highly excited projectile spectators emerging from peripheral collisions of two Au-nuclei at 600 A·MeV was determined by measuring fragments and neutrons of its decay. This analysis will be reviewed and preliminary results obtained from the decay also of the target spectator will be presented.

1 Introduction

An interesting aspect of highly excited nuclei is the prediction of a liquid-gas phase transition [1,2,3]. The decay channel which appears most promising for its verification is multifragmentation, a simultaneous break-up into several fragments of intermediate mass, where the coexistence of nuclear fragments and free nucleons can be seen as nuclear vapor.

In order to heat a nucleus up to the excitation energy where the phase transition is expected [4] - of the order of the total binding energy - two different classes of experiments are possible:
- central collisions of two equally sized nuclei at intermediate energies or
- peripheral collisions at much higher energies.

Whereas in the first case the overlap region is investigated, in the second case the spectator nuclei emerging from the primary collision will have the

---

†Present address: NSCL, Michigan State University, East Lansing, MI 48824
‡Present address: Nuclear Science Division, LBNL, Berkeley, CA 94720
§Present address: Max Planck Institut für Kernphysik, Heidelberg, Germany

excitation energies of interest.

Using the ALADIN spectrometer at GSI together with the Large Neutron Detector LAND we have investigated the reaction Au + Au at an incident energy of 600 A·MeV. From the measured fragments and neutrons the caloric curve [5] of the projectile spectator has been determined.

After a short review of the characteristic properties of spectator nuclei, I will discuss the determination of the caloric curve, including preliminary results from a new experiment, where also decay products of the target spectator have been measured.

2 Projectile Source

Projectile spectators from peripheral heavy ion collisions at high energies offer favorable conditions for the search for signatures of the predicted liquid-gas phase transition:

- At sufficiently high beam energies ($\geq 600 A \cdot MeV$) the decay products are focussed into a narrow angular region around the beam axis. They maintain to a large extent the initial (beam) rapidity and thus are well separated from contributions from the interaction region. The primary projectile spectator emerging from the abrasion phase is therefore a well-defined source which can be reconstructed with high accuracy on an event-by-event basis.

- Because of the high source velocity heavy fragments can also be measured and detector thresholds are negligible. On the other hand the high velocities render a precise determination of the momenta in the decaying system difficult. This can be done better for the target spectator.

- The spectator source shows a high degree of equilibration (thermalization?). Without any selection criteria the transverse and longitudinal widths of fragments are found to be equal [6]. Only in the spectra of very light particles small preequilibrium contributions are found, which show up as a small enhancement towards lower rapidities, but still much closer to the projectile rapidity than to midrapidity. This distinction between preequilibrium contributions in the spectator decay and the interaction region is only possible at high incident energies where the different sources are well separated in rapidity.

- Depending on impact parameter very different excitation energies are transferred into the spectator nuclei. Therefore a whole excitation function can be measured simultaneously [7]. A disadvantage is, that in order to reach high excitation energies, rather central collisions have to be selected for which the mass of the remaining spectator contains as little as 25% of the initial projectile [5].

- For a fixed impact parameter the excitation energy of the projectile spectator increases with increasing beam energy and/or target mass [8]. Despite of this, a universal behavior independent of beam energy or target mass is found if the events are selected according to the mass of the projectile spectator (more precisely according to Z_{bound} being defined as the charge sum of all projectile fragments excluding H-isotopes). This universality of the fragmentation process [9] is not only observed for the multiplicity of intermediate mass fragments (IMF, $3 \leq Z_{IMF} \leq Z_{proj}/3$) but also for the correlations between different fragments [10]. Obviously the process of fragment formation is independent of the details of the entrance channel dynamics, suggesting that the system passes through an equilibrated intermediate state for which the concept of temperature is meaningful.

- During the primary collision spectators undergo no or at most very little compression and the collective flow observed in the subsequent expansion is small (about 1 A·MeV) [11,12]. As a consequence up to excitation energies of about 8 A·MeV the expanding system may reach a turning point [13] leading to a long time span before the system breaks up and consequently a high degree of thermalization can be reached.

- Although the spectator decay as a function of Z_{bound} is independent of target mass, sufficiently heavy targets have the advantage that the cross section $d\sigma/dZ_{bound}$ is rather constant over nearly the whole range of Z_{bound}-values, whereas in case of light targets or low incident energies it drops steeply towards small Z_{bound}-values (see fig. 2). In this region fluctuations in the decay may reduce the sensitivity of the event characterizing variable Z_{bound} to the initial excitation energy.

3 Determination of the Caloric Curve

In order to determine the caloric curve for the excited projectile spectator not only its temperature and excitation energy must be reconstructed, but also its mass.

3.1 Masses

The ALADIN-spectrometer is well suited for a complete reconstruction of the projectile spectator, because at sufficiently high energies (≥ 600 A·MeV) nearly all IMF's are focussed into its acceptance and only a small number (≤ 0.5 per event) is found in the surrounding Si-CsI hodoscope. The rapidity distributions observed in the ToF-wall are therefore the true distributions in case of fragments originating from the projectile spectator decay and distortions due to the angular and also the rigidity acceptance of the spectrometer would only show up for those of the fireball. Because of the high temperature of this source the latter yield is anyhow very small.

For He-fragments, and somewhat stronger for neutrons and H-isotopes, small preequilibrium contributions are seen [14] leading, however, even in the worst case of rather central collisions to an uncertainty in the projectile mass of at most 6 mass units.

Mass and charge of the primary projectile spectator were determined by summing up all fragments and nucleons assigned to this source. For fragments, where only the atomic charge has been measured, the corresponding masses were taken from an EPAX parametrization with parameters adjusted in a previous experiment [12]. For the first analysis [5] the multiplicity of H-isotopes had been deduced from the assumption that the N/Z-ratio of the projectile source (after abrasion) remained that of the initial projectile. This has now been confirmed by measuring protons from the target spectator.

3.2 Excitation Energies

The total excitation energy transferred to the projectile spectator can be determined by adding the kinetic energies of all particles transformed into the moving system to the Q-value of the reaction (the difference between the sum of the final masses and the mass of the primary fragment after the abrasion phase). For fragments and neutrons the energies were measured in the experiment. For H-isotopes those of neutrons (measured at 600 A·MeV) increased by the Coulomb energies had been assumed. At 1000 A·MeV considerably higher energies were measured for both species, the energies of fragments, however, remained constant. This difference is not yet understood.

A possible increase in the excitation energies will, however, partly be compensated if preequilibrium contributions and directed flow are taken into account.

3.3 Temperatures

Following the work of Albergo et al. [15] we have deduced the temperature of the primary projectile spectator from the yields of different light isotopes. For a thermally equilibrated system at low densities the temperature can be derived from a double ratio of the yields of two isotope pairs, differing by the same number of protons or neutrons:

$$R = \frac{Y_1/Y_2}{Y_3/Y_4} = \alpha \cdot e^{[(B_1 - B_2) - (B_3 - B_4)]/T} \tag{1}$$

B_i denotes the binding energies of the fragments and α can be calculated from their spins and masses. As can be seen the sensitivity increases with the difference between the binding energy differences of the two pairs. This makes ^3He and ^4He an ideal pair. A further requirement is that the fragments must be produced with sufficient yield over a wide range of excitation energies from the liquid to the gaseous phase. This excludes the use of much heavier fragments. For the projectile source H-isotopes had not been measured. Therefore ^3He and ^4He were combined with ^6Li and ^7Li and the resulting temperature was denoted as $T_{He,Li}$. For the target spectator we have also determined $T_{H,He}$ by measuring the d/t - ratio. Furthermore we have extracted $T_{Li,Be}$ from the ratio $\frac{^9Be/^8Li}{^7Be/^6Li}$.

Eq. 1 is valid as long as the measured yield ratios are not modified by contributions from either γ-decay of excited states or from feeding from heavier isotopes. Possible corrections were determined for each isotope thermometer individually by Quantum Statistical Model [16] calculations, a model which includes secondary decays. Over a wide range of densities (ρ/ρ_0 between 0.1 and 0.5) the reconstructed temperatures show a linear dependence with the initial ones. The same result was obtained when the temperatures were deduced from GEMINI-calculations for $\rho = \rho_0$. Since in all cases the temperatures deduced from the isotope yields were about 20% smaller than the initial ones, all isotopic temperatures have been scaled by a factor of 1.2.

4 Caloric Curve

From the reconstructed excitation energies and masses together with the isotopic temperature the caloric curve of the projectile spectator shown in fig. 1 has been obtained [5]. It has a remarkable resemblence with e.g. that of water. Within the region from 3 to 10 MeV per nucleon a nearly constant value of about 5 MeV is observed, followed by an increase towards higher excitation energies which can be related to the transition from the vapour to the gas phase. The broad plateau at 5 MeV is in agreement with a series of experiments carried out over a wide range of incident energies where almost identical temperatures have been deduced from the population of excited states in He and Li. An often given explanation for this constancy of the emission temperatures is that fragment emission occurs when the initially hot nucleus has cooled down by the emission of lighter particles. In this case, however, the multiplicity of fragments should only depend on the size of the system at that time and not on the initial temperature. For the projectile spectator going from $Z_{bound} = 70$ to $Z_{bound} = 40$ the initial mass drops by 25% (the masses after light particle emission even more), but the fragment multiplicity increases

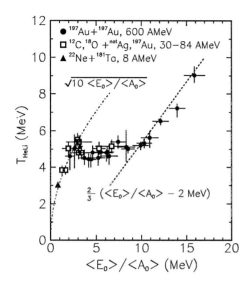

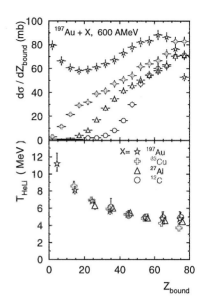

Figure 1: Caloric curve of excited spectator nuclei determined by the dependence of the isotopic temperature $T_{He,Li}$ on the excitation energy[5].

Figure 2: Top: Differential cross section as a function of Z_{bound}[17] for different targets. Bottom: $T_{He,Li}$ as a function of Z_{bound} for Au-interactions with different targets.

by a factor of three. This excludes the above explanation.

The final analysis of excitation energies (see the above discussion) is not yet finished. Therefore the temperatures obtained with different targets (see fig. 2) and those deduced from different isotope ratios (see fig. 3) will be presented as a function of Z_{bound}. Since large excitation energies correspond to small values of Z_{bound}, the temperature rise towards the gas phase in this presentation occurs towards the origin.

In fig. 2, bottom part, the dependence of $T_{He,Li}$ on Z_{bound} is shown for Au-interactions with different targets. Within the error bars all targets lead to identical distributions, confirming the universality of spectator decay. With the light targets ^{12}C and ^{27}Al this comparison is, however, limited to Z_{bound} values larger than 35 and 25, respectively, because for lower values of Z_{bound} the cross sections (see fig. 2, upper part) are vanishing.

A similar agreement is found for the different isotopic temperatures $T_{H,He}$, $T_{He,Li}$ and $T_{Li,Be}$ shown in fig. 3. Although this cannot be seen as a general proof for the validity of temperatures deduced from isotope yields, the agree-

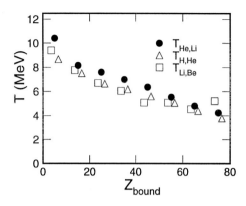

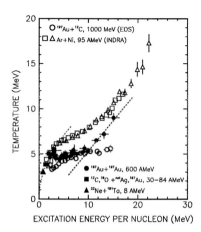

Figure 3: Temperatures determined from different isotope combinations for the reaction Au+Au at 600 A·MeV as a function of Z_{bound} [17].

Figure 4: Comparison of caloric curves obtained by the ALADIN-, the EOS-[18] and the INDRA [19]-collaboration.

ment between $T_{H,He}$ and $T_{Li,Be}$ gives another indication that light particles and fragments see the same source temperatures, resp. have similar emission times. A further conclusion is that the $^3He/^4He$-ratio can be used for temperature determination, despite of the modification of the α-yield by evaporation.

5 Comparison with other Experiments and Conclusion

In fig. 4 the caloric curve measured by ALADIN [5] is compared to preliminary results from the INDRA [19]- and the EOS [18]-collaboration, for which $T_{He,Li}$ has also been determined.

The EOS-result was obtained for Au-interactions with a C-target at 1000 A·MeV and therefore should be directly comparable with our measurement. The agreement in the plateau is very good, but the second rise towards vaporisation is not seen, as in our result with a C-target. The vanishing cross sections observed for small Z_{bound} values indicate that at 1000 A·MeV even in central collisions C is too light in order to vaporize the Au-projectile completely. The fact that the plateau in the two experiments extends to different energies may be related to the different bombarding energies as discussed above.

The *preliminary* result of the INDRA-collaboration was obtained for the Ar+Ni system at 95 A·MeV, where only that half of projectile fragments emitted in forward direction has been analyzed. While the deduced temperatures are about 2 MeV higher, the qualitative shape of the ALADIN-result is reproduced. A possible explanation for the difference could be the N/Z-ratio or the small system size.

The determination of the caloric curve of highly excited spectators opens up a new way to investigate the expected liquid-gas phase transition in nuclear matter. The close resemblance between the caloric curves of nuclei and that of water is surprising, since nuclei are closed systems and neither the pressure nor the volume remains constant when the energy is increased. Although we believe that spectator decays are in so far optimal as nearly no flow is observed during expansion, it is not clear whether the dynamics of the decay can be completely neglected [13]. This will certainly be the focus for further experiments in the future.

Acknowledgments

This work was supported in part by the European Community under contract ERBCHGE-CT92-0003 and ERBCIPD. J.P. and M.B. acknowledge the financial support of the Deutsche Forschungsgemeinschaft under contract Po256/2-1 and Be1634/1-1.

References

1. D.Q.Lamb et al., Phys. Rev. Lett. **41**, 1623 (1978)
2. H.Jaqaman et al., Phys. Rev. C **27**, 2782 (1983)
3. P.J.Siemens, Nature **305**, 410 (1983)
4. J.Bondorf et al., Nucl. Phys. A **444**, 460 (1985)
5. J.Pochodzalla et al., Phys. Rev. Lett. **75**, 1040 (1995)
6. U.Lynen et al. in 11th Winter Workshop on Nuclear Dynamics (1994).
7. J.Hubele et al., Z. Phys. **340**, 263 (1991)
8. W.D.Kunze, PhD-thesis, University Frankfurt (1996)
9. C.A.Ogilvie et al.,Phys. Rev. Lett. **67**, 1214 (1991)
10. P.Kreutz et al., Nucl. Phys. A **556**, 672 (1993)
11. G.J.Kunde et al. Phys. Rev. Lett. **74**, 38 (1995)
12. V.Lindenstruth, PhD-thesis, University Frankfurt (1993)
13. G.Papp and W.Nörenberg, 22nd Winter Workshop, Hirschegg (1994).
14. A.Schüttauf, PhD-thesis, University Frankfurt (1996)
15. S.Albergo et al., Nuovo Cimento A **89**, 1 (1985)
16. D.Hahn and H.Stöcker, Nucl. Phys. A **476**, 718 (1988)
17. T.Moehlenkamp PhD-thesis, Dresden (1996)
18. M.L.Tincknell et al., 12th Winter Workshop on Nuclear Dynamics (1996)
19. G.Auger et al., 1st Catania Rel. Ion Studies, Acicastello (1996)

New vistas on multifragmentation

P.B. Gossiaux[1], Ch. Hartnack, Rajeev K. Puri[2], J. Aichelin[3]

SUBATECH

(Laboratoire de Physique Subatomique et des Technologies Associées)

University of Nantes/ CNRS/ Ecole des Mines de Nantes

4 rue Alfred Kastler, F-44072 Nantes, Cedex 03, France

[1] *Present address: Universität Heidelberg, Heidelberg, Germany*

[2] *Present address: Panjab University, Chandigarh, India*

[3] *Invited speaker*

We study the dynamics of fragment production at intermediate energies using two different methods, one using 'standard' MST and the other one using a simulated annealing technique. We find a rather early formation of the fragments. Our analysis indactes that at least for central collisions the fragments are formed due to preserved correlations between the nucleons and fluctuation which are build up in the violent phase of the reaction.

1 Introduction

Why does a nucleus shatter into to several (up to a dozen) intermediate mass fragments (IMF's) if hit by a projectile nucleus? Is this only a statistical process and hence microcanonical phase space models are the proper tool for its description [1] - [2] or is this a dynamical process as also conjectured [3] - [8]?

Despite of extensive efforts of several experimental groups [9] - [14], this question is not finally decided yet. However, the results of the recent and presently most complete experiments by the FOPI collaboration [10] can up to now not be reconciled with the predictions of statistical models. This raises the question of what - if not phase space - is the driving force for multifragmentation. Since the situation of a heavy ion reactions is too complicated to allow an approach starting from first principles the only means at hand for this search are the dynamical models which simulate heavy ion reactions on an event by event basis by following the time evolution of the nucleons. It is all but easy to extract from the complex n body dynamics the underlying physical process. But this is presently the only possibility to identify the physics behind the multifragmentation. Therefore our contribution will present – after a comparison of QMD results with present experimental data – an analysis of the time evolution of simulated events concerning fragmentation. Our question will be:

when will be the first time we can identify the fragments?

2 Comparison with experimental data

Before any conclusions from the simulation models can be drawn one has to verify that they reproduce the experimental results in a quantitative way. For the Quantum Molecular Dynamics (Q.M.D.), which will be used in this contribution, this has been done extensively [7,10]. As an example we present here the multiplicity of intermediate mass fragments (IMF's) as a function of the number of charged particles Nc . The system Kr + Au has been measured at MSU [11], the system Xe + Sn has been investigated by the INDRA collaboration[12]. For all systems we present the experimental results, the Q.M.D. results if all particles were detected and the Q.M.D. results filtered by the experimental acceptance filter.

We see that the average IMF multiplicity as a function of Nc (and hence as a function of the impact parameter) is well reproduced. This is as well the case for the average number of IMF's as a function of the beam energy as presented in the lower right panel. Here one sees clearly that a beam of 100 A MeV is most suited to investigate the process of multifragmentation.

A more detailed investigation of the reaction at 50 AMeV is available for the asymmetric system Fe + Au where the angular distribution of the fragments have been measured [14]. The comparison with the Q.M.D. calculation is displayed on the right hand side of fig.2. For the Q.M.D. events we have summed over all fragments with charges in between 5 and 25 in order to obtain sufficient statistics whereas the experiment has measured the angular distribution for each fragment charge. We see that the angular distribution of the fragment yield is quite nicely reproduced. Also the absolute value is in reasonable agreement.

On the left hand side of this figure we present the Erat $(= \frac{<p_\perp^2>}{<p_\parallel>})$distribution (calculated with the Q.M.D. model by the FOPI collaboration) as compared with experiment. The reproduction of the results for central collisions (the calculation stopped at b = 7 fm) demonstrates that the excitation energy of the system is properly described in the Q.M.D. model. The acceptance cuts of the detector lower the Erat value as compared to the unfiltered Q.M.D. events. However, the general statement that Erat = 2, and hence thermal equilibrium, is achieved only in rare events is independent of the acceptance.

210

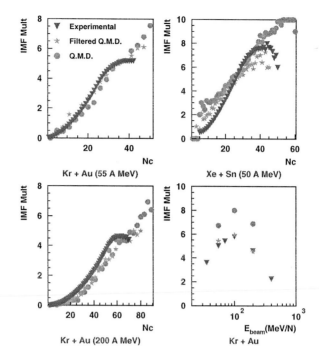

Figure 1: Multiplicity of IMF's as a function of the number of charged particles. The system Kr + Au has been measured at MSU, the system Xe + Tn represents a result from the INDRA collaboration. The lower right panel shows the average IMF multiplicity as a function of the beam energy

3   Early fragment recognition

This verification of the Q.M.D. model allows to employ this model to proceed further towards a physical understanding of the production process by taking advantage of the fact, that it contains the time evolution of the n-body phase space. The first step towards an understanding of the multifragmentation process is the identification of the time point, at which the fragments are formed. This allows then to investigate the environment in which the formation takes place. This requires a fragment identification already at an early stage of the reaction.

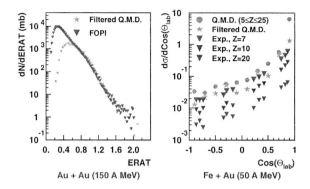

Figure 2: ERAT-distribution of the system Au+Au at 150 AMeV energy and angular distribution for the system Fe+Au at 50 AMeV.

3.1 The basic principles of MST and SACA

Before one can study the origin of fragments on has to identify the fragments. Up to now the fragments have been identified by a minimum spanning tree (MST) procedure which has also been used to obtain the results displayed in fig.1 and fig.2.

One first simulates the reaction for about 200 fm/c, using Q.M.D.. Then the spatial distance of all nucleons is checked. A nucleon is part of a fragment if there is another one within a distance of $r_{min} = 3$ fm[6]. This procedure yields stable results, i.e. gives the same fragment pattern for times later than 200 fm/c but cannot be used for earlier times because it only makes sense when the system is very dilute [7].

Recently we have developed a new approach which defines the fragments in phase-space. There nucleons can form a fragment if the total fragment energy/nucl. ζ_i is below a minimum binding energy:

$$\zeta_i = \frac{1}{N^f} \left[\sum_{i=1}^{N^f} \frac{(\vec{p}_i - \vec{p}_{cm})^2}{2m} + \frac{1}{2} \sum_{i \neq j}^{N^f} V_{ij} \right] < E_{Bind}, \qquad (1)$$

We take for $E_{Bind} = $ -4.0 MeV if $N \geq 3$ and $E_{Bind} = 0$ otherwise. In this equation, N^f is the number of the nucleons in a fragment, $\vec{p}_{cm}$ is the center-of-mass momentum of the fragment. This new definition has the advantages that the requirement of a minimum binding energy excludes loosely bound fragments which will decay later. It modifies the definition of Ref.[15], where nucleons can be bound even if the binding energy of the fragment is extremely small. We employ a simulated annealing [16] mechanism to find the most bound configuration and dubbed this approach simulated annealing cluster algorithm (SACA).

3.2 Description of central collisions

The result for the reaction Au + Au 400 A MeV, b = 3 fm is presented in fig.3. The first row shows the time evolution of the collision rate and of the mean density

$$< \rho(t) >= \frac{1}{N} \sum_{i,j=1,i\neq j}^{N} \frac{1}{2\pi L} e^{(x_i(t)-x_j(t))^2/2L} \tag{2}$$

where $2L = 2.16fm^2$ and the x_i are the centroids of the gaussian wave function of the nucleons. The density reaches its maximum at about 30 fm/c, whereas the collision rate has its peak at about 60-70 fm/c. With increasing time the collision rate becomes negligible whereas the mean density stays constant at about 0.4 ρ_0. This value is an average over all nucleons. It includes the free nucleons which will finally have see a density equal zero as well as the fragment nucleons which feel the density of their fellow nucleons.

The second row shows that SACA finds the heaviest fragment rather early, at a time when the system is still quite dense and interactions among the nucleons are still continuing. This gives an indication that the heaviest fragment is formed from the spectator matter which is correct, as we will see later. Note that the MST needs as long as 200 fm/c to find the surviving heaviest fragment. As a consequence SACA predicts also already the asymptotic single particle multiplicity at around 60 fm/c, much earlier than MST. There the heaviest fragment emits continuously loosely bound nucleons until 250 fm/c. These nucleons are still around the heaviest fragment and therefore MST counts them as belonging to the heaviest fragment but are very loosely bound and therefore the most stable configuration is obtained if one consider them no longer as part of the heaviest fragment.

From the third row we see that also the light fragments $2 \leq A \leq 4$ are formed quite early before they are separated visibly in coordinate space. SACA finds the stable pattern at about 50 fm/c, whereas MST identifies them at

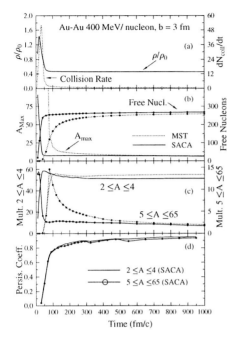

Figure 3: Evolution of the collision of Au-Au at 400 MeV/nucl. and at an impact parameter b = 3 fm. From top to bottom the rows display the time evolution of the density and of the collision rate, of the size of the heaviest fragments and of the number of emitted nucleons, of the multiplicity of fragments with mass $2 \leq A \leq 4$ and with mass $5 \leq A \leq 65$ and of the persistence coefficient.

about 100 fm/c. The formation of intermediate mass fragments ($5 \leq A \leq 65$) is finished after 50 fm/c as well. The MST, however, needs very long (about 300-400 fm/c) until it can identify the final fragments. This is again due to loosely bound fragment nucleons. Their relative velocity with respect to the fragment momentum is small, therefore they rest for a long time in the vicinity of the fragments until they get eventually sufficient energy to escape.

In order to quantify the change of the nucleon content of the fragments between two successive time steps we introduce the persistent coefficient [15,17]. It is one when during a time step the fragment neither looses nor gains a nucleon and 0 when it disintegrates completely. The average persistence coefficient for the fragments $2 \leq A \leq 4$ and $4 \leq A \leq 65$ is displayed in the last row.

One can conclude that the final fragments are formed as early as at 50 fm/c. The persistent coefficient reaches its asymptotic value later due to the (mainly potential) interaction between fragments and between fragments and free nucleons. This interaction changes the details but not the general structure of the fragmentation pattern.

We can conclude that as soon as the violent phase of the reaction is over the fate of the final fragments is already determined. There are first indications from FOPI analyses that the fragment pattern can be formed before the interactions ceases to exits [9].

3.3 Analysis of Peripheral collisions

Let us now turn to more peripheral collisions. Fig. 4 displays the same quantities as fig. 3 for Au-Au at 600 MeV/nucl. and at impact parameter b = 8 fm. This reaction has recently raised a lot of interest because the Aladin collaboration has conjectured [18] that the results present direct evidence for a liquid gas phase transition.

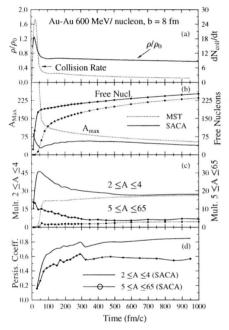

Figure 4: Same as preceeding figure but for Au on Au at 600 MeV/nucl. and an impact parameter of b=8fm.

We observe that the mean density as well as the number of free nucleons is finally about the same as at the central collisions at 400 MeV/nucl. However, this value is obtained very slowly by emission of nucleons from clusters. Comparing this result with fig. 3 we see that despite of the higher beam energy the reaction takes much longer until the finite distribution is obtained. This is due to the small energy transfer to the spectator matter which is consequently little excited and hence the emission time scale is very long and comes even to that of a compound nucleus.

The nucleons continue to collide up to the end of the reaction. These are, however, soft collisions inside the clusters. In contradistinction to the central collision SACA is not able to identify the size of the largest fragment before 150 fm/c, however, the MST need still much longer. SACA seems to fail completely for the small clusters as compared to the minimum spanning tree. It takes as well very long ($> 300 fm/c$) before a persistent coefficient of .8 is obtained. Before that time there is a strong exchange of nucleons between the fragments. The number of IMF's increases due to the fact the largest fragment falls finally into this mass bracket.

How can this result be interpreted? The fact that the minimum spanning tree recognizes the final fragments $5 \leq A \leq 50$ already at 80 fm/c means that at this time these fragments are clearly separated from the rest of the system and each other in coordinate space. The additional fragments observed by SACA are not separated in coordinate space but are only obtained because this configuration gives the lowest binding energy. Shortly after the interaction between projectile and target the both large remnants (the spectator matter) are perturbed in a way that dividing them into small fragments is energetically favourable. Therefore also the size of the largest fragment is below the asymptotic value. The nucleons entrained in these by SACA detected fragments continue to interact (as can be inferred as well from the low value of the persistence coefficient) and smoothen finally the perturbation in a way that some nucleons get emitted and the rest of the system forms a single large fragment.

Is this realistic? Between the time steps analyzed by SACA the nucleons propagate on trajectories as calculated in the QMD program. Despite of the local interaction due to the Gaussian form of the wave function the range of the interaction is finite and different from zero even at large distances between the nucleons. This is of no importance if the excitation energy of the system is large. Here, however, we deal with excitation energies of a couple of MeV/nucl. were these details may play a role. Thus one may conjecture that the equilibration of the spectator matter which takes place at a time larger than 60 fm/c is not realistic.

Following this conjecture one may assume that the realistic fragment multiplicity is that detected by SACA at the moment where the size of the largest fragment is minimal. If one compares this number with the experimental results one finds agreement (Z_{bound} for b= 8 fm is about 35) [13]. We have furthermore analysed the time evolution of several dynamical observables (stopping, flow, etc.) for different fragment classes and found agreement between an analysis performed very early (at 50 fm/c) and very late (at 1000 fm/c). This indicates as well that the observables of the reaction reach their asymptotic values already quite early. For a detailed discussion we refer to ref. [17].

If the conjecture that the late equilibration of the spectator matter in QMD is artificial is true one can take directly the dynamical observables of the fragments determined by SACA at 60 fm/c and can compare this result with experiment. Indeed and in agreement with the results of the Aladin collaboration we observe that the average rapidity of the projectile like fragments is independent of the mass of the fragment.

3.4 Is there realm for thermal models?

Recently is has been found [20] that the fragment distribution of the ALADIN experiment 600 A MeV Au + Au can be quite nicely reproduced by a statistical model. In this experiments one observes the fragmentation of one of the two nuclei. Most of the fragments come from the projectile spectator matter and have lost very little of their initial momentum. The collaboration conjectured later that this may be an evidence for an equilibration of the system.

To verify this conjecture we made simulations for semiperipheral reactions (b = 8 fm) where one observes a peak in the fragment multiplicity distribution. We find that in the average 192 nucleons entrained in fragments $A > 5$. The smaller fragments as well as nucleons can be less clearer attached to either projectile or target spectator as the intermediate mass fragments. Due to the magnitude of possible exit channels it is not possible in the context of Q.M.D. calculations to investigate in detail whether a distribution expected from a statistical model for a finite particle number is realized. However we can verify whether some necessary but not sufficient conditions for thermalization are fulfilled:

- the average velocity of the fragments has to be independent of the fragment mass, i.e. the fragments may come from a common source

- the variance Δ of the rapidity distribution of the fragments and a possible "temperature" T of the system are connected by

$$\Delta^2 \cdot A \cdot m = T_{eff}^{dyn}$$

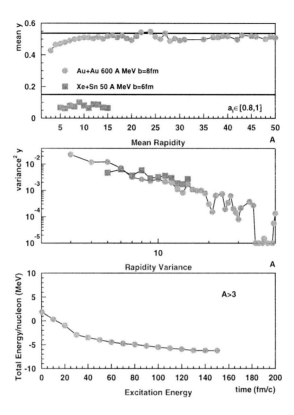

Figure 5: Mass dependence of several observables as obtained from simulations of the reaction 600 MeV/N Au + Au b=8 fm. We observe a nearly constant average rapidity but a mass dependence of the rapidity width which is much stronger than expected in a thermalized system. The average excitation energy/nucleon in the fragments is quite small and comes to about 1 MeV at the end of our simulation.

- the excitation energy E in the system is related with the possible "temperature" by the relations

$$T_{eff}^{ex} = \sqrt{\frac{E}{a}}$$

where a is the level density parameter of about A / (8. MeV^{-1}). and both values of the "temperature" have to be the same.

In fig.5 we investigate these conditions. We see from top to bottom the mean rapidity and the squared variance of the rapidity distribution of the fragments as a function of their mass number, as well as the total energy per nucleon as a function of time.

The mean rapidity of the fragments is slightly below the beam rapidity of 0.54. The heavier fragments approach the beam rapidity whereas the lighter ones are slightly below. Hence the first condition is almost fulfilled.

The squared variance Δ^2 of the rapidity distribution is plotted in a double logarithmic graph. In this graph the second condition results in straight line which is indeed observed for the low mass fragments. The slope of the *variance*^{2} is however quite close to -2 and hence quite different from the expected -1 dependence. It is, however, in agreement with data[21], although more precise experiments would be certainly of value. The latest experimental analysis of the Aladin collaboration find a similar dependence of the variance on the fragment mass[23]. The values extracted for T_{eff}^{dyn} are of the order of 15 MeV. This value is compatible with that found experimentally in[22] and agrees also with the preliminary results of the ALADIN collaboration[23]. As mentioned above the mass yield of these results has been successfully described in the framework of a statistical model, however with a much lower "temperature".

This obvious discrepancy in the "temperature" values renders any statement about a possible thermalization premature before a detailed analysis of the momentum distribution of the fragments has been performed.

The energy of the system, i.e. the total energy of those nucleons finally be entrained in projectile (resp.) targetlike fragments with $A > 3$ is displayed in the last row. Initially this energy is high due to some projectile nucleons which become part of the targetlike fragments and vice versa. After the projectile spectator has been disentangled from the fireball and before the spectator matter is disassembled into fragments we observe an excitation energy of about 3 A MeV. At 130 fm/c the system is completely deexcited and the difference between the observed value of - 6.5 A MeV and the value for finite nuclei of - 8 A MeV is partially due to the relative motion between the fragments. T_{eff}^{ex} is about 4 MeV, in clear disagreement to the value of T_{eff}^{dy} extracted from the spectrum.

In order to understand the creation of a large 'temperature' let us now regard the time evolution of the mean rapidity and mean transverse velocity and the corresponding variances as it is depicted in fig. 6 for fragments with A=5 and A=12-14. If we inspect the mean rapidity (circles) and transverse velocity (triangle up) of all fragments ('all nuc') we see some changes of the values in

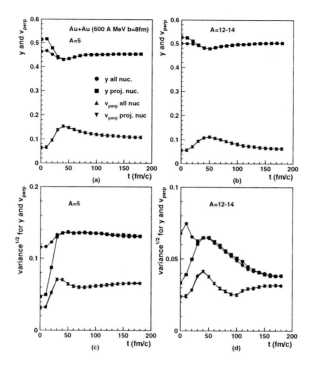

Figure 6: Time evolution of the mean rapidity and transverse momentum and the respective variances for fragments with A=5 and A=12-14 for the system Au(600 AMeV)+Au at b=8fm.

the early time of the violent reaction and a constant behaviour afterwards. The variance of the rapidity already starts with an very high value and increases (for A=5) afterwards only modestly (and descends for A=12-14). If we now, however, only regard those particles in the cluster which originally belonged to the projectile (i.e. we reduce eventually the mass of the A=5, A=12-14 cluster and change its CM momentum) we see hardly a change in the transverse velocities (triangle down) but visible effects on the rapidities (squares). The mean rapidity shows an enhanced initial value which falls down rapidly. The variance of the rapidity starts at a rather moderate value comparable to the variance of the transverse velocity and increases rapidly afterwards. From this we conclude that the large variance in rapidity is mainly due to the mixing of particles coming from projectile and from target which leads to an enhancement of the fragments relative momenta.

4 Conclusion

In conclusion we have demonstrated that QMD is able to reproduce the available data in the energy range of $50\,\mathrm{AMeV} \leq E_{lab} \leq 400\,\mathrm{AMeV}$ to a degree that it most probably contains the proper physics. We have used an extended algorithm for cluster recognition based on the simulated annealing method which is able to recognize the stable structure of fragments in reactions between heavy ions at a very early time in central collisions. We have shown that in central collisions the fragments are formed during the high density stage. This early recognition of fragments gives us new strong evidence that the fragments are formed just due to preserved correlations between the nucleons and fluctuation which are build up in the violent phase of the reaction.

In peripheral reactions in addition to the fragments formed at the time when projectile and target are in contact with each other the SACA algorithm also finds that it is energetically preferable to subdivided the spectator matter into smaller fragments. During the further time evolution the QMD propagation equilibrizes the spectator matter. Hence finally only one large fragment is left. Whether this equilibrium is an artificial QMD feature has to be investigated. For the studies of peripheral collisions there may be a realm of thermodynamics. However, the use of thermal models may imply that the temperature should be rather regarded as a statistical parameter than as a physical observable. The measurement of energy spectra for fragments $d^2\sigma/dE\,dZ$ will shurly help to answer this question.

Acknowledgements

The authors acknowledge valuable support by Dr. C. Dorso. This work was supported by the french IN2P3.

References

1. J.P. Bondorf et al., Nucl. Phys. A **443**, 321 (1985).
   D.H.E Gross, Rep. Prog. Phys. **53**, 605 (1990), and references therein.

2. L.G. Moretto and G.J. Wozniak, Annual Reviews of Nuclear and Particle Science, J.D. Jackson, ed., **43** , 379 (1993).

3. W. Bauer, G.F. Bertsch and H. Schulz, Phys. Rev. C **69**, 1888 (1992).

4. W.A. Friedman and W.G. Lynch, Phys. Rev. C **28**, 556 (1983).

5. J. Aichelin and H. Stöcker, Phys. Lett. B**163**, 59 (1986).

6. G. Peilert, H. Stöcker, A. Rosenhauer, A. Bohnet, J. Aichelin and W. Greiner, Phys. Rev. **C 39**, 1402 (1989).

7. J. Aichelin, Phys. Rep. **202**, 233 (1991).

8. P.B. Gossiaux and J. Aichelin, SUBATECH preprint 95-4, submitted to Phys. Rev. C

9. M. Petrovici et al., Phys. Rev. Lett. **74**, 5001 (1995).

10. S.C. Jeong et al., Phys. Rev. Lett. **72**, 3468 (1994)

11. B. Tsang, Proceedings of the International Workshop XXII on Gross Properties of Nuclei and Nuclear Excitations, Hirschegg, Austria, January 17 -22, 1994 edt by H. Feldmeier and W. Nörenberg

12. INDRA Collaboration, private communication

13. M. Begemann-Blaich et al., Phys. Rev.**48**, 610 (1993).

14. T.C. Sangster, private communication

15. C. Dorso, and J. Randrup, Phys. Lett. B **301**, 328 (1991).
    C. Dorso , and P.E. Balonga, Phys. Rev. C **50**, 991 (1994).
    C. Dorso and, J. Aichelin, Phys. Lett. B**345** , 197 (1995).

16. N. Metropolis, A.W. Rosenblut, M.N. Rosenblut, A.H. Teller, and E. Teller, J. Chem. Phys. **21**, 1087 (1953).
    P.J. M. Laarhoven, and E.H. L. Aarts, Simulated Annealing: theory and applications (Reidel, Dordrecht, 1987).

17. R.K. Puri, Ch. Hartnack and J. Aichelin, to be submitted to Nucl. Phys. A

18. J. Pochodzalla et al., Phys. Rev. Lett. **75**, 1040 (1995)

19. M. Morjean et al, Nouvelles du Ganil 56, p. 33 (1995)

20. A. S. Botvina *et al.*, GSI 94-36 (1994).

21. F.P. Brady et al., Phys. Rev. C50 (1994) R525

22. J. Dreute et al., Phys. Rev. **C44** (1991) 1057

23. W. Trautmann et al., private communication

Strangeness Production and Flow
in Relativistic Heavy Ion Collisions

N. Herrmann, FOPI Collaboration
Gesellschaft für Schwerionenforschung, Darmstadt, Germany
and Physikalisches Institut der Universität Heidelberg, Heidelberg, Germany

Strangeness production is studied with the FOPI detector at GSI in the reaction
Ni+Ni at 1.93 AGeV. K^+ and Λ distributions are compatible with the assumption
of kinetic equilibrium with the baryons. The extrapolated production yields are
in variance with chemical equilibrium. Effects of possible in-medium modification
of the strange meson masses on the experimental observables are discussed. The
directed sideward flow of Kaons is used as an additional probe of the in-medium
properties.

1 Introduction

Strangeness production is considered to be a promising tool to investigate the
properties of hot and dense nuclear matter that is eventually created in the
course of relativistic heavy ion collisions. Originally, it was hoped that the
Kaon yield can be directly related to the nuclear matter equation-of-state
(EOS), since depending on how much energy is stored in the compression,
a different amount of the total energy is available for particle production [1].
There are, however, competing processes that influence the Kaon yield. Most
importantly, as it was shown experimentally [2], Kaons are produced in multiple
collisions via intermediate resonances. The production is therefore sensitive
not only to the EOS but also to the number of excited resonances, their mutual interaction and possibly on changes of the particle masses in the nuclear
medium that may be caused by a partial restoration of chiral symmetry [3,4,5,6].

To achieve a better understanding on the relevant processes and in order to
sort out the different contributions, more complete measurements are needed.
This paper focuses on the attempt made by the FOPI collaboration of GSI
to measure strange particles and presents the first results for the reaction
Ni+Ni at 1.93 AGeV. The results are based on a central event sample of 420 mb
selected by means of charged particle multiplicity.

2 Strange Particle Identification with FOPI

For polar laboratory angles $30° < \Theta_{lab} < 145°$ charged particles are identified
in the central drift chamber of the FOPI detector via the correlation of their
specific energy loss and the curvature of the reconstructed tracks that are

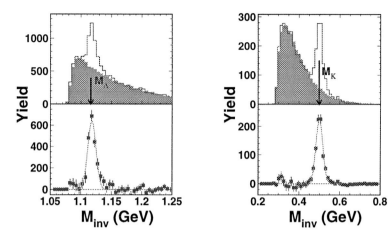

Figure 1: Invariant Mass Distributions for $\pi^- p$ (left panel) and $\pi^- \pi^+$ pairs (right panel) in the reaction Ni+Ni at 1.93 AGeV

bent by the magnetic field of a solenoid of $B = 0.6T$. π^+ are separated from Protons up to laboratory momenta of 600 MeV/c and Kaons from both Pions and Protons up to momenta of 350 MeV/c. For higher momenta the Time-of-Flight information from the plastic scintillator barrel is necessary to achieve the separation. Currently K^+ can be identified in the interval $100 MeV/c < p_{lab} \leq 500 MeV/c$ for laboratory angles of $\Theta_{lab} > 38°$.

Since all the emitted particles are registered the primary vertex can be reconstructed with an accuracy of approximately 1 mm. Tracks not originating from the primary vertex are used to reconstruct long lived resonances. Combining π^- - tracks that have in the plane perpendicular to the beam direction a minimum distance of 2 cm to the primary vertex with either Proton (left panel) or π^+ (right panel) track candidates results in the invariant mass spectra of fig.1. Both spectra show prominent structures which are attributed to Λ and K°, respectively. The background is determined quantitatively by mixing tracks from different events. Normalization can be either done by adjusting to the structurefree parts of the spectra or on the absolute number of events and gives the same results. Subtracting the background yields symmetric mass peaks with a resolution of $\Delta m \approx 5 MeV$ around the nominal particle mass. A total of 3000 Λs and 600 K° has been identified so far and is used for the further analysis.

The same technique of combining well identified tracks and reconstructing invariant masses was applied to $K^+ - K^-$ pairs originating from the primary

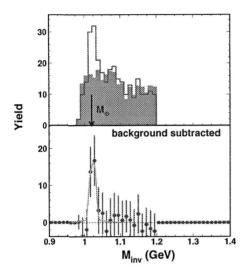

Figure 2: Invariant Mass Distribution of K$^+$-K$^-$ pairs

The top panel shows the measured distribution as well as the one obtained from a mixed event analysis. The lower panel is obtained after subtraction of the uncorrelated background. The spectrum is artificially cut off at 1.2 GeV.

vertex. The results are shown in fig.2. A statistically significant enhancement is found close to the phase space boundary. Background subtraction in the same fashion as described above leaves a peak with a centroid at 1.02 GeV. It is therefore tentatively assigned to the Φ- meson with a free mass of $1019.6 MeV$ and a width of $\Gamma = 4.4 MeV$ decaying with a probability of 49.1 % into a $K^+ - K^-$ pair. To learn more about Φ- meson production in this energy regime clearly more statistics is needed, but the accuracy obtained within the available statistics of $7 \cdot 10^7$ events shows that this will be doable in the future. Implications of the overall yield estimated from the current statistics of 30 ± 8 events will be discussed in section 4.

3   Phase Space Distributions

In order to understand the strangeness production process as complete distributions as possible are necessary. The current status of FOPI data is shown in fig.3. The transverse mass distributions of K^+ are shown for various rapidity intervals. The mean values of the rapidity bins is given on the right side of the figure in terms of the normalized center-of-mass rapidity $y^{(0)} \equiv y/y_{projectile}$. The abscissa is chosen in a way that a thermal spectrum corresponds to a single exponential function, i.e. a straight line in this representation. The Kaon spectra in each of the rapidity intervals can be described well by single exponential functions. The slope parameter is found to decrease from rapidities

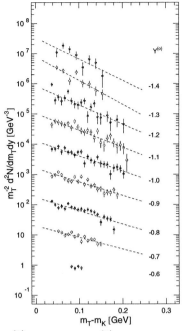

Figure 3: Transverse mass spectra at Kaons for various rapidity intervals for the reaction Ni+Ni 1.93 AGeV

The spectra for successive $Y^{(0)}$ bins are offset by a factor of 10.

of $y^{(0)} = -0.7$ to $y^{(0)} = -1.3$. This is expected for an isotropically emitting source and should follow a cosh(y) dependence.

The same procedure is undertaken for Protons and Λ-particles. Within the acceptance covered by the experiment the spectra of the baryons are exponential as well. The relative comparison of the resulting fit parameter (slope T_B and integrated yield $dN/dy^{(0)}$) as function of the normalized rapidity $y^{(0)}$ is shown in fig.4. On the left panel the slope parameters of all three particle species are shown to be very similar within a given rapidity interval. Slight differences are, however, expected for collectively expanding sources. The proton and deuteron transverse mass spectra at midrapidity have been fitted assuming a radially expanding source [7,10] that is described by a collective expansion velocity β and a temperature T. For such a source the dependence of T_B and $dN/dy^{(0)}$ on rapidity can be predicted (smooth lines in fig. 4). The variation of the slope parameter is reasonably well described for the Kaons. For Lambdas and Protons the slopes at target rapidity are significantly smaller than the expectations. This can be related to the existence of spectator matter and the strong rescattering of the baryons therein.

For the rapidity density distributions on the right side of fig.4 the corresponding effects are observed: Proton and Lambda distributions are more

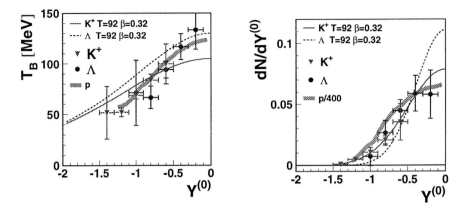

Figure 4: Slope parameter and rapidity distributions of strange particles in comparison to protons for the reaction Ni+Ni at 1.93 AGeV

elongated in longitudinal direction with respect to an isotropic source while K^+ mesons can be described reasonably well by the model assumptions.

4 Production Yields of Strange Particles

Since there is no consistent picture of the reaction one has to be careful when trying to extract overall 4π production yields. The blast model description seems to work well for the produced mesons while it underpredicts the rapidity density of baryons around target rapidity due to the lack of spectators. For this reason the summary of particle production yields in fig.5 is restricted to strange mesons, where in addition to the K^+ and the Φ-meson also preliminary K^- yields and data from the KaoS [9] and the TAPS [11] collaboration are included. Note that the Φ efficiencies were derived by simulating the detector response and reconstructing procedure for an initial distribution that was generated with the parameters shown in fig.4. The systematic error that results from uncertainties of this basic assumption is difficult to assess and is not included in the error bar.

The representation used in fig.5 was suggested by Metag [12] and takes into account the available phase space for the different particles by normalizing the beam energy per nucleon to the production threshold energy in nucleon-nucleon collisions. The trivial size dependence of the various reactions is taken care of by normalizing to the number of participating nucleons. For the current FOPI measurement those where extracted from the cross section of the event sample. As reference the solid line depicts the pion production probabilities as

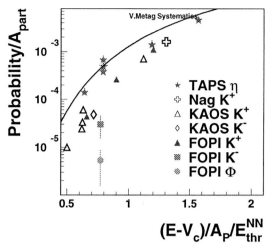

Figure 5: Extrapolated Production Probabilities of Strange Mesons

it describes all known pion data within a factor of 2. The production probabilities for the charged Kaons are consistent among the different experiments. The Kaon production is clearly lower than the non-strange meson production. Thus the available phase space is not fully populated ruling out a chemical equilibrium. In addition the strange particles are not behaving consistently: η and Φ mesons (both with hidden strangeness) differ by a factor of 100 for the same available energy. One other remarkable observation is the similar production yields of K^+ and K^- for the same available energy. This is surprising since in the elementary production process K^- production is suppressed by one order of magnitude[9]. In addition K^- - mesons are much stronger absorbed and their rate should be even further suppressed.

The production yields support the speculations that because of the partial restoration of chiral symmetry, particle masses might change in the nuclear medium. Especially the K^- would loose rapidly its mass[3,5,6] and could therefore be produced much more frequently. The theoretical considerations involve the existence of scalar and vector potentials that besides modifying the particle mass also would influence the particle propagation[13].

5 Sideward Flow of Strange Particles

With the rapidity coverage of FOPI the emission probabilities of Kaons and Lambdas can be examined with respect to the reaction plane. The reaction plane resolution within the event sample is $\Delta\phi = 40°$. The Λ - particles follow

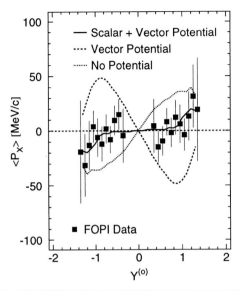

Figure 6: Sensitivity of K^+ sideward flow to in-medium potentials according to Li and Ko

the Proton flow quantitatively[8]. The result for the positively charged Kaons is shown in Fig.6 where the projection of the Kaon transverse momenta onto the reaction plane $< p_x >$ vs. rapidity is displayed. The data are compared to RBUU transport model predictions that make use of different versions for the in-medium Kaon potential [14]. Note that the data are reflected around midrapidity. The errors on the data are mostly of statistical origin and will be decreased by further measurement. The effects of event selection and reaction plane resolution are taken into account in an approximate way. Nevertheless, already at this stage, the comparison to the model prediction reveals sensitivity to the existence of a potential. The *no-potential* version is excluded as well as the *vector-potential only* option. Only the sum of both potentials bring the calculations in agreement with the data.

6 Conclusion

Strangeness production in the near threshold region shows a rich phenomenology. Strange particles seem to reach a kinetic equilibrium with the surrounding matter as evidenced by the slope parameters and the rapidity distributions. A chemical equilibrium is, however, not obtained. Rather, the comparison of the yields of the different strange particle species might call for some fundamental changes of their properties, e.g. dropping masses. The flow properties of strange particles offer an independent way of cross checking the ideas that are introduced to reproduce the Kaon yields. To find out whether those ideas are correct or not is a challenging task for the future that promises insights into fundamental properties of QCD.

Acknowledgments

I would like to thank all members of the FOPI collaboration of SIS/GSI:
B. Hong[4], J. Ritman[4], D. Best[4], A. Gobbi[4], K. D. Hildenbrand[4], Y. Leifels[4], C. Pinkenburg[4], W. Reisdorf[4], D. Schüll[4], U. Sodan[4], G. S. Wang[4], T. Wienold[4], J. P. Alard[3], V. Amouroux[3], N. Bastid[3], I. Belyaev[7], L. Berger[3], J. Biegansky[5], A. Buta[1], R. Čaplar[11], N. Cindro[11], J. P. Coffin[9], P. Crochet[9], R. Dona[9], P. Dupieux[3], M. Dzelalija[11], M. Eskef[6], P. Fintz[9], Z. Fodor[2], A. Genoux-Lubain[3], G. Goebels[6], G. Guillaume[9], E. Häfele[6], S. Hölbling[11], F. Jundt[9], J. Kecskemeti[2], M. Kirejczyk[4,10], Y. Korchagin[7], R. Kotte[5], C. Kuhn[9], D. Lambrecht[3], A. Lebedev[7], A. Lebedev[8], I. Legrand[1], C. Maazouzi[9], V. Manko[8], J. Mösner[5], S. Mohren[6], W. Neubert[5], D. Pelte[6], M. Petrovici[1], F. Rami[9], V. Ramillien[3], C. Roy[9], Z. Seres[2], B. Sikora[10], V. Simion[1], K. Siwek-Wilczyńska[10], V. Smolyankin[7], L. Tizniti[9], M. Trzaska[6], M. A. Vasiliev[8], P. Wagner[9], D. Wohlfarth[5] and A. Zhilin[7]
[1] IPNE, Bucharest, [2] CRIP Budapest, [3] LPC Clermont-Ferrand, [4] GSI, Darmstadt, [5] FZ Rossendorf, Dresden, [6] Universität Heidelberg, [7] ITEP Moscow, [8] KI Moscow, [9] CRN and Université Strasbourg, [10] Warsaw University, [11] RBI Zagreb

References

1. J.Aichelin, C.M. Ko PRL 55, 2661 (1985)
2. D.Miskowiec et al.: PRL 72, 3650 (1994)
3. D.B. Kaplan and A.E. Nelson, Phys.Lett. B 175, 57 (1986)
4. S.Klimt et al., Phys. Lett. B 249, 386 (1990)
5. G.E.Brown et al., Phys.Rev. C43, 1881 (1991)
6. J.Schaffner et al., Phys.Lett. B 334 (1994) 268
7. P.J.Siemens, J.O. Rasmussen, PRL 42, 880 (1979)
8. J. Ritman et al.: Z. Phys. A 352 (1995) 355
9. P.Senger et al., Proceedings of the Meson'96 Workshop, Cracow (1996)
10. B.Hong et al. (FOPI), to be published
11. R.Averbeck, private communication
12. V.Metag, Prog.Part.Nucl.Phys. 30,75 (1993)
13. G.Q.Li et al., PRL 74, 235 (1995)
14. G.Q.Li, C.M. Ko, private communication

FRAGMENT PRODUCTION, COLLECTIVE FLOW, AND THE PROPERTIES OF EXCITED NUCLEAR MATTER

JENS KONOPKA AND HORST STÖCKER

Institut für Theoretische Physik
Johann Wolfgang Goethe-Universität
Postfach 11 19 32
D–60054 Frankfurt am Main, Germany
e-mail: konopka@th.physik.uni-frankfurt.de

Quantum Molecular Dynamics (QMD) calculations of central collisions between heavy nuclei are used to study fragment production and the creation of collective flow. It is shown that the final phase space distributions are compatible with the expectations from a thermally equilibrated source, which in addition exhibits a collective transverse expansion. However, the microscopic analyses of the transient states in the reaction stages of highest density and during the expansion show that the system does not reach global equilibrium. Even if perfect equilibration is assumed, the measurable final state is not suited to determine macroscopic parameters, e.g. the temperature, of the hypothetical transient equilibrium state unambiguously.

1 Introduction

The only possibility, how excited nuclear matter can be probed in the laboratory are nucleus–nucleus reactions[1,2]. In particular when two heavy ions like Au or Pb collide most centrally, the combined system forms a zone of high density and high random agitation of the involved constituents. However, it is still an open question, to which extent the system equilibrates and hence allows for the application of thermodynamical concepts.

From the experimental point of view it is clear that the measurable final state has to be compatible with an equilibrium configuration. Two basically different approaches are commonly used: i) analysis of the final spectra in terms of emission from a thermally equilibrated source, i.e. they should fall off as $\propto \exp(-E/T)$ [3,4,5,6] and ii) analysis of the final fragment composition in terms of a chemically equilibrated source, i.e. the population of a state j is $\propto \exp(\mu_j/T)$, where the chemical potentials of all states are connected via some Gibbs equilibrium conditions[7,8,9,10].

For the theory there are many other "observables", which may indicate the degree of equilibration reached. In this article we use the Quantum Molecular Dynamics model (QMD) to analyze the complete space time history of heavy ion collisions on a microscopic basis. The conclusions, which are drawn from the final state of the reaction only are confronted with that information, which is obtained from the knowledge about the intermediate reaction stages. We demonstrate that it is impossible to extract nuclear temperatures unambiguously from heavy ion induced spectra.

2 Quantum Molecular Dynamics with Pauli-potential

Quantum Molecular Dynamics (QMD) [11,12,13,14,15] is a semi-classical model which calculates the trajectory of a heavy ion collision in the entire many-body phase-space. It simulates the many-body dynamics due to the real and imaginary part of the optical potential by merging two transport theoretical approaches: The real part of the potential is treated via a phenomenological nucleon-nucleon interaction, whereas the effect of the imaginary part can be translated into a collision term of the Boltzmann type [13] which is supplemented by the Uehling-Uhlenbeck correction. QMD therefore contains a classical molecular dynamics section and a collision term, which performs a Monte-Carlo integration of the local collision kernel in a similar manner as in intranuclear cascade models [16]. Using a product ansatz

$$\Psi(\boldsymbol{x}_1, \boldsymbol{x}_2, ..., \boldsymbol{x}_A) = \prod_{j=1}^{A} \varphi_j(\boldsymbol{x}_j) \tag{1}$$

of gaussian wavefunctions

$$\varphi_j(\boldsymbol{x}_j) = \left(\frac{2\alpha}{\pi}\right)^{\frac{3}{4}} \exp\left\{-\alpha\left(\boldsymbol{x}_j - \boldsymbol{r}_j(t)\right)^2 + \frac{i}{\hbar}\boldsymbol{p}_j(t)\boldsymbol{x}_j\right\} \tag{2}$$

the quantummechanical variation principle [17] leads to equations of motion for the time-dependent parameters $\boldsymbol{r}_j$ and $\boldsymbol{p}_j$ that are equivalent to the classical trajectories for the center of gravity of the gaussians.

Special attention has been given to the fermionic character of the nucleons in the molecular dynamics part as well. For an antisymmetrized state of two gaussian wavepackets the expectation value of the kinetic energy operator reads

$$E_{\text{kin}} = \frac{p^2}{2\mu} + \frac{3\alpha\hbar^2}{4\mu} + \frac{\alpha\hbar^2}{2\mu} \frac{\alpha r^2 + \frac{p^2}{\alpha(\hbar c)^2}}{\exp\left\{\alpha r^2 + \frac{p^2}{\alpha(\hbar c)^2}\right\} - 1}, \tag{3}$$

where the second term on the right hand side is due to the zero point energy of the wavepackets. Since the gaussian width parameter α is time-independent and the corresponding energy cannot be transformed in other forms of energy this constant term is neglected. The third term can be interpreted as a coordinate- and momentum-dependent potential between the two gaussians [18]. However, in our case we took another functional form of the Pauli-potential [19]

$$V_{\text{Pauli}} = V_0^{\text{Pauli}} \left(\frac{\hbar}{p_0 q_0}\right)^3 \exp\left\{-\frac{(\boldsymbol{x}_j - \boldsymbol{x}_k)^2}{2q_0^2} - \frac{(\boldsymbol{p}_j - \boldsymbol{p}_k)^2}{2p_0^2}\right\} \delta_{\tau_j \tau_k} \delta_{\sigma_j \sigma_k}, \tag{4}$$

whose parameters have been adjusted to the temperature- and density-dependence of the kinetic energies of a free Fermi-gas [14]. The delta-functions indicate that this potential acts only between particles with identical spin and isospin projection. Taking into account Fermi momenta in such a manner allows for a selfconsistent determination of ground-states by searching for that configuration, which binds $A - Z$ neutrons and

Z protons and minimizes the total energy. Necessary conditions for this minimum are

$$0 = \frac{\partial H^A}{\partial p_j} = \dot{r}_j \quad \text{and} \quad 0 = \frac{\partial H^A}{\partial r_j} = -\dot{p}_j, \qquad j = 1, \ldots, A. \tag{5}$$

From the last equation we can conclude that these model nuclei are absolutely stable, because none of the nucleons moves nore there are forces acting on the constituents.

3 Macroscopic temperature measurement from particle spectra

As an idealized test case, we performed QMD calculations[11,12,14] of a heavy system, namely Au (150 MeV/nucleon, b=0) + Au. Such reactions are supposed to yield the largest accessible participant region, where the nucleons undergo a rapid sequence of binary collisions[20,22]. Consequently, equilibrium, with a large number of nucleons involved, is more likely established in central collisions than in peripheral reactions or collisions of lighter partners.

In the present analysis no information other than mass, charge and four-momentum of the emitted products will be used for extraction of further information on the properties of the hot and dense nuclear matter formed in the transient state. Thus such an analysis would be applicable to experimental data as well[4,5].

After a full QMD propagation for 300 fm/c, involving all nuclear interactions and hard binary scatterings, fragments are calculated via a 3 fm configuration space coalescence. Then, in addition, all charged products are propagated on their mutual Coulomb-trajectories for another 10^6 fm/c $\approx 3 \cdot 10^{-18}$ s.

Having obtained the full triple-differential spectra in such a manner for each charge independently, the spectra were fitted by a transversally expanding, thermally equilibrated source[20]. This implies that the transverse spectrum is composed of a transverse flow and a random component, which can be associated with a temperature. The strength of the collective component is assumed to increase linearly with the transverse distance; the shape of the source is a homogeneous sphere. Nonrelativistically the double-differential distribution reads

$$\frac{d^2 f(T, p_{max})}{dp_l \, dp_t} = N \frac{2}{p_{max}} \int_0^{p_{max}} dp_0 \, \frac{p_0}{p_{max}} \left(\frac{\eta}{\pi}\right)^{\frac{3}{2}} 2\pi \, p_t$$
$$\times \exp\{-\eta(p_t^2 + p_l^2 + p_0^2)\} \, I_0(2\eta p_t p_0) \tag{6}$$

Here η is an abbreviation for $\frac{A}{2m_N T}$. With these formulas the average flow momentum per nucleon is $\frac{2}{3} p_{max}$ and the averaged flow kinetic energy amounts to $p_{max}/4m_N$.

In order to get information on the thermal collective energy sharing, we restrict our analysis for the moment to the transverse momentum spectra only. This procedure is motivated by the fact that transverse momenta are newly created and are not directly connected to the initial (longitudinal) beam momentum. A possible collective expansion in longitudinal direction cannot unambiguously be related to the properties of the hot and dense reaction zone because the system may have memory about its history, in particular the incident momentum of the beam.

Fig. 1 shows transverse momentum spectra of various charged fragments obtained with QMD for the system Au (150 MeV/nucleon, b=0) + Au (symbols) together with fits to these calculated data, which are based on the assumptions from above. In fact, the corresponding count rates have been fitted, rather than the invariant distributions, which are displayed. All spectra are compatible with temperatures between 20 and 25 MeV and averaged collective flow velocities of 0.1–0.13 c.

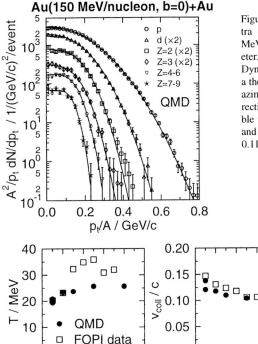

Figure 1: Invariant transverse velocity spectra of various reaction products of Au (150 MeV/nucleon) + Au at vanishing impact parameter. The predictions of the Quantum Molecular Dynamics model (symbols) have been fitted with a thermally equilibrated source, which expands azimuthally symmetric towards the transverse direction (lines). Note, all spectra are compatible with temperatures between 20 and 25 MeV and an averaged transverse collective velocity of 0.11–0.13 c.

Figure 2: Temperature and flow velocities for fragments. The QMD results (closed circles) deduced from the fits to the transverse momentum spectra in Fig. 1 are compared to data of the FOPI-collaboration [4] (open squares).

The strength of the collective expansion exhibits a weak decrease with increasing fragment mass, which is displayed in Fig. 2. The results on the averaged collective transverse momentum and temperature from the QMD calculation (full symbols) compare reasonably well to a experimental compilation of available data on central Au+Au collisions[4], which have been analysed with the very same method. Similar collective flow velocities and temperatures have also been reported from the EOS collaboration[5]. On the contrary these rather high temperatures are in variance with the expectations ($T \approx 8\,\mathrm{MeV}$ at $150\,\mathrm{MeV/nucleon}$) from a quantumstatistical analysis of the large number of intermediate mass fragments observed in the very same system[8].

The collectivity is more emphasized in the spectra of the heavy clusters, since the thermal energy drops as $1/A$. This is expressed in Fig. 3, where the averaged kinetic energy in the center of mass frame and the thermal energy per nucleon is plotted as a function of the fragment mass. The thermal energies obtained from the fit to the spectra show the expected fall off, whereas the total energy is shifted roughly by a constant amount. Note, for light ejectiles like protons the collective energy amounts to only 30% of the total energy, for the heavy fragments the collective–thermal energy sharing is reversed.

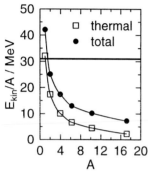

Figure 3: Averaged kinetic energies in the center of mass frame of central Au+Au collisions at 150 MeV/nucleon bombarding energy. The total energy (closed circles) clearly shows a non-thermal behaviour. The horizontal line indicates the kinetic energy per nucleon in the final state averaged over all fragments.

Another interesting aspect of the many-body dynamics can be read off Fig. 3. The average available kinetic energy per nucleon in the center of mass after the reaction, which is indicated by the horizontal line, is considerably lower than the averaged kinetic energies of free nucleons. This excess energy can be associated with the heat released due to the formation of heavy fragments, which move with much lower kinetic energy per nucleon compared to the average. The description of this behaviour is out of scope of one-body models like BUU/VUU[21], since they lack the many-body correlations, which are responsible for the formation of complex fragments. Thus these models also underpredict the averaged proton kinetic energies at low bombarding energies.

4   Temperature determination from microscopic phase space densities

Since QMD provides full information about the entire collision history it allows not only for the calculation of final state properties but also – what is even more important – for a detailed analysis of the intermediate reaction stages. In the following we again consider head on collisions of two Au nuclei at 150 MeV/nucleon. For a study of the local phase space distribution we divide configuration space by a cylindrical grid with 1 fm extension in longitudinal as well as in transverse direction. By superimposing many events of the same kind the local velocity distribution can be calculated within each of these cells. Its mean corresponds to that velocity the entire cell is collectively moving with. The spread reflects the thermal agitation inside the cell. Fig. 4 shows a typical snapshot of the expansion phase of a heavy ion reaction Density contours are shown together with local collective expansion velocities. A strong correlation between configuration and velocity space is observed. Inside any of these cells the velocity distribution exhibits a gaussian shape.

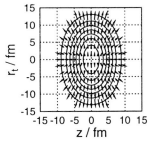

Figure 4: Snapshot of the expansion after 50 fm/c starting from a 5 fm separation of projectile and target surfaces. Density contours are at 0.1–0.7 ρ_0. The arrows indicate the direction and the absolute velocity (proportional to the length) of the collective motion of the individual fluid cells.

An example for one particular cell is shown in Fig. 5. However, the spreads in the different coordinate directions are not equal, thus even locally there remains a non-equilibrium component in the spectrum! In the following, this will be illustrated through reference of a longitudinal, a transversal, and a azimuthal temperature respectively.

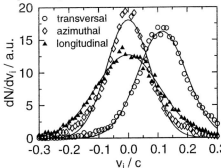

Figure 5: Example of a local velocity distribution in longitudinal (closed triangles), transversal (open circles), and azimuthal direction (open diamonds). The results correspond to the cell centered at $r_t = 5$ fm and $z= 0$ after 50 fm/c. Note, how the transverse velocity distribution is shifted due to the collective motion of the entire cell.

These "local temperatures" vary only slightly over large volumes in real space (see Fig. 6), which supports a common analysis of a larger number of cells. For this purpose the central reaction volume is defined as the sum of all cells in which the local density exceeds half the maximum density at this instant, which implies a time-dependent volume. In Fig. 7 the properties of the excited nuclear matter inside this volume are displayed as a function of time. Even in the late stages of the expansion this zone still contains $\approx 1/3$ of the mass of the entire Au+Au system. The evolution of the spreads of the local velocity distributions suggests a rapid cooling, which goes in line with a strong density decrease. The different temperatures tend to converge in the late stages only.

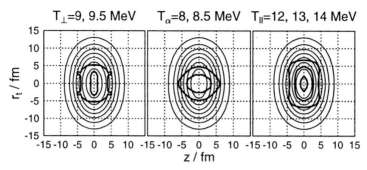

Figure 6: Distribution of local temperatures (heavy contours) during the expansion phase. The T parameters have been calculated from the widthes of the local velocity distributions.

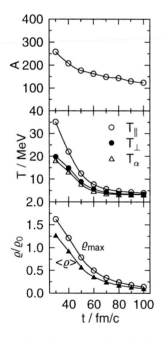

Figure 7: Thermodynamics in the central reaction zone, i.e. the volume where the local density is at least half of the maximum value. Mass content a), spreads of the local velocity distributions b), and maximum as well as averaged density c) are displayed as a function of time.

At densities around 0.1–0.5 ϱ_0, where the freeze-out of fragments is supposed to happen, the corresponding temperatures have dropped below 10 MeV, which is no longer in vast disagreement with the chemical temperatures needed for the understanding of the large intermediate mass fragment multiplicities[8].

5 Impossibility of temperature measurements

We have shown in the two previous sections that final state spectra, although they seem to be perfectly discribed by an equilibrated, expanding source, suggest too high temperatures even if collective flow effects are taken into account. Locally much lower temperatures are observed in the very same class of events.

In turn the question arises what was wrong with the assumptions which underlied the global fit. Assuming an overall thermally equilibrated source which disintegrates not only due to the thermal pressure but also due to some additional collective expansion, the probability of finding a particle which has been submitted some collective velocity $v_{coll.}$, i.e. $dN/dv_{coll.}$ is essentially unknown. It can be expressed as

$$\frac{dN}{dv_{coll.}} = \frac{dN}{dr} \cdot \frac{dr}{dv_{coll.}} = \frac{dN}{dr} \cdot \left(\frac{dv_{coll.}}{dr}\right)^{-1}. \tag{7}$$

Here the density distribution as well as the flow velocity profile enters as the first and second term on the right hand side of the preceding equation. For the fit to the spectra in Fig. 1 a homogeneous density with a sharp cut-off and a linearly increasing velocity profile have been used (dotted lines in Fig. 8).

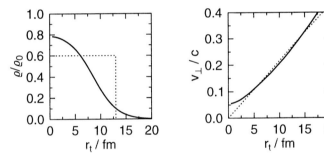

Figure 8: Local density and collective velocities as a function of the transverse distance in the $z = 0$ plane. The Figure corresponds again to the situation after 50 fm/c. The calculated shape (solid lines) are compared to the assumptions, which entered the global fit (dotted lines). The longitudinal as well as the azimuthal collective velocity vanish. The reason why $v_\perp$ does not reach 0 for $r_t = 0$ is due to the fact that the innermost cell is not symmetric around the symmetry axis.

As it is expressed in Fig. 8 the shape of the local flow profile is in accord with the principles which underlied our fit, the density however, is not at all constant. Therefore the high transverse momentum components of the particle spectra do not correspond to the high momentum tails of a hot source, but to fastly moving cells with a considerably lower temperature. The microscopic analysis suggests that the combinations of density and temperatures, which a traversed in the course of the reaction, are in agreement with the expectations from the quantum statistical analysis of the fragment distributions in the final state[8].

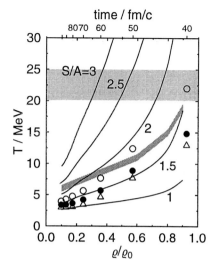

Figure 9: Temperatures versus densities for central Au+Au collisions at 150 MeV/nucleon. The microscopically determined local temperatures (symbols as in Fig. 6) are considerably lower than the simple macroscopic fit to the spectra (light grey shaded area). However, they compare reasonably well with curves of constant entropy from QSM and with experimental data from a quantum statistical analysis of the fragment distributions (dark grey shaded area)[8].

Moreover, it is shown in Fig. 9 that the correlation of averaged densities and microscopic temperatures compares reasonably well with the quantumstatistical analysis of multifragmentation data as well as with isentropes in the temperature-density plane. On the contrary, the temperature deduced form the global fit to the spectra is independent of the density, see Eq. 7. The nature of the expansion process is presently under intense investigation[24].

6 Summary

We have presented QMD transport calculations for head on collisions of Au+Au. Temperatures were deduced by fitting final state particle spectra as well by analysis of local velocity distributions in the intermediate reaction stages. The latter are in agreement with the temperatures which are necessary to explain the large fragment abundances in terms of a chemical equilibrium.

In summary, any temperature obtained from particle spectra – even if collective flow has been taken into account – can only serve as an upper estimate for the conditions at freeze-out. The main reason is that the configuration space distribution is essentially unknown. In view of these reasons fireball model estimates for ultrarelativistic heavy ion collisions may well overestimate the thermal energy reached in such kind of reactions substantially. This finding has severe impact also on temperatures extracted from e.g. slopes of pion and light particle spectra at higher and highest incident energies[23].

Acknowledgments

This work has been supported by the Bundesministerium für Bildung und Forschung (BMBF), the Deutsche Forschungsgemeinschaft (DFG), and the Gesellschaft für Schwerionenforschung mbH (GSI).

References

1. H. Stöcker and W. Greiner, Phys. Rep. **137**, 277 (1986).
2. Proceedings on the NATO Advanced Study Institute Programme on Hot and Dense Nuclear Matter, eds. W. Greiner, H. Stöcker, and A. Gallmann, NATO ASI Series B 335, Plenum, New York (1994).
3. H. Stöcker, A.A. Ogloblin, and W. Greiner, Z. Phys. A **303**, 259 (1981).
4. J.P. Coffin et al, Nucl. Phys. **A 583** 567c (1995).
5. M.A. Lisa et al., Phys. Rev. Lett. **75**, 2662 (1995).
6. G.J. Kunde et al., Phys. Rev. Lett. **74**, 38 (1995).
7. D. Hahn and H. Stöcker, Nucl. Phys. **A 476**, 718 (1988).
8. C. Kuhn et al., Phys. Rev. C **48**, 1232 (1993).
9. J. Pochodzalla et al., Phys. Rev. Lett. **75**, 1040 (1995).
10. J. Stachel and P. Braun-Munzinger, Phys. Lett. **B216**, 1 (1989).
11. J. Aichelin et al., Phys. Rev. C **37**, 2451 (1988).
12. G. Peilert et al., Phys. Rev. C **39**, 1402 (1989).
13. J. Aichelin, Phys. Rep. **202**, 233 (1991).
14. G. Peilert et al., Phys. Rev. C **46**, 1457 (1992).
15. A. Faessler, Prog. Part. Nucl. Phys. **30**, 229 (1993).
16. Y. Yariv and Z. Fraenkel, Phys. Rev. C **20**, 2227 (1979).
17. A.K. Kerman and S.E. Koonin, Annals of Phys. **100**, 332 (1976).
18. L. Wilets et al., Nucl. Phys. **A282**, 341 (1977).
19. C. Dorso et al., Phys. Lett. B **188**, 287 (1987).
20. J. Konopka et al., Nucl. Phys. **A 583**, 357c (1995) and to be published.
21. G.F. Bertsch, H. Kruse, and S. Das Gupta, Phys. Rev. C **29**, 673 (1984).
22. P. Danielewicz, Phys. Rev. C **51**, 716 (1995).
23. R. Mattiello et al., Phys. Rev. Lett. **74**, 2180 (1995).
24. J. Konopka, Horst Stöcker, and Walter Greiner, Proceedings of the 1st Catania Relativistic Ion Studies, Acicastello (Italy) 1996 and to be published.

Investigation of the radial expansion of nuclear matter in collisions of heavy nuclei at SIS energies [†]

Ch. Hartnack and J. Aichelin

SUBATECH

(Laboratoire de Physique Subatomique et des Technologies Associées)
University of Nantes - IN2P3/CNRS - Ecole des Mines de Nantes
4 rue Alfred Kastler, F-44072 Nantes, Cedex 03, France
[†] *Contributed paper*

The Isospin Quantum Molecular Dynamics Model (IQMD) is used to investigate the transverse radial expansion of nuclear matter. A linear velocity profile is found at nearly all energies. A strong influence of the potential part of the interaction on the asymptotic transverse velocities is reported. The energy and impact parameter dependence is investigated. An influence of the nuclear equation of state on the transverse velocities could not be established.

1 Introduction

A major goal in the study of heavy ion collisions is the investigation of the properties of dense and hot nuclear matter. Various activities have been started in order to enhance our understanding of matter under extreme conditions. For an overview we refer to [1] and [2].

Besides the investigation of the production of secondary particles like pions and kaons and besides the study of anisotropies in the momentum distribution like the transverse flow of nuclear matter in plane (bounce-off) and out of plane (squeeze-out) also the examination of isotropic components like the radial flow has recently achieved large interest (see e.g.. [3]).

Our aim is to investigate the radial expansion of fragments in collisions of heavy nuclei at SIS energies. Hereby we will only regard the direction transverse to the beam. In the following the transverse velocity v_t will denote the absolute value of the velocity component perpendicular to the beam and the radial velocity v_r will denote the radial component of the velocity vector in cylindrical coordinates r, φ, z.

2 QMD and its different flavours

The Quantum Molecular Dynamics Model (QMD) [4] is a microscopical model for the description of heavy ion collisions on a n-body level. It has been frequently used for the description of fragmentation processes (e.g. in [5,6,7]). For a detailed description of the model we refer to [2].

We would only like to recall briefly that each particle in the model is represented by its Wignerfunction

$$f_i(\vec{r}, \vec{p}, t) = \frac{1}{\pi^2 \hbar^2} e^{-(\vec{r} - \vec{r}_{i0}(t))^2 \frac{1}{2L}} e^{-(\vec{p} - \vec{p}_{i0}(t))^2 \frac{2L}{\hbar^2}}$$

which is described as a Gaussian in coordinate and momentum space. The Gaussian width L is a free parameter of the theory and corresponds to the interaction range of the particles.

The nucleons are propagated according to equations of motion which look similar to the classical Hamilton equations:

$$\dot{p}_i = -\frac{\partial H}{\partial q_i} \quad , \qquad \dot{q}_i = \frac{\partial H}{\partial p_i} \quad .$$

The expectation value of the Hamiltonian is given by

$$H = T + V = \sum_i \frac{p_i^2}{2m_i} + \sum_i \sum_{j>i} \int f_i(\vec{r}, \vec{p}, t) \, V^{ij} f_j(\vec{r}', \vec{p}', t) \, d\vec{r} \, d\vec{r}' \, d\vec{p} \, d\vec{p}'$$

The particles interact via two-body interactions of local (Skyrme-) type, Yukawa type, Coulomb type, momentum dependent interactions and proton-neutron asymmetry interactions

$$V^{ij} = V^{ij}_{loc} + V^{ij}_{Yuk} + V^{ij}_{Coul} + V^{ij}_{mdi} + V^{ij}_{asym}$$

with

$$V^{ij}_{loc} = t_1 \delta(\vec{r}_i - \vec{r}_j) + \sum_{k>j>i} \int f_k(\vec{r}'', \vec{p}'') t_2 \delta(\vec{r}_i - \vec{r}_j) \delta(\vec{r}_i - \vec{r}_k) d\vec{r}'' d\vec{p}''$$

$$V^{ij}_{Yuk} = t_3 \frac{\exp\{|\vec{r}_i - \vec{r}_j|/a\}}{|\vec{r}_i - \vec{r}_j|/a}$$

$$V^{ij}_{Coul} = \frac{Z_i Z_j e^2}{|\vec{r}_i - \vec{r}_j|}$$

$$V^{ij}_{mdi} = t_4 \ln^2(1 + t_5(\vec{p}_i - \vec{p}_j)^2) \delta(\vec{r}_i - \vec{r}_j)$$

$$V^{ij}_{asym} = 100 \text{ MeV} \frac{1}{\varrho_0} T_{3i} T_{3j} \delta(\vec{r}_i - \vec{r}_j)$$

where T_3 is the component of the isospin vector and $t_1, \ldots t_5$, a and δ are parameters which have to be adjusted to the properties of normal nuclear matter.

Additionally a collision term is introduced which includes elastic and inelastic collisions of nuclei respecting the Pauli principle and the production of deltas and pions.

There exist different numerical realizations of the QMD model. Two of the most prominent ones are BQMD [2,8] and IQMD [9]. BQMD was mainly applied to the description of fragmentation [6,7] while IQMD was mainly dedicated to the studies of dynamical variables and particles production. A first comparison of different QMD versions with VUU and BUU was published in [10] a second detailed comparison especially of BQMD and IQMD has recently been performed [11]. However both programs are destined to be superseded by a novel unified QMD version [12].

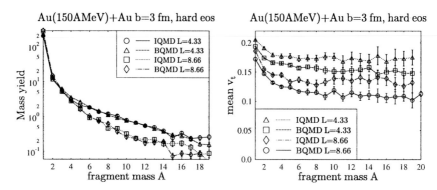

Figure 1: Comparison of the fragment yields and the transverse velocity distributions for the system Au(150 AMeV)+Au at b=3fm calculated with BQMD and IQMD using $L = 4.33\text{fm}^2$ and $L = 8.66\text{fm}^2$.

It was found that BQMD failed in the description of nucleonic flow while IQMD in its default version shows a to weak production of fragments. A detailed analysis will be found in [11]. It should only be noted that BQMD and IQMD yield similar fragment yields for central collisions if the same interaction range L is used. This can be seen in Fig. 1 where the fragment yield of Au(150 AMeV)+Au at b=3fm is compared between BQMD and IQMD both with $L = 4.33\text{fm}^2$ and $L = 8.66\text{fm}^2$. IQMD and BQMD using $L = 4.33\text{fm}^2$ yield more fragments than the calculations with $L = 8.66\text{fm}^2$. Standard BQMD uses as a default value $L = 4.33\text{fm}^2$ while IQMD uses $L = 8.66\text{fm}^2$.

It can furthermore be seen that L also influences the transverse velocity of the fragments. The calculations with $L = 4.33\text{fm}^2$ reach higher asymptotic values of v_t than the corresponding calculation with $L = 8.66\text{fm}^2$. It is also found that using the same interaction range L IQMD shows slightly enhanced radial flow as compared to BQMD.

The effect of the better description of fragment numbers using $L = 4.33\text{fm}^2$ motivated us to investigate the velocity of fragments in IQMD using this value. Therefore all results presented in the following are obtained by IQMD using $L = 4.33\text{fm}^2$.

3 Investigation of velocity profiles

Let us first study the reaction Au(1 AGeV)+Au at b=1fm. For our analysis we used a calculation with a soft equation of state with momentum dependent interactions (soft mdi). However, similar results are achieved if another equation of state (eos) is applied.

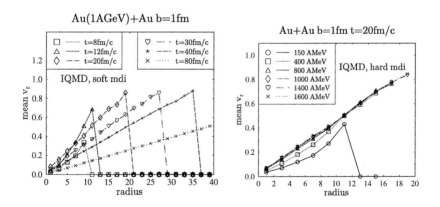

Figure 2: Velocity profiles found in Au+Au collisions at b=1fm. Left at 1 AGeV for different times, right for different energies taken after 20 fm/c.

Fig 2 shows on the left side the time evolution of the velocity profile (v_r as a function of the radius r). We see that very early a linear profile is built up. During the expansion the profile remains linear. This linear profile can be found at different energies. On the right side of Fig 2 we show (here using a hard eos with mdi) the velocity profile of the same system at different energies taken after 20 fm/c. We see (with some exception of the 150 MeV case) a very similar behaviour. The profiles even show the same slopes. Only the point of maximum velocity (and thus the maximum radius) is different. A comparison of hard eos and soft eos (left figure with right figure) shows a very similar behaviour.

Let us now look for the slope of the profiles. For this we sorted all particles according to size of the fragment to which it will finally belong and compared the mean value of v_r/r at different times. As we learn from the left part of Fig.

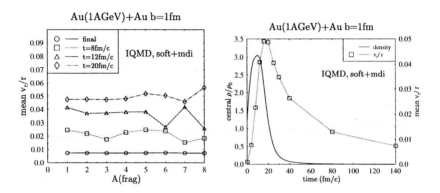

Figure 3: Left: Slope parameter v_r/r of the velocity profiles taken at different time as a function of the size of the final fragment. Right: Time evolution of the density and of the slope parameter v_r/r.

3 the particles of all (final) fragments show the same slope parameters over the whole time evolution which indicates that all particles show a similar velocity profile. Thus it is sufficient to describe the global profile with one parameter, its slope. The right hand side of Fig. 3 therefore shows the time evolution of the average value of v_r/r (squares). Additionally the time evolution of the density in the center of the reaction zone has been included. We see a strong increase of the slope v_r/r with time up to a time of about 16 fm/c. At this time the maximum compression has already been reached at the system is already expanding. After about 20 fm/c the slope decreases. Its time evolution for $t \geq 20$fm/c can be fit to the following formula

$$\frac{\vec{v} \cdot \vec{r}}{r^2} = \frac{v_r}{r} = \frac{a(t - t_0)}{1 + a(t - t_0)^2} \qquad a = \frac{T_0}{m\Delta^2}$$

This formula taken from [13] describes the expansion of a Gaussian shaped fireball of the width Δ^2 with a temperature T_0 at $t = t_0$. The fit yields $a = 0.014$ and $t_0 = 4\text{fm/c}$.

From this similarity to a free adiabatic expansion for $t \geq 20$ fm/c we assume that the driving forces of a radial expansion should have had their major contribution before a time of 20 fm/c for the 1 AGeV case.

4   Influence of the potential part

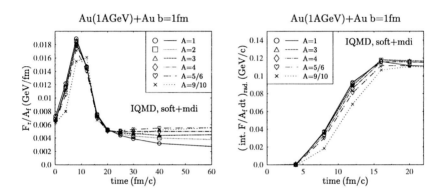

Figure 4: Time evolution of the mean radial component of the forces F_r (left) and of the mean radial component of the momentum gain forced by the potential (right) for different fragment sizes.

In order to check this assumption we show in Fig. 4 the average value of the radial component of the forces acting on the particles. The values are taken as force per nucleon. We see a strong increase of the forces during the compression and an decrease afterwards. After about 20 fm/c the forces saturate. The forces on the particles depend only slightly on the mass of the final fragment. Single particles show only slightly stronger forces at the maximum. Therefore it is not astonishing that the momentum transferred by the potential to one nucleon is rather independent of the fragment size. This is demonstrated on the right side of Fig. 4 which shows the radial component of the average momentum transfer per nucleon caused by the potential. The common value obtained at 20 fm/c is about $0.11 - 0.12$ GeV/c.

Let us now investigate the time evolution of the transverse velocities v_t for

Au(1AGeV)+Au b=1fm

Au(1AGeV)+Au b=1fm

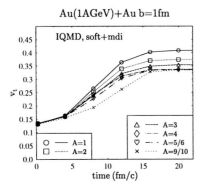
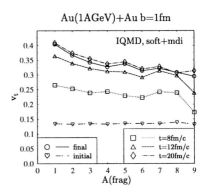

Figure 5: Time evolution of the transverse velocity v_t of different fragment types (left) and fragment mass dependence of v_t at different time steps (right)

different sizes of the final fragment. As we obtain from Fig. 5 all fragments start with the same initial value of $v_t \approx 0.14c$ which correspond to the Fermi momentum of the initial distribution. The curves start to deviate from each other quite early at a time of about 5-10 fm/c. After 20 fm/c the velocities remain constant.

This can be proven by regarding the right side of Fig. 5 which shows the fragment dependence of v_t at different time. The final pattern and the distribution at 20 fm/c look identical. The first indication of higher velocities for light particles can be seen at about 8-12 fm/c.

We find an asymptotic velocity of about $v_t \approx 0.28c$. In a pure thermal expansion the mean kinetic energy should be the same for all fragment species depending on the temperature. This would cause an asymptotic value (for large A_f) of 0. The obtained asymptotic value can be mainly attributed to the initial Fermi momentum ($v_t \approx 0.14$ initially) and the gain due to the potentials which gives for 1 AGeV about $0.12 - 0.13c$. There is some remaining rest which might correspond to the pressure caused by the collisions.

5 Energy and impact parameter dependence

Let us now finally study the dependence of the asymptotic transverse velocity on the impact parameter. Fig 6 shows on the left hand side this dependence for the system Au+Au at 1 AGeV for calculations with hard and soft eos

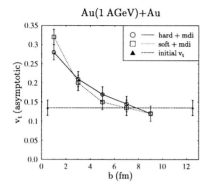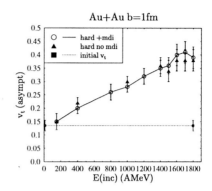

Figure 6: Impact parameter dependence of the asymptotic velocity for Au(1 AGeV)+Au using a hard and a soft eos with momentum dependent interactions (left) and excitation function for the same system with $b = 1$fm using a hard eos with and without mdi.

both with momentum dependent interactions (mdi). We see a very similar behaviour for both equations of state. The asymptotic velocity decrease rapidly with increasing impact parameter. At $b \approx 7$fm the asymptotic velocity reaches about the initial value. It may still decrease below the initial value since the particles loose kinetic energy to compensate the binding energy when they are set free.

The right hand side presents the excitation function of the asymptotic velocity for the system Au+Au at $b = 1$fm. Calculations with a hard eos (and as well with a soft eos which was not included in the figure) with and without momentum dependent interactions yield the same result. We see a continuous rise of v_t up to an energy of about 1600 AMeV. Whether the flow still rises afterwards or whether it reaches a saturation is not yet clear.

6   Conclusion

We used the IQMD model to investigate the transverse radial expansion of nuclear matter. A linear velocity profile is found at nearly all energies. A strong influence of the potential part of the interaction on the asymptotic transverse velocities is reported. Together with the initial velocity cause by the Fermi momentum it is the dominant contribution to the asymptotic velocity. The asymptotic velocity shows a strong increase for very central collisions. A continuous rise of the asymptotic velocity with energy up to beam energies of 1600 AMeV has been found. An influence of the nuclear equation of state on the transverse velocities could not be established.

References

1. H. Stöcker and W. Greiner, Phys. Rep. 137 (1986).
2. J. Aichelin, Phys. Rep. **202**, 233 (1991).
3. S.C. Jeong et al., Phys. Rev. Lett. 72 (1994) 3468.
   M.A. Lisa et al. , Phys. Rev. Lett. 75 (1995) 2662.
   W. Reisdorf for the FOPI collaboration, submitted to Z. Phys. A
   N. Herrmann for the FOPI collaboration, to be published in proceedings on Quark Matter 96, Heidelberg (Germany), 1996.
4. J. Aichelin and H. Stöcker, Phys. Lett. B**163**, 59 (1986).
5. G. Peilert, H. Stöcker, A. Rosenhauer, A. Bohnet, J. Aichelin and W. Greiner, Phys. Rev. **C 39**, 1402 (1989).
6. P.B. Gossiaux and J. Aichelin, SUBATECH preprint 95-4, submitted to Phys. Rev. C
7. P.B. Gossiaux, R.K. Puri, Ch. Hartnack and J. Aichelin, contribution to this proceeding
8. A. Bohnet et al., Phys. Rev. C44 (1991) 2111.
9. C. Hartnack, H. Stöcker and W. Greiner, Proc. Gross Properties of Nuclear Matter XVI, Hirschegg, Austria, 1988, 138.
   C. Hartnack, Li Zhuxia, L. Neise, G. Peilert, A. Rosenhauer, H. Sorge, J. Aichelin, H. Stöcker, W. Greiner, Nucl. Phys. A495 (1989) 303c.
   Ch. Hartnack, GSI report 93-05
   S.A. Bass, C. Hartnack, H. Stöcker and W. Greiner, Phys. Rev. C51 (1995) 3343.
10. J. Aichelin, C. Hartnack, A. Bohnet, Li Zhuxia, G. Peilert, H. Stöcker, W. Greiner, Phys. Lett. B224 (1989) 34.
11. Ch. Hartnack et al., manuscript in preparation
12. S.A. Bass et al., this proceeding
    L.A Winckelmann et al., to be published in the proceedings on Quark Matter 96, Heidelberg (Germany), 1996.
13. J. Aichelin, Nucl. Phys. A 411 (1983) 474.

PRODUCTION OF ANTI-PROTONS AND MESONS IN NUCLEUS-NUCLEUS COLLISIONS FAR BELOW THE THRESHOLD

P. KIENLE and A. GILLITZER

Technical University Munich, James-Franck-Str., 85748 Garching, Germany

Abstract:

Experiments are described in which below threshold forward production of $\bar{p}$, K^- and π^- have been studied in Ne-NaF and Ni-Ni collisions at bombarding energies up to 2AGeV. Strong evidence for multistep production with nucleon resonances as energy storing intermediate states is shown by the data. Indications of a reduction of $\bar{p}$ and K^- effective mass in the medium are extracted from excitation functions and spectral distributions. Pion production shows anomalous high yields of high energy pions and strong spectral deviations from thermal distributions not understood so far. An approximate scaling of particle production as function of the difference of the available and required energy has been found.

1. Introduction

The work presented here is devoted to Walter Greiner in honour of his 60. birthday. It is complementary to his predictive theoretical approach to understand nature, because it followed a purely method oriented experimental scheme with the hope to discover something new. Our meeting place will be the discussion and understanding of the new results.

The original motivation of this work was to make use of the fragment separator FRS [1] at GSI, built for the separation and identification of medium energy projectile fragmentation products, for studies of forward production of $\bar{p}$, K^- and π^- in nucleus-nucleus collisions far below their respective N-N-collision thresholds. Our hope was to learn something new about high momentum components and excited hadrons in the collision zone, created in multistep processes. Particular questions had to do with the effects of creation of Δ-matter and chiral symmetry restoration in the dense, hot medium on particle production in central nucleus-nucleus collisions at energies around 2 AGeV, which lead to densities in the participant zone of $\rho \sim 3 \, \rho_o$ with $\rho_o = 0.17 \, \text{fm}^{-3}$ and temperatures of up to 100 MeV. Why study forward production of particles in view of the obvious experimental difficulties caused by the background of scattered particles? A kinematical advantage is provided by the Lorentz boost, which leads to forward focusing and detectability of short lived particles with rather low energy. A more speculative feature has to do with the possibility of detection of coherent effects in forward particle production. In section 2, we will discuss the experimental set up of the „Antiproton Experiment" at the FRS [2] for the $\bar{p}$-, K^-- and π^-- detection in forward direction. New results for particle spectral distributions and excitation functions in Ne-NaF and Ni-Ni collisions will be reviewed in section 3. Finally we discuss the results in the framework of transport and chiral symmetry restoration models in section 4.

2. Experimental set up at the FRS

The main experimental problems in the study of forward production of particles in nucleus-nucleus collisions is, first the suppression of scattered beam particles and break up products, second the large cross section differences if one wants to measure $\bar{p}$-, K^- and π^-- production at 2AGeV, with characteristic cross section ratios of 10^7 : $3 \cdot 10^3$: 1 respectively. In order to have a detectable antiproton production yield one has to run the experiment with π^-- rates up to 10^6/s. Furthermore a good particle identification scheme is required. The solution was to use the 72 m long FRS as a magnetic guide for negatively charged particles emitted in forward direction and measure along the guide their velocities with various techniques. This method is specially adapted for the identification of stable particles, such as antiprotons, but works also for π^-- and K^-- mesons to certain extent. By the adoption of a special optical setting of the FRS, its solid angle acceptance could be increased to $\Omega = 3$msr at a momentum acceptance $\Delta p/p = \pm 3$ % which yields a factor of 10 gain compared to the standard setting.

Fig. 1 shows a schematic presentation of the experimental set up. ^{20}Ne and ^{58}Ni-beams with energies up to 1.95 AGeV and 1.85 AGeV and intensities up to 3 x 10^{10} and 3 x 10^8 particles per spill respectively (spill time 1s, repetition rate 1/5 Hz) hit various targets (NaF, Cu, Sn, Bi and Ni) with 10% interaction thickness. In the focal planes S_1, S_2, S_3 and S_4 of the FRS, scintillation detectors Sc were placed to measure the flight times repeatedly. In addition various threshold Cerenkov detectors with thresholds of $\beta > 0.90$, 0.96 and 0.98 respectively provided further particle identification.

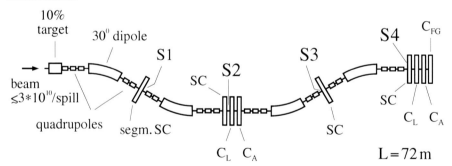

Fig. 1: Schematic view of the FRS including the detector installations used for the „Antiproton Experiment". S_{1-4} are FRS focal planes, SC scintillation counters for time of flight measurements, C_L lucite Cerenkov detectors, C_A aerogel Cerenkov detectors C_{FG},/glass/freon Cerenkov detector.

Fig. 2 shows the effectiveness of the particle identification in 1.94 AGeV ^{20}Ne-Sn collisions with p_{lab} = 1.5 GeV/c, making use of multiple time of flight determination and Cerenkov thresholds ($\beta < 0.96$ and < 0.90) as indicated.

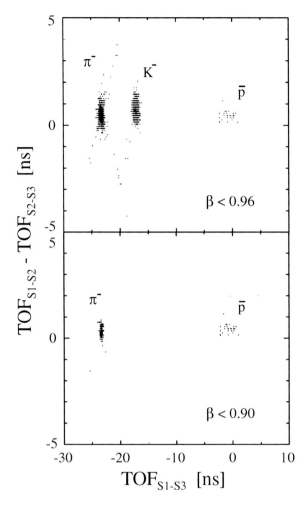

Fig. 2: Time of flight identification spectra of negatively charged particles in the 1.94 AGeV ^{20}Ne+Sn reaction at p_{lab} = 1.5 GeV/c. The time of flight (TOF) differences between S_1-S_2 and S_2-S_3 are plotted versus the time of flight between S_1-S_3 for particles with β < 0.96 and β < 0.90 respectively.

3. Experimental results for $\bar{p}$ -, K^-- and π^-- production

3a spectral distributions

Fig 3a and 3b show invariant antiproton production cross sections as function of the cm kinetic energies of the $\bar{p}$'s for Ne-NaF (a) and Ni-Ni (b) collisions at the bombarding energies indicated. The spectral slope parameters are about T=100 MeV at the highest bombarding energies in both collision systems with a slight decrease

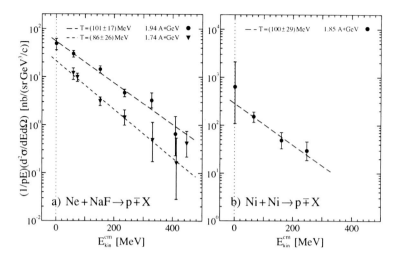

Fig.3a and b: Invariant $\bar{p}$ production cross sections as function of the $\bar{p}$ cm energy, E_{kin}^{cm}, for the reactions Ne+NaF $\to$ $\bar{p}$ +X(a) and Ni+Ni $\to$ $\bar{p}$ +X(b) at bombarding energies indicated. The dashed lines are exponential fits to the data points with slope parameters T as indicated.

indicated in the Ne-NaF system at lower bombarding energies. Fig. 4 shows the ratio of the invariant $\bar{p}$-production cross sections for Ni-Ni and Ne-NaF collision as function of the laboratory momenta, scaled to identical bombarding energies, indicating an experimental value for σ(Ni-Ni)/σ(Ne-NaF) of about 9. This is three

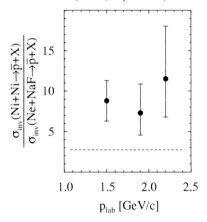

Fig. 4: Ratio of the invariant $\bar{p}$ -production cross sections for the reactions Ni+Ni and Ne+NaF as function of the $\bar{p}$ momenta p_{lab}. The dashed line represents the ratio of participants.

times higher than the ratio of participants, which is a strong indication for the presence of multistep processes, such as Δ-resonance contributions.

The spectral distributions for the K^--production as shown in Fig. 5a and 5b for both collision systems respectively shows similar characteristics as the $\bar{p}$ spectra with slope parameter ranging between 100 and 80 MeV and a similar ratio of about 10 for $\sigma(Ni\text{-}Ni)/\sigma(Ne\text{-}NaF)$.

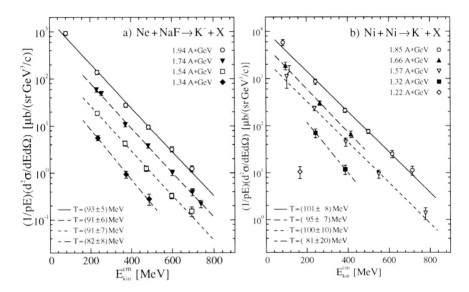

Fig. 5a,b: Invariant K^--production cross sections as function of the K^-- cm energy, E_{kin}^{cm}, for the reactions Ne+NaF $\rightarrow K^- + X$(a) and Ni+Ni $\rightarrow K^-$+X at bombarding energies indicated. The lines are exponential fits to the data points with slope parameters T as indicated.

π^- production has been observed in the momentum range from $p_{lab} = 0.4$ to 2.75 GeV/c and $p_{lab} = 0.4$ to 4.5 GeV/c for Ni-Ni and Ne-NaF collisions, Fig. 6a and 6b, respectively, which means from production above threshold to far below threshold for the high momentum pions. Distinct deviations from exponential slopes were observed and will be discussed in section 4.

In our new data set, we paid special attention to determine excitation functions for particle production, because for p_{cm} = const one expects no strong distortions due to absorption effects, which are difficult to handle. Fig. 7 shows a summary of our data on $\bar{p}$ excitation functions from Ni-Ni-collisions at $p_{cm} = 0.34$ GeV/c and Ne-NaF collision at $p_{cm} = 0.34$ GeV/c and 0.56 GeV/c as function of the lab. beam energy. The slopes are rather flat and within the large errors similar in both collision systems and antiproton momenta.

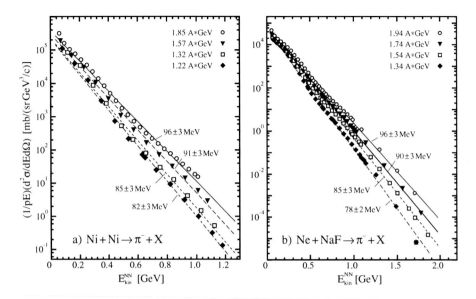

Fig. 6a,b: Invariant π^--production cross sections as function of the π^--cm energy E_{kin}^{NN} for the reactions Ni+Ni $\rightarrow$ π^-+X(a) and Ne+NaF $\rightarrow$ π^-+X(b) at bombarding energies indicated. The lines are exponential fits to the data points with slope parameters T as indicated.

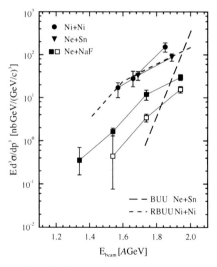

Fig. 7: Excitation functions for $\bar{p}$-production for Ni-Ni, Ne-Sn and Ne-NaF collisions. The Ni-Ni and Ne+NaF data are taken at p_{cm} = 0.34GeV/c ($\bullet$,$\blacksquare$) and 0.56 GeV/c ($\square$) respectively. The Ne-Sn data are at p_{lab} = 1.5GeV/c. The data points are connected with solid lines. The dashed lines represent BUU and RBUU calculations respectively [6].

4. Discussion of the results and comparison with transport models

From the ratio of the particle production cross sections for Ni-Ni and Ne-NaF collisions we found already strong indications for a multistep particle production process with dominant contributions from Δ and N^* resonances, which also act as an energy storage [3]. This basic feature of the production process is quantitatively predicted by various transport models [4], [5], [6], [7], [8], such as BUU, QMD, RBUU and RQMD.

Before we go into a more detailed comparison of our observations with transport models, we like to point out some observations on π^--production which are not well understood.

While the low to medium energy pion emission is greatly influenced by the strong πN coupling leading to resonance formation and preferential emission following the high density phase, the high energy pions, observed with cross section 20-40 times larger [9] than $\bar{p}$ with corresponding energy transfers, must originate from very high momentum components of the participants, characteristic for the high density phase of the collision. This most interesting observation deserves a more detailed theoretical treatment. An other interesting observation is the distinct deviation of the observed π^-- spectra from a thermal distribution as shown in Fig. 8a and 8b for the Ni-Ni (a) and Ne-NaF (b) collision system respectively [9]. Whereas the enhanced emission at cm energies of about 300 MeV may be attributed to an enhancement of π^--emission in the Δ-resonance region, the increased π^- intensity at energies of $\sim$ 1 GeV is more difficult to understand.

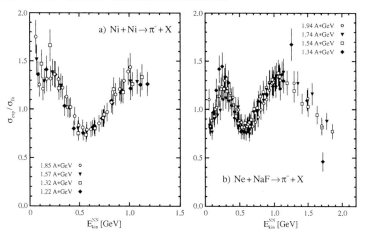

Fig. 8a,b: Ratio of the experimentally observed pion production cross sections and the exponential fits shown in fig. 6a,b as function of the cm energies E_{kin}^{NN} for the reactions Ni-Ni(a) and Ne-NaF(b) at bombarding energies indicated.

In the following we will compare some of our observation for $\bar{p}$- and K⁻-production with predictions of transport models.

As has been shown by Teis et al. [6], transport models can predict the high cross sections of $\bar{p}$-production in nucleus-nucleus collisions at bombarding energies of about 2AGeV if one takes into account multistep processes with strong contributions from NR and RR collisions, where R stands for resonances (mainly Δ and N^*) created in the collision. In addition the strong absorption of $\bar{p}$ in matter and in medium modification of the $\bar{p}$ mass have to be considered. Fig. 9 compares our results for $\bar{p}$-production in Ni+Ni collisions with RBUU predictions and various assumptions for the absorption and an effective field contribution of - 150 MeV. This effective mass reduction of $\bar{p}$ together with the absorption contribution taken from data on antiproton absorption seem to fit the data rather well. But the main

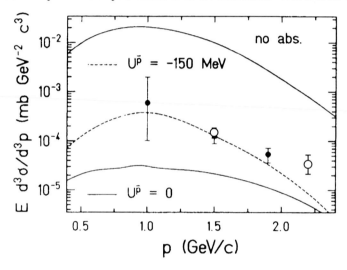

Fig. 9: Invariant $\bar{p}$ production cross section in Ni+Ni $\rightarrow$ $\bar{p}$ +X reactions at 1.85 AGeV in comparison with theoretical RBUU model prediction with effective $\bar{p}$ fields $U_{\bar{p}}$ = -150MeV and O MeV without absorption (solid line) and $U_{\bar{p}}$ = -150 MeV including absorption (dashed line) [6].

uncertainty is contained in the estimate of the absorption effects. One way to avoid this difficulty is a comparison of transport models with excitation functions. This has been done by the Gießen group [6] for our new Ni-Ni$\rightarrow \bar{p}$ + X excitation functions shown in fig. 7. Using an RBUU-model with $M < M_o$ a result in agreement with the data is obtained, whereas a BUU calculation [6] with $M=M_o$ overestimates the slope of the excitation function appreciable. Thus we do have a strong indication that the effective mass of the antiproton in dense matter is reduced. Similar conclusions have been derived by comparison with other transport models [5], [8].

The data on K⁻-production have obtained special attention, because one expects an attractive effective interaction for K⁻ in the medium, where as for K⁺ the interaction should be repulsive. The first indication for an attractive effective interaction of K⁻-mesons in medium was gained by comparison of our earlier ⁵⁸Ni-⁵⁸Ni → K⁻ + X data at 1.85 AGeV bombarding energy with predictions of a relativistic transport model by Li et al. [10] taking account the mass reduction in the framework of a chiral symmetry model, as shown in Fig. 10.

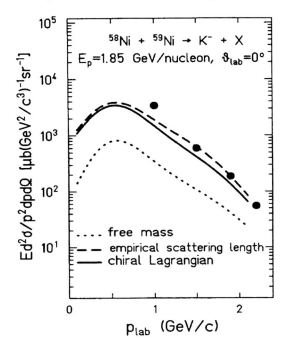

Fig. 10: Invariant K⁻-production cross-sections in Ni+Ni → K⁻+X reactions at 1.85 AGeV in comparison with a transport model prediction [10] using the $\bar{p}$ free mass, the empirical scattering length and a chiral Lagrangian respectively.

A strong evidence for the effective mass reduction of K⁻ comes from a comparison of the K⁻- and K⁺-production data in Ni-Ni collisions as shown in Fig. 11, which contains K⁺- and K⁻- data from the KAOS collaboration by Grosse et al.

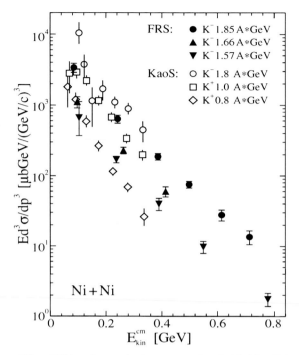

Fig. 11: Comparison of K⁻ and K⁺ invariant production cross sections in Ni+Ni collisions as function of cm kinetic energies at the bombarding energies indicated. Data marked with FRS are our data, these marked by Kaos are from Grosse et al.

[11] taken at 44° and our new K⁻-data which extends up to cm energies of 0.8 GeV. The K⁺-data are taken at equivalent energies, by scaling with the lower threshold. Note the higher K⁻ cross section, despite the lower absorption of the K⁺-mesons and its higher elementary production cross section, is again a further strong indication of changes of the effective mass in media. In fact it is predicted that at high densities, the effective mass of the K⁻ mesons becomes so low [12], that absorption by strong interaction will be forbidden energetically resulting in a trapping of K⁻ mesons in the high density phase. One should observe them at very small cm kinetic energies, after escaping at the end of the collision. The study of forward production of K⁻-mesons seems to us an excellent tool to search for this effect. Unfortunately our present data does not extend to low enough momenta to detect trapping. It is proposed to extend our experiments to study forward production of K⁻-meson to as low as possible energies.

We like to end our discussion with a presentation of all our data on particle production in Fig. 12 in the frame of a generalized scaling with the energy needed to produce a particle $E_{prod} = E_{thre}^{cm} + E_{kin}^{cm}$ and the energy available in the collision $E_{avail} = \sqrt{S_{NN}} - 2 M_N$ according to the scaling law

$$\sigma_{inv} \propto \exp\left(\frac{\left(E_{avail} - E_{prod}\right)}{E_0}\right)$$

The approximate scaling extends over 10 orders of magnitude in the invariant cross sections. But note distinct deviations from scaling. As pointed out in the presentation of the high energy π^- data, one notes that the π^--cross-sections are more than an order of magnitude larger than the scaling equivalent K^- and $\bar{p}$- cross section. Such deviations may have their origin in different absorption and/or elementary production cross sections.

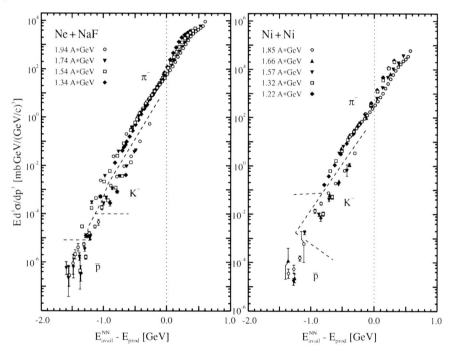

Fig. 12: Scaling behaviour of invariant particle production cross sections (π^-, K^-, $\bar{p}$) as function of the difference of the available energy and the energy required for the production in the cm systems for Ne+NaF and Ni+Ni collisions respectively.

In summary, we were able to measure forward production cross sections of $\bar{p}$, K^- and π^- in nucleus-nucleus collisions down to - 1.6 GeV in cm below the N-N production thresholds and found an approximate scaling for all particles with the difference of the available energy and the energy required for production in the cm

260

systems. The dominant contribution is due to high momentum components and resonances of the participating hadrons in the collision zone, populated in multistep collisions. The comparison of the measured cross sections with results of transport models indicates a reduction of the $\bar{p}$ and the K^- mass in the medium, with the latter being supported by a comparison of K^- and K^+ production at similar kinematic conditions. Pion production shows anomalous high yields of high energy pions pointing at very high momentum components of the participants and strong spectral deviations from thermal distribution.

One of the authors (P.K.) owes many discussions and advices on infinite number of matters in physics and management to Walter Greiner and wishes to thank him for the strong interaction.

References:

[1] H. Geissel et al., Nucl. Instrum. Methods B70, 286 (1992)
[2] A. Schröter et al., Z. Phys. A350, 101 (1994)
[3] see: Mosel U., Annual Rev. Nucl. Part. Sci 41, 29 (1991)
[4] G. Batko et al., Phys. Lett B256, 331 (1991)
[5] S.W. Huang et al., Nucl. Phys. A547, 653 (1992)
[6] S. Teis et al., Phys. Lett. B319, 47 (1993) and Phys. Rev. C50, 388 (1994)
[7] C. Spieles et al., Mod. Phys. Lett. A8, 27 (1993)
[8] G. Q. Li and C.M. Ko, Phys. Rev. C49, 1139 (1994) and Phys. Rev. C50, 1725 1994)
[9] A. Gillitzer et al., Z. f. Phys. A354, 3 (1996)
[10] G.Q. Li, C.M. Ko, X.S. Fang Phys. Lett. B329, 149 (1994)
[11] Grosse priv. comm. and R. Barth et al., GSI Scientific Rep. 1995, GSI-96-1,
[12] G.E. Brown et al., Z. Phys. A341, 301 (1992) and G.E. Brown and Manque Rho Phys. Lett. B338, 301 (1994) and priv. communication.

Pion and Kaon Production as a Probe for the Hot and Dense Nuclear Matter

H. Oeschler for the KaoS Collaboration*

Institut für Kernphysik, Technische Hochschule Darmstadt, D - 64289 Darmstadt, Germany

Using a magnetic spectrometer pions, kaons and protons were detected with laboratory angles between 40 and 48 degrees in mass-symmetric heavy ion reactions (Ne+NaF, Ni+Ni, Au+Au, Bi+Pb) at incident energies between 0.8 and 1.8 AGeV. The center-of-mass pion spectra deviate from a Boltzmann distribution. Trends of the inverse slope parameters of the high-energetic pions with system mass and incident energy are discussed. Results are presented indicating that high-energy pions are emitted at an early stage of the collision using (i) the centrality dependence of the yield and (ii) a comparison of π^+ and π^- spectra.

1 Introduction

In central collisions of heavy ions at relativistic energies the colliding nuclei are expected to be stopped leading to dense and highly excited nuclear matter in their collision zone. The investigation of particle production is a well established method to explore the properties of this hot and excited dense nuclear matter [1,2,3]. Pions are the most abundant secondary particles in the incident energy regime around 1 AGeV. They can be easily produced in individual nucleon-nucleon collisions. Pions interact strongly with nuclear matter and therefore, are expected to leave the collision zone in a late stage of the collision when the system has expanded and cooled down [4].

In this paper, emphasis is put on the properties of high-energy pion, i.e. pions with a total energy above the available energy in free nucleon-nucleon collisions. Those pions can be called "subthreshold" particles. Results are presented evidencing that high-energy pions are emitted at an early stage of the collision [5,6,7]. Before concentrating on this aspect, general trends of pion spectra are discussed.

2 Pion Spectra

The experiments had been performed with the double-focussing QD magnetic Kaon Spectrometer [8] installed at SIS/GSI. Details of the experiments and the data analysis are described in Ref. [6].

Data on pion spectra up to laboratory momenta of 1400 MeV/c had been measured in mass symmetric systems from A = 20 to A = 209 and at incident

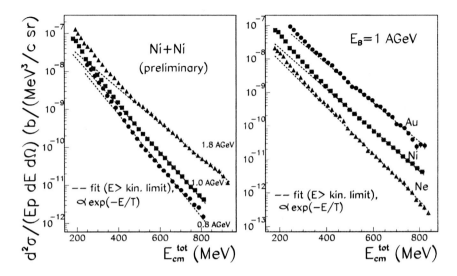

Figure 1: π^+ c.m. spectra in a Boltzmann representation for Ni+Ni (preliminary data) at 0.8, 1.0 and 1.8 AGeV incident energy (left) and for Au+Au, Ni+Ni, Ne+NaF at 1 AGeV incident energy (right).

energies from 0.8 to 1.8 AGeV. As a selection, Fig. 1 shows double differential inclusive cross sections of positively charged pions in the Boltzmann representation $1/(pE)d^2\sigma/(dEd\Omega)$ for Ni+Ni collisions at three different energies (left side) and for three different mass systems at 1 AGeV incident energy (right side). The spectra are measured at laboratory angles of 44 ± 4 degrees corresponding to a center-of-mass angular range within 90 ± 30 degrees. All spectra exhibit a concave, non-thermal shape. Straight lines fitted to the high-energy tail, i.e. to kinetic energies above the corresponding free NN kinematical limit, are shown. The variation of these inverse slope parameters with the mass of the collision system is rather weak, increasing from 64 ± 3 MeV for Ne+NaF to 74 ± 3 MeV for Au+Au for high-energy pions. However, proton spectra measured close to midrapidity show a much stronger increase of the inverse slope parameters with mass, i.e. 79 ± 5 MeV for Ne+NaF to 103 ± 5 MeV for Au+Au. The inverse slope parameters of the high-energy pions in Ni+Ni collisions (Fig. 1, right) increase rather strongly with incident energy rising from 68 ± 3 MeV at 0.8 AGeV to 95 ± 3 MeV at 1.8 AGeV. The corresponding proton spectra exhibit a weaker increase (from 90 ± 5 MeV to 103 ± 5 MeV). For details see Ref. [6].

In a recent paper, we study a scenario where all pions are originating from decaying baryonic resonances which are embedded in the flow of nuclear matter. Therefore, the pion spectra are governed by the decay kinematics of the baryonic resonances and due to their large Q-values the influence of the motion of the resonances on the shape of the pion spectra is weak. This explains the rather weak dependence with the mass of the colliding system. The thermal properties of the collision zone influence the pion spectra via the population of baryonic resonances. The population of e.g. the Δ_{33} resonance is not that of the free one but weighted with the Boltzmann factor and thus depends on temperature. Further, with increasing temperature resonances with masses higher than the Δ_{33} are increasingly populated. Hence, both the pion and the proton spectra exhibit an increase of the inverse slope parameters with increasing incident energy. These arguments are given in the frame of a thermal picture. The validity of such a concept, however, is not proven.

3 On the Origin of High-Energy Pions

In this chapter two independent arguments demonstrate that high-energy pions are emitted during an early stage of the collision. The first one is based on a comparison of the centrality dependence of high-energy pions with that of positive kaons[6]. The second argument is based on a comparison of spectra of positive and negative pions. This difference is attributed to the known influence of the different isospins and to the oppositely acting Coulomb field which is extracted from the data. From this, effective source radii are established for high kinetic pion energies. Further, it is shown that pions with lower energies are emitted from a source with a weaker Coulomb field[7].

3.1 High-energy pions and kaons

Figure 2 (left top) shows the ratio $(d\sigma_{\pi+}/d\Omega_{CM})$ / $(d\sigma_p/d\Omega_{CM})$ (labeled $\sigma_{\pi+}/\sigma_p$) as a function of the average number of participating nucleons for the heavy mass systems at 0.8 and 1 AGeV beam energy. Parameterizing $(d\sigma_{\pi+}/d\Omega_{CM})$ / $(d\sigma_p/d\Omega_{CM}) \propto A_{part}^{\alpha-1}$ (solid lines in Fig. 2), an exponent α of 1.04 ± 0.13 (1.05 ± 0.13) is obtained at 1.0 (0.8) AGeV incident energy. This result demonstrates that the number of pions at midrapidity, dominated by the low-energy part of the spectra, exhibits a linear increase with the number of participating nucleons as already reported in Refs.[5,6,9,10] using the assumption that the number of high-energy protons emitted close to midrapidity scales linearly with A_{part}. Absorption is expected to play a minor role on these trends as only the spatial distribution of the mass varies. That is the advantage of studying the A_{part} dependence of one mass system only and not comparing

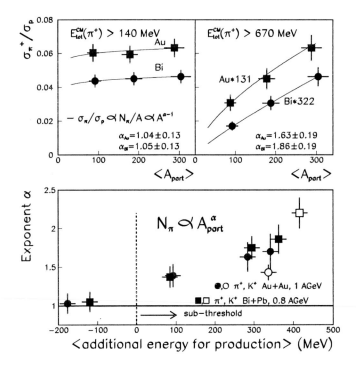

Figure 2: Upper part: The ratio π^+/p for all (left) and high-energy (right) pions from Au+Au (1 AGeV) and Bi+Pb (0.8 AGeV) as a function of the average number of participating nucleons. Lines represent a fit $\propto A^{\alpha-1}$. Lower part: Exponent α for positive pions and kaons as a function of the "average additional production energy" (see text for definition).

different mass systems.

A different trend in σ_{π^+}/σ_p (Fig. 2, right top) is observed when studying the high-energy part of the spectra alone. This ratio increases with the size of the reaction zone resulting $\alpha = 1.63\pm0.19$ ($\alpha = 1.86\pm0.19$) for 1.0 (0.8) AGeV incident energy (lines in Fig. 2). Here, only those pions are taken into account which have a total energy above 671 MeV in the center-of-mass frame. This value has been chosen to compare with the results from positive kaon production and it represents the minimum energy needed for the production of positive kaons $(m_K + (m_\Lambda - m_N))$, see below.

In the lower part of Fig. 2 the dependence of pion production on the beam energy and on the kinetic energy of the pions is combined: We study the dependence of the exponent α on the energy which is available to produce a particle in a nucleon-nucleon collision, corrected for the kinematical limit in free nucleon-nucleon collisions ($E_{kin}+E_{threshold}$-($\sqrt{s_{NN}}$-$2m_N$)). The resulting quantity which is called "additional energy needed for production", is defined to be positive for subthreshold production.

Figure 2 (bottom) evidences an exponent $\alpha \approx 1$ for particles produced above threshold and a continuous increase of the exponent α with the energy which is needed to produce a particle with a given kinetic energy. For positive kaons a non-linear dependence of the yield with the number of participating nucleons has been observed, too [11,12]. To include these results in Fig. 2 the corresponding full kaon spectra were integrated and the average kinetic energy was determined. To obtain the average energy needed for the production of these kaons, the minimum energy to produce a positive kaon of 671 MeV was added. Within the error bars no significant difference between particle species is observed in Fig. 2. The similarity in production characteristics of the positive kaons and the high-energy pions, characterized by the exponent α, can be understood by the fact that the same total energy is needed for their production. As mentioned above, a key in this representation is that absorption likely cancels and only the production properties are seen. The observed trend suggests that the more energy is needed to produce a particle the more secondary collisions during the course of the heavy ion reaction have to take place to accumulate the necessary energy. For positive kaons such a suggestion has already been made from a theoretical point of view [13,14,15]. Secondary collisions happen more frequently in central collisions of heavy nuclei during the hot and dense stage of the reaction.

Our observations indicate that both kaons as well as high-energy pions are produced and emitted at the same early stage of the reaction, whereas the majority of pions are emitted in a later and cooler stage.

3.2 Comparison of positively and negatively charged pions

In this section, we present a comparison of positively and negatively charged pions emitted in ^{197}Au+^{197}Au collisions at 1.0 AGeV incident kinetic energy. The observed difference in the π^- and π^+ spectra is attributed to the different isospins and to the oppositely acting Coulomb field. At the incident energy of 1 AGeV the π production occurs essentially via the formation of the Δ_{33} resonance. Hence, the influence of the isospin can be calculated. Further, only the one-pion decay is relevant. For details see Ref. [7].

Some data on π^-/π^+ ratios has been measured in nucleus-nucleus collisions in the last 15 years [16,17,18]. In Ref. [16] a strong peak has been found for

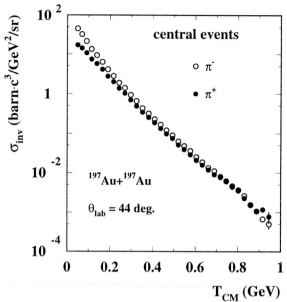

Figure 3: Invariant production cross section of negative and positive pions from the central collisions of the reaction system ^{197}Au+^{197}Au at an incident beam energy of 1.0 AGeV.

^{20}Ne+NaF at 400 AMeV in the π^-/π^+ ratio for laboratory angles around 0 degrees and was interpreted as caused by the strong Coulomb interaction with the beam spectator fragments. For our study, we have selected central collisions to reduce the influence of spectator matter.

Figure 3 shows the invariant production cross section as a function of the kinetic energy for both reaction systems at mid-rapidity and for central reactions representing 14% of the total reaction cross section.

At low kinetic energies of the pions, the π^- yield exceeds the π^+ yield and approaches it at higher energies. It can be seen clearly that both pion species exhibit different slopes with a more pronounced concave shape of the π^-.

The energy integrated π^-/π^+ ratio $R_{exp}^{tot} = (d\sigma(\pi^-)/d\Omega)/(d\sigma(\pi^+)/d\Omega)$ is determined by extrapolating the energy distribution to $T_{CM} = 0$ describing the spectra with the sum of two Maxwell-Boltzmann distributions (see also[5,6]). The experimental value of 1.94±0.05 agrees rather well with the ratios derived from an isospin decomposition using the parametrization given in Ref.[19] (1.90) and with the nearly identical values using the isospin formulas corresponding to a formation purely via the Δ_{33} resonance (1.95). This agreement motivates the assumption that the global π^--π^+ difference is given by the isobar model and that the observed energy dependence is caused by the oppositely acting

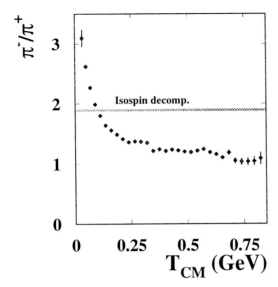

Figure 4: π^-/π^+ ratio as a function of the pion kinetic energy. The horizontal bar gives the value of the isospin decomposition (see text).

Coulomb field. In the following, we propose a method to extract the acting Coulomb force and from that freeze-out radii.

For this purpose, we show in Fig. 4 the ratios of π^- to π^+ as a function of the kinetic energy of the pions. At low pion energies, it exceeds the values from the isospin decomposition and at higher energies it drops below this value.

Following the ideas in Ref. [20], the Coulomb force disturbs the pion spectra by modifying the kinetic energies of the particles and the available phase space. As a first attempt to deduce the Coulomb energy V_{coul} we neglegt the phase-space factor. To construct the energy spectra of the pions prior to the Coulomb distortion, the slopes of π^- and π^+ are slope-averaged giving values very close to π^0 spectra obtained in the same collision systems by the TAPS collaboration [9].

In our strategy we divide the π^- spectra by 1.95, correcting thus for the influence of the isospin, and the remaining difference is then due to the Coulomb force. Now it turns out that a constant energy shift which reflects twice the Coulomb potential, cannot describe the measured results. A Coulomb-energy shift varying with pion energy is needed. At kinetic energies above 0.6 GeV a Coulomb potential of 26 MeV is required to describe the data. Yet, for the low-energy part of the pion spectrum a much smaller V_{Coul} is extracted (≈ 8 MeV).

From the extracted values of the Coulomb potential of the high-energy part we estimate the radius of the pion emitting source from the extracted Coulomb potential. For central collisions the number of participating charges Z_{part} has been measured to 110±8. In a simple assumption of an emission of pions from the surface of a charged sphere, the Coulomb potential ($V_{coul} = Z\alpha\hbar c/r_{eff}$) yields an effective radius of $r_{eff} \approx 6.3$ fm. Our procedure to determine source dimensions constitutes an alternative way to the well-known Hanbury-Brown and Twiss correlation experiments [21].

We have observed that the Coulomb field which acts on low-energy pions, is weaker than the field acting on high-energy pions. This indicates a more dilute charge distribution at freeze-out for low-energy pions. Similar conclusions were obtained from pion-correlation studies [22,23].

4 Summary

Using the Kaon Spectrometer at SIS/GSI Darmstadt pions and protons have been measured with high statistics and up to high momenta for mass-symmetric collision systems (Ne+NaF, Ni+Ni, Au+Au, Bi+Pb) at energies of 0.8, 1.0 and 1.8 AGeV. The shape of the pion spectra are interpreted as originating from decaying baryonic resonances embedded in the flow of the nulear matter.

Studying the yield of high-energy pion and kaons as a function of centrality, similar behaviour is found when comparing at the same total energy needed for their production indicating that both K^+ and high-energy pions are produced and emitted at the same early stage of the collision.

Comparing the shapes of π^- and π^+ spectra, it is demonstrated that pions with high kinetic energies are emitted from a stronger Coulomb field than low pion energies. This agrees with the idea that high-energy pions are emitted early while low-energy pions are emitted from a more dilute system.

This work was supported by the German Federal Minister for Research and Technology (BMBF), by the Gesellschaft für Schwerionenforschung (GSI) and by the Polish Committee of Scientific Research.

* The members of the **KaoS Collaboration**:

C. Müntz[a], A. Wagner, P. Baltes, H. Oeschler, S. Sartorius, C. Sturm, (*Institut für Kernphysik, Technische Hochschule Darmstadt, D - 64289 Darmstadt, Germany*),

C. Bormann, D. Brill, Y. Shin, J. Stein, H. Ströbele (*Johann Wolfgang Goethe-Universität, D-60325 Frankfurt/Main, Germany*),

W. Ahner, R. Barth, M. Cieslak, M. Debowski, E. Grosse, W. Henning, P. Koczon, M. Mang, D. Miskowiec, R. Schicker, P. Senger (*Gesellschaft für Schwerionenforschung, D-64220 Darmstadt, Germany*),

B. Kohlmeyer, H. Pöppl, F. Pühlhofer, J. Speer, K. Völkel (*Philipps-Universität, D-35037 Marburg, Germany*),

W. Walus (*Jagiellonian University, PL-30-059 Kraków, Poland*)

References

1. S. Nagamiya et al., *Phys. Rev.* C **24**, 1981 (971).
2. R. Brockmann et al., *Phys. Rev. Lett* **53**, 1984 (2012).
3. R. Stock, *Phys. Rep.* **135**, 1986 (259).
4. S. Nagamiya, *Phys. Rev. Lett* **49**, 1982 (1383).
5. C. Müntz et al., *Z. Phys.* A **352**, 1995 (175).
6. C. Müntz et al., submitted to *Z. Phys. A*.
7. A. Wagner et al., to be published.
8. P. Senger et al., *Nucl. Instrum. Methods* A **327**, 1993 (393).
9. O. Schwalb et al., *Phys. Lett.* B **321**, 1994 (20).
10. J.W. Harris et al., *Phys. Rev. Lett* **58**, 1987 (463).
11. D. Miśkowiec et al., *Phys. Rev. Lett* **72**, 1994 (3650).
12. M. Cieślak , PhD Thesis (1995), Jagiellonian University Cracow, Poland; and to be published.
13. J. Aichelin et al., *Phys. Rev. Lett* **58**, 1987 (1926).
14. W. Cassing et al., *Phys. Rep.* **188**, 1990 (363).
15. C. Hartnack et al., *Nucl. Phys.* A **580**, 1994 (643).
16. W. Benenson et al., *Phys. Rev. Lett* **43**, 1979 (683).
17. J. Miller et al., *Phys. Rev. Lett* **58**, ((1)987)2408.
18. J.W. Harris et al., *Phys. Rev.* C **41**, 1990 (147).
19. B. J. Ver West, R. A. Arndt, *Phys. Rev.* C **25**, 1982 (1979).
20. M. Gyulassy and S.B. Kauffmann, *Nucl. Phys.* A **362**, 1981 (503).
21. D. l'Hôte, *Nucl. Phys.* A **545**, 1992 (381c); and references therein.
22. D. Beavis et al., *Phys. Rev.* C **43**, 1986 (757).
23. R. Bock et al., *Z. Phys.* A **333**, 1989 (193).

---

[a]now at Brookhaven National Laboratory, Upton, NY

CHARGED PION PRODUCTION IN HEAVY ION COLLISIONS AT *SIS* ENERGIES

D. PELTE

Physikalisches Institut der Universität Heidelberg, Philosophenweg 12,
D-69120 Heidelberg, Germany

THE *FOPI* COLLABORATION
Gesellschaft für Schwerionenforschung mbH, Postfach110552,
D-64220 Darmstadt, Germany

Charged pion production is studied at different energies in two sytems, $Ni + Ni$ and $Au + Au$, using the $FOPI$ detector at GSI. The pion production probability is found to be reduced for the heavier system. The rescattering of pions from spectator matter is observed under all conditions, it is responsible for the non-thermal shape of the rapidity spectra. The excited Baryon resonances are shown to be a major source of the observed pions, higher resonances than the $\Delta(1232)$ are responsible for pions with large transverse momenta.

1 Introduction

The production of charged pions $\pi^{\pm}$ was studied at bombarding energies ranging from 1 to $2 AGeV$ with the 4π detector $FOPI$ at the SIS accelerator of the GSI/Darmstadt. The following symmetric projectile - target systems were measured:

$$Au\,1: \quad {}^{197}Au +{}^{197}Au \text{ at } E = 1.06 AGeV,$$
$$Ni\,1: \quad {}^{58}Ni + {}^{58}Ni \text{ at } E = 1.06 AGeV,$$
$$Ni\,2: \quad {}^{58}Ni + {}^{58}Ni \text{ at } E = 1.45 AGeV,$$
$$Ni\,3: \quad {}^{58}Ni + {}^{58}Ni \text{ at } E = 1.93 AGeV.$$

At the time of these experiments the $FOPI$ detector was not yet completed. It consisted of the forward detectors PLA and ZER, which covered the lab. region from $1.2^0 < \vartheta < 30^0$ with full azimuthal coverage, and of the central drift chamber CDC at angles $30^0 < \vartheta < 150^0$, which was installed inside a solenoidal magnet field of $0.6T$ strength. Also the CDC provided full azimuthal coverage. Charged pions could only be identified in the CDC since the PLA and ZER detectors only allow nuclear charge identification.

Because of the CDC boundary at $\vartheta = 30^0$ only 30% of the pion phase space in the forward hemisphere of the cm frame was measured. To extent the coverage of the total phase space to approximately 90% the symmetry relation $f(\Theta, \Phi) = f(\pi - \Theta, \pi + \Phi)$ was employed which holds when target and projectile are identical.

To present the data we use a rapidity normalized to the cm rapidity
$$Y^{(0)} = (y - y_{cm})/y_{cm},$$
and a normalized transverse momentum
$$p_t^{(0)} = p_t/A \cdot (P_{proj}/A_{proj})^{-1}.$$
In general quantities in the cm frame will be presented by capital letters.

In this contribution the forward detector PLA is only used for trigger purposes, i.e. it defined central, intermediate and peripheral collisions by means of the selected particle multiplicity n_{PLA}. Minimum bias collisions correspond to no trigger selection, i.e. all multiplicities are accepted.

In the following sections we present results on

- the pion multiplicity n_π as function of the participants A_{part},

- the pion phase space distributions,

- the identification of Baryon$(S = 0)$ resonances in the $(p, \pi^\pm)$ channel.

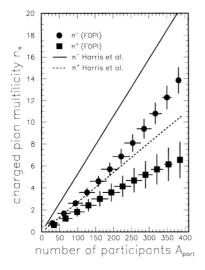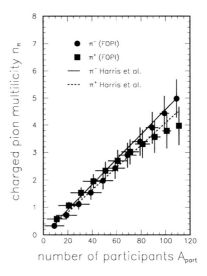

Figure 1: Pion multiplicities as functions of A_{part} for the reactions $Au\,1$(left) and $Ni\,1$(right).

2   Charged pion multiplicities

The number of participants A_{part} is not directly measurable but has to be deduced from other observables x. For x one may choose any quantity which is a monotonous function of the impact parameter b, like the forward particle multiplicity n_{PLA}, or the total particle multiplicity n_{TOT}, or the baryonic charge Z_{CDC}^{bar} measured with the CDC. The method to relate x with b and then A_{part} is to compare the measured cross section $\sigma(x)$ with the calculated cross section $\sigma_{geo}(b) = 2\pi \int_b^{b_{max}} \rho^2(r/2)r\,dr/\rho_0^2$, and then to obtain A_{part} from the geometrical and b dependent overlap of 2 identical nuclei with Saxon - Woods density profile $\rho(r) = \rho_0(1 + exp\{(r - r_0)/a\})^{-1}$. In accordance with electron scattering we chose for the nuclear radius $r_0 = 1.2 \cdot A^{1/3}$, and for the surface parameter $a = 1.0 \pm 0.6 fm$. The error in a was used to obtain an estimate for the uncertainty in determining A_{part}. The $\pi^{\pm}$ multiplicities are shown in Fig.1 as functions of A_{part} for the $Au\,1$(left) and $Ni\,1$(right) reactions. The π^- multiplicity as function of A_{part} was studied previously by Harris et al. [1] in the light systems $Ar + Ca$ and $La + La$. These results, when extrapolated to the $Au\,1$ and $Ni\,1$ reactions, are shown as straight lines in Fig.1. Two conclusions are obvious: The number of charged pions n_π increases with A_{part}, and the data are in close agreement with the Harris extrapolation for the $Ni\,1$, and also the $Ni\,2$, $Ni\,3$ reactions, but they are smaller than predicted by the Harris extrapolation in case of the $Au\,1$ reaction. The observed reduction of the pion multiplicities in this reaction was also observed by the $KaoS$ and $TAPS$ collaborations [2]. To obtain the total pion yield per A_{part} the nonlinearities with

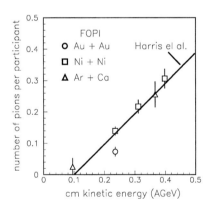

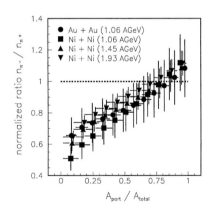

Figure 2: Pion production probability for different systems as function of energy.

Figure 3: π^- to π^+ ratio for different systems as function of A_{part}.

A_{part} have to be taken into account. From $n_\pi = a_1 \cdot A_{part} + a_2 \cdot A_{part}^2 + a_3 \cdot A_{part}^3$ it follows that $\frac{<n_\pi>}{<A_{part}>} = a_1 + \frac{a_2}{2} \cdot A_0 + \frac{a_3}{3} \cdot A_0^2$, where A_0 is the total number of nucleons in the system. The results, obtained with the $FOPI$ detector, are plotted in Fig.2, the π^0 contribution was included using the isobar model[3].

The nonlinearities with A_{part} also influence the π^- to π^+ ratio R_π. If these ratios are normalized to the values expected from the isobar model, i.e. $R_\pi = 1.95$ for $Au + Au$ and $R_\pi = 1.12$ for $Ni + Ni$, the increase of the normalized ratios with increasing A_{part}/A_0 is similar for all reactions studied, as shown in Fig.3.

3 The Pion Phase Space Distributions

The phase space distributions of charged pions under minimum bias condition are shown in Fig.4 for the $Ni\,1$ reaction. Increasing grey shades correspond to the increase of the invariant cross section $\frac{1}{p_t} \cdot \frac{d^2\sigma}{dp_t\,dy}$ on a logarithmic scale. A close inspection reveals that the phase space distributions are non-thermal.

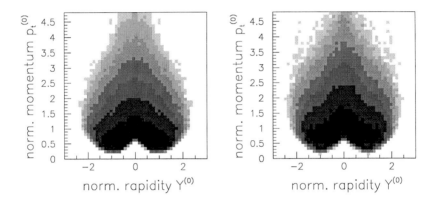

Figure 4: Phase space distributions of negative(left) and positive(right) pions.

We have chosen to demonstrate this behaviour by projecting from these distributions into the pion kinetic energies for different cm angles Θ, and into the pion rapidities with $p_t^{(0)} > 1$ in case of $Au + Au$ and $p_t^{(0)} > 0.8$ in case of $Ni + Ni$ collisions.

3.1 The Kinetic Energies

In all reactions studied the shape of the kinetic energy spectra $\frac{1}{p_t E} \cdot \frac{d^2\sigma}{dE\,d\Omega}$ is concave, suggesting that the assumption of 2 temperatures, $T_{l,\pi}$ and $T_{h,\pi}$ has to be made if these spectra are to be fitted by exponentials. The deduced temperatures vary for angles $|90^0 - \Theta| > 45^0$ whereas for angles around $\Theta = 90^0$

Table 1: Temperatures T and yields $R_T = \frac{I(T_l)}{I(T_h)}$ from fits to the pion energy spectra

	T_{l,π^-}	T_{l,π^+}	$T_{l,\pi}$	$R_T(\pi^-)$	$R_T(\pi^+)$
$Au\,1$	42.2 ± 2.7	49.4 ± 2.3	96.4 ± 5.1	0.74 ± 0.28	0.75 ± 0.43
$Ni\,1$	47.8 ± 1.3	52.5 ± 4.3	93.1 ± 6.6	1.54 ± 0.57	1.40 ± 0.98
$Ni\,2$	51.7 ± 2.4	56.4 ± 2.7	99.5 ± 4.0	1.17 ± 0.63	1.53 ± 0.67
$Ni\,3$	49.0 ± 1.0	56.6 ± 2.5	101.6 ± 2.0	0.52 ± 0.18	0.53 ± 0.14

the temperatures remain reasonably constant. The average temperatures in this angular range are listed in Table 1 together with their contribution to the complete $\frac{d\sigma}{dE}$ spectrum. It is evident that the temperatures are almost independent of system mass or energy, whereas the relative contribution of the high-temperature component increases with system mass and energy.

3.2   The Rapidity Distributions

The projected rapidity distributions $\frac{d\sigma}{dY^{(0)}}$ of the $Au\,1$ reaction are shown in Fig.5 under minimum bias condition. The dotted curve displays the expected

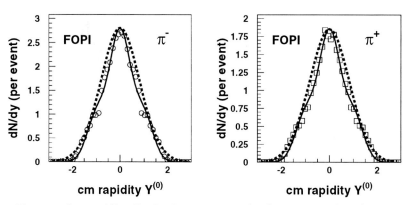

Figure 5: $Au\,1$ rapidity distributions of negative(left) and positive(right) pions.

distribution if the pions were emitted from thermal sources at midrapidity, which have 2 temperatures $T_{l,\pi}$, $T_{h,\pi}$. The data deviate from these expectations and suggest that pion emission is enhanced at rapidities $|Y^{(0)}| > 1$. This finding is corroborated by the pion angular distributions which are enhanced at forward and backward angles in the cm frame. The $Ni + Ni$ reactions yield similar results. The rapidity spectra were fitted by 3 Gaussians centered at

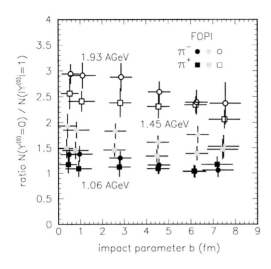

Figure 6: Impcat parameter dependence of the ratio R_Y for $Ni + Ni$ reactions

$Y^{(0)} = 0$ and $|Y^{(0)}| = 1$, the latter representing the contributions of pions rescattered from target and projectile spectators. The relative contribution R_Y of pions from the participant region to pions from the spectator regions is displayed in Fig.6. For the $Ni + Ni$ reactions R_Y does not depend strongly on the impact parameter b, but it increases with energy. This is mainly due to an increase of the width of $\frac{d\sigma}{dY^{(0)}}|_{y=0}$, whereas the rapidity widths of pions from the spectators remain constant.

The Identification of Baryon(S=0) Resonances

Pions are assumed to originate from Baryon resonances which were excited by the nucleon-nucleon collisions in the participant. The $\Delta(1232)$ resonance dominates this process:
$$N + N \rightleftharpoons N + \Delta \qquad \Delta \rightleftharpoons N + \pi$$
The observed independence of the pion temperature $T_{l,\pi}$ from system mass and energy supports this hypothesis. The concave shapes of the pion kinetic energy spectra, which is equivalent to the existence of a second higher temperature $T_{h,\pi}$, is interpreted as due to the emission of 'direct' pions or the contribution from higher Baryon resonances.

The transient existence of Baryon resonances in the participant ought to be seen in the invariant mass spectra of correlated $(p, \pi^{\pm})$ pairs. Because of the large amount of uncorrelated protons, the probablity to find a correlated pair amounts to $r = \frac{1}{n_p}\alpha\epsilon_p\epsilon_f \approx 0.01$, where n_p is the proton multiplicity, α accounts for isospin conservation, ϵ_p is the detector acceptance for protons, and ϵ_f takes care of the rescattering of protons by nuclear matter. The background of uncorrelated pairs S_b, which amounts to approximately 99% of all pairs S_t, is reconstructed using the technique of event mixing.

The invariant mass spectrum of the Baryon resonances is obtained from
$$S = S_t - (1 - r) \cdot S_b,$$
with ϵ_f as a free parameter. All p, π correlations other than those due to the resonance decay, are assumed to be of identical strength in S_t and S_b, in order to make this procedure applicable. For this purpose the azimuthal angle is measured with respect to the reaction plane, and only those events are mixed which have the identical n_p values. The effects caused by the pion rescattering in spectator matter are not well understood.

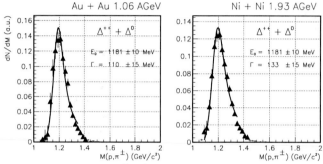

Figure 7: Invariant mass specrum of (p, π) pairs with pion transverse momenta $100 < p_t < 200 \, MeV/c$ for $Au\,1$(left) and $Ni\,3$(right) reactions.

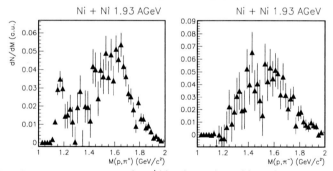

Figure 8: Invariant mass spectrum of (p, π^+)(left) and (p, π^-)(right) pairs with pion transverse momenta $400 < p_t < 600 \, MeV/c$ for $Ni\,3$ reactions.

The background corrected invariant mass spectrum of $(p, \pi^{\pm})$ pairs depends on the pion transverse momentum p_t. It is shown for the $Au\,1$ and $Ni\,3$ reactions in Fig.7 where $100 < p_t < 200 MeV/c$. The curves show the fitted Cugnon [4] parametrization of the Δ resonance, which yields a value for the Δ mass which is lower than accepted for the free Δ resonance. Notice that (p, π^-) and (p, π^+) pairs were combined in Fig.7. The Fig.8 displays the invariant mass spectra separately for (p, π^-) and (p, π^+) pairs from the $Ni\,3$ reaction where now $400 < p_t < 600 MeV/c$. The spectra are shifted to larger mass values and they are different for the two types of pairs. This difference is expected when higher Baryon resonances become involved, since the (p, π^+) channel can only be populated by the decay of the Δ resonances, whereas the (p, π^-) channel can also be populated by the decay of N^* resonances.

Conclusion

The main conclusions from our investigation of the $\pi^\pm$ production at 1 to 2 $AGeV$ energy are:
The number of pions n_π per participants A_{part} is for the system $Ni + Ni$ in agreement with the earlier results from Harris et al.[1], which were obtained with systems of total mass $A_0 < 200$. For the $Au + Au$ system with $A_0 \approx 400$ the $\frac{<n_\pi>}{<A_{part}>}$ value is reduced by a factor 0.65 when compared to the results of Harris et al. The pions from the $Ni + Ni$ and $Au + Au$ systems are rescattered by spectator matter, which causes the pion rapidity distributions to become non-thermal. The concave shape of the pion kinetic energy spectra near midrapidity is generated by the decay of Baryon resonances. The contribution of resonances with masses larger than $\Delta(1232)$ increases with the pion transverse momentum, and also with the system energy. Equivalently high-energy pions become more abundant without a considerable change of the apparent pion temperature.

Acknowledgments

This work was supported by the Bundesministerium für Bildung und Forschung under contract no. 06 HD 525I (3), and by the Gesellschaft für Schwerionenforschung mbH under contract no. HDPEK

References

[1] J.W. Harris et al., *Phys. Rev. Lett.* **58**, 463 (1987)
[2] P. Senger in *Multiparticle Correlations and Nuclear Reactions*, ed. J. Aichelin and D. Ardouin (World Scientific, 1988).
[3] R. Stock, *Phys. Rep.* **135**, 259 (1986).
[4] J. Cugnon and M. Lemaire, *Nucl. Phys.* A **489**, 781 (1988).

HYPERON-RICH MATTER

Jürgen Schaffner

The Niels Bohr Institute, Blegdamsvej 17, DK-2100 Copenhagen

The phase diagram of nuclear matter offers fascinating features for heavy ion physics and astrophysics when extended to strangeness and antimatter. We will discuss some regions of this phase diagram as strange matter at zero temperature, hypermatter at high densities and the antiworld at high densities and strangeness fraction.

1 Strange Matter and Hypermatter

Relativistic heavy ion collisions provide a promising tool for studying the physics of strange quark and strange hadronic matter (see recent review [1]). Fig. 1 shows schematically the phase diagram of hot, dense and strange matter.

Perhaps the only unambiguous way to detect the transient existence of a quark gluon plasma (QGP) might be the experimental observation of exotic remnants, like the formation of strange quark matter (SQM) droplets. First studies in the context of the MIT-bag model predicted that sufficiently heavy strangelets might be metastable or even absolutely stable. The reason for the possible stability of SQM is related to a third flavour degree of freedom, the strangeness. As the mass of the strange quark is smaller than the Fermi energy of the quarks, the total energy of the system is lowered by adding strange quarks. According to this picture, the number of strange quarks is nearly equal to the number of massless up or down quarks and saturated SQM is nearly charge neutral. This simple picture does not hold for small baryon numbers. Finite size effects shift unavoidably the mass of strangelets to the metastable regime. Moreover strangelets can have very high charge to mass ratios for low baryon numbers. This behaviour is well known from normal nuclei. Therefore, instead of long-lived nearly neutral objects, strangelet searches in heavy ion experiments have to cope with short-lived highly charged objects [2] !

On the other hand, metastable exotic multihypernuclear objects (MEMOs) consisting of nucleons and hyperons have been proposed [3] which extend the periodic system of elements into a new dimension. MEMOs have remarkably different properties as compared with known nuclear matter as e.g. being negatively charged while carrying a positive baryon number! Even purely hyperonic matter has been predicted [4]. These rare composites would have a very short lifetime, at the order of the lifetime of the Λ. Central relativistic heavy ion collisions provide a prolific source of hyperons and hence, possibly, the way

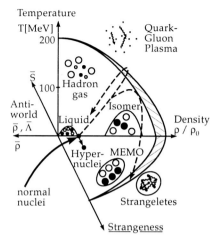

Figure 1: Sketch of the nuclear matter phase diagram with its extensions to strangeness and to the antiworld.

for producing MEMOs. Again, heavy ion experiments looking for these exotic composites have to deal with very short-lived highly charged objects!

2   Hyperon-rich Matter in Neutron Stars

Strangeness, in form of hyperons, appears in neutron star matter at a moderate density of about $2 - 3$ times normal nuclear matter density $\rho_0 = 0.15$ fm^{-3} as shown by Glendenning within the Relativistic Mean Field (RMF) model [5]. These new species have considerable influences on the equation of state and the global properties of neutron stars.

On the other side, much attention has been paid in recent years to the possible onset of kaon condensation as the other hadronic form of strangeness in neutron stars. Most recent calculations based on chiral perturbation theory [6] show that kaon condensation may set in at densities of $(3-4)\rho_0$. Nevertheless, these calculations do not take into account the presence of hyperons which may already occupy a large fraction of matter when the kaons possibly start to condense [7].

Below I present new results from our recent paper [8], where the properties of neutron matter with hyperons were studied in detail. We use the extended version of the relativistic mean field (RMF) model and constrain our parameters to the available hypernuclear data and to the kaon nucleon scattering lengths.

2.1 The Model with Hyperons

The implementation of hyperons within the RMF approach is straightforward. SU(6)-symmetry is used for the vector coupling constants and the scalar coupling constants are fixed to the potential depth of the corresponding hyperon in normal nuclear matter [4]. We choose

$$U_\Lambda^{(N)} = U_\Sigma^{(N)} = -30 \text{ MeV} \quad , \qquad U_\Xi^{(N)} = -28 \text{ MeV} \quad . \tag{1}$$

Note that a recent analysis [9] comes to the conclusion that the potential changes sign in the nuclear interior, i.e. being repulsive instead of attractive. In this case, Σ hyperons will not appear at all in our calculations.

The observed strongly attractive $\Lambda\Lambda$ interaction is introduced by two additional meson fields, the scalar meson $f_0(975)$ and the vector meson $\phi(1020)$. The vector coupling constants to the ϕ-field are given by SU(6)-symmetry and the scalar coupling constants to the σ^*-field are fixed by

$$U_\Xi^{(\Xi)} \approx U_\Lambda^{(\Xi)} \approx 2U_\Xi^{(\Lambda)} \approx 2U_\Lambda^{(\Lambda)} \approx -40 \text{ MeV} \quad . \tag{2}$$

Note that the nucleons are not coupled to these new fields.

2.2 Neutron Stars with Hyperons

Fig. 2 shows the composition of neutron star matter for the parameter set TM1 with hyperons including the hyperon-hyperon interactions.

Up to the maximum density considered here all effective masses remain positive and no instability occurs. The proton fraction has a plateau at $(2-4)\rho_0$ and exceeds 11% which allows for the direct URCA process and a rapid cooling of a neutron star. Hyperons, first Λ's and Σ^-'s, appear at $2\rho_0$, then Ξ^-'s are populated already at $3\rho_0$. The number of electrons and muons has a maximum here and decreases at higher densities, i.e. the electrochemical potential decreases at high densities. The fractions of all baryons show a tendency towards saturation, they asymptotically reach similar values corresponding to spin-isospin and hypercharge-saturated matter. Hence, a neutron star is more likely a giant hypernucleus!

2.3 Kaon Condensation ?

In the following we adopt the meson-exchange picture for the KN-interaction simply because we use it also for parametrizing the baryon interactions. We start from the following Lagrangian

$$\mathcal{L}_K' = D_\mu^* \bar{K} D^\mu K - m_K^2 \bar{K} K - g_{\sigma K} m_K \bar{K} K \sigma - g_{\sigma^* K} m_K \bar{K} K \sigma^* \tag{3}$$

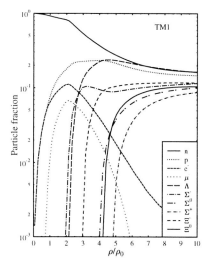

Figure 2: The composition of neutron star matter with hyperons which appear abundantly in the dense interior.

with the covariant derivative

$$D_\mu = \partial_\mu + i g_{\omega K} V_\mu + i g_{\rho K} \vec{\tau} \vec{R}_\mu + i g_{\phi K} \phi_\mu \quad . \tag{4}$$

The coupling constants to the vector mesons are chosen from SU(3)-relations. The scalar coupling constants are fixed by the s-wave KN-scattering lengths. We have found that this leads to an $\bar{K}$-optical potential around $U^{\bar{K}}_{\text{opt}} = -(130 \div 150)$ MeV at normal nuclear density for the various parameter sets used. This is between the two families of solutions found for Kaonic atoms [10]. The onset of s-wave kaon condensation is now determined by the condition $-\mu_e = \mu_{K^-} \equiv \omega_{K^-}(k = 0)$.

The density dependence of the K and $\bar{K}$ effective energies is displayed in Fig. 3. The energy of the kaon is first increasing in accordance with the low density theorem. The energy of the antikaon is decreasing steadily at low densities. With the appearance of hyperons the situation changes dramatically. The potential induced by the ϕ-field cancels the contribution coming from the ω-meson. Hence, at a certain density the energies of the kaons and antikaons become equal to the kaon (antikaon) effective mass, i.e. the curves for kaons and antikaons are crossing at a sufficiently high density. At higher densities the energy of the kaon gets even lower than that of the antikaon! Since the electrochemical potential never reaches values above 160 MeV here

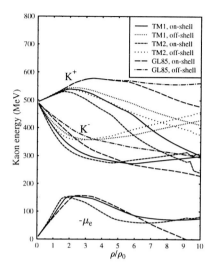

Figure 3: The effective energy of the kaon and the antikaon. and the electrochemical potential. Kaon condensation does not occur over the whole density region considered.

antikaon condensation does not occur at all. We have checked the possibility of antikaon condensation for all parameter sets and found that at least 100 MeV are missing for the onset of kaon condensation in contrast to previous calculations disregarding hyperons [6].

3 Strange Antiworld

There are evidences for strong scalar and vector potentials in nuclear matter. Already in 1956 Dürr and Teller proposed a relativistic model with strong scalar and vector potentials to explain the saturation of nuclear forces [11] and found a scalar potential of $U_s = a m_N \phi$ where ϕ is a scalar field and a is a coupling constant. This was the first version of the RMF model discussed above, where $U_s = g_\sigma \sigma$, and its extension, the Relativistic Brückner-Hartree-Fock (RBHF) [12] calculations, where the scalar potential is the scalar part of the self-energy of the nucleon $U_s = \Sigma_s(p_N)$. In the chiral $\sigma\omega$ model [13] one finds $U_s = m_N \sigma / f_\pi - m_N$ which incorporates the (approximate) chiral symmetry at the underlying QCD Lagrangian. Here f_π is the pion decay constant. Besides these models based on a hadronic description there exists effective models dealing with constituent quarks, like the Nambu–Jona-Lasinio (NJL) model and models based on QCD sum rules [14]. These models can be linked to hadronic observables in the dense medium by using the low density expansion of the quark condensate which gives $U_s = -\frac{m_N \sigma_N}{m_\pi^2 f_\pi^2} \rho_N$, where $\sigma_N \approx 45$ MeV is the pion-nucleon sigma term.

Astonishingly, *all* these approaches come to the same conclusion, namely that the scalar potential is as big as

$$U_s = -(350 \div 400) \text{ MeV } \rho_N/\rho_0 \tag{5}$$

for moderate densities! This strong scalar attraction has to be compensated by a strong repulsion to get the total potential depth of nucleons correct. Hence, one finds for the vector potential

$$U_v = (300 \div 350) \text{ MeV } \rho_N/\rho_0 \quad . \tag{6}$$

These big potentials are in fact needed to get a correct spin-orbit potential. The idea of Dürr and Teller [11] was that the antinucleons feel the difference of these two potentials, i.e.

$$U_{\bar{N}} = U_s - U_v = -(650 \div 750) \text{ MeV } \rho_N/\rho_0 \tag{7}$$

which is already comparable to the mass of the nucleon. Note that the extrapolation to high densities is quite dangerous as effects nonlinear in density might get important. It is already known from RMF models that the scalar potential saturates at high densities instead of growing steadily. RBHF calculations show that this might be also true for the vector potential. With this in mind one can extrapolate to higher densities and finds that the field potentials get overcritical at $\rho_c = (3 - 7)\rho_0$ which was first pointed out by Mishustin [15]. At this critical density the potential felt by the antinucleons is equal to $U_{\bar{N}} = 2m_N$, the negative energy states are diving in the positive continuum and this allows for the spontaneous nucleon-antinucleon pair production. This has certain parallels to the spontaneous e^+e^- production proposed by Pieper and Greiner [16]. Assuming SU(6)-symmetry one gets for Λ's

$$U_v^\Lambda = (200 \div 230) \text{ MeV } \rho_N/\rho_0 \tag{8}$$

and combining with hypernuclear data this gives then for the total $\bar{\Lambda}$ potential

$$U_{\bar{\Lambda}} = U_s^\Lambda - U_v^\Lambda = U_\Lambda - 2U_v^\Lambda = -(430 \div 500) \text{ MeV } \rho_N/\rho_0 \quad . \tag{9}$$

In the hyperon-rich medium additional fields will enhance this potential. Assuming again SU(6)-symmetry one can estimate the vector potential coming from the ϕ meson

$$V_v^\Lambda = \frac{2}{9} \frac{m_\omega^2}{m_\phi^2} U_v \cdot f_s \approx 40 \text{ MeV } \rho_B/\rho_0 \cdot f_s \tag{10}$$

where f_s is the total strangeness fraction. The corresponding strange scalar potential is in principle unknown but definitely higher than the strange vector potential to explain the strongly attractive $\Lambda\Lambda$ interaction seen in double Λ hypernuclei. Hence one gets at least an additional $\bar{\Lambda}$ potential of

$$V_{\bar{\Lambda}} = V_s^{\Lambda} - V_v^{\Lambda} \approx -120 \text{ MeV } \rho_B/\rho_0 \cdot f_s \tag{11}$$

in the hyperon-rich medium.

These strong antibaryon potentials will have certain impacts for heavy ion reactions. Proposed signals for antiprotons are: enhanced subthreshold production[15], change of the slope of the excitation function[15], apparent higher temperatures[17], which have indeed been measured at GSI[18]. Nevertheless, a recent analysis indicates that the antiproton potential might be quite shallow at normal nuclear density, around $U_{\bar{p}} = -100$ MeV[19]. Possible other signals include: enhanced antihyperon production[15], strong antiflow of antibaryons[20], cold baryons from tunnelling[15], cold kaons from annihilation in the medium (the phase space of the reaction $\bar{\Lambda}+p \to K^+ + \pi's$ is reduced by $U_{\bar{\Lambda}}+U_N-U_K \approx -600$ MeV ρ_N/ρ_0 compared to the vacuum), enhanced pion production due to the abundant annihilation processes which would also enhance the entropy. Definitely, more elaborate work is needed to pin down the possible signals from the critical phenomenon of the antiworld.

We conclude this section with a brief comment concerning the limitations of the RMF model. This is clearly an effective model which successfully describes nuclear phenomenology in the vicinity of the ground state. On the other hand, this model does not respect chiral symmetry and the quark structure of baryons and mesons. Also negative energy states of baryons and quantum fluctuations of meson fields are disregarded. These deficiencies may affect significantly the extrapolations to high temperatures, densities or strangeness contents.

Acknowledgements

This paper is dedicated to Prof. Walter Greiner on the occasion of his 60th birthday. I am indebted to him for guiding me to the fascinating field of hypermatter and antimatter and his continuous support. I thank my friends and colleagues A. Diener, C.B. Dover, A. Gal, Carsten Greiner, and especially I.N. Mishustin and H. Stöcker for their help and collaboration which made this work possible.

References

1. C. Greiner and J. Schaffner, in *Quark-Gluon Plasma 2*, Ed. R.C. Hwa (World Scientific, Singapore, 1995), p. 635
2. C. Greiner, A. Diener, J. Schaffner, H. Stöcker, Nucl. Phys. **A566**, 157 (1994)
3. J. Schaffner, C. Greiner, H. Stöcker, Phys. Rev. **C46**, 322 (1992)
4. J. Schaffner, C.B. Dover, A. Gal, C. Greiner, H. Stöcker, Phys. Rev. Lett. **71**, 1328 (1993) and Ann. of Phys. (N.Y.) **235**, 35 (1994)
5. N.K. Glendenning, Astrophys. J. **293**, 470 (1985)
6. G.E. Brown, C.-H. Lee, M. Rho, V. Thorsson, Nucl. Phys. **A567**, 937 (1994)
7. J. Schaffner, A. Gal, I.N. Mishustin, H. Stöcker, W. Greiner, Phys. Lett. **B334**, 268 (1994)
8. J. Schaffner and I.N. Mishustin, Phys. Rev. **C53**, 1416 (1996)
9. J. Mares, E. Friedman, A. Gal, B.K. Jennings, Nucl. Phys. **A594**, 311 (1995)
10. E. Friedman, A. Gal, C.J. Batty, Phys. Lett. **B308**, 6 (1993); Nucl. Phys. **A579**, 518 (1994)
11. H.P. Dürr and E. Teller, Phys. Rev. **101**, 494 (1956)
12. L.S. Celenza, A. Pantziris, C.M. Shakin, W.D. Sun, Phys. Rev. **C45**, 2015 (1992)
13. J. Boguta, Phys. Lett. **120B** 34 (1983)
14. T.D. Cohen, R.J. Furnstahl, D.K. Griegel, Phys. Rev. Lett. **67**, 961 (1991)
15. I.N. Mishustin, Yad. Fiz. (Sov. J. Nucl. Phys.) **52**, 1135 (1990), J. Schaffner, I.N. Mishustin, L.M. Satarov, H. Stöcker, W. Greiner, Z. Phys. **A341**, 47 (1991), I.N. Mishustin, L.M. Satarov, J. Schaffner, H. Stöcker, W. Greiner, J. Phys. **G19**, 1303 (1993)
16. W. Pieper and W. Greiner, Z. Phys. **A218** 327 (1969)
17. V. Koch, G.E. Brown, C.M. Ko, Phys. Lett. **B265**, 29 (1991)
18. A. Schröter, E. Berdermann, H. Geissel, P. Kienle, W. König, A. Gillitzer, J. Homolka, F. Schumacher, H. Ströher, B. Povh, Nucl. Phys. **A553** (1993) 775c
19. St. Teis, W. Cassing, T. Maruyama, U. Mosel, Phys. Rev. **C50**, 388 (1994)
20. A. Jahns, C. Spieles, H. Sorge, H. Stöcker, W. Greiner, Phys. Rev. Lett. **72**, 3464 (1994)

DILEPTON PRODUCTION
FROM HADRONIC PROCESSES

U. MOSEL

Institut für Theoretische Physik,Universität Giessen
D-35392 Giessen, Germany

Abstract

I discuss elementary processes for production of dileptons from hadronic interactions and emphasize that a quantitative understanding of
heavy-ion dilepton data at relativistic and ultrarelativistic energies requires an accurate understanding of the elementary processes. Taking
the bremsstrahlungs- and Δ-decay contributions as an example I emphasize the dependence of the expected cross-sections on the (unknown)
electromagnetic formfactor of the nucleon in the time-like regime and
point out that proton- and photon-induced reactions can provide important information on the validity of vector meson dominance for the
nucleon.

1 Introduction

Since a few years data on dilepton production from relativistic heavy-ion collisions exist, taken at the BEVALAC in Berkeley [1]. These data have been
analyzed by using a semiclassical transport theory; in Fig. 1 I show the results
of a very recent calculation that explains the published data quite well. The
transport theory used here incorporates the lowest-lying mesons and nucleon
resonances; it contains not only the primary interactions, but also all possible
secondary collisions. For example, at 1 GeV/A the strong Δ contribution in
the $^{40}Ca + ^{40}Ca$ data is primarily due to secondary ΔN collisions and the
dominant shoulder in the vector meson mass region is to a large part caused
by $\pi^+\pi^-$ annihilation. Note that a major part of the yield were missing if
these secondary processes were not taken into account.

Data taken very recently in the ultrarelativistic regime [2] exhibit exactly
the same sources in the low invariant mass regime shown here; they are usually
analyzed in terms of a first collision only ("cocktail plot"). Very recent transport theoretical calculations for this energy regime have shown, however, that
the secondary collisions there are as essential as they are in the 1 - 2 GeV/A
experiments at Berkeley [3].

It is thus obvious that besides the reaction history the elementary processes
that make up the total observed yield have to be well under control if one
wants to look for any 'exotic' effects in such spectra. In this paper I discuss

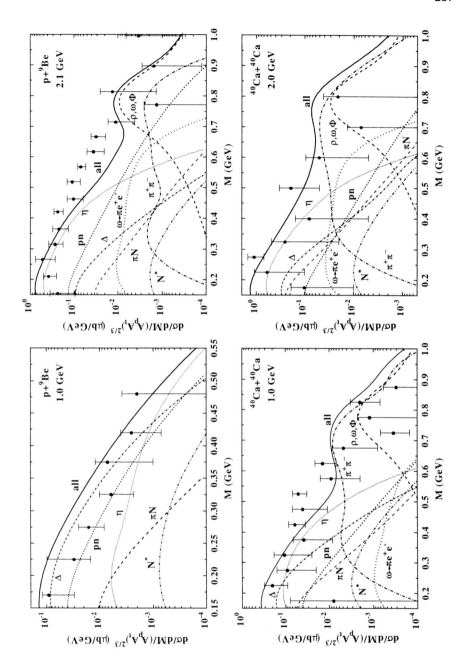

Figure 1: Theoretical analysis of the DLS dilepton data. The various contributions to the total yield are indicated (from [4]). According to a private communication from the authors of ref.[1] the ^{40}Ca data contain a normalization error and should be higher by about a factor of 6.

one process in detail, namely the radiation from the nucleon and the Δ in bremsstrahlung-like processes. This article is based on 3 recent publications which contain all the details of the theory and the calculations [5,6,7]. The present article therefore summarizes only the main results without going into any of the technical details which can be found in the references just given.

2   Dilepton Bremsstrahlung

As Fig. 1 shows dilepton spectra measured in heavy-ion collisions are dominated by two components, the η Dalitz decay and the $\pi^+\pi^-$ annihilation [8,9]. Below these two dominant components lies, however, a strong background of dileptons from nucleon-nucleon bremsstrahlung and Δ decay that has a rather strong bombarding energy dependence [8]. In proton reactions on nuclei these two components become dominant for higher invariant masses of the dileptons. It is, therefore, of interest to study these components in a more refined model than the commonly employed long-wavelength approximation.

In ref.[5] we have therefore studied the dilepton production in nucleon-nucleon collisions using an effective T matrix that is based on a One-Boson-Exchange (OBE) model; the T matrix is determined by fitting the nucleon-nucleon elastic scattering. The calculation treats the dilepton emission from nucleon lines and from Δ lines coherently, whereas in standard BUU simulations these two decay channels do not interfere. The interference turns out to be quite important at the higher invariant masses and can describe the observed mass- and bombarding energy dependence of the pd/pp ratio in the dilepton yield [5]. The overall yield was dominated by the radiation from the Δ line. This important conclusion has now been put on a firmer basis in a recent calculation in which we used a state-of-the-art NN potential [10] instead of the phenomenological OBE potential used in ref.[5].

All these calculations were done without any formfactor for the nucleons because the needed half-off-shell formfactor of the nucleon in the time-like region is experimentally not known. One way to proceed would be to postulate vector meson dominance (VMD).

VMD is a theoretically well developed concept for the description of the coupling between hadrons and photons. While the original idea of Sakurai [11] relied to a large extent on the equality of the quantum numbers of the ρ meson and the photon and the similarity of the Lagrangians describing both particles, the present-day picture is that of charged quark-antiquark loop insertions in the photon propagator which also can change the interaction vertices.

This picture, appealing as it is, is experimentally well established only for the pion where the measured formfactor shows a clear resonance at the ρ meson mass and can be described by a monopole fit [12]. For other vector mesons [13] and for the nucleon the picture is much less clear. For example, for the nucleon the formfactor in the spacelike region exhibits a well-established dipole form, and the time-like region below a momentum of around $2\,GeV/c$, i.e. the vector meson pole region, is experimentally not accessible [14].

In our calculation of ref. [7] we have simply assumed that the electric formfactor has, first, a VMD shape and, second, no off-shell dependence. Under these assumptions the calculations, which are gauge-invariant, give a strong vector meson signal at the ρ meson mass. There are, so far, no published data on dilepton production in pp collisions, but under the assumption that a proton-induced reaction on a light target proceeds mainly through independent pp and pn collisions, the calculations can – properly weighted with the correct number and type of nucleon-nucleon collisions – be compared to the $p + {}^9Be$ data obtained by the DLS group at Berkeley [5]. While the absolute magnitude of the cross-section is reproduced reasonably well, the data show no special effect in the vector meson mass region [5].

3 Off-Shell Formfactors

If this result holds up for the $p + p$ data, then there is obviously a discrepancy between experiment and data, and the question for its reason arises. We have, therfore, studied the electromagnetic formfactor of the nucleon in a model that combines a loop expansion for the π, N and Δ with vector meson dominance for the pion [6]. In this picture the photon couples to the nucleon only through the pion cloud which, in turn, is dominated by VMD in its coupling to the photon. The model allows to study also half-off-shell vertices as they are important in the bremsstrahlungs studies.

For the half-off-shell vertex in a bremsstrahlungs-process the photon vertex has the following general structure

$$
\begin{aligned}
\Gamma(p,p') \;=\; & e\Lambda^+(p)\left(F_1^{++}\gamma_\mu + F_2^{++}\frac{i\sigma_{\mu\nu}q^\nu}{2m_N} + F_3^{++}q_\mu\right)\Lambda^+(p') \\
+ \;& e\Lambda^+(p)\left(F_1^{+-}\gamma_\mu + F_2^{+-}\frac{i\sigma_{\mu\nu}q^\nu}{2m_N} + F_3^{++}q_\mu\right)\Lambda^-(p') \;.
\end{aligned}
\tag{1}
$$

Here p and p' are the on-shell and off-shell momenta of the nucleon, respectively. The formfactors F_i depend on the two variables p'^2 and q^2 and the $\Lambda^\pm$ are projection operators on to states of positive (+) or negative (-) energy.

In ref.[5] we have shown that F_3 can be removed from further consideration by exploiting the Ward-Takahashi identities, so that we are left with 4 independent formfactors. All of these, and in particular their off-shell dependence, we have calculated in ref.[6].

The main result of this study, that reproduces the on-shell electromagnetic formfactor of the nucleon in the spacelike region quite well, is that the time-like formfactors $F_{1,2}^{+\pm}$, are dominated by the well-established VMD for the pion, and that this VMD-effect is remarkably stable as a function of the off-shellness of the nucleon; the off-shell dependence of the electric formfactor is thus negligible in the time-like sector. Therefore, experiments that are sensitive to half-off-shell timelike electromagnetic processes should be able to give important information on these formfactors.

On the other hand, the space-like formfactors are strongly off-shell dependent, except for the formfactor F_1^{++} which is constrained by the electric charge. It thus seems to be necessary to include these off-shell effects in any analysis of data on $p + p$ bremsstrahlung. Such work, employing realistic potentials, is now in progress.

One complication in the interpretation of dilepton production experiments in NN collisions is the presence of distinct sources. For example, we have seen already above that dileptons can be produced both from a nucleon line and from Δ-decay. At higher bombarding energies, above about $1.2 \ GeV$, for example η-mesons can be produced. These η-mesons can also contribute to the $e^+ e^-$ yield through their Dalitz decay. An interesting possibility to distinguish between the various sources has recently been proposed by Bratkovskaya et al.[4]. It amounts to measuring the angle between the relative momentum of the dilepton pair and its total momentum. The corresponding angular distribution is sensitive to the production mechanism. A measurement of this distribution can, therefore, help to disentangle this various $e^+ e^-$ sources and thus put more stringent limits on theoretical predictions.

4   Compton Scattering into the Timelike Region

The off-shellness of the nucleon in nucleon-nucleon collisions is governed by the strong interaction T matrix, so that the dilepton data are due to an interplay of strong and electromagnetic vertices. This makes it difficult to extract the information on the electromagnetic vertex in an unique way.

A "cleaner" experiment is the photon-induced dilepton production $\gamma + p \rightarrow p + e^+ e^-$ that - at least in the s and u channels - is free of any strong interaction vertex. The t channel, which involves a photon-photon-meson vertex and a meson-nucleon vertex, does contribute, but only at forward angles between the incoming and outgoing photon because of the low mass of the intermediate virtual pion (in this energy range there are no other physical mesons

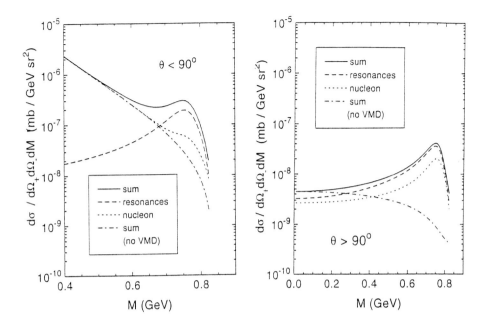

Figure 2: Dilepton invariant mass spectrum for $\gamma + p$ at 1.2 for forward (left) and backward (right) emission angles of the dilepton pair (from [7]).

with an appreciable photon-vector meson coupling strength). As we will see below the forward direction is, however, unsuitable for any dilepton production experiments, because here the (in this context uninteresting) Bethe-Heitler contribution dominates.

Our calculation [7], therefore, starts from a pole fit to Compton scattering data over a wide energy and angular range [15]. These data are described by the nucleon Born terms (including exchange) plus the s-channel resonance terms where a coherent summation over as many as 26 nucleon resonances is performed.

In ref.[7] we have generalized this semi-empirical description to the case of outgoing massive, time-like photons; in this way we make sure that the calculations reproduce the data when the outgoing photon approaches the on-shell point. The coupling to the new longitudinal degree of freedom of the massive photon is determined by explicitly demanding current conservation and gauge invariance.

The analysis by Wada et al. involves momentum-dependent vertex factors for the photon-nucleon vertex. We interpret this dependence (on the three-momentum) as a half-off-shell dependence of the formfactors for a real photon; in this way we have effectively determined the half-off-shell behaviour of the formfactors from experiment. In order to check the consequences of an additional VMD-like dependence on the photon momentum wethen multiply the outgoing vertex factor by a VMD formfactor that peaks at the vector meson mass. This procedure assumes that the nucleon off-shell effects decouple from the photon off-shell effects on the formfactor.

The results of these calculations are shown in Fig. 2. The left part of Fig. 2 gives the dilepton invariant mass spectrum for forward emission angles of the dilepton (small opening angles of the pair with respect to the direction of the incoming photon momentum). It is evident that the spectrum is dominated over a wide range by the Bethe-Heitler contribution. Nevertheless the spectrum shows a shoulder-like structure in the vector meson mass region, which is dominated by the resonance channels. For large opening angles, i.e. backward emission (right part of Fig. 2) the overall cross section is smaller, but the resonance structure is more pronounced; again the resonances contribute somewhat more to it than the nucleon Born graph. Thus an experiment of this sort will determine a formfactor that is averaged over the nucleon and its excited states.

5 Conclusions

In this paper I have emphasized that the low-invariant-mass region of heavy-ion induced dilepton spectra is made up from exactly the same hadronic sources both at relativistic and at ultrarelativistic energies. In both energy regimes a quantitative understanding of the heavy-ion dilepton spectra necessitates an equally quantitative understanding of the elementary hadronic processes *and* of secondary interactions. While the treatment of the latter is now under reasonable control (through the development of quantitatively reliable transport theories), the former stil pose many interesting and fundamental questions.

As an example, I have discussed the electromagnetic formfactor of the nucleon in the timelike regime which is experimentally not accessible in on-shell processes. I have shown that data on proton induced dilepton production on protons or light nuclei may yield important information on this fundamental property. Based on results of a microscopic model I have argued that the nucleon's VMD is essentially determined by that of the well established VMD of the pion em formfactor. An experiment that would give direct access to the timelike electromagnetic formfactor without the strong interaction complications is Compton scattering into the timelike region. I have presented calculations of the cross section for this process and have shown that the dilepton spectra expected from such a reaction are quite sensitive to the electromagnetic formfactor.

Acknowledgements

This work was supported by GSI Darmstadt and BMBF. The result described in this paper were obtained in collaboration with H.C. Dönges, F. de Jong and M. Schäfer.

1. For a recent summary see: H.S. Matis et al, Nucl. Phys. A583 (1995) 617c
2. G. Agakichiev et al, Phys. Rev. Lett. 75 (1995) 1272
3. W. Cassing, W. Ehehalt, C.M. Ko, Phys. Lett. 363 (1995) 35; W. Cassing, W. Ehehalt and I. Kralik, Phys. Lett. B (1996) in print
4. E.L. Bratkovskaya et al., Phys. Lett. B348 (1995) 283; Phys. Lett. B348 (1995) 325; Phys. Lett. B (1996), in press
5. M. Schäfer, H.C. Dönges, A. Engel, U. Mosel, Nucl. Phys. A575 (1994) 429
6. H.C. Dönges, M. Schäfer, U. Mosel, Phys. Rev. C51 (1995) 950
7. M. Schäfer, H.C. Dönges, U. Mosel, Phys. Lett. B342 (1995) 13
8. Gy. Wolf, W. Cassing, U. Mosel, Nucl. Phys. A552 (1993) 549
9. L. Xiong, Z.G. Wu, C.M. Ko, J.Q. Wu, Nucl. Phys. A512 (1990) 772
10. T.-S.H. Lee, Phys. Rev. C29 (1984) 195
11. J.J. Sakurai, Currents and Mesons, University of Chicago Press, 1969
12. T. Ericson and W. Weise, Pions and Nuclei, Clarendon Press, Oxford, 1988
13. L.G. Landsberg, Phys. Rep. 128 (1985) 301
14. S. Dubnicka, Nuovo Cimento 103A (1990) 1417; P. Mergell, U.-G. Meißner, D. Drechsel, hep-ph/9506375; H.-W. Hammer, U.-G. Meißner, D. Drechsel, hep-ph/9604294
15. Y. Wada et al., Nucl. Phys. B247 (1984) 313

DILEPTONS FROM BREMSSTRAHLUNG: GOING BEYOND THE SOFT PHOTON APPROXIMATION

H.C. EGGERS

Department of Physics, University of Stellenbosch, 7600 Stellenbosch, South Africa

C. GALE[1], R. TABTI[1], and K. HAGLIN[2]

[1] *Department of Physics, McGill University, Montréal QC, Canada H3A 2T8*
[2] *National Superconducting Cyclotron Laboratory, Michigan State University East Lansing, MI 48824-1321, USA*

The traditional calculation of dilepton yields from bremsstrahlung relies on the assumption that electromagnetic and strong processes factorize. We argue that this assumption, embodied by the soft photon approximation, cannot hold true for invariant mass spectra on very general grounds. Deriving a formula for the dilepton cross section for pion-pion scattering that does not rely on such factorization, we formulate the problem exactly in terms of three-particle phase space invariants. Using a simple one boson exchange model for comparison, we find that dilepton cross sections and yields are generally overestimated by the soft photon approximation by factors 2–8. In extreme cases, overestimation up to a factor 40 is possible.

1 Introduction

Interest in the use of dileptons as a probe of the hot and early phases of heavy ion collisions is fed by the desire of finding new physics such as the vaunted quark gluon plasma and generally probing the behavior of nuclear matter under extreme conditions[1]. Recent data by HELIOS and NA38[2] and discrepancies found by CERES[3] have provided impetus to hopes that new physics is finally in sight. Realising such hopes, however, requires detailed understanding of background processes: the contribution from each must be calculated quantitatively.

While quantitative calculations are challenging already on a technical level, it is a much harder problem to identify untested assumptions that enter such calculations and to quantify their effects. We here aim to show, by example of bremsstrahlung from pion-pion scattering, that the "soft photon approximation" (SPA) represents such an untested assumption[4].

2 Why the SPA must fail

Conventional dilepton yields at very low invariant masses $M < 500$ MeV are dominated by Dalitz decays and bremsstrahlung. Calculations of such bremsstrahlung yields rely heavily on the soft photon approximation because it is simple to use. This simplicity is achieved mainly through the assumption that the electromagnetic and strong processes factorize.

In order for this assumption to be valid, two conditions [5] must be met: First, the photon energy q_0 must be much smaller than the energy E_i of any one of the hadrons participating in the scattering, $q_0/E_i \ll 1$; secondly, the hadronic and electromagnetic scales must be sufficiently different to permit separate treatment. This translates into the condition $q_0 \ll m_Y |\mathbf{p}_i|/E_i$, where m_Y is the mass of the exchange boson. For the case where two hadrons of equal mass m collide, these equations read in their cms

$$q_0 \ll \sqrt{s}/2, \tag{1}$$

$$q_0 \ll m_Y \sqrt{1 - 4m^2/s}. \tag{2}$$

Implicit in these equations is, of course, a specific Lorentz frame with respect to which the energies are measured. In a simple bremsstrahlung experiment, these limits are easily satisfied by selecting only photons or dileptons of low energy in the laboratory frame. In the complex multiparticle systems formed in the course of nucleus-nucleus collisions, however, there are many binary collisions, and their respective cms frames do not generally coincide either with one another or with the overall nucleus-nucleus cms frame.

In such a situation, it is better to look at relativistically invariant quantities, such as the dilepton invariant mass M. Looking at invariant masses means that q_0 is no longer fixed but must vary over its full kinematic range, which for our example of colliding equal-mass hadrons is given by

$$M \leq q_0 \leq \frac{s - 4m^2 + M^2}{2\sqrt{s}}. \tag{3}$$

In Figure 1, we show the three functions (1) and (2) and (3) for the case $m = m_\pi = 140$ MeV, $m_Y = m_\sigma \simeq 500$ MeV and dilepton invariant masses $M = 10$ and 300 MeV. It is immediately clear that the assumptions underlying the SPA are not fulfilled even for small M: the kinematic range accessible to q_0 is never much smaller than the limits set by the SPA. The situation becomes even worse for larger M.

3   Cross sections: approximate and exact

In order to quantify what effect the use of the SPA has on yields, it is necessary to compare approximate cross sections to an exact formulation. All such dilepton cross sections can be written [6] as the product of a purely leptonic prefactor

$\kappa \equiv (\alpha/3\pi)\left[1 + (2\mu^2/M^2)\right]\sqrt{1 - 4\mu^2/M^2}$ (with μ the lepton rest mass) times the cross section for production of a virtual photon γ^* with mass M,

$$d\sigma_{hh\ell^+\ell^-} = \kappa \, d\sigma_{hh\gamma^*} . \tag{4}$$

Now the traditional procedure has been to factorise the virtual photon cross section $d\sigma_{hh\gamma^*}$ into electromagnetic and strong pieces, writing it in terms of a

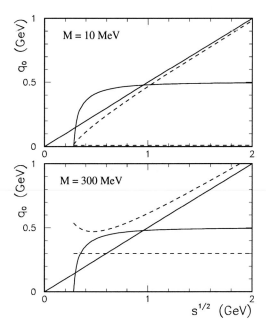

Figure 1: The region between the dashed lines is the domain of integration for q_0 when calculating cross sections as a function of dilepton invariant mass. The SPA is valid only when q_0 is much smaller than the two solid lines shown. This is clearly not the case.

current J^μ and three-particle phase space $dR_3 \equiv \delta^4(p_a + p_b - p_1 - p_2 - q)(d^3p_1/2E_1)(d^3p_2/2E_2)(d^3q/2q_0)$,

$$d\sigma_{hh\gamma^*} = 4\pi\alpha \frac{dM^2}{M^2}(-J^\mu J_\mu)|\mathcal{M}_h|^2 \frac{dR_3}{(2\pi)^5 F} , \tag{5}$$

where F is the incoming flux. It turns out that such factorization is unnecessary. A re-derivation yields exactly, without factorization [4],

$$d\sigma_{hh\gamma^*} = \frac{dM^2}{M^2}\left[-\sum_{mn}(\mathcal{M}_m)^\mu(\mathcal{M}_n^*)_\mu\right]\frac{dR_3}{(2\pi)^5 F} , \tag{6}$$

where m, n run over all diagrams of the reaction $\pi\pi \to \pi\pi\gamma^*$, including emission by the central blob.

Since people do not like 3-particle phase space, dR_3 is usually simplified by neglecting q in the delta function, permitting reduction to 2-phase space $dR_3 \simeq dR_2 \, (d^3q/2q_0)$, with $dR_2 \equiv \delta(p_a + p_b - p_1 - p_2)(d^3p_1/2E_1)(d^3p_2/2E_2)$. Eqs. (4), (5) then simplify to the Rückl form [7,8]

$$\frac{d\sigma_{hh\ell^+\ell^-}^{\text{Rückl}}}{dM^2} = \frac{3}{2}\frac{\kappa}{M^2}\frac{\alpha}{\pi}\int d\sigma_{hh}\int_M^A dq_0\sqrt{q_0^2 - M^2}\int \frac{d\Omega_q}{4\pi}(-J^\mu J_\mu)\,, \quad (7)$$

where $d\sigma_{hh}$ is the on-shell cross section for the purely hadronic reaction $\pi\pi \to \pi\pi$. The current is first angle-averaged over the photon solid angle $d\Omega_q$ and then integrated over the kinematic range $M \le q_0 \le A = [s + M^2 - 4m^2]/2\sqrt{s}$ of the photon energy q_0. At this point an ad hoc factor to correct for the factorization of dR_3 is usually also included.

A better derivation by Lichard [6] led to a form similar to Eq. (7) but without the $(3/2)$ prefactor and with the inclusion of q and M in the current.

In order to use the exact formulation (6), by contrast, it is necessary to formulate $d\sigma_{hh\gamma^*}$ in terms of *three-particle phase space* invariants, which for the schematic reaction $a + b \to 1 + 2 + 3$ are defined [9] as $s = (p_a + p_b)^2$, $t_1 = (p_1 - p_a)^2$, $s_1 = (p_1 + p_2)^2$, $s_2 = (p_2 + p_3)^2$, and $t_2 = (p_b - p_3)^2$. The final-state phase space integral is then given by [9]

$$dR_3(s) = \frac{\pi}{4\lambda^{1/2}(s, m_a^2, m_b^2)}\int \frac{dt_1\,ds_2\,ds_1\,dt_2}{\sqrt{B}}\,, \quad (8)$$

weighted by the 6×6 Cayley determinant B and where $\lambda(x, y, z) = (x - y - z)^2 - 4yz$. One then obtains exactly

$$\frac{d\sigma_{hh\ell^+\ell^-}^{\text{exact}}}{dM^2} = \frac{4\pi\alpha}{(2\pi)^5 M^2}\frac{\kappa(M^2)\pi}{8\lambda(s, m_a^2, m_b^2)}\int \frac{dt_1\,ds_2\,ds_1\,dt_2}{\sqrt{B}}\left[-\sum_{mn}(\mathcal{M}_m)^\mu(\mathcal{M}_n^*)_\mu\right],$$
$$(9)$$

where all terms $(\mathcal{M}_m)^\mu(\mathcal{M}_n^*)_\mu$ are written in terms of the five invariants.

As a by-product of the 3-particle phase space language, one can define an intermediate approximation which, while still factorising the electromagnetic part out of $d\sigma_{hh\ell^+\ell^-}$, writes the current J in terms of its invariants [4]:

$$\frac{d\sigma_{hh\ell^+\ell^-}}{dM^2} \simeq \frac{4\pi\alpha}{(2\pi)^5 M^2}\frac{\kappa\,\pi}{8\lambda(s, m_a^2, m_b^2)}\int dt_1\,|\mathcal{M}_h(s, t_1)|^2$$
$$\times \int \frac{ds_2\,ds_1\,dt_2}{\sqrt{B}}\left[-J^2(s, t_1, s_2, s_1, t_2)\right]\,. \quad (10)$$

4   One boson exchange model

Equation (9) may be exact in terms of the matrix elements $\mathcal{M}$, but it does not specify what $\mathcal{M}$ should be. For a quantitative comparison of the approximations (7) and (10) to the exact cross section (9), we must therefore turn to a simple microscopic model of pion-pion interactions [4]. We use a simple one boson exchange model with the σ, ρ and $f(1270)$ mesons included. The hadronic part of the lagrangian

$$\mathcal{L} = g_\sigma\, \sigma\partial_\mu\boldsymbol{\pi}\cdot\partial^\mu\boldsymbol{\pi} + g_\rho\, \boldsymbol{\rho}^\mu\cdot\boldsymbol{\pi}\times\partial_\mu\boldsymbol{\pi} + g_f\, f_{\mu\nu}\partial^\mu\boldsymbol{\pi}\cdot\partial^\nu\boldsymbol{\pi} \tag{11}$$

is fitted to elastic $\pi^+\pi^-$ collision data to fix the constants, then supplemented through minimal substitution by the corresponding electromagnetic interaction lagrangian. We work at tree level only. Monopole strong form factors are included for t and u-channels; no electromagnetic form factors are needed at present. Virtual photon emission is implemented gauge-invariantly for pion and ρ emission as well as the $\pi\pi\sigma$, $\pi\pi f$ and $\pi\pi\rho$ vertices.

This model is probably far from a perfect description of the pion-pion interaction, so that results obtained below can serve merely as a good pointer towards answering the question: How different are the results when we use the SPA or the exact cross section?

5   Numerical results

To quantify the differences between the approximations and the exact formulation, we have studied the five distinct pion-pion reactions. Writing $(+-) \to (+-)$ as shorthand for the reaction $(\pi^+\pi^- \to \pi^+\pi^-\ell^+\ell^-)$ and so on, we have calculated within our OBE model cross sections for $(+-) \to (+-)$, $(++) \to (++)$, $(+-) \to (00)$, $(00) \to (+-)$ and $(+0) \to (+0)$. Numerical results were checked by performing a number of consistency checks, including testing for gauge invariance in the σ, ρ and f sectors separately.

Figures 2–4 show the cross sections $d\sigma_{hh\ell+\ell-}$ for dielectron production from the five distinct reactions as functions of $\sqrt{s}$, for $M = 10$ MeV and 300 MeV respectively. Final-state symmetrization factors were included where appropriate. Initial-state symmetrization was also included for $(++) \to (++)$ and $(00) \to (+-)$ in order to facilitate use within a thermal pion gas environment. Because they are identical in structure to their charge-conjugate versions, cross sections for the reactions $(+0) \to (+0)$ and $(++) \to (++)$ were doubled. All cross sections were computed using the same OBE model and parameter values.

We see that there is a complete hodgepodge of curves, with a few under-estimating the "exact" cross sections (solid lines) but most overestimating the exact curves by factors 1–5. The largest discrepancy between approximations and the exact result occur for the reaction $(++) \to (++)$ (Figure 4): for the Rückl approximation, factors 3 (for 10 MeV) to 40 (for 300 MeV) arise, while the Lichard approximation yields corresponding overestimation factors of 1.9 and 14–20.

In Figure 5, we attempt to cast some light on the physical origin of the discrepancies observed. Plotted are the cross sections for reactions $(++) \to (++)$ and $(+-) \to (00)$. The upper lines correspond, as before, to the Rückl, Lichard and 3-phase space current approximations, while the solid line again represents the exact result. The lowest dash-dotted line, on the other hand, represents the exact result but excluding all internal [a] diagrams and their cross terms with external ones. The difference between this lower line and the exact result (solid line) therefore represents the contribution of the internal diagrams; while the difference between the lower dash-dotted line and the upper lines (approximations) represents the change in cross section due to inclusion/exclusion of q in the *external*-emission diagrams. We see that the contribution of internal emission is not all that large, albeit nonnegligible. By far the most important effect on $d\sigma/dM$ is the neglect of the dependence of $\mathcal{M}$ on the photon momentum q. We believe that this neglect is at the heart of the considerable differences between approximations and exact results.

Finally, Figure 6 shows the dilepton production rates per unit spacetime summed over all seven reactions, calculated using the Boltzmann formula [10]

$$\frac{dN_{\ell+\ell-}^{\text{Boltz}}}{d^4x \, dM^2} = \frac{1}{32\pi^4} \int ds \, \lambda(s, m^2, m^2) \frac{K_1(\sqrt{s}/T)}{(\sqrt{s}/T)} \frac{d\sigma_{hh\ell+\ell-}}{dM^2} \tag{12}$$

for temperatures $T = 100$ and 200 MeV respectively. Right-hand panels show the corresponding ratios, obtained by dividing a given approximate by the "exact" result. Again, the Rückl approximation is the worst, as expected; overestimation factors range from 2 to 4 for $T = 100$ MeV, and 2–8 for $T = 200$ MeV. The Lichard and 3-phase space current approximations overestimate by factors 1.4–4, depending on temperature and M.

Note that none of the approximations approaches the exact result for small values of M: even for the smallest value shown ($M = 10$ MeV), the discrepancy is still above 40% for the Lichard and current approximations and larger than a factor 2 for the Rückl approximation.

---

[a] "Internal" diagrams are those for which the photon is emitted at a hadronic vertex or by the exchange boson.

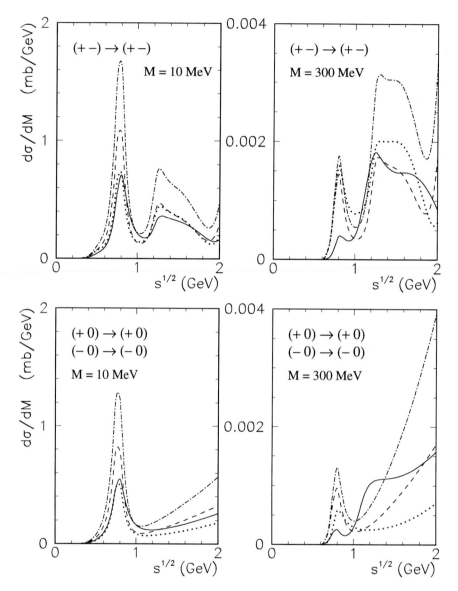

Figure 2: Cross sections for $\pi^+\pi^- \to \pi^+\pi^- e^+ e^-$ and $\pi^+\pi^0 \to \pi^+\pi^0 e^+ e^-$ for fixed invariant masses $M = 10$ and 300 MeV. Solid line: exact OBE calculation Eq. (9). Dash-dotted line: Rückl approximation (7). Dashed line: Lichard approximation. Dotted line: 3-phase space current (10).

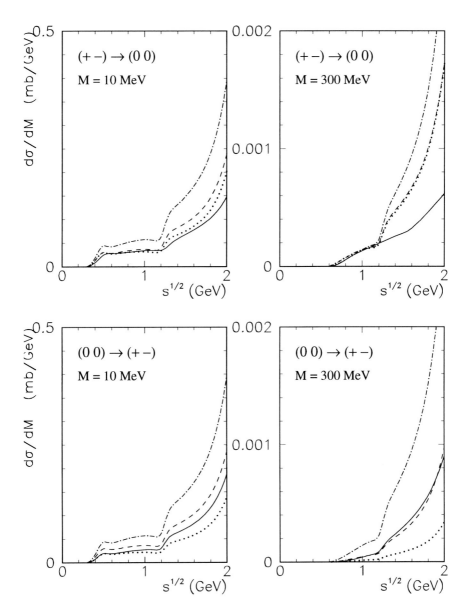

Figure 3: Same as Figure 2, for the reactions $\pi^+\pi^- \to \pi^0\pi^0 e^+e^-$ and $\pi^0\pi^0 \to \pi^+\pi^- e^+e^-$.

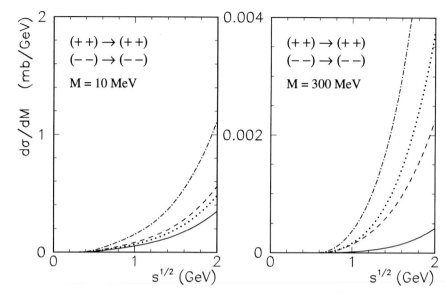

Figure 4: Same as Figure 2, for the reaction $\pi^+\pi^+ \to \pi^+\pi^+ e^+ e^-$.

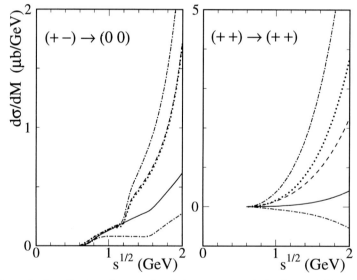

Figure 5: Contribution of external-emission vs. external-plus-internal diagrams for the reactions $(+-) \to (00)$ and $(++) \to (++)$ for $M = 300$ MeV. Lines are as in Figs. 3 and 4. The new dash-dotted line below the (solid line) exact calculation represents contributions arising solely from emission of γ^* by an external pion line, but taking q into account in $\mathcal{M}$, in contrast to the approximations (upper lines) which neglected q.

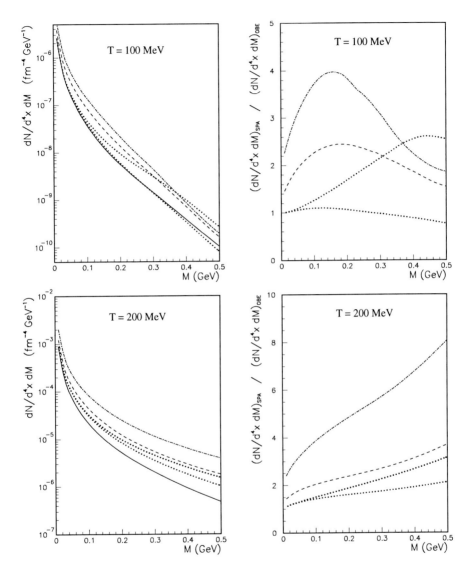

Figure 6: Left: Total bremsstrahlung yields of e^+e^- for all seven pion-pion reactions, for a Boltzmann gas with temperatures $T = 100$ MeV and $T = 200$ MeV. Right: Ratios of SPA approximation calculations divided by exact OBE rate. Solid line: exact OBE calculation, Dash-dotted line: Rückl approximation, Dashed line: Lichard approximation, Dotted line: 3-phase space current, including (upper) and excluding (lower) the mass of the virtual γ^* in the current.

We hence believe that the SPA is flawed when used in a heavy ion context except in special situations. The suppression of the "exact" rates compared to traditional calculations shown here would imply that bremsstrahlung dielectrons cannot make up for the discrepancies between measured dielectron rates and the cocktail of reactions used for comparison. Greater attention will have to be paid to the $\pi\pi$ annihilation, η Dalitz and other channels contributing at low M. The suppression we find would also have a bearing on calculations of the Landau-Pomeranchuk effect [11].

Acknowledgments

This work was supported in part by the Austrian Fonds zur Förderung der wissenschaftlichen Forschung (FWF), the Natural Sciences and Engineering Research Council of Canada, the Québec FCAR fund, a NATO Collaborative Research Grant, and the National Science Foundation.

References

1. J.W. Harris and B. Müller, *The Search for the Quark-Gluon Plasma*, hep-ph/9602235, to appear in *Ann. Rev. Nucl. Part. Sci.*.
2. HELIOS-3 Collaboration, T. Åkesson et al., *Z. Phys.* C **68**, 47 (1995); NA38 Collaboration, C. Lourenco et al., in *Proc. 11*[th] *Int. Conf. on Ultrarelativistic Nucleus-Nucleus Collisions*, Monterey, Jan. 9–13, 1995, *Nucl. Phys.* A **590**, 1c (1995).
3. NA45/CERES Collaboration, G. Agakichiev et al., *Phys. Rev. Lett.* **75**, 1272 (1995).
4. H.C. Eggers, R. Tabti, C. Gale and K. Haglin, hep-ph/9510409, *Phys. Rev.* D **53**, 4822 (1996).
5. F. Low, *Phys. Rev.* **110**, 974 (1958).
6. P. Lichard, *Phys. Rev.* D **51**, 6017 (1995).
7. R. Rückl, *Phys. Lett.* B **64**, 39 (1976).
8. K. Haglin, C. Gale and V. Emel'yanov, *Phys. Rev.* D **47**, 973 (1993).
9. E. Byckling and K. Kajantie, *Particle Kinematics*, (Wiley, London, 1973).
10. K. Kajantie, J. Kapusta, L. McLerran and A. Mekjian, *Phys. Rev.* D **34**, 2746 (1986).
11. J. Cleymans, V.V. Goloviznin and K. Redlich, *Phys. Rev.* D **47**, 989 (1993).

PROBING HADRON STRUCTURE BY REAL AND VIRTUAL PHOTONS [*]

HARTMUTH ARENHÖVEL

Johannes Gutenberg-Universität, D-55099 Mainz, Germany

The importance of subnuclear degrees of freedom in photon absorption and electron scattering by nuclei is discussed. After a short introduction into the basic properties of one-photon processes and a very brief survey on the nuclear response in the various regions of energy and momentum transfers, the particular role of subnuclear degrees of freedom in terms of meson and isobar degrees of freedom is considered. Their importance is illustrated by several reactions on the deuteron which are currently under study at c.w. electron machines like MAMI in Mainz.

1 Introduction

Since the beginning of nuclear physics, when the existence of the atomic nucleus was deduced by Rutherford from the famous experiments of α-particle scattering on a gold foil by Geiger and Marsden up to the deep inelastic scattering experiments at Stanford by Friedman, Kendall and Taylor revealing the parton substructure of nucleons, the electromagnetic probe has always played an outstanding role in the study of nucleon and nuclear structure [1].

One of the major goals of present day research in this field is to clarify the role of subnuclear degrees of freedom (d.o.f.) in the structure of nuclei as well as their relation to the underlying quark-gluon dynamics of QCD. In this talk I will concentrate on the manifestation of of meson and isobar degrees of freedom in electromagnetic processes where they contribute as two-body meson exchange currents (MEC) and via nuclear isobar configurations (IC). The latter are nuclear wave function components containing internally excited nucleons (isobars).

2 Properties of the Electromagnetic Probe

The special role of the electromagnetic interaction in unravelling the microstructure of the world is due to the fact that (i) its properties as a classical field are well known, (ii) the electromagnetic interaction fulfills the basic requirements of a fundamental interaction incorporating relativity and representing the simplest case of a gauge theory, and (iii) the electromagnetic interaction is weak enough to allow lowest order perturbation treatment resulting in simple

---

[*]Dedicated to Walter Greiner on the occasion of his 60th birthday

and unique interpretations of experimental results. However, this weakness constitutes also a disadvantage since the cross sections for photoreactions and electron scattering are considerably smaller than for pure hadronic reactions.

The lowest order processes are the one-photon processes like photon absorption (or emission) and electron scattering which is governed by the exchange of one virtual photon. The main differences between photoabsorption and electron scattering or real and virtual photon processes, respectively, are:

(i) For real photons one has a fixed relation between energy and momentum transfer $(\vec{q}^2 = \omega^2)$ while for the exchange of a virtual photon in electron scattering the four momentum is space-like $(\vec{q}^2 \geq \omega^2)$ allowing an independent variation of energy and momentum transfer within the spacelike region.

(ii) Real photons have only transverse polarization so that only the transverse current density contributes while virtual photons have both, transverse and longitudinal polarization allowing also the charge density to contribute.

Qualitatively one may distinguish different regimes of the nuclear response to photo absorption. At low energies below particle emission threshold, one finds sharp resonances corresponding to the excitation of bound excited nuclear states. Above particle emission threshold for photon energies of about 10 to 30 MeV the dominant feature is the giant dipole resonance. It is a collective nuclear mode which exhausts almost one classical TRK-sum rule. It can be described in a phenomenological two-fluid model as an oscillation of the proton against the neutron fluid or in a microscopic description as a coherent superposition of one-particle one-hole excitations. Increasing the energy, one enters the domain of short-range correlations or the so-called quasi-deuteron region, where the leading process is the absorption by a correlated two-nucleon pair emitted mainly back to back. Here, the major contributions come from MEC. Above pion production threshold, the dominant mode of absorption is the isobar excitation of a nucleon to a Δ or to higher nucleon resonances.

In electron scattering one can explore the same dynamical regions but with the additional possibiliy of independent variation of the momentum transfer. It allows, for example, in the elastic process "cum grano salis" the determination of ground state charge and magnetization densities.

3 Subnuclear Degrees of Freedom in Deuteron Break-up

Now I will turn to the discussion of subnuclear d.o.f. in deuteron disintegration by photons and electrons. These subnuclear d.o.f. are described in terms of meson exchange currents (MEC) and isobar configurations or currents (IC). The special role of the electromagnetic deuteron break-up is a consequence of (a) the simple structure of the two-body system allowing exact solutions at least in the nonrelativistic domain, and (b) the specific features of the electromagnetic probe as discussed in the preceding section. The continued interest in this process has persisted over more than sixty years because the two-body system is in fact a unique laboratory, in particular for the study of subnuclear d.o.f. [2]. One may summarize the evidence for MEC and IC in $\gamma^{(*)} + d \to p + n$ as follows:

(i) The strongest MEC contributions appear in $E1$ in $d(\gamma, N)N$, but they are mostly covered by the Siegert operator.

(ii) A clear signature of MEC is furthermore observed in $M1$ in $d(e, e')np$ near break-up threshold at higher momentum transfers.

(iii) In the Δ-region one finds a strong manifestation of IC.

3.1 MEC and relativistic effects in electrodisintegration

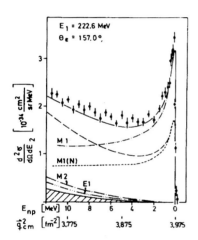

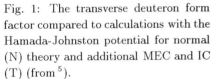

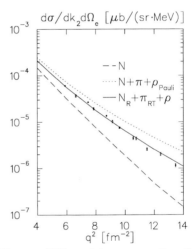

Fig. 1: The transverse deuteron form factor compared to calculations with the Hamada-Johnston potential for normal (N) theory and additional MEC and IC (T) (from [5]).

Fig. 2: Differential cross section for $d(e, e')np$ near threshold at backward angles. Dashed: normal theory without MEC; dotted: with nonrelativistic MEC; full: with RC (from [6])

First, I will consider the inclusive process $d(e, e')pn$, where no analysis is made of the hadronic final state. The threshold region is dominated by the excitation of the antibound $^1S_0$-resonance in NN scattering at very low energies. It is the inverse process to thermal n-p radiative capture, which proceeds via $M1$ transition and where MEC and IC give about a 10 percent enhancement[3,4]. The advantage of electron scattering in having the momentum transfer at one's disposition becomes now apparent since the relatively small effect of subnuclear d.o.f. in the real photon process can be amplified considerably. The reason for this lies in the fact that with increasing momentum transfer the one-body contribution drops rapidly due to a destructive interference of S- and D-wave contributions and thus the distribution of the momentum transfer onto both nucleons via the two-body operators becomes more favourable. This can be seen in Fig. 1 where the transverse form factor is shown as obtained from the Rosenbluth separation [5]. While the longitudinal form factor is well described by the classical theory supporting the Siegert hypothesis, that the charge density is little affected by exchange effects, one finds a large discrepancy in F_T which is only resolved if MEC and IC are added.

The situation for higher momentum transfers is shown in Fig. 2 where the inclusive cross section is plotted at backward angles for moderate momentum transfers. One readily sees the strong enhancement due to MEC, but in addition one notes sizeable relativistic contributions. It is clear from this result that already at low excitation energies relativistic effects may become important and have to be considered in a quantitative comparison of theory with experiment. One has to keep in mind that relativistic contributions appear both in the current operators and in the wave functions. For the latter case one may distinguish between (i) the internal dynamics of the rest frame wave function and (ii) the boost operation transforming the rest frame wave function into a moving frame. It is obvious that for a conclusive interpretation one has to include all corrections of the same order consistently. In this work, presented in Fig. 2, where we have adopted a $(p/M)^2$-expansion starting from a covariant approach [7], all relativistic terms are included consistently.

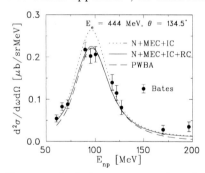

Fig. 3: Inclusive cross section for $d(e, e')$ for PWBA and with FSI, MEC, IC, and RC (from [8]).

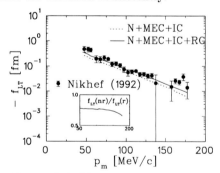

Fig. 4: Coincidence structure function f_{LT} as function of the missing momentum without and with relativistic corrections (from [9]).

In Fig. 3 we show for higher energy and momentum transfer another comparison of experimental data for $d(e, e')$ with a realistic calculation including FSI, MEC, IC and the dominant relativistic contributions (RC) and a comparison to the pure plane wave Born approximation (PWBA). One notes that the relativistic contributions improve considerably the agreement between theory and experiment. Furthermore, the comparison to the PWBA shows that off the quasi-free region interaction effects are important.

As last example of relativistic effects, I show in Fig. 4 the longitudinal-transverse interference structure function f_{LT} appearing in the exclusive process $d(e, e'p)n$. The comparison with recent experimental data is significantly improved if relativistic contributions are included. The relative size of the RC is shown in the inset as ratio of nonrelativistic (nr) to relativistic (r) result.

3.2 Signature of a $\Delta\Delta$ component in the deuteron

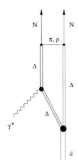

Fig. 5: Contribution of the $\Delta\Delta$ component to the longitudinal structure function f_L

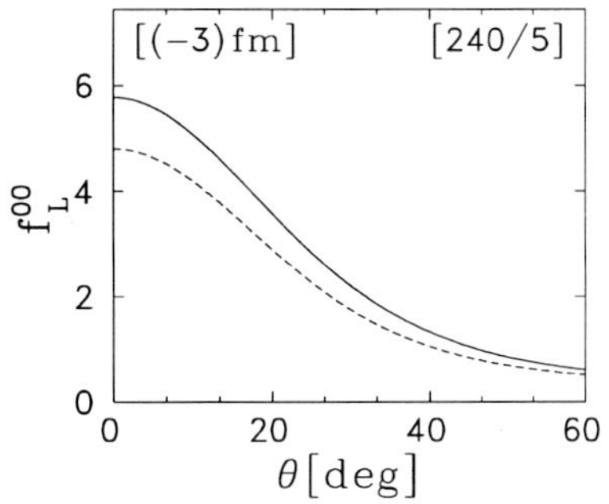

Fig. 6: Longitudinal structure function in the Δ region for $E_{np} = 240$ MeV and $\vec{q}^2 = 5$ fm^2 without (dashed) and with (full) $\Delta\Delta$ component.

One important consequence of the internal nucleon dynamics is the presence of small wave function components where one or more nucleons are internally excited into an isobar state [10]. These isobar configurations (IC) are rather small and thus their presence is difficult to detect. In any case, evidence for them will only be indirect since they are not directly observable. However, under certain favourable conditions they may lead to sizeable effects. I would like to show one example for the double Δ-component in the deuteron.

In view of the fact, that the charge excitation of a Δ which has to proceed via $C2$ is largely suppressed, the only contribution of IC to the longitudinal structure function f_L comes from the double Δ-component as sketched in the diagram of Fig. 5. At low momentum transfer, the contribution is negligible. However, with increasing momentum transfer its relative importance is enhanced considerably because of the much shorter ranged structure of the isobar configurations compared to the normal ones. This behavior is shown in Fig. 6. It also illustrates nicely the advantage of varying the momentum transfer independently in electron scattering.

3.3 $N\Delta$ dynamics in $d(\gamma, p)n$ in a coupled channel approach

Deuteron photodisintegration is fairly well understood at low energies in the framework of nucleon, meson and isobar d.o.f.[11] whereas at higher photon energies between 200 and 450 MeV, where the Δ-excitation dominates the reaction, the theoretical description is much less well settled[2]. Often the Δ-excitation is treated in the impulse approximation (IA). It turns out, however, that it is important to include the $N\Delta$ dynamics in a coupled channel approach[12,13]. Here I will present an improved coupled NN-$N\Delta$ channel approach which includes in addition explicit pion d.o.f.[14]. In this calculation, the dominant magnetic dipole excitation of the Δ has been fitted to the experimental $M_{1+}(3/2)$ multipole of pion photoproduction on the nucleon. Details can be found in Ref.[14].

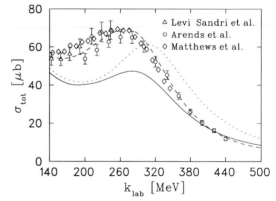

Fig. 7: Total cross section for $\gamma d \to pn$ in comparison with experiment. Complete calculation (full), IA (dotted) and complete calculation with modified $\gamma N\Delta$-coupling (dashed) as described in the text (from[14]).

I first show in Fig. 7 the total cross section. In comparison with the IA, the full calculation leads to a strong reduction of the cross section above 260 MeV and in addition to a shift of the maximum position towards lower energies in accordance with the experimental energy dependence, whereas the IA peaks close to the resonance position of the free $N\Delta$ system at 320 MeV. However, below 340 MeV the cross section becomes definitely too small. Since the $\gamma N\Delta$-coupling is weaker than the effective one used in[12] we have also considered a modified $\gamma N\Delta$-coupling, which was determined from the elementary amplitude under the assumption of vanishing nonresonant contributions to the $M_{1+}(3/2)$ multipole. Using this coupling, we achieved a good agreement of the total cross section with experiment over the whole energy range as demonstrated in Fig. 7. Since in this case the Born terms are effectively incorporated in the modified coupling we are led to the conclusion that the framework of static π-exchange currents, containing in principle the Born terms, gives a poor description of them in this energy region.

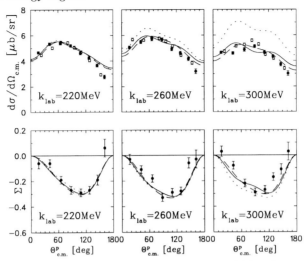

Fig. 8: Differential cross section and photon asymmetry Σ for three energies: complete calculation with modified $\gamma N \Delta$-coupling (full), IA (dotted) and complete calculation with modified $\gamma N \Delta$-coupling in M1 only (dashed) (from [14]).

Differential cross sections and photon asymmetries for three energies are shown in Fig. 8. In addition to the IA and the complete calculation with modified coupling, we show results where the $\gamma N \Delta$-coupling has been modified in the M1 multipole only. Despite the good description of the total cross section, problems in the angular distributions still remain. At the lower energy our calculation does not show such a strong decrease at the backward angles as the data. Furthermore, a dip around $90°$ appears, in particular at the highest energy, which clearly contradicts the experimental shape. The occurrence of this dip structure is also reported in [13]. The photon asymmetry in Fig. 8 is quite satisfactory although at the higher energies it appears to be more asymmetrical around $90°$ than the experiment. At even higher energies this trend continues so that the model clearly fails to reproduce the data.

3.4 Unitary ambiguity in the resonance multipoles for $\gamma + N \to \Delta$

From the foregoing it is clear that for a reliable description of Δ excitation in nuclei one needs to know their multipoles fairly well. Consequently, there is considerable experimental effort in measuring the corresponding E_{1+} and

M_{1+} isospin $3/2$ multipole amplitudes for photoproduction of pions on the nucleon. However, all realistic pion photoproduction models show that both multipoles, in particular $E_{1+}^{3/2}$, contain nonnegligible nonresonant background contributions. Unfortunately, their presence complicates the isolation of the resonant parts. Basically two different approaches are used in order to extract these transition amplitudes. The first one is an effective lagrangian approach, in which the πN scattering is not treated dynamically and thus different unitarization schemes are used. These introduce some model dependence.

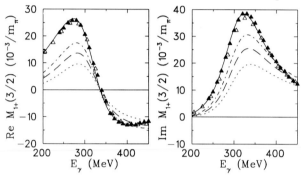

Fig. 9: Real and imaginary parts of the $M_{1+}^{3/2}$ multipole as function of the photon laboratory energy E_γ. Dashed, dotted and dash-dotted curves are bare multipoles corresponding to transformation angles $\tilde{\alpha} = 0°$, $10°$ and $-10°$, respectively. The solid curves show the total multipoles which are representation independent (from [15]).

In the second approach, the πN interaction is treated dynamically and thus unitarity is respected automatically. However, also the latter approach contains considerable model dependence of the resonance properties because of an inherent unitary ambiguity. The reason for this unitary freedom is that the separation of a resonant Δ contribution corresponds to the introduction of a Δ component into the πN scattering state which vanishes in the asymptotic region. It is known that the explicit form of a wave function depends on the chosen representation, and can be changed by means of unitary transformations. Consequently, the probability of a certain wave function component is not an observable, since it depends on the representation, whereas any observable is not affected by a change of the representation. We have demonstrated this fact in a simple model where the unitary transformation mixes resonant and background πN interactions [15]. I show in Fig. 9 the effect of such unitary transformations (governed by a parameter $\tilde{\alpha}$) on the $M_{1+}^{3/2}$ multipole. One notices a considerable variation of the bare resonance multipole with $\tilde{\alpha}$ while the total one is unchanged.

With this I would like to close the brief review in which I could give only a very cursory survey on the many interesting facets of present day research in current electromagnetic physics programs at various c.w. electron machines.

Acknowledgments

The collaboration of J. Adam, G. Beck, H. Göller, F. Ritz, Th. Wilbois and P. Wilhelm is gratefully acknowledged. This work has been supported by the Deutsche Forschungsgemeinschaft (SFB 201).

References

1. H. Arenhövel, Lecture Notes in Physics, Vol. **426**, 1 (1994)
2. H. Arenhövel, M. Sanzone, Few-Body Syst., Suppl. **3**, 1 (1991)
3. D.O. Riska, G.E. Brown, Phys. Lett. **38B**, 193 (1972)
4. H. Arenhövel, Lecture Notes in Physics, Vol. **137**, 136 (1981)
5. G.G. Simon et al., Nucl. Phys. **A324**, 277 (1979)
6. F. Ritz, diploma thesis, University Mainz 1995
7. J. Adam, Jr., E. Truhlik, D. Adamova, Nucl. Phys. **A492**, 556 (1989)
8. T. Wilbois, G. Beck, H. Arenhövel, Few-Body Syst. **15**, 39 (1993)
9. G. Beck, T. Wilbois, H. Arenhövel, Few-Body Syst. **17**, 91 (1994)
10. H.J. Weber, H. Arenhövel, Phys. Rep. **36C**, 277 (1978)
11. K.-M. Schmitt, P. Wilhelm, H. Arenhövel, Few-Body Syst. **10**, 105 (1991)
12. W. Leidemann, H. Arenhövel, Nucl. Phys. **A465**, 573 (1987)
13. H. Tanabe, K. Ohta, Phys. Rev. **C40**, 1905 (1989)
14. P. Wilhelm, H. Arenhövel, Phys. Lett. **B318**, 410 (1993); Few-Body Syst., Suppl. **7**, 235 (1994)
15. P. Wilhelm, T. Wilbois, H. Arenhövel, preprint MKPH-T-96-1, Mainz 1996

CRYSTALLINE STRUCTURE IN THE CONFINED-DECONFINED MIXED PHASE: NEUTRON STARS AS AN EXAMPLE

N. K. GLENDENNING

Nuclear Science Division and Institute for Nuclear & Particle Astrophysics
Lawrence Berkeley National Laboratory, University of California, Berkeley,
California 94720

We review the differences in first order phase transition of single and multi-component systems, and then discuss the crystalline structure expected to exist in the mixed confined deconfined phase of hadronic matter. The particular context of neutron stars is chosen for illustration. The qualitative results are general and apply for example to the vapor-liquid transition in subsaturated asymmetric nuclear matter.

1 Introduction

First order phase transitions are very familiar only in one-component substances such as water. The transition in multi-component substances is very different [1,2]. In such systems the common pressure and all other properties of each phase varies with the proportion of the phases in equilibrium. This has unique consequences in certain situations, such as in the presence of a gravitational field. More than that, when one of the independent components is electrically charged, the two phases in equilibrium may form a Coulomb lattice of the rare phase immersed in the dominant one [2]. These are quite general theorems and do not depend on details of models. They will hold for diverse systems such as in the gas-liquid transition in nuclear physics, in the confined-deconfined transition in nuclear collisions and the same transition in neutron stars as well as in chemical systems. However in the first two cases mentioned, the ideal equilibrium situation is probably never achieved, and the observed results will depend on such things as how central or peripheral the collision was, how much of the colliding nuclei thermalized and had time to reach equilibrium in every sense, and so on. Nevertheless, it is important to keep in mind, even for such systems as the gas-liquid nuclear transition, what the ideal transition would look like. *There is no plateau* as is often referred to in the ideal case *except* for isospin symmetric systems.

Our aim here is to briefly recapitulate the physical reason for the different behavior of a first order phase transition in single- and multi-component substances, and then to compute the varying geometry of the crystalline structure as a function of proportion of the phases in equilibrium. We shall do this in the context of the confined-deconfined phase transition in neutron star matter—matter that is charge neutral and in equilibrium with respect to all baryon and quark species. The results would be qualitatively similar for the liquid-vapor transition in sub-saturated nuclear matter if equilibrium were achieved.

2   Degrees of Freedom in Multi-component System

Consider a substance composed of two conserved 'charges' or independent components— Q of one kind, B of the other. In the case of a neutron star, these could denote the net electric charge number (in units of e) and baryon charge number. Let the substance be closed and in a heat bath. Define the concentration by $c = Q/B$. The ratio is fixed *only* as long as the system remains in one pure phase or the other! When in the mixed phase the concentration in each of the regions of one phase or the other may be different and they are restricted *only* by the conservation on the total numbers,

$$c_1 = Q_1/B_1, \quad c_2 = Q_2/B_2, \qquad (Q_1 + Q_2 = Q, \quad B_1 + B_2 = B). \tag{1}$$

If the internal forces can lower the energy of the system by rearranging the concentration, they will do so. The essential point is that conservation laws in chemical thermodynamics are global, not local. The concentrations, c_1, c_2 will be optimized by the internal forces of the system at each proportion of the phases, so that the properties of the phases are *not* constant in the mixed phase. This contrasts with the liquid-vapor transition in water.

The mathematical proof of the above properties is not nearly so illuminating as the physical verbal proof above, but we give it for completeness. The Gibbs condition for phase equilibrium is that the chemical potentials μ_b, μ_q corresponding to B and Q, temperature T and the pressures in the two phases be equal,

$$p_1(\mu_b, \mu_q, T) = p_2(\mu_b, \mu_q, T) \tag{2}$$

As discussed, the condition of *local* conservation is stronger than required. We apply the weaker condition of *global* conservation,

$$< \rho > \equiv (1 - \chi)\rho_1(\mu_b, \mu_q, T) + \chi \rho_2(\mu_b, \mu_q, T) = B/V, \tag{3}$$

$$(1 - \chi)q_1(\mu_b, \mu_q, T) + \chi q_2(\mu_b, \mu_q, T) = Q/V, \quad \chi = V_2/V. \tag{4}$$

Given a temperature, the above three equations serve to determine the two independent chemical potentials and V for a *specified* volume fraction χ of phase '2' in equilibrium with phase '1'. We note that the condition of global conservation expressed by (3) and (4) is compatible, together with (2), with the number of unknowns to be determined. It would *not* be possible to satisfy Gibbs conditions if *local* conservation were demanded, for that would replace (4) by *two* equations, such as $q_1(\mu_b, \mu_q, T) = Q_1/V_1$, $q_2(\mu_b, \mu_q, T) = Q_2/V_2$, and the problem would be over determined. Note also that the solution of the equations is different for each proportion of the phases χ so that all properties, including the common pressure, vary through the mixed phase.

The internal force that can exploit the degree of freedom opened by *global* charge neutrality and which is closed to one in which *local* neutrality is artificially enforced, is the isospin restoring force experienced by the confined phase of hadronic matter. The hadronic regions of the mixed phase can arrange to be more isospin symmetric than in the pure phase by transferring charge to the quark phase in equilibrium with it. Symmetry energy will be lowered thereby at only a small cost in rearranging the quark Fermi surfaces. Thus the mixed phase region of the star will have *positively* charged regions of nuclear matter and *negatively* charged regions of quark matter.

3 Structure in the Mixed Phase

The Coulomb interaction will tend to break the regions of like charge into smaller ones, while this is opposed by the surface interface energy. Their competition will be resolved by *forming a lattice of the rare phase immersed in the dominant one whose form, size and spacing will minimize the sum of surface and Coulomb energies.* In other words, a crystalline lattice will be formed.

Consider a Wigner-Seitz cell of radius R containing the rare phase object of radius r and an amount of the dominant phase that makes the cell charge neutral. The whole medium can be considered as made of such non-interacting cells. The solution to problems involving a competition between Coulomb and surface interface energies is universal. The radius of the rare phase immersed in the other and the minimum of the sum of Coulomb and surface energies, in the case of three geometries, slabs, rods and drops are

$$\frac{1}{r^3} = \frac{4\pi[q_H(\chi) - q_Q(\chi)]^2 e^2 f_d(x)}{\sigma d}, \qquad d = 1, 2, 3, \qquad (5)$$

$$\frac{E_C + E_S}{V} = 6\pi x \left(\frac{[\sigma d(q_H(\chi) - q_Q(\chi))e]^2 f_d(x)}{16\pi^2} \right)^{1/3}, \qquad (6)$$

where, q_H, q_Q are the charge densities of hadronic and quark matter (in units of e) at whatever proportion χ being considered. We have denoted the volume fraction of quark matter V_Q/V by χ. The ratio of droplet (rod, slab) to cell

Figure 1: The charge densities in the mixed phase carried by regions of quark and hadronic matter, as well as leptons which permeate all regions in our approximation. Multiplied by the respective volumes occupied, the total charge adds to zero.

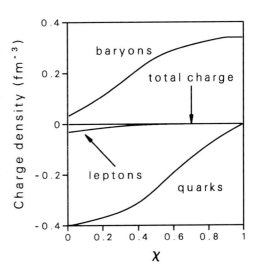

volume is called $x = (r/R)^d$. It is related to χ by $\chi = x$ when hadronic matter is the background (ie. dominant) phase. The quark droplets (rods, slabs) have radius r and the spacing between centers is R. In the case of drops or rods, r is their radius and R the half distance between centers while for slabs, r is the half thickness. In the opposite situation where quark matter is the background, $1 - \chi = x$ is the fraction of hadronic matter which assumes the above geometric forms.

The function $f_d(x)$ is given in all three cases by,

$$f_d(x) = \frac{1}{d+2}\left[\frac{1}{(d-2)}(2 - dx^{1-2/d}) + x\right],\qquad(7)$$

where the apparent singularity for $d = 2$ is well behaved in the limit.

The surface tension should be self-consistent with the two models of matter, quark and hadronic, in equilibrium with each other. The densities of each phase change as their proportion does[2,1]. So the surface energy is not a constant. Gibbs studied the problem of surface energies, and as a gross approximation, one can deduce that it is given by the difference in energy densities of the substances in contact times a length scale typical of the surface thickness [3], in this case of the order of the strong interaction range, $L = 1$ fm. In other words, the surface interface energy should depend on the proportion of phases in phase equilibrium, just as everything else does.

$$\sigma = \text{const} \times [\epsilon_Q(\chi) - \epsilon_H(\chi)] \times L.\qquad(8)$$

318

Figure 2: Diameter (lower curves) and Spacing (upper curves) of rare phase immersed in the dominant as a function of the proportion of quark phase. Geometries are identified as drops, rods, slabs, and composition as q (quark) or h (hadronic). Dots are a continuous dimensionality interpolation of the discrete shapes.

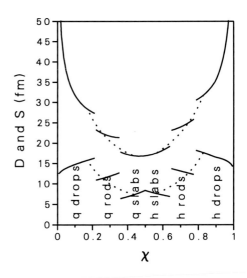

Following our deduction that a Coulomb lattice should exist in the mixed hadron-quark phase[2,1], Heiselberg, Pethick and Staubo investigated the dependence of the geometrical structure on the surface tension[4]. They adopted a selection of values from various sources, none of them computed self-consistently. The constant in (8) should be chosen so that the structured phase lies below the unstructured one. Heiselberg et al found their energy difference to be about 10 MeV. We choose the constant accordingly.

The sum of surface and Coulomb energies scale with the surface energy coefficient as $\sigma^{1/3}$ independent of geometry. Therefore the location in the star where the geometry changes from one form to another is independent of σ.

4   Bulk Description of the Phases

The geometrical structure of the mixed phase occurs against the background of the bulk structure. The energy and pressure are dominated by the bulk properties of matter. The description of the hadronic phase of neutron star matter has been discussed in detail elsewhere [5]. We include all the charge states of the lowest baryon octet, $(p, n, \Lambda, \Sigma^+, \Sigma^-, \Sigma^0, \Xi^-, \Xi^0)$.

The nuclear properties that define the values of the couplings are the binding, $B/A = -16.3$ MeV, saturation density, $\rho_0 = 0.153$ fm^{-3}, and symmetry energy coefficient, $a_{\text{sym}} = 32.5$ MeV, compression modulus $K = 240$ MeV and nucleon effective mass at saturation $m^\star_{\text{sat}}/m = 0.78$. The hyperon couplings can be constrained by levels in hypernuclei, the binding of the Λ in nuclear

Figure 3: Boundaries of the various crystalline structures and radius of the star are shown for stars of mass shown on the y-axis. Detail of the dotted boxed area is shown in Fig. 4.

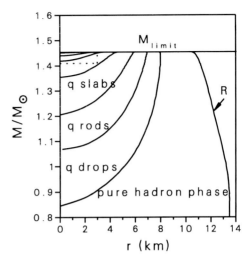

matter, and from neutron star masses [6]. We shall assume that all hyperons in the octet have the same coupling as the Λ. They are expressed as a ratio to the above mentioned nucleon couplings, $x_\sigma = g_{H\sigma}/g_\sigma$, $x_\omega = g_{H\omega}/g_\omega$, $x_\rho = g_{H\rho}/g_\rho$. The first two are related to the Λ binding by a relation derived in [6]; the third can be taken equal to the second by invoking vector dominance. We adopt the value of $x_\sigma = 0.6$ and corresponding x_ω taken from [6].

To describe quark matter we use a simple version of the bag model with $B^{1/4} = 180$ MeV and quark masses $m_u = m_d = 0$, $m_s = 150$ MeV [2]. Because of the long time-scale, strangeness is not conserved in a star. The quark chemical potentials for a system in chemical equilibrium are therefore related to those for baryon number and electron by $\mu_u = \mu_c = (\mu_n - 2\mu_e)/3$, $\mu_d = \mu_s = (\mu_n + \mu_e)/3$.

5 Varying Crystalline Structure

To illustrate the rearrangement of the electric charge concentration between the quark and baryonic regions of the mixed phase, we show the charge density in each region, and the electron charge density, assumed to be uniform throughout the Wigner-Seitz cell, as functions of the proportion of quark matter in Fig. 1. Quark matter, which in the absence of baryonic matter ($\chi = 1$) is charge neutral in a star, carries a high negative charge density when there is little of it and it is in equilibrium with baryonic matter. The latter acquires an ever increasing density as the quantity of quark matter with which it can

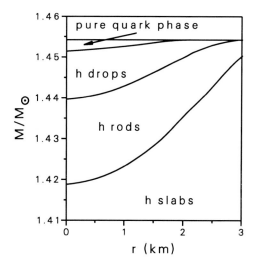

Figure 4: Detail of the dotted boxed area of Fig. 3

balance electric charge grows. This illustrates how effectively the symmetry driving force acts to optimally rearrange charge.

As shown above, because one of the conserved quantities is the electric charge, having long range, an order will be established in the mixed phase, the size of the objects of the rare phase and their spacing in the dominant one, being determined by the condition for a minimum sum of Coulomb and surface energy. In Fig. 2 the diameter D and spacing S is shown by the lower and upper curves as a function of proportion of quark phase. It is noteworthy that at the limit of the pure phases corresponding to $\chi = 0$ or 1, the spheres of rare phase are of finite diameter, but spaced far apart. The size of the objects is between 7 and 15 fm. As noted previously the location in χ of the geometries is independent of σ, but the size and spacings scale as $\sigma^{1/3}$.

6 Crystal Structure in the Cores of Neutron Stars

In the environment of a particular mass star χ is a function of radial coordinate. The various structures occur at various radial locations in stars of different mass. In Fig. 3 the radial location of the boundaries are shown for the stellar masses shown on the y-axis. Stars of low mass have no mixed phase, as expected. With increasing mass, quark drops appear and at still higher mass the full complement of geometries appear in both quark and hadronic form. For stars very close to the limiting mass an inner sphere of pure quark matter appears having a radius of 2 km at the mass limit. Generally the region spanned by the crystalline structures is of the order of 8 km. We note the transition from pure hadronic to mixed phase occurs at the rather low density of about $2\rho_0$, as was found also by several other authors [4,7].

7 Summary

It is almost certain that a solid region in a pulsar will play a role in the period glitch phenomenon, which is highly individualistic from one pulsar to another. We have suggested that this high degree of individual behavior may be due to the extreme sensitivity on stellar mass of the radial extent of the solid region [8]. The sensitivity arises because of the rather flat radial profile of the pressure and energy density in neutron stars, so that a small change in central density and therefore a small change in stellar mass, moves a transition pressure a considerable distance in the radial direction in the star.

Acknowledgements: This work was supported by the Director, Office of Energy Research, Office of High Energy and Nuclear Physics, Division of Nuclear Physics, of the U.S. Department of Energy under Contract DE-AC03-76SF00098..

1. N. K. Glendenning, Nuclear Physics B (Proc. Suppl.) **24B** (1991) 110.
2. N. K. Glendenning, Phys. Rev. D, **46** (1992) 1274.
3. W. D. Myers, W. J. Swiatecki and C. S. Wang, Nucl. Phys. **A436** (1985) 185.
4. H. Heiselberg, C. J. Pethick, and E. F. Staubo, Phys. Rev. Lett. **70** (1993) 1355.
5. N. K. Glendenning, Astrophys. J. **293** (1985) 470.
6. N. K. Glendenning and S. A. Moszkowski, Phys. Rev. Lett. **67** (1991) 2414.
7. V. R. Pandharipande and E. F. Staubo, in *International Conference on Astrophysics*, Calcutta, 1993, ed. by B. Sinha, (World Scientific, Singapore).
8. N. K. Glendenning and S. Pei, Phys. Rev. C **52** (1995) 2250.

QUARK MATTER, MASSIVE STARS AND STRANGE PLANETS

F. WEBER, CH. SCHAAB, M. K. WEIGEL

Institute for Theoretical Physics, University of Munich,
Theresienstr. 37, 80333 Munich, Germany

N. K. GLENDENNING

Nuclear Science Division & Institute for Nuclear and Particle
Astrophysics, Lawrence Berkeley National Laboratory,
Berkeley, CA 94720, USA

This paper gives an overview of the properties of all possible equilibrium sequences of compact strange-matter stars with nuclear crusts, which range from strange stars to strange dwarfs. In contrast to their non-strange counterparts, –neutron stars and white dwarfs–, their properties are determined by two (rather than one) parameters, the central star density and the density at the base of the nuclear crust. This leads to stellar strange-matter configurations whose properties are much more complex than those of the conventional sequence. As an example, two generically different categories of stable strange dwarfs are found, which could be the observed white dwarfs. Furthermore we find very low-mass strange stellar objects, with masses as small as those of Jupiter or even lighter planets. Such objects, if abundant enough in our Galaxy, should be seen by the presently performed gravitational microlensing searches. Further aspects studied in this paper concern the limiting rotational periods and the cooling behavior of neutron stars and their strange counterparts.

1 Introduction

The theoretical possibility that strange quark matter may be absolutely stable with respect to iron (energy per baryon below 930 MeV) has been pointed out by Bodmer[1], Witten[2], and Terazawa[3]. This so-called strange matter hypothesis constitutes one of the most startling possibilities of the behavior of superdense nuclear matter, which, if true, would have implications of greatest importance for cosmology, the early universe, its evolution to the present day, astrophysical compact objects, and laboratory physics[4]. Unfortunately it seems unlikely that QCD calculations will be accurate enough in the foreseeable future to give a definitive prediction on the absolute stability of strange matter, such that one is left with experiments and astrophysical tests, as performed here, to either confirm or reject the hypothesis.

One striking implication of the hypothesis would be that pulsars, which are conventionally interpreted as rotating neutron stars, almost certainly would be rotating strange stars (strange pulsars). Part of this paper deals with an

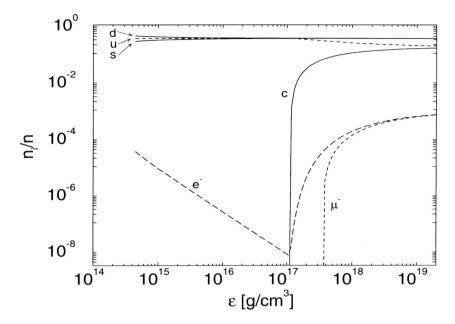

Figure 1: Relative densities of quarks and leptons in absolutely stable strange-quark-star matter versus density. n_i and n denote partial and total densities, respectively.

investigation of the properties of such objects. In addition to this, we develop the complete sequence of strange stars with nuclear crusts, which ranges from the compact members, with properties similar to those of neutron stars, to white-dwarf-like objects (strange dwarfs) and discuss their stability against acoustical vibrations[5]. The properties with respect to which strange-matter stars differ from their non-strange counterparts are discussed.

2   Quark-lepton Composition of Strange Matter

The relative quark–lepton composition of quark-star matter at zero temperature is shown in Fig. 1[6]. All quark flavor states that become populated at the densities shown are taken into account. Since stars in their lowest energy state are electrically charge neutral to very high precision, any net positive quark charge must be balanced by leptons. In general, as can be seen in Fig. 1, there is only little need for leptons, since charge neutrality can be achieved essentially among the quarks themselves. The concentration of electrons is largest at the lower densities of Fig. 1, due to the finite strange-quark mass which leads

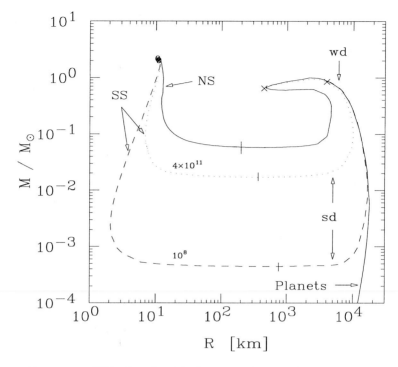

Figure 2: Neutron star (NS)–white dwarf (wd) sequence (solid line). The dotted and dashed curves refer to strange star (SS)–strange dwarf (sd) sequences with inner crust densities as indicated (in g/cm^3). Vertical bars mark minimum mass stars, crosses mark the termination points of the strange star sequences.

to a deficit of net negative quark charge, and at densities beyond which the charm-quark state becomes populated which increases the net positive quark charge.

3   Nuclear Crusts on Strange Stars and Equation of State

The presence of electrons in strange quark matter is crucial for the possible existence of a nuclear crust on such objects. As shown in Refs.[6,7], the electrons, because they are bound to strange matter by the Coulomb force rather than the strong force, extend several hundred fermi beyond the surface of the strange star. Associated with this electron displacement is a very strong electric dipole layer which can support – out of contact with the surface of the strange star – a crust of nuclear material, which it polarizes. The maximal possible density at the base of the crust (inner crust density) is determined by the neutron drip density ($\epsilon_{drip} = 4.3 \times 10^{11}$ g/cm^3), at which neutrons begin to drip out of the nuclei and form a free neutron gas. Being electrically charge neutral, the neutrons do not feel the repulsive Coulomb force and hence would gravitate toward the quark matter core, where they become converted, via hypothesis, into strange quark matter. So neutron drip sets a strict upper limit on the crust's maximal inner density.

The somewhat complicated situation of the structure of a strange star with crust can be represented by a proper choice of equation of state[8], which consists of two parts. At densities below neutron drip it is represented by the low-density equation of state of charge-neutral nuclear matter, for which we use the Baym-Pethick-Sutherland equation of state. The star's strange-matter core is described by the bag model. The graphical illustration of such an equation of state can be found in Ref.[8].

4   Complete Sequences of Strange-Matter Stars

Since the nuclear crusts surrounding the cores of strange stars are bound by the gravitational force rather than confinement, the mass-radius relationship of strange-matter stars with crusts is qualitatively similar to the one of purely gravitationally bound stars –neutron stars and white dwarfs–, as illustrated in Fig. 2[5,9]. The strange-star sequences are computed for the maximal possible inner crust density, $\epsilon_{crust} = \epsilon_{drip}$, as well as for an arbitrarily chosen, smaller value of $\epsilon_{crust} = 10^8$ g/cm^3, which may serves to demonstrate the influence of less dense crusts on the mass-radius relationship[5]. From the maximum-mass star (dot), the central density decreases monotonically through the sequence in each case. The neutron-star sequence is computed for a representative model for the equation of state of neutron star matter, the relativistic Hartree-Fock equation of state of Ref.[10], which has been combined at subnuclear densities with the Baym-Pethick-Sutherland equation of state. Hence the white dwarfs shown in Fig. 2 are computed for the latter. The gravitationally bound stars with radii $\lesssim$ 200 km or $\gtrsim$ 3000 km represent stable neutron stars and white dwarfs, respectively.

The fact that strange stars with crusts tend to possess somewhat smaller radii than neutron stars leads to smaller rotational mass shedding (Kepler) periods P_K for the former, as is indicated classically by $P_K = 2\pi\sqrt{R^3/M}$. Of course the general relativistic expression,

$$P_K \equiv 2\pi/\Omega_K\,, \quad \text{with} \quad \Omega_K = \omega + \frac{\omega'}{2\psi'} + e^{\nu-\psi}\sqrt{\frac{\nu'}{\psi'} + \left(\frac{\omega'}{2\psi'}e^{\psi-\nu}\right)^2}\,, \quad (1)$$

which is to be applied to neutron and strange stars, is considerably more complicated. It is to be computed simultaneously in combination with Einstein's equations (see Ref.[5] for details and further references),

$$\mathcal{R}^{\kappa\lambda} - \frac{1}{2} g^{\kappa\lambda} \mathcal{R} = 8\pi T^{\kappa\lambda}(\epsilon, P(\epsilon)), \qquad (2)$$

However the qualitative dependence of P_K on mass and radius as expressed by the classical expression remains valid. So one finds that, due to the smaller radii of strange stars, the complete sequence of such objects (and not just those close to the mass peak, as is the case for neutron stars) can sustain extremely rapid rotation[5]. In particular, our model calculations indicate for a strange star with a typical pulsar mass of $\sim 1.45\,M_\odot$ Kepler periods as small as $0.55 \lesssim P_K/\mathrm{msec} \lesssim 0.8$, depending on crust thickness and bag constant[8,5]. This range is to be compared with $P_K \sim 1$ msec obtained for neutron stars of the same mass.

The minimum-mass configurations of the strange-star sequences in Fig. 2 have masses of about $\sim 0.017\,M_\odot$ (about 17 Jupiter masses) and $10^{-4}\,M_\odot$, depending on inner crust density. More than that, for inner crust densities smaller than 10^8 g/cm^3 we find strange-matter stars that can be even by orders of magnitude lighter than Jupiters. If abundant enough in our Galaxy, all these light strange stars could be seen by the gravitational microlensing searches that are being performed presently. Strange stars located to the right of the minimum-mass configuration of each sequence consist of small strange cores ($\lesssim 3$ km) surrounded by a thick nuclear crust, made up of white dwarf material. We thus call such objects strange dwarfs. Their cores have shrunk to zero at the crossed points. What is left is an ordinary white dwarf with a central density equal to the inner crust density of the former strange dwarf[5,9]. A detailed stability analysis of strange stars against radial oscillations[5] shows that all those strange-dwarf sequences that terminate at stable ordinary white dwarfs are stable against radial oscillations. Strange stars that are located to the left of the mass peak of ordinary white dwarfs, however, are unstable against oscillations and thus cannot exist stably in nature. So, in sharp contrast to neutron stars and white dwarfs, the branches of strange stars and strange dwarfs are stably connected with each other[5,9].

Finally the strange dwarfs with 10^9 g/cm$^3 < \epsilon_{\mathrm{crust}} < 4 \times 10^{11}$ g/cm^3 form entire new classes of stars that contain nuclear material up to $\sim 4 \times 10^4$ times denser than in ordinary white dwarfs of average mass, $M \sim 0.6\,M_\odot$ (central density $\sim 10^7$ g/cm^3). The entire family of such strange stars owes its stability to the strange core. Without the core they would be placed into the unstable region between ordinary white dwarfs and neutron stars[9].

5   Thermal Evolution of Neutron Stars and Strange Stars

The left panel of Fig. 3 shows a numerical simulation of the thermal evolution of neutron stars. The neutrino emission rates are determined by the modified and direct Urca processes, and the presence of a pion or kaon condensate. The baryons are treated as superfluid particles. Hence the neutrino emissivities are suppressed by an exponential factor of $\exp(-\Delta/kT)$, where Δ is the width of the superfluid gap (see Ref. [11] for details). Due to the dependence of the direct Urca process and the onset of meson condensation on star mass, stars that are too light for these processes to occur (i.e., $M < 1\,M_\odot$) are restricted to standard cooling via modified Urca. Enhanced cooling via the other three processes results in a sudden drop of the star's surface temperature after about 10 to 10^3 years after birth, depending on the thickness of the ionic crust. As one sees, agreement with the observed data is achieved only if different masses for the underlying pulsars are assumed. The right panel of Fig. 3 shows cooling simulations of strange quark stars. The curves differ with respect to assumptions made about a possible superfluid behavior of the quarks. Because of the higher neutrino emission rate in non-superfluid quark matter, such quark stars cool most rapidly (as long as cooling is core dominated). In this case one does not get agreement with most of the observed pulsar data. The only exception is pulsar PSR 1929+10. Superfluidity among the quarks reduces the neutrino emission rate, which delays cooling[11]. This moves the cooling curves into the region where most of the observed data lie.

Subject to the inherent uncertainties in the behavior of strange quark matter as well as superdense nuclear matter, at present it appears much too premature to draw any definitive conclusions about the true nature of observed pulsars. Nevertheless, should a continued future analysis in fact confirm a considerably faster cooling of strange stars relative to neutron stars, this would provide a definitive signature (together with rapid rotation) for the identification of a strange star. Specifically, the prompt drop in temperature at the very early stages of a pulsar, say within the first 10 to 50 years after its formation, could offer a good signature of strange stars[12]. This feature, provided it withstands a more rigorous analysis of the microscopic properties of quark matter, could become particularly interesting if continued observation of SN 1987A would reveal the temperature of the possibly existing pulsar at its center.

6   Summary

This work deals with an investigation of the properties of the complete sequences of strange-matter stars that carry nuclear crusts. The following items

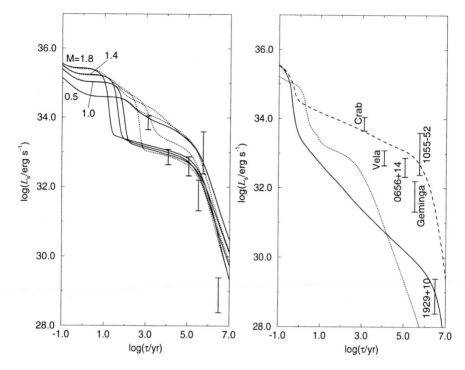

Figure 3: Left panel: Cooling of neutron stars with pion (solid curves) or kaon condensates (dotted curve). Right panel: Cooling of $M = 1.8\,M_\odot$ strange stars with crust. The cooling curves of lighter strange stars, e.g. $M \gtrsim 1\,M_\odot$, differ only insignificantly from those shown here. Three different assumptions about a possible superfluid behavior of strange quark matter are made: no superfluidity (solid), superfluidity of all three flavors (dotted), and superfluidity of up and down flavors only (dashed). The vertical bars denote luminosities of observed pulsars.

are particularly noteworthy:

1. The complete sequence of compact strange stars can sustain extremely rapid rotation and not just those close to the mass peak, as is the case for neutron stars!

2. If the strange matter hypothesis is correct, the observed white dwarfs and planets could contain strange-matter cores in their centers. The baryon numbers of their cores are smaller than $\lesssim 2 \times 10^{55}$!

3. The strange stellar configurations would populate a vast region in the mass-radius plane of collapsed stars that is entirely void of stars if strange

quark matter is not the absolute ground state!

4. If the new classes of stars mentioned in (2) and (3) exist abundantly enough in our Galaxy, the presently performed gravitational microlensing experiments could see them all!

5. Due to the uncertainties in the behavior of superdense nuclear as well as strange matter, no definitive conclusions about the true nature (strange or conventional) of observed pulsar can be drawn from cooling simulations yet. As of yet they could be made of strange quark matter as well as of conventional nuclear matter.

Of course, there remain various interesting aspects of strange pulsars, strange dwarfs and strange planets, that need to be worked out in detail. From their analysis one may hope to arrive at definitive conclusion about the behavior of superdense nuclear matter and, specifically, the true ground state of strongly interacting matter.

References

1. A. R. Bodmer, Phys. Rev. D **4** (1971) 1601.
2. E. Witten, Phys. Rev. D **30** (1984) 272.
3. H. Terazawa, INS-Report-338 (INS, Univ. of Tokyo, 1979); J. Phys. Soc. Japan, **58** (1989) 3555; **58** (1989) 4388; **59** (1990) 1199.
4. *Strange Quark Matter in Physics and Astrophysics*, Proceedings of the International Workshop, Aarhus, Denmark, 1991, ed. by J. Madsen and P. Haensel, Nucl. Phys. B (Proc. Suppl.) **24B** (1991).
5. N. K. Glendenning, Ch. Kettner, and F. Weber, Astrophys. J. **450** (1995) 253.
6. Ch. Kettner, F. Weber, M. K. Weigel, and N. K. Glendenning, Phys. Rev. D **51** (1995) 1440.
7. C. Alcock, E. Farhi, and A. V. Olinto, Astrophys. J. **310** (1986) 261.
8. N. K. Glendenning and F. Weber, Astrophys. J. **400** (1992) 647.
9. N. K. Glendenning, Ch. Kettner, and F. Weber, Phys. Rev. Lett. **74** (1995) 3519.
10. F. Weber and M. K. Weigel, Nucl. Phys. **A505** (1989) 779.
11. Ch. Schaab, F. Weber, M. K. Weigel, and N. K. Glendenning, "Thermal Evolution of Compact Stars", March 1996, submitted to Nuclear Physics *A*, (astro-ph/9603142).
12. P. Pizzochero, Phys. Rev. Lett. **66** (1991) 2425.

THE EQUATION OF STATE, FINITE NUCLEI, HYPERNUCLEI and NEUTRON STAR PROPERTIES IN THE RELATIVISTIC APPROACH

H. Huber, F. Weber, and M. Weigel

Sektion Physik, Universität München, Am Coulombwall 1, D-85748 Garching

The microscopic relativistic approach for nuclear systems is investigated in the framework of the relativistic Brueckner–Hartree–Fock–approximtion (RBHF). The bulk parameters of symmetric and asymmetric nuclear matter are in rather good agreement with the semiempirical values. Properties of finite nuclei and hypernuclei were calculated in a relativistic density–dependent Hartree–scheme, using the self–energies of nuclear matter as an input. The agreement of this approach with the data is also satisfactory. Furthermore neutron star matter is calculated (parameterfree) in the RBHF for densities up to twice nuclear matter density. The extension to higher densities with more baryons is performed in the relativistic Hartree–Fock–scheme, adjusted to the outcome of the RBHF–calculations for lower densities. The resulting equations of state are used in "neutron star" calculations for rotating configurations. So far, the present NS–data are in accordance with a pure baryon/lepton composition.

1 Introduction

The primary goal of every microscopic many–body theory is to "understand" the properties of the system in tems of the interactions between its constituents (so–called parameter–free theories). In nuclear physics the relativistic approach has some advantages in comparison with the nonrelativistic treatment, as, for instance, the natural incorporation of the spin–orbit force and a new saturation mechanism, which favors the repulsion with increasing density.[1,2] One basic attempt towards the microscopic relativistic description is the implementation of dynamical two–body correlations using modern one–boson exchange potentials (OBE–potentials) adjusted to the two–nucleon problem only. In order to test the capability of such an approach, one should require that one can reproduce the parameters of the Bethe-Weizsäcker formula (BWF) and/or the properties of finite nuclei. Both questions are related, but the determination of the parameters of the BWF has a conceptional advantage, since they can be studied in idealized systems, as, for instance, infinite nuclear matter (INM) etc. A further benefit of such a treatment is based on its microscopic origin, which makes it more reliable for extensions to higher densities. A famous example is neutron–star–matter (NSM), where one has to rely for the equation of state almost on theoretical and/or speculative extrapolations.

2 Theory

Since the basic ingredients of the theory are outlined in greater detail in several investigations (see, for instance, Refs. 2-6), we repeat only the general scheme.

The dynamics of the system is governed by a Lagrange density

$$\mathcal{L} = \mathcal{L}_N + \sum_M (\mathcal{L}_M + \mathcal{L}_{MN}) \quad , \tag{1}$$

where the three terms describe the noninteracting nucleons, the free meson fields and the interaction (i.e. OBE–potential).

By restriction to dynamical two–body correlations one obtains a coupled system of the Dyson equation for the Green's function G and the Bethe–Salpeter equation for the effective scattering matrix T in matter:

$$\left\{ (G_1^0)^{-1}(1,2) - \Sigma(1,2) \right\} G_1(2,1') = \delta(1,1') \quad , \tag{2}$$

$$< 12|T|1'2' > = < 12|v|1'2' - 2'1' > + < 12|v|34 > \Lambda(34,56) < 56|T|1'2') \ , \tag{3}$$

where the self–energy is given by

$$\Sigma(1,2) = -i < 14|T|52 > G_1(5,4) \quad . \tag{4}$$

The Hartree – or Hartree – Fock approximatin is defined by $T = v$ and $T = v_{AS}$, respectively. For the intermediate $p - p$ propagator a standard choice is the Brueckner propagator Λ_B, on which we will concentrate in this contribution (for more details, see Refs. 3,4). The relativistic Hartree – (RH) and Hartree – Fock – approximations (RHF) are not suited, since no binding is obtained for realistic OBE – potentials in these schemes.[4]

3 Calculations and Results

3.1 Symmetric and asymmetric nuclear matter

By calculating the properties of symmetric and asymmetric infinite nuclear matter (INM, AINM) one can determine the volume parameters of the BW–formula, defined via[3] density of

$$e(\rho, \delta) = (a_v + K_v \epsilon^2 + \cdots) + \delta^2 (J + \frac{1}{3} L \epsilon + \frac{1}{18} K_{sym} \epsilon^2 + \cdots). \tag{5}$$

The calculations for these systems were performed in the RBHF–approximation (for more details, see Refs.[3,5] using three different OBE–potentials – constructed by Brockmann and Machleidt – denoted by A, B, and C, which differ mainly by the strength of the tensor force increasing from A to C. Since the standard procedure of solving the RBHF–equations in the Dirac space of positive energy–spinors only can lead to nonunique self–energies (for details, see Ref.[7]), we have performed the calculations in the full Dirac space, which renders the calculations more complicated.

A detailed comparison (see Table III of Ref. 3) shows that the values for E/A, ρ_{00}, K_v, and J agree very well with the semiempirical values and even L and K_{sym} follow nicely the expected pattern. The EOSs for AINM are given in Fig. 1 for the potential B (for A one obtains rather similar EOSs). Furthermore it turns out that the quadratic approximation is fulfilled in the RBHF–approach. However the symmetry energy bends much less a function of the density as in non–relativistic approaches (for details, see Ref.[3]), which has a significant impact on the composition of the EOS in neutron stars.

3.2 Finite Nuclei

The RBHF–treatment for finite nuclei is too complicated. In order to test the compliance of the RBHF–approach with the properties of finite nuclei, we follow a proposal of Brockmann and Toki[8], extended by Lenske and Fuchs[9] by inclusion of rearrangement contributions, in which one uses the RH–method with density–dependent couplings adjusted in LDA–approximation to RBHF–self–energies. Due to the density dependence of the interaction

$$\mathcal{L}_{int} = \bar{\psi}\Gamma_\sigma(\rho_0)T\psi - \bar{\psi}\gamma_\mu\Gamma_\omega(\rho_0)\omega^\mu\psi + \cdots \quad , \tag{6}$$

one obtains via the Euler–Lagrange equation

$$\frac{\partial \mathcal{L}_{int}}{\partial \bar{\psi}} = \frac{\partial \mathcal{L}_{int}}{\partial \bar{\psi}} + \frac{\partial \mathcal{L}_{int}}{\partial \rho_0}\frac{\delta \rho_0}{\delta \bar{\psi}} \tag{7}$$

a second (new) term, which leads to the so–called rearrangement contributions. A comparison with the experimental data, (binding energies, charge raddi etc.) shows that the agreement with the experimental data is satisfactory for this simple parameterfree treatment, where only the results of symmetric INM were used (with no adjustment of g_ρ). The inclusion of rearrangement improves the agreement significantly. Also the charge density distributions are close to the experimental values (for more details, see Ref.[10])

3.3 Hypernuclei

The same approach was also used for hypernuclei, where we used the ad hoc assumption that we can use the same density–dependence for the coupling constants as for nucleons, only the relative coupling strengths were chosen according to the quark model. Surprisingly the agreement with the data for $\Lambda-, \Sigma-$ and $\Xi-$ nuclei is almost as good as in phenomenological treatments (for more details, see Ref.[11])

3.4 Neutron Star Matter

For NSs the density stretches from zero at the star's edge to several times the density of normal nuclear matter at its core. Starting from the two outer crusts with crystalline structures, one has to treat in the uniform matter region ($\rho \geq 4 \times 10^{14} g/cm^3$) a system of interacting baryons (i.e. p, n, hyperons, possibly $\Delta's$) and/or quarks that are in generalized β–equilibrium with leptons (e^-, μ^-). Constraints are zero total electric charge and baryon number conservation, but strangeness is not conserved. Here we will restrict ourselves to the case, for which NSM consists of hadron/lepton matter solely.

α) RBHF – calculation for NSM (p, n, e^-, μ^-):
Since the onset of hyperon population happens at densities around twice nuclear matter density, we implement first the ARBHF – scheme for uniform matter, where only protons, neutrons, electrons, and muons (in β–equilibrium) are present in NSM. The task is even more complicated than in AINM, since β–equilibrium plus charge constraint has additionally to be invoked.
Due to the new degrees of freedom the EOS becomes softer as the EOS of pure neutron matter. This behavior is illustrated for different EOSs in Fig. 2.

β) Extension to higher densities:
Since for higher densities with the onset of the hyperon population the RBHF – approach is not tractable, we have selected for higher densities the extension via the RHF – approximation, since the formal and the OBEP – structure is similar as in RBHF. In this scheme one can include more baryons (16 baryons). The dynamics, i.e. the Lagrangian has to be determined such that for lower densities agreement with the RBHF – NSM is obtained. The new and much more difficult problem is, that in comparison with former treatments, which are adjusted only to the properties of INM, one faces now the problem to reproduce the properties of NSM (i.e. pressure, nucleon/lepton composition etc.). As can be seen from Fig. 3, two parameter sets, which fit the properties of symmetric INM, lead to different predictions for asymmetric matter with fixed asymmetries. Such deviations occur also for NSM, a typical example is

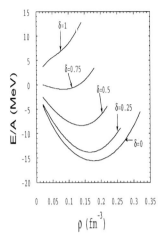

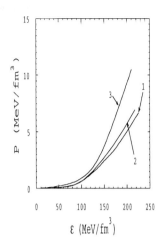

Figure 1: EOSs for asymmetrical INM in the ARBHF for different asymmetries (Potential B).

Figure 2: Pressure versus energy-density for NSM (potential B) in the RBHF: curve 1: p, n, e^- and μ^-; curve 2: p, n, e^-; curve 3: pure neutron matter.

given in Fig. 4, where we show the agreement of the baryon/lepton composition. For the extension we use then the parametrization of the RHF – Lagrangian, which gives the "best" agreement (pressure, baryon/lepton composition, chemical potentials etc.) with the RBHF – NSM calculations (for more details, see Ref.[12]).

4 Neutron Stars

Since the obtained EOSs for NSM are based on a microscopic and relatively parameterfree description it seems worthwhile to use these EOSs in NS–calculations. For a given EOS, one can determine Einstein's curvature tensor via

$$G_{\mu\nu} = 8\pi T_{\mu\nu}(\epsilon, P(\epsilon)) \quad . \tag{8}$$

In order to determine the properties of (rapidly) rotating NSs and their limiting rotational periods, one has to generalize the standard Schwarzschild–metric to

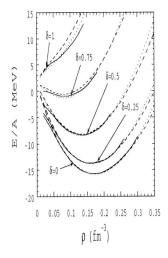 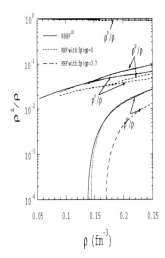

Figure 3: Comparison of the RBHF–EOSs (solid curves) for AINM with its RHF parametrizations (pot. B).

Figure 4: Nucleon/lepton composition of NSM in β–equilibrium for RBHF and RHF (potential A).

a generalized Schwarzschild metric for rotating, axially symmetric configurations (for details, see Ref.[13]).

Of special interest – besides the standard properties of NSs – is the problem of stability of such objects for rapid rotations. Since general criteria do not exist in GR, we used as criteria the mass shedding at the star's equator at the Kepler frequency Ω_K and the onset of gravitation–radiation instabilities, respectively.[13] β–equilibrium, The limiting rotational periods are given in Fig. 5. It turns out that the onset of gravitational radiation determines the limit of stable rotation. The rectangle covers the range of observed periods and masses. Since similar limitations were obtained for many EOSs (≈ 20 EOSs) with baryon/lepton composition, one can conclude that the observations so far are not in contradiction with the assumption of a baryon/lepton composition.[14] However one has to keep in mind that technical limitations in radioastronomy do not permit – at present – to measure faster rotations. Detection of periods considerably smaller than 1 ms would lead (inevitably) to another (new) scenario (strange stars? ; see Refs.[15,16]).

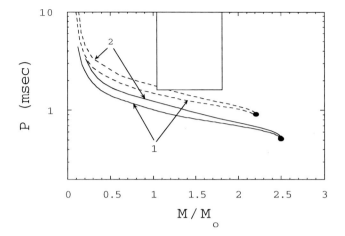

Figure 5: Limiting rotational periods P of pulsars. The solid curves correspond to NS rotating with their Kepler frequencies Ω_K, the dashed curves give the limits due to instability of gravitation–radiation. The curves with label 2 correspond to the case of pure NM. The rectangle covers the range of observed periods and masses. Therefore a pure hadronic/leptonic composition is so far in compliance with the data.

5 Conclusions

In the frame of the RBHF – theory with modern OBE – potentials one obtains a rather satisfactory description of the properties of (a-) symmetric nuclear matter and of finite systems. The treatment is "parameterfree", since only the two–body interaction enters. Also for hypernuclei one gets a description similar in quality as phenomenological approaches. The resulting (softer) EOS in β – equilibrium, relevant for NS, can explain the present data with respect to masses, radii, etc. and rotation frequencies. Periods smaller then 1 ms would contradict such a baryon/lepton composition and require another scenario[15,16].

References

1. B. D. Serot and J. D. Walecka, Adv. Nucl. Phys. **16**, 1 (1986), and references therein.

2. L. S. Celenza and C. M. Shakin, Relativistic Nuclear Physics (World Scientific, Singapore 1986), and references therein.
3. H. Huber, F. Weber, and M. K. Weigel, *Phys. Rev.* C **51**, 1790 (1995), and references therein.
4. P. Poschenrieder and M. K. Weigel, *Phys. Rev.* C **38**, 471 (1988), and references therein.
5. B. ter Haar and R. Malfliet, *Phys. Rev. Lett.* **59**, 1652 (1990).
6. R. Brockmann and R. Machleidt, *Phys. Rev.* C **42**, 1965 (1990).
7. C. Nuppenau, Y. J. Lee, and A. D. MacKellar, *Nucl. Phys.* A **504**, 839 (1989), and references therein.
8. R. Brockmann and F. Toki, *Phys. Rev. Lett.* **68**, 3408 (1992).
9. H. Lenske and C. Fuchs, *Phys. Lett.* B **345**, 355 (1995).
10. F. Ineichen, M. K. Weigel, and D. Von-Eiff, "Nuclear structure calculations in the density–dependent Hartree theory", Journal *Phys. Rev.* C (in press, May 1996).
11. F. Ineichen, D. Von-Eiff, and M. K. Weigel, "A Density–Dependent Relativistic Hartree Approach for Hypernuclei", preprint 1995, Sektion Physik, Universität München.
12. H. Huber, F. Weber, and M. K. Weigel, *Nucl. Phys.* A **596**, 684 (1996).
13. F. Weber, N. K. Glendenning, and M. K. Weigel, ApJ **373**, 579 (1991), and references therein.
14. H. Huber, F. Weber, and M. K. Weigel, *Phys. Rev.* C **50**, R1287 (1994).
15. Ch. Kettner, F. Weber, M. K. Weigel, and N. K. Glendenning, *Phys. Rev.* D **51**, 1440 (1995).
16. F. Weber, Ch. Schaab, M. K. Weigel, and N. K. Glendenning, "Quark Matter, Massive Stars, and Strange Planets", elsewhere, and references therein.

1. G. Münzenberg, R. Mattiello and H. Stöcker preparing for a walk.

2. Fritz Hahne and his crew: Shann Wyngaardt, Brandon van der Ventel, Gregory Hillhouse, and Hans Eggers.

3. J. Schukraft, G. Röpke, Oda Stöcker and J. Aichelin.

1. On the road: G. Röpke, H. C. Britt, Donna Britt, Sybille Mersmann, and W. Scheid at the end.

2. J. Nagle, S. Bass, N. Cindro, Elizabeth Cindro, Nelly Eisenberg and J. Eisenberg.

4. Ultrarelativistic Heavy Ion Collisions, Hot Matter versus Quark-Gluon Plasma

On Hadronic Particle Ratios and Flow in Ultra-Relativistic Nucleus-Nucleus Collisions

P. Braun-Munzinger[1] and J. Stachel[2]

[1] *Gesellschaft für Schwerionenforschung, Darmstadt, Germany*
[2] *Physikalisches Institut der Universität Heidelberg, Germany*

We discuss recent data on particle production and directed flow with emphasis on the degree of thermal and chemical equilibration achieved, and on the dependence of the directed flow signal on particle species.

1 Introduction

Collisions between ultra-relativistic heavy nuclei create matter of high initial energy and particle density. This translates, for the experimental program currently underway at the fixed target machines, to regions of very high initial baryon density and temperatures of the order of 150 - 200 MeV. Under such conditions it is expected that a phase of deconfined quarks and gluons, where also chiral symmetry is restored, is more stable than hadronic matter.

An experimental heavy ion program has been underway for nearly ten years at the Brookhaven AGS and the CERN SPS. In a number of experiments many hadronic observables have been measured, often over a large fraction of the full solid angle [1,2]. With these experimental data the final freeze-out stage has been well characterized in terms of the hadronic observables at least for the relatively light Silicon and Sulfur beams and similar data for the heaviest beams, Gold and Lead, are rapidly becoming available. In addition, detailed data on directed flow in Au+Au collisions at AGS energy have recently been reported by the E877 collaboration [3,4,5,6] . We want to briefly survey here to what extent the final hadronic stage can be characterized in terms of the concepts of thermal and hadrochemical equilibrium, *i.e.* to what extent we can talk about a thermodynamic phase. Although one might think that thermodynamic equilibrium at freeze-out implies that there is no information in hadronic observables on the hot and dense phase formed in the collision, we will argue with the example of directed flow that this is not the case.

2 Thermodynamical Analysis of Particle Production Ratios

As a starting point for the present discussion we use results from recent numerical solutions of QCD on the lattice. For zero net baryon density the critical temperature for the transition between the hadronic and the quark-gluon phase has been obtained by several groups[7]: $T_c = 145 \pm 10$ MeV. However, systematic uncertainties in the calculation of T_c with dynamical fermions might still allow a range of up to 200 MeV[8]. Lattice QCD calculations do not, up to now, shed light on the properties of the phase transition at finite baryon density. In order to continue the lattice results into the region of finite baryon density relevant for present fixed target experiments we construct[9,10] the phase boundary by equating chemical potential μ_B and pressure P of a hadron resonance gas with the equivalent quantities of a non-interacting quark-gluon plasma. The hadron resonance gas contains all known baryons and mesons up to a mass of 2 and 1.5 GeV, respectively. Interactions among the baryons are approximately taken into account by an excluded volume correction[10]. The plasma phase contains massless gluons, u and d quarks and strange quarks with $m_s = 150$ MeV. A bag constant of B = 262 MeV/fm^3 is used to insure that the transition for $\mu_B = 0$ takes place at $T_c = 145$ MeV, consistent with the lattice QCD results.

With this simple thermal model we demonstrate in the following to what accuracy the hadron abundances from AGS and SPS experiments with Si and S beams can be described in terms of the two thermodynamic variables T and μ_B. The comparison of the thermal model predictions with experiment is shown in Figure 1 and Figure 2. The details of the analysis are described in [9,10]. The resulting values are $\mu_B = 0.54$ GeV, $120 \leq T \leq 140$ MeV at AGS energy and $\mu_B = 0.17(0.18)$ GeV, $T = 160(170)$ MeV. The overall agreement between the measured particle ratios and the thermal model predictions is rather impressive, on the 20% level or better in most cases. Note that, because of the large baryon chemical potential and relatively low temperature, the particle ratios at AGS energy vary over six orders of magnitude, while the ratios measured at the SPS are much more bunched together, consistent with a relatively small μ_B value and higher temperature.

We note, however, that the SPS data are not yet as complete as the AGS data. In particular, extrapolation of the particle ratios to those expected from a 4π measurement is not yet possible for most particle species. Consequently, flow effects, which to first order cancel out for particle ratios integrated over the full solid angle, may still influence somewhat the values displayed in Figure 2. Furthermore, the K/π ratios are not included in Figure 2 because final values are not yet available. Preliminary results[11] indicate values of $K^+/\pi^+ \approx 0.15$ for the systems S+Ag and Pb+Pb. This would be nearly 30 % smaller than what is predicted by the thermal model. Whether this indicates a difference between thermal and chemical freeze-out for pions and kaons, or has some other dynamical origin, is an interesting question to be followed up once final data are available.

The baryon density at freeze-out for both the AGS and the SPS data is

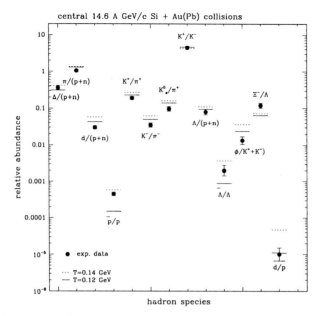

Figure 1: Comparison of thermal model prediction to experimental data for hadron abundance ratios at AGS energy. For details see text.

$\rho_B \approx 0.05\text{-}0.08/\text{fm}^3$, while the corresponding pion densities are $\rho_\pi \approx 0.08/\text{fm}^3$ (AGS, $T = 120$ MeV) and $\rho_\pi \approx 0.30/\text{fm}^3$ (SPS, $T = 160$ MeV). This may reflect the fact that freeze-out at the AGS, where the pion/nucleon ratio is near unity, is determined by the pion-nucleon and nucleon-nucleon dynamics and the associated large cross sections (~ 100 mb) while the small $\pi\pi$ cross section (~ 10 mb) is the relevant quantity at SPS energies where pions dominate nucleons by about 5 to 1 in the central region. To within the accuracy of present data this analysis shows that the hadronic freeze-out configuration is close to thermal and chemical equilibrium. In particular, the strangeness suppression which is well known for data from nucleon-nucleon collisions is not observed in nucleus-nucleus collisions. Rather, strangeness degrees of freedom are close to the values expected for a hot hadronic system in chemical equilibrium.

If chemical equilibration of the strangeness degrees of freedom at freeze-out is confirmed for the heavy systems such as Au+Au at the AGS and Pb+Pb

346

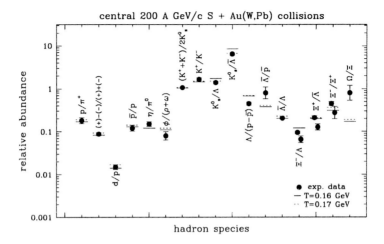

Figure 2: Comparison of thermal model prediction to experimental data for hadron abundance ratios at SPS energy. The notation (+) or (-) refers to the density of positively or negatively charged hadrons. For more details see text.

at the SPS, this could be considered evidence, albeit indirect, that either the phase boundary has been crossed or that at least partial restoration of chiral symmetry with a concomitant reduction of hadron masses has been achieved in the hot and dense fireball created in the collision.

Similar conclusions may be drawn from Figure 3 , where we have compared the above determined experimental freeze-out parameters with the calculated position of the phase boundary. The surprising result is that, even at freeze-out, the fireball parameters are close to those expected for a the quark-gluon plasma. Although the system is, by definition, purely hadronic at freeze-out, its proximity to the phase boundary suggests that the boundary was reached or even crossed during the course of the evolution of the fireball towards freeze-out. The arrows indicate where the systems should evolve from if one takes predictions by cascade models such as RQMD as guideline for the system parameters at the time of maximum compression and temperature.

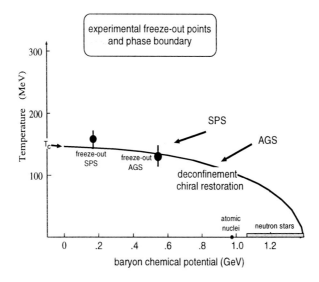

Figure 3: Comparison of experimentally determined freeze-out parameters at AGS and SPS energy with the phase boundary calculated in the present approach.

3 Flow

As was discussed in the previous section, particle ratios integrated over the full solid angle are close to what one expects for thermal and chemical equilibration. On the other hand, analysis of particle spectra[9,10,12] shows that the hot and compressed fireball expands both longitudinally and transversely. For Si+Au at AGS energy the expansion velocity is on average 52 % and 33 % of the speed of light in longitudinal and transverse direction, respectively. Corresponding numbers for S+Au at the SPS are 75 % and 27 %. In addition to this expansion, the system, at least for collisions between heavy nuclei such as Au+Au, also preserves a memory of the impact parameter direction, as shown by the analysis of the azimuthal distribution of transverse energy backward and forward of midrapidity[3].

Using more detailed data from the E877 calorimeters and multiplicity detectors, and taking advantage of the particle identification capabilities of the

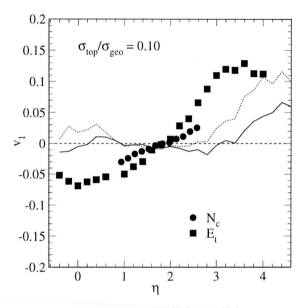

Figure 4: Comparison of the flow parameters $v_1^{(N_c)}$ and $v_1^{(E_t)}$ (solid symbols) for the centrality range 5-15 %. Also shown are the equivalent parameters extracted from RQMD events (solid and dashed lines).

E877 forward spectrometer, the E877 collaboration has recently managed to provide much more detailed information on the flow [4,5]. Applying a Fourier decomposition of the distribution of charged particle multiplicity and transverse energy with respect to the reaction plane one can determine the Fourier coefficients v_i for different multipolarities i as function of centrality and pseudorapidity. In practice, the $i = 1$ term is by far the dominant one, implying that, in the plane perpendicular to the beam direction, the two-dimensional distributions $d^2N/dp_x dp_y$ are not centered at $(p_x = 0, p_y = 0)$ but are shifted along or opposite to the direction of the impact parameter vector, depending on the rapidity interval studied. This dipole term v_1, unfolded for effects due to the finite resolution of the reaction plane or impact parameter vector measurement, is shown in Figure 4 for transverse energy E_t and charged particle multiplicity N_c distributions from Au+Au collisions at a beam energy of 10.8 A GeV/c and intermediate centrality.

The data now cover the full pseudorapidity range $-0.5 < \eta < 4$ and exhibit the typical S-shaped behavior observed in flow studies at lower energies. Note that the experimental anisotropies of the charged particle and of the transverse energy distributions are significantly different: at a given pseudorapidity the v_1 values for E_t are nearly a factor of two larger than those for N_c. Since both distributions are dominated by pions and nucleons whose relative weights are quite different for E_t and N_c distributions, one may attribute this difference to different flow patterns for pions and nucleons.

The observed experimental dipole anisotropies cannot be reproduced by calculations using the cascade code RQMD [13]. The predicted flow parameters are significantly smaller than what is observed, both for E_t and N_c distributions. Looking into RQMD in more detail [5] this is traced, in part, to the fact that in this model pions and nucleons show opposite flow behavior, leading to large cancellations for E_t and N_c distributions.

Clearly this result makes it very important to separate experimentally the flow due to pions from that of nucleons. We have achieved this with the E877 calorimeter and charged particle multiplicity data in two different ways. First, we assume that the observed flow effect in the global observables E_T and N_c is a linear superposition of the anisotropy of nucleons and pions, thereby neglecting the contribution from other particle species. We denote the corresponding flow parameters by $v_1^{(N_c)}$ and $v_1^{(E_T)}$ and further differentiate between coefficients $v_1^{(N_c,n)}$, $v_1^{(N_c,\pi)}$, $v_1^{(E_T,n)}$, and $v_1^{(E_T,\pi)}$ for nucleons and pions, respectively.

Under the above assumptions the dipole anisotropy of the two global observables can then be written in the following way:

$$v_1^{(N_c)} = \frac{dN_c^\pi/d\eta \cdot v_1^{(N_c,\pi)} + dN_c^n/d\eta \cdot v_1^{(N_c,n)}}{dN_c^\pi/d\eta + dN_c^n/d\eta}, \tag{1}$$

$$v_1^{(E_T)} = \frac{dE_T^\pi/d\eta \cdot v_1^{(E_T,\pi)} + dE_T^n/d\eta \cdot v_1^{(E_T,n)}}{dE_T^\pi/d\eta + dE_T^n/d\eta}. \tag{2}$$

These equations can be solved for the flow parameters of pions and nucleons if one knows: (i) the relative contribution of pions and nucleons to the charged particle and transverse energy pseudorapidity distributions, and (ii) the ratio of the flow parameters measured using particle or E_T azimuthal distributions, i.e. $v_1^{(N_c)}/v_1^{(E_T)}$.

Using the by now rather complete information on particle distributions (averaged over the reaction plane orientation) from AGS experiments E866 and E877 one can indeed disentangle the two distributions [6]. The results are somewhat model dependent but cover the full pseudorapidity range. As a second alternative one can determine, as before, the reaction plane orientation

350

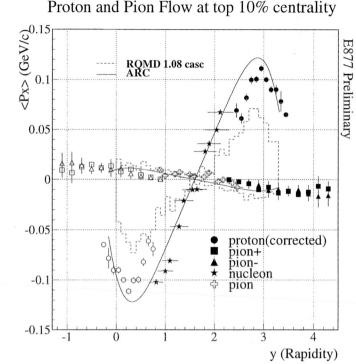

Figure 5: Flow parameters for charged pions and protons as a function of (pseudo-) rapidity for Au+Au collisions at AGS energy.

in the E877 calorimeters, but measure the distributions relative to the reaction plane, in the E877 forward spectrometer. This has the advantage of full particle identification. On the other hand, the results are limited to the acceptance of the E877 spectrometer, *i.e.* about one unit of rapidity or more forward of the center-of-mass. The results of both methods are shown in Figure 5. In this Figure we have plotted the mean transverse momentum in the reaction plane $\langle p_x \rangle$ *vs* rapidity y. The data labelled proton, pion+ and pion- are from the spectrometer analysis, corrected for the acceptance of the spectrometer. The data labelled nucleon and pion are from the calorimeter analysis described above. What was already suspected before becomes very apparent here: the nucleons are the main carriers of flow, while the pion distributions exhibit only a weak flow effect, opposite in direction to that of the nucleons. For the nucleons, a maximum values of $\langle p_x \rangle \approx 110$ MeV/c is reached near beam rapidity ($y_{beam} = 3.15$). The data are compared to the predictions from RQMD and ARC. For the pions both event generators reproduce the measured shapes very well, while there are significant differences for the nucleons. Predictions for protons from RQMD without inclusion of mean field effects [13] have roughly the observed shape, but the maximum $\langle p_x \rangle$ value is only about $1/2$ of that measured. Inclusion of mean field effects leads to an increase in flow [14] and we look forward to a detailed comparison of such model predictions with our data. The ARC calculation, where a parameterized energy dependence of the nucleon-nucleon interaction is used [15], shows a flow pattern which is close to, although somewhat above the data. This may be an indication that the nucleon-nucleon interaction is less repulsive at high density. These comparisons indicate that the recent flow data from the E877 collaboration can be used to test and hopefully improve the assumptions underlying such codes with the ultimate hope to gather information about the equation of state.

4 Conclusion

We have shown that data on hadron production in ultrarelativistic nucleus-nucleus collisions exhibit many features expected for systems which are in thermal and chemical equilibrium. In addition, there is now convincing evidence in the data for the presence of hydrodynamical flow effects. If we will succeed, as it is planned, to determine experimentally the flow also for produced particles such as $K^{\pm}$ and antiprotons, the analysis of such flow patterns should bring us a significant step further in the quest for the determination of the equation of state.

References

1. see, *e.g.* Proc. Quark Matter '95 Conference, Nucl. Phys. **A590** (1995) and references there.
2. see, *e.g.* Proc. Quark Matter '96 Conference, Nucl. Phys. **A610** (1996), in print.
3. J. Barrette et al., E877 collaboration, Phys. Rev. Lett. **73**, 2532(1994).
4. J. Barrette et al., E877 collaboration, ref. 1, p. 259c.
5. T. Hemmick et al., E877 collaboration, ref. 2.
6. J. Barrette et al., E877 collaboration, preprint Sep. 1996.
7. C. DeTar, in Proc. "Lattice '94", Nucl. Phys. **B42** (Proc. Suppl.), 73(1995).
8. E. Laermann, ref. 2.

9. P. Braun-Munzinger, J. Stachel, J.P. Wessels, and N. Xu, Phys. Lett. **B344**, 43(1995).

10. P. Braun-Munzinger, J. Stachel, J.P. Wessels, and N. Xu, Phys. Lett. **B365**, 1(1996).

11. Reinhard Stock, private communication.

12. U. Heinz, Nucl. Phys. **A566**, 205c(1994); J. Sollfrank and U. Heinz, preprint HU-TFT-95-27, to be published in "Quark-Gluon Plasma 2", R.C. Hwa, Editor, World Scientific.

13. R. Mattiello, A. Jahns, H. Sorge, H. Stöcker, and W. Greiner, Phys. Rev. Lett. **74**, 2180(1995) and refs. therein.

14. R. Mattiello, Dissertation, Frankfurt, May 1995.

15. D.E. Kahana, Y. Pang, and E. Shuryak, preprint nucl-th/9604008.

Recent Results from the WA93 Experiment

I. Langbein[1], S. K. Nayak[2], Y. P. Viyogi[3], H. H. Gutbrod[4]
and the WA93 Collaboration

[1] *Gesellschaft für Schwerionenforschung mbH, Darmstadt*
[2] *Institute of Physics, Bhubaneswar 751005, India*
[3] *Variable Energy Cyclotron Centre, Calcutta 700064, India*
[4] *SUBATECH, Ecole des Mines, 44070 Nantes, France*

Abstract

The WA93 experiment at CERN SPS employed for the first time a large preshower detector with a high granularity for the event-by-event measurement of multiplicity and rapidity- and angular distribution of photons produced in 200 AGeV S + Au collisions [1]. The data have been analysed to investigate multi photon correlations leading to photon clusters containing up to 30 photons. First results on the frequency of the cluster size are compared with predictions of Monte-Carlo simulations using the VENUS event generator and also with a purely random distribution of hits.

1 Introduction

With the ultrarelativistic heavy ion experiments at the CERN-SPS one tries to achieve through nuclear collisions energy densities and temperatures in nuclear matter which are close to those of the Early Universe. It is assumed that at a given energy density and temperature the quarks and gluons will be deconfined and build a quark-gluon plasma. During the expansion a phase transition into the hadronic world occurs. Photons are emitted reflecting the black body temperature at each given state. Unfortunately the distribution of these photons is dominated by that of the photons from the decay of e.g. π^0s and ηs. But since these hadrons carry also information on the dynamics of the reaction, we investigate the distribution of photons per event.

Hwa et al. [2] studied the formation of hadron clusters in heavy ion collisions in view of cluster formation. He assumes that hadrons in the mixed phase are surrounded by quarks and build a condensation core for the hadronisation of the quarks, similar to the condensation in thermodynamics. These hadrons may stick together and could form clusters. It is predicted that the growth of the clusters will be of a non trivial behaviour in the case of a first order phase transition and that will be reflected in the distribution of the cluster size.

In the mixed phase a cluster can appear randomly. The probability P(S) finding clusters of a given size S is assumed to be constant over a certain range and will rapidly fall to zero after attaining the maximum size. On the other hand, if the non linear growth process is limited by the size of the total system and the duration of the mixed phase, a stochastic growth process will dictate the behaviour of P(S). In this case a power law

$$P(S) \propto S^a \tag{1}$$

is expected for a certain range of S. Such a behaviour can be taken as a signature of a phase transition.

Measurements of the distribution of charged hadrons in high energy physics have been performed since a long time. Two particle correlations, i.e. Bose-Einstein correlation (HBT) are a well established tool to determine the dynamics and the geometry of a heavy ion collision. However, these correlations are subject to strong Coulomb corrections due to the Coulomb repulsion. Correlations of neutral pions do not have to be corrected for the Coulomb repulsion. However, due to their short life time the detection of neutral pions is not possible. Only the photons from their decay can be detected. If correlations of neutral pions can be studied at large momenta as in the case of S + Au reactions at 200 AGeV where measurements are done in the forward hemisphere of the centre of mass system, then we may consider that the distribution of the photons reflect the spacial distribution of the neutral pions. Bose condensation and pion clusters, if they occur, can be studied through the investigation of the photon pattern.

The investigation of photon clusters should also lead to a non trivial distribution of the cluster size in the case of a first order phase transition. Two different methods to define clusters have been employed and will be described in chapter 3. The results of both methods are compared with Monte-Carlo simulations using the VENUS event generator as well as with the cluster production in a random distribution of photons.

2 Experimental Setup

The experiment was performed at the CERN-SPS using the WA93 set up which is described in detail in [3]. The multiplicity, the rapidity- and angular distribution of photons produced in 200 AGeV S + Au collisions were measured with the Photon Multiplicity Detector (PMD), a highly granular preshower detector. Two calorimeters were used to obtain the centrality of an event. Photon multiplicities up to $N_\gamma \sim 250$, rapidity densities of up to dN/dy ~ 200

in central events and the angular distributions have been measured. The Zero-Degree-Calorimeter (ZDC) measures the forward energy distribution ($\eta > 6$) which is mainly determined by projectile spectators. The total transverse energy is measured with the Mid-Rapidity-Calorimeter (MIRAC). It was shown that the total number of photons, the mean values of the ZDC energy and the total transverse energy are strongly correlated [4].

The PMD used in this experiment has a 1.6 m × 2.0 m large lead converter of three radiation length thickness where the photons are converted into an electromagnetic shower. The shower particles generate light in the subsequent sensitive medium of the detector, 7600 scintillator pads. Other charged particles also generate light in the scintillators, but the signal is much lower than the signal from the showers. The light is collected via wavelength shifting (WLS) fibers and is then guided to the entrance window of the readout chains.

After the hit reconstruction $\gamma - like$ clusters are selected by applying a threshold to the ADC values to suppress signals from charged particles. The chosen threshold at three times the signal given by a Minimum Ionising Particle (MIP) passing through the scintillator gives a reasonably high photon counting efficiency in the range of 65 % to 75 % depending on the centrality. One can find a more detailed description of the detector in reference [3].

3 Analysis Methods

These photons are investigated to search for correlations and for emission pattern which could be related to critical phenomena such as the phase transition from Quark-Gluon-Plasma (QGP) to hadrons. Due to their origin from the decay of higher resonances photons may appear in groups, reflecting the decay branch. For example approximately 30% of the η's produced in the nuclear collision decay into three π^0 and subsequently into six photons. Since the relative momentum is much smaller than the boost, these six photons will stay together and may be detected as a group. The largest fraction of photons results from the π^0 decay and will therefore reflect the spatial distribution of the π^0s. If the emission of π^0s is isotropic or random one should see this in the spatial distribution of the photons. Different methods are developped to study the photon emission pattern with regard to the search of clusters and multi photon correlation. The questions to answer are, wether the photon distribution is totally random or wether the distribution of the cluster size shows a behaviour which leads to the assumption that there are large resonances or hadron clusters decaying into photons. Does the probability distribution of the cluster size follow a power law and will this be a clear signature of the existence of a mixed phase with hadron clusters during the quark-hadron phase transition?

3.1 Cluster definition with the Minimal-Spanning-Tree Method

One of the methods for the definition of clusters is based on the construction of a Minimal Spanning Tree (MST) among all hits in the pattern of an event from the PMD. All distances from one point to all other points are determined and can be defined in different metrices. The following definition was used:

Metric (η - ϕ space): $\sqrt{\Delta\eta^2 + \Delta\phi^2}$

Δ η and Δ ϕ are normalized to the geometrical range of the PMD which is used for the analysis. For each event all hits are connected in such a way that the sum of all connections in the tree attains a minimum. The median is found from the distances which span the tree and is used as the splitting criterion. After eliminating all connections larger than this given length groups of connected hits remain which are called clusters. This splitting criterion always defines clusters in any distribution and was already successfully used by Dorfan to find three-jet events at PEP and PETRA energies [5]. Characteristic numbers of these clusters have been studied on an event-by-event basis for measured, simulated and mixed events. A characteristic parameter of a cluster is for example the number of photons contained in a cluster which is called the cluster size S. As the Minimal Spanning Tree Method finds clusters in all kinds of distributions of points, a comparison of the distribution of the cluster size from real data events to the distribution resulting from random or simulated events is required.

3.2 Cluster definition with cells in Δ y and Δ ϕ

For a further method the spacial distribution of the photon hits of the PMD is overlapped by a grid defined in polar coordinates with its origin at the center of the PMD. The cells of the grid are then described in units of rapidity Δ y and the azimuthal angle Δ ϕ. A cluster is now identified as an island of connected cells, containing at least one hit, in the y - ϕ space surrounded by one layer of empty cells. The total number of hits in this island is then defined as the cluster size S. We have studied the cluster size in terms of cell size by changing Δ y and Δ ϕ to obtain an idea of the cluster structure.

In the present analysis the smallest rapidity bin size was 0.15 and the azimuthal bin size was 0.1 radian. This binning is larger than our two track resolution at all rapidity values.

4 Results and Discussion

If the size distribution of photon clusters is influenced by the assumed hadron clusters in the mixed phase, it will follow a power law as predicted for the hadrons by Hwa et al. like $f(S) \propto S^\alpha$. The cluster size distribution will mainly reflect the distribution of π^0s and the underlying two photon correlation from the π^0 decay. A small fraction of the photons comes from the decay of higher resonances which may result in an enhancement of larger clusters compared to a totally random distributed sample of hits.

Random events were generated once by distributing hits randomly over the detector surface and once by mixing hits of real events thus destroying underlying correlations, but keeping effects from detector inefficiencies.

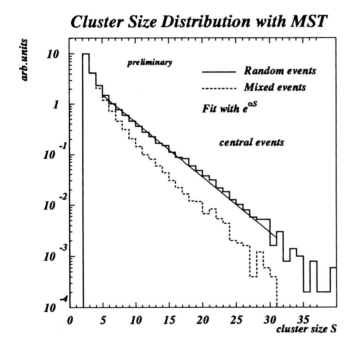

Figure 1: Distributions of the cluster size from mixed events and from random events together with an exponential function.

For both types of events the Minimal Spanning Tree method was used to define clusters. The resulting distributions of the cluster sizes are shown in figure 1. An exponential fit is made in the range of $5 < S < 30$. The reduction of large clusters in mixed events compared to random events may be attributed to the detector inefficiencies.

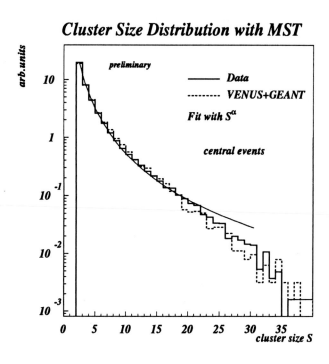

Figure 2: Distributions of the cluster size from real data and from simulated VENUS events filtered through the GEANT package and corrected for the readout inefficiency

In figure 2 the cluster size distribution resulting from real data is compared to simulated events, generated with the VENUS event generator and filtered through the GEANT package for the detector response. In the range of small clusters containing up to 20 photons, the slopes of the simulated and measured data agree well with the slope from a power law (solid line). As can be seen from the plot, events simulated with the event generator VENUS and the GEANT package, result in the same distribution of the cluster size as the real events.

Applying the second method to the same input events gives a slightly different shape of the cluster size distribution. In figure 3 the distribution of the cluster size is shown for data, simulated and mixed events for that kind of a cluster definition.

Cluster Size Distribution with Cells in $\Delta y - \Delta\phi$

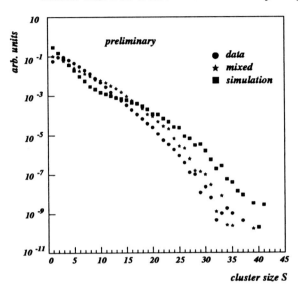

Figure 3: Distribution of the cluster size for the method where the clusters are defined via cells in y - ϕ space.

The distribution of the cluster size S can be described neither with an exponential function nor with a power law.

The distributions of the size of clusters which are defined with the MST method in real data and simulated events, can be fitted with a power law only in the range of small S. The differences at larger clusters may be due to the finite number of photons in one event and the limited acceptance of the PMD. In additon, in the MST method the found cluster size is always significantly smaller than the multiplicity in the event because of the use of the median as the splitting parameter. This may also be the reason for the different range in the frequency of the cluster sizes found with the two different methods.

The calculated and the measured distribution of photons are a superposition of photons coming from different sources with a major contribution from the π^0 decay and the neutral decay modes of ηs. The decay of other resonances may also result in groups of photons. All those resonance decays are taken into consideration for the generation of the events with the VENUS event generator. After applying the detector simulation by using the GEANT package, the resulting cluster size distribution from simulated events has the same slope as the cluster size distribution from real measured data.

References

[1] Proposal for WA93, WA93 Collaboration, CERN/SPSC/90 14, SPSC/P-252(1990)

[2] Hwa et al., Cluster Production in Quark-Hadron Phase Transition in Heavy-Ion Collisions, Phys.Rev.Lett. 72(1994)

[3] M.M.Aggarwal et al., WA93 Collaboration, A Preshower Photon Multiplicity Detector for the WA93 Experiment, Nucl.Instr.Meth A372(1996)143

[4] M.M.Aggarwal et al., WA93 Collaboration, Multiplicity and Rapidity Distribution of Photons in S + Au Reactions at 200 A GeV, to be published

[5] J.Dorfan, A Cluster Algorithm for the Study of Jets in High-Energy Physics, Z.Phys.C, Particles and Fields 7(1981)

HADRONIC RATIOS IN Si-Au COLLISIONS

J. CLEYMANS, D. ELLIOTT
Department of Physics, University of Cape Town,
Rondebosch 7700, South Africa

H. SATZ
Fakultät für Physik, Universität Bielefeld,
D-33501 Bielefeld, Germany

R.L. THEWS
Department of Physics, University of Arizona,
Tucson, AZ85721, USA

The most abundantly produced hadron species in $Si-Au$ collisions at the BNL-AGS (nucleons, pions, kaons, antikaons and hyperons) are shown to be in accord with emission from a thermal resonance gas source. Within the uncertainties of the present data, two freeze-out points are possible. The best agreement is obtained for a temperature $T \simeq 110$ MeV and a baryochemical potential $\mu_B \simeq 540$ MeV, corresponding to about 1/3 standard nuclear density. Another possible point lies at about twice nuclear density, with $T \simeq 160$ MeV and $\mu_B \simeq 620$ MeV. Our analysis takes the isopin asymmetry of the initial state fully into account.

1 Introduction

A recent comprehensive analysis of BNL-AGS data from $Si-Au$ collisions [1,2] provides evidence that the most abundantly observed hadrons are from a thermal source. If supported by further results from $Au-Au$ collisions at the AGS and at higher energies from the CERN-SPS, this conclusion would be a decisive step in showing that strongly interacting *matter* can be produced by nuclear collisions. In view of the importance of such a result, we find it useful to analyze the present AGS data independently in a self-contained hadrosynthesis approach. Our analysis will be based solely on particle ratios since many effects (transverse flow, excluded volume) cancel out there. As the initial state contains more neutrons than protons, it is necessary to assure overall charge conservation, which leads, e.g., to a π^-/π^+ ratio different from unity. The analysis presented below takes such deviations from an isospin-symmetric initial state fully into account. Since the neutron to proton and the π^-/π^+ ratios deviate from one by about ten to twenty percent, the results from this analysis will generally differ from the isospin-symmetric case by this order of magnitude. Estimates of the effects of a charge chemical potential have been presented previously [3,4].

2 Hadronic Ratios.

The measured hadron production ratios listed in Table 1 form the basis of our analysis. If the hadrons are emitted from a source in full chemical equilibrium, each measured ratio determines a range of T–μ_B values with which it is compatible. This range can be calculated on the basis of an ideal gas of all observed hadrons and hadronic resonances, requiring overall charge and strangeness conservation and correct resonance decays. The method has been presented in e.g. reference [5,6]. We include all hadrons listed in the latest Particle Data Compilation [7], excluding charm and bottom resonances; decays are taken into account with their experimental branching ratios, with an educated guess being made whenever the information on decays is not complete. The resulting table contains 479 entries. The partition function then determines all thermal properties of the system in terms of the four parameters T, μ_B, μ_S and μ_Q. The chemical potential for the strangeness, μ_S, can be fixed by requiring the overall strangeness of the system to vanish. Similarly, the chemical potential for the charge, μ_Q, is fixed by requiring the charge/baryon ratio of the final state to be equal to that of the initial state. If there is a one-stage freeze-out of all thermal hadrons, then all production ratios (π/p, K/π, $K/\bar{K}$, ρ/π, ϕ/π, Y/p, ...) are given in terms of the two remaining parameters T, and μ_B. For example, we can use the measured π^+/p and K^+/π^+ ratios to fix the values of T and μ_B; all other ratios are then predicted, if the different particle species are indeed present according to their equilibrium weights.

To fix the overall charge/baryon, we thus have to estimate the number of interacting protons and neutrons in the initial state. For a central A–B collision ($A \ll B$), the number N_{part} of participant nucleons is the sum of the nucleons in A and those in a tube of radius R_A through nucleus B,

$$N_{\text{part}} \simeq [A + (\pi R_A^2)n_0 2R_B] \simeq [A + 1.5A^{2/3}B^{1/3}]. \tag{1}$$

Here $n_0 = 0.17$ fm^{-3} is standard nuclear density; $R_A = 1.12A^{1/3}$, and similarly for B. For Si–Au collisions, we thus obtain $N_{\text{part}} \simeq 108$. With $Z/A = 0.5$ for Si and $Z/A = 0.4$ for Au, this is made up of $N_p \simeq 46$ protons and $N_n \simeq 62$ neutrons. We thus have to fix the overall charge/baryon of the final state at 46/108 by suitably adjusting the charge potential μ_Q at the temperature T and baryochemical potential μ_B obtained from particle ratios. As noted, the strangeness potential μ_S is fixed to give the final state a vanishing overall strangeness.

We begin with the most abundantly observed hadron species, p, $\pi^\pm$, $K^\pm$ and

Λ, since these are most likely to be thermalised. What we denote by Λ will always include the Σ^0, since the two are experimentally not separable. From the production rates of these six hadron species, we get five independent ratios. As seen in Fig. 1, four of these in fact cross in a common region in the plane of temperature and baryochemical potential, so that pions, nucleons, kaons, antikaons and lambdas are observed according to their thermal weights. The

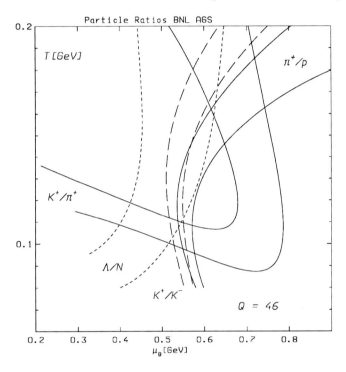

Figure 1: The $T - \mu_B$ regions determined by the indicated particle ratios (including experimental errors). The charge is kept fixed at 46, corresponding to 46 participant protons and 62 participant neutrons.

freeze-out parameters thus obtained are

$$T_F = 110 \pm 5 \text{ MeV} \qquad \mu_B^F = 540 \pm 20 \text{ MeV}. \qquad (2)$$

The fifth ratio, which can be taken either as the K^-/π^- or the π^-/π^+ ratio, is also in agreement with the above values. In the isospin symmetric case, with $\mu_Q = 0$, the ratios narrowly fail to define a common region. If we fix the total charge at $Q = 46$, rather than the corresponding chemical potential

Table 1: Ratios of Hadron Species in Si-Au Collisions at the AGS (Thermal parameters: $T = 110 \pm 5$ MeV, $\mu_B = 540 \pm 20$ MeV)

Ratio	Experimental	Thermal
π^+/p	0.80 ± 0.08	0.87 ± 0.15
K^+/π^+	0.19 ± 0.02	0.21 ± 0.02
K^+/K^-	4.40 ± 0.40	4.51 ± 0.62
Λ/p	0.20 ± 0.04	0.16 ± 0.02
K^-/π^-	$0.035 \pm .005$	0.038 ± 0.006
Ξ^-/Λ	$(1.2 \pm 0.2) \times 10^{-1}$	$(4.9 \pm 0.5) \times 10^{-2}$
ϕ/π^+	$(4.5 \pm 1.2) \times 10^{-3}$	$(4.6 \pm 1.3) \times 10^{-3}$
$\bar{p}/p$	$(4.5 \pm 0.4) \times 10^{-4}$	$(7.2 \pm 6.3) \times 10^{-5}$
$\bar{\Lambda}/\Lambda$	$(2.0 \pm 0.8) \times 10^{-3}$	$(3.4 \pm 3.0) \times 10^{-4}$

μ_Q, for all values of the temperature and the baryon chemical potential, the results remain essentially unchanged. In Table 1, we list all thermal ratios as determined by these values, together with the measured ratios. It is evident that the emission rates of nucleons, pions, kaons, antikaons and hyperons are correctly described by thermal composition. The corresponding baryon density is $n_B = 0.055 \pm 0.025$ fm^{-3} and hence about 1/3 standard nuclear density, indicating considerable expansion before freeze-out. The freeze-out values (2) agree with those determined in [1]; our temperature is slightly lower. In [1], the temperature range for freeze-out was fixed by a study of Δ production. However, the determination of the Δ yield is quite delicate, based on pion transverse momentum distributions at small p_T and/or reconstruction over a partially known background [9]. To avoid the uncertainties this leads to, we use the directly measured hadron ratios. For the freeze-out parameters (2), we get an overall thermal ratio $\Delta/N = 0.25 \pm 0.02$.

Given the initial state baryon number of 108, we can extract from the ratios of Table 1 the thermal abundances of the different species; they are listed in Table 2. Also shown there are the experimental abundances, obtained in the

Table 2: Abundances of Hadron Species in Si-Au Collisions at the AGS (Thermal parameters: $T = 110 \pm 5$ MeV, $\mu_B = 540 \pm 20$ MeV)

Species	Experimental	Thermal
nucleons	94	94
pions	120	133
kaons	14	17
hyperons	14	12
antikaons	3	4
Ξ's	2	1
ϕ's	2×10^{-1}	2×10^{-1}
antinucleons	4×10^{-2}	6×10^{-3}
antihyperons	3×10^{-2}	4×10^{-3}

same way from the experimental ratios.

Returning now to Fig. 1, we note that three of the four ratios shown have another crossing region at higher temperature and slightly higher μ_B. The fourth, K^+/K^-, misses this crossing only slightly. In view of the fact that the K^+/π^+ ratio does show a considerable increase in going from $Si - Au$ to $Au - Au$ data[10], one should not discount such a higher freeze-out point before analyzing the $Au - Au$ data. We therefore note

$$T_F = 160 \pm 5 \text{ MeV} \qquad \mu_B^F = 620 \pm 20 \text{ MeV}. \qquad (3)$$

as an alternative point of hadrosynthesis. It would correspond to a baryon density $n \simeq 0.33$ fm^{-3} and thus to a much denser and smaller freeze-out system.

Next we turn to less copiously produced hadron species. In Table 1, we also list their measured ratios together with the corresponding predictions from thermal emission at the freeze-out parameters (2) and (3). We see that the value for ϕ production [11] at the lower freeze-out point agrees well with the thermal prediction, although there are only about 0.1 ϕ's per event; at the higher point, it is a factor two off. Multiply strange baryons (Ξ) seem experimentally somewhat more abundant than thermally predicted at either point. A definite disagreement with the lower thermal predictions is found for antibaryons ($\bar{p}$ and $\bar{\Lambda}$): these are produced at least an order of magnitude more copiously than the thermal predictions based on the parameter set (2). The high temperature freeze-out point, on the other hand, reproduces the anti-baryons reasonably well, with predictions which are higher than the data by factors 1.5 (antiprotons) and three (anti-lambdas). We shall return to possible sources for difficulties with anti-baryon rates somewhat later on.

3 Thermalization.

Thermalization in the AGS energy range can lead to an increase or to a decrease of hadron abundances relative to those measured in $p - p$ or $p - A$ interactions.

In view of the general evolution towards thermalization, it seems misleading to single out enhanced kaon/pion or hyperon/nucleon ratios as "strangeness enhancement". There are fewer pions per nucleon in nucleus-nucleus collisions than in $p - p$ or $p - A$ interactions: this alone would drive the K/π ratio up, even for constant kaon production. Moreover, there are strange hadron species (e.g., antikaons) whose thermal production rate is smaller than that in hadron-hadron collisions, and there are non-strange species (e.g., Δ's) with enhanced thermal production.

We return now briefly to the discrepancy between the experimental antibaryon results and the thermal prediction. The point we wish to make is that such a disagreement cannot be used to rule out the parameters (2) The thermal rates are determined by the dependence on the baryochemical potential μ_B: increasing μ_B at fixed temperature decreases the ratio $\bar{p}/p$ as $\lambda_B^{-2} = \exp(-2\mu_B/T)$. However, such a suppression presupposes that the antibaryons can really experience the thermal medium, and that appears not clear. The fate of antiprotons in nuclear matter has been studied in detail in low energy $p - A$ interactions [18]. Here one also expects enhanced annihilation, particularly in the production of very slow $\bar{p}$'s, which spend a long time in the medium. In contrast, one finds that over a large momentum range down to 0.5 GeV/c, the antiprotons apparently do not interact with the medium. This has been interpreted in terms of a very long formation time for baryon-antibaryon pairs in nuclear collisions, so that antiprotons emerge as well-defined particles only after leaving the nucleus [19]. It has also been considered as the effect of a specific screening of antiprotons in nuclear matter [20]. In any case, until the considerable transparency of nuclear matter for antiprotons is understood, it is not clear what role they will play in the medium produced in nucleus-nucleus collisions. It thus seems safest to exclude them from thermal considerations of AGS data, which are in the energy range studied in the mentioned $p - A$ collisions [18].

4 Summary

We have here addressed the question of chemical equilibrium at freeze-out in a self-contained fashion, including the isospin asymmetry of the initial state. Our conclusion, in full agreement with that of[1], is that in $Si - Au$ collisions at the AGS all copiously produced hadron species are emitted in accord with their thermal weights, as calculated for an ideal gas of hadrons and hadron resonances. The freeze-out parameters thus determined still are not really unique, with the two possible sets (2) and (3) both accounting quite well for the presently available data. Some less frequently produced species also agree with this. Our conclusion is thus supported by six or seven independent and directly measured hadron production ratios, so that it seems quite well-founded. It will be interesting to see if the forthcoming data from $Au - Au$ collisions lead to a unique freeze-out point.

In closing we note one difficulty with the low temperature solution (2). The freeze-out temperature determined in hadrosynthesis can in principle be counter-checked by the measured transverse momentum spectra, provided we know the expansion pattern. In the absence of any transverse nuclear effects, we would have for light hadrons ($m_\pi \simeq 0$)

$$\langle |p_T| \rangle \simeq 2T. \tag{4}$$

For the freeze-out set (2), this is definitely too small, while for set (3) it leads to the canonical transverse momentum of around 0.32 GeV. As noted in[1], a disagreement between the hadrosynthesis temperature and the inverse p_T-slope could be taken as an indication of predicted flow effects[8]. However, to establish conclusively the presence of such effects requires a more complete analysis, comparing in particular the change in transverse momentum spectra in going from $p - A$ to $A - B$ collisions. Although of great interest, that is beyond the scope of the present note.

Acknowledgment

Stimulating discussions with M. Gazdzicki, D. Kharzeev, R. Stock and in particular with J. Stachel are gratefully acknowledged. One of the authors (J.C.) thanks the organizers for the stimulating atmosphere of the conference. We extend our warmest congratulations to Walter Greiner for his 60th birthday.

References

1. P. Braun-Munziger, J. Stachel, J. P. Wessels and N. Xu, *Phys. Lett.* B **344**, 43 (1995).
2. P. Braun-Munziger and J. Stachel, in *Hot Hadronic Matter: Theory and Experiment*, J. Letessier, H. Gutbrod ands J. Rafelski (Eds)., Plenum Press, New York 1995, p. 451.
3. M.I. Gorenstein, H. Miller, R.M. Quick and Shin Nan Yang, *Phys. Rev.* C **50**, 2232 (1994).
4. J. Letessier, A. Tounsi, U. Heinz. J. Sollfrank and J. Rafelski, *Phys. Rev.* D **51**, 3408 (1995).
5. J. Cleymans and H. Satz, *Z. Phys.* C **57**, 135 (1993).
6. J. Cleymans, K. Redlich, H. Satz and E. Suhonen, *Z. Phys.* C **58**, 347 (1993).
7. *Review of Particle Properties Phys. Rev.* D **50**, 1177 (1994).
8. U. Heinz, *Nucl. Phys.* A **566**, 225c (1994).
9. T. K. Hemmick et al., *Nucl. Phys.* A **566**, 431c (1994).
10. T. Abbott et al., *Nucl. Phys.* A **544**, 237c. (1992).
11. Y. Wang, *Nucl. Phys.* A **566**, 379c (1994); Y. Wang, in *Heavy Ion Physics at the AGS*, G. S. F. Stephans et al. (Eds)., MITLNS-2158, p. 239, 1993.
12. W. A. Zajc, *Nucl. Phys.* A **544**, 237c (1992); M. Gonin, in *Heavy Ion Physics at the AGS*, G. S. F. Stephans et al. (Eds)., MITLNS-2158, p. 184, 1993.
13. G. Odyniec, in *Hot Hadronic Matter: Theory and Experiment*, J. Letessier, H. Gutbrod ands J. Rafelski (Eds)., Plenum Press, New York 1995, p. 399.
14. G.S.F. Stephans, *Nucl. Phys.* A **566**, 269c (1994).
15. J. Stachel, these proceedings.
16. B. Shiva Kumar, in *Heavy Ion Physics at the AGS*, G. S. F. Stephans et al. (Eds)., MITLNS-2158, p. 144, 1993.
17. M. Gazdzicki, private communication.
18. A. O. Vaisenberg et al., *JETP Lett.* **29** 661 (1979).
19. B. Z. Kopeliovich and F. Niedermayer, *Phys. Lett.* B **151**, 437 (1985).
20. S. Kahana et al., *Phys. Rev.* C **47**, 1356 (1993).

MEASUREMENT OF DIRECT PHOTONS
IN CERN EXPERIMENTS WA80/WA98

F. PLASIL

Physics Division, Oak Ridge National Laboratory
Oak Ridge, TN 37831-6372, USA

Presented for the WA80 and WA98 Collaborations

Results from measurements of photons in reactions of 200-GeV/nucleon ^{32}S ions with Au nuclei are presented. A photon excess corresponding to 5% of the total inclusive photon yield is observed in central collisions with a systematic error of 5.8%. Upper limits on the yield for direct photon production at the 90% confidence level are presented. Preliminary results from photon measurements in reactions of Pb+Pb at 158-GeV/nucleon are shown.

At the Quark Matter '93 Conference, we reported preliminary results from our measurements of photons in reactions of 200·A GeV ^{32}S ions with Au nuclei.[1] For central collisions we reported a small, but statistically significant, excess of photons over those that can be accounted for by the two-photon decay branches of π^0 and η mesons and by the small photon contributions from other radiative decays. Here we present the final WA80 single "direct" photon results,[2] and we discuss their implications for various theoretical calculations. We also present very preliminary results from photon measurements of Pb+Pb at 158 GeV/nucleon.

As is pointed out frequently, photons, because of their low interaction probability, are considered to be a very promising probe suitable for studies of the early phases of hot and dense matter produced in relativistic nucleus-nucleus collisions. In particular, directly-radiated single photons may reflect the thermal properties of hot and dense matter, whether it be a hadron gas, a quark-gluon plasma (QGP), or a mixture of both. The possibility that photons are radiated from the QGP had been considered as early as 1978.[3] Expected contributions to the observed photon yield from the QGP would very probably include $q - \bar{q}$ annihilation and the QCD equivalent of the Compton process in which a quark interacts with a gluon to produce a photon.

It was pointed out only relatively recently that a hot hadron gas is expected to radiate photons with emission rates that are similar to those that are expected from a QGP.[4] The dominant production mechanisms in a hot hadron gas are expected to be $\pi^+\pi^- \rightarrow \rho\gamma$ and $\pi\rho \rightarrow \pi\gamma$, with a further increase in the production rate in the latter case due to the formation of the intermediate broad A_1 (1260) resonance.[5] The pion annihilation mechanism is expected to dominate below about $p_T = 0.7$ GeV while the $\pi - \rho$ channel dominates at higher values of transverse momentum. Since the mean free path of the produced photons is considerably larger than the size of the nuclear volume associated with the reactions, photons produced during all stages of the reaction will be observed, and should the QGP be formed, the total yield will be the result of time integration over three phase regions: pure QGP, mixed QGP-hadron phase, and pure hadron-gas phase.

In a thermalized system, when attempts are made to estimate contributions to the observed photon yield from the QGP, on the one hand, and from the hadron gas, on the other hand, it is useful to consider the number of degrees of freedom involved in each case. Given a fixed energy density produced in a given collision, a higher temperature is attained in a thermalized system that has a lower number of degrees of freedom. For a pure pion gas consisting only of bosons, the total number of degrees of freedom is 3, due to the isospin involved. In the case of the QGP, considering spin/polarization, isospin, and color, the number of degrees of freedom is 12 for the quarks and 16 for gluons. This large difference in the total number of degrees of freedom involved in the two cases leads to very different predictions for temperatures of thermalized systems and, hence, to the predicted, resulting photon yields. Unfortunately, whether the photons are estimated to originate from the QGP or from a hadron gas, their yield is predicted to be low relative to the yield of photons resulting from the decay of copiously-produced hadrons such as π^0 and η. This makes reliable measurements of direct photons extremely difficult. Of course, a hadron gas will not consist only of pions. Other hadrons and resonances need to be taken into account. However, even when these are considered, the predicted temperatures, and hence the photon yields, associated with hadron-gas scenarios tend to be significantly higher than those associated with possible QGP production.

There are, generally speaking, three experimental approaches to the determination of direct-photon yields. In the first, sometimes used in hadron-hadron reactions, a search is made for isolated photons that do not have a partner identified as resulting from hadron decay (e.g., from π^0 or η). This method requires low multiplicities and wide experimental coverage. It is not applicable to the high-multiplicity environment encountered in nucleus-nucleus collisions. An alternative approach is the measurement of a quantity that is related to the inclusive photon yield (e.g., conversion electrons) combined with model-dependent estimates of photons resulting from hadronic decays. This approach suffers from large uncertainties associated with the model assumptions. We have chosen a third approach which is appropriate for high-multiplicity environments and in which model assumptions are minimized. In our method we measure p_T spectra of identified photons simultaneously with the p_T spectra of π^0 and η mesons, which are the primary sources of decay photons. We then obtain the direct-photon yield on a statistical basis, as a function of p_T, by comparing the total photon yield to that which can be attributed to all long-lived decays. While the analysis of the data is very complicated and requires numerous corrections and careful estimates of systematic errors, it does lead to results that are sensitive at the 5% level of inclusive photons.

The WA80 experimental setup was described earlier.[1] The photon measurement was performed with a finely-segmented electromagnetic calorimeter consisting of 3798 lead-glass modules. The detector coverage ranged from one-tenth to one-half of ϕ in the rapidity range between 2.1 and 2.9. Two layers of a streamer tube charged-particle veto (CPV) detector were located immediately in front of the lead-glass array.

The preliminary WA80 S+Au results presented earlier[1] showed no significant excess of observed photons over those that can be accounted for by decays of known hadrons in peripheral collisions, while an excess at about the 2σ level was observed in central collisions. The preliminary results were presented in the form of plots of γ/π^0 ratios as a function of the transverse momentum. The data as depicted in this form appeared to be of little interest to the theory community. However, when the data were presented at an informal meeting in the form of a p_T spectrum of the excess photons, theoretical interest picked up immediately and, within a period of less than two years, five papers appeared in which calculations were compared to the preliminary WA80 results.[6-10] The calculations involved essentially standard thermal-model QGP formation assumptions, and comparisons were made for various scenarios involving QGP formation, hadron gas formation, or a combination of both, including, in some cases, contributions from a mixed QGP-hadron gas phase. In general, scenarios in which a QGP did not form and in which photons were radiated only from a hadron gas led to overpredictions of the photon excess, as can be expected on the basis of the "degrees-of-freedom" arguments presented above.

The surge of theoretical interest, together with the difficulties associated with the extraction of the small direct photon emission component from data that are dominated by expected background photons originating from hadronic decays, has led to the decision to reanalyze the data by a WA80 team that was not intimately involved with the original analysis. As was mentioned above, an important feature of the WA80 measurements is that π^0 and η yields are obtained simultaneously with the inclusive photon yield, γ^{obs}, in the same p_T and rapidity region for each event class considered. This leads to a minimization of systematic errors resulting from the known centrality dependence of the meson p_T spectra. The single photon excess is determined, in principle, on a statistical basis from $\gamma^{excess} = \gamma^{obs} - \gamma^{bkgd}$, where γ^{bkgd} is calculated based on the measured π^0 and η yields. (A small contribution of the order of 2% due to photons originating from other radiative decays is estimated based on m_T scaling, which has been shown to hold for π^0 and η mesons[11] in the reactions considered here. In practice, however, results are usually presented in the form ratios $(\gamma/\pi^0)^{obs}$ and $(\gamma/\pi^0)^{bkgd}$, since these ratios are less sensitive to systematic errors. Here we go a step further and present results in terms of the ratio of ratios, $(\gamma/\pi^0)^{obs}/(\gamma/\pi^0)^{bkgd}$, which gives the fraction of photons observed relative to the expected decay background. This ratio will have a value of 1 if there are no excess photons. It has the advantages that both statistical and systematic errors are approximately constant over the entire p_T region and that systematic errors arising from acceptance corrections are minimized.

Due to the relatively low expected yield of direct photons, very accurate estimates of systematic errors are required. In many cases, the available data themselves can be used for systematic error estimates. Thus, for example, evaluation of the energy-dependence of the π^0 mass peak leads to estimates of energy nonlinearity. Another approach is to carry out the analysis as a function of different criteria and then compare the results to each other. This was done in the crucial case of photon identification, where different criteria lead to significantly different photon and π^0 identification efficiencies and to different background corrections. In order to determine which observed separated showers in the WA80 lead-glass detector are due to impinging photons, the following criteria of varying degree of restriction were applied: all observed showers; only showers with small lateral profile; only showers without an overlapping hit in either of the two charged-particle veto (CPV) layers located in front of the lead-glass array; and only showers without an overlapping hit in both CPV layers. In a consistent analysis, all of the above identification methods should give the same result, and the variation of the results gives an indication of systematic errors associated with both the single-photon and the π^0 yield determinations.

A list of all known sources of systematic error in the determination of the direct-photon yield is given in Table 1 for peripheral and central collisions and for two different regions of p_T. The errors are expressed as a percentage of $(\gamma/\pi^0)^{obs}/(\gamma/\pi^0)^{bkgd}$. Note that the largest systematic errors are associated with the determination of the π^0 yield and efficiency. This yield extraction is particularly difficult since the large, total multiplicity leads to centrality-dependent modifications of the π^0 mass peak due to shower overlap and to very unfavorable (as low as 0.1) peak-to-background ratios resulting from the huge two-photon combinatorial background.

Table 1: Various sources of systematic error in the WA80 200·A GeV ^{32}S+Au direct photon analysis specified as a percentage of $(\gamma/\pi^0)^{obs}/(\gamma/\pi^0)^{bkgd}$. The dependence of the errors on p_T is indicated.

Centrality:	Periph. coll. (31% σ_{mb})		Central coll. (7.4% σ_{mb})	
p_T range (defined below):	a	b	a	b
Source of error				
γ reconstruction efficiency	1.0	1.0	2.0	2.0
π^0 yield extraction and efficiency	2.0	3.0	4.0	5.0
Detector acceptance^c	0.5	0.5	0.5	0.5
Energy nonlinearity^c	2.0	1.0	2.0	1.0
Binning effects^c	1.0-0.0	0.0	1.0-0.0	0.0
Charged vs. neutral shower sep.^{c}	1.0	1.0	1.0	1.0
γ conversion correction^c	1.0	1.0	1.0	1.0
Neutrons	1.5	0.5	1.5	0.5
Other neutrals, e.g., $\bar{n}, K_L^0$	1.0	0.5	1.0	0.5
η/π ratio, m_T scaling	1.5	1.5	1.5	1.5
Other radiative decays	0.5	0.5	0.5	0.5
Total: (quadratic sum)	4.2	4.0	5.7	5.9

$^a p_T < 1.5$ GeV/c
$^b p_T > 1.5$ GeV/c
^{c}Centrality independent

The ratio $(\gamma/\pi^0)^{obs}/(\gamma/\pi^0)^{bkgd}$ is shown in Fig. 1 as a function of p_T for peripheral and central collisions. The final result was obtained with the π^0 yield determined on the assumption that all identified showers are due to photons and is depicted by filled circles. Results of analyses using other criteria (see above) are shown by means of open symbols. The scatter of the points provides an indication of the level of systematic errors that can be attributed to the π^0 yield extraction. A further indication of systematic errors is provided by yet another independent analysis which was carried out without making use of the CPV data and in which independent methods were used to extract yields, efficiencies, and backgrounds. Results of this analysis are shown in Fig. 1 by means of light-shaded squares. A fit to the final ratios with a constant value over the range 0.5 GeV/c $\leq p_T \leq$ 2.5 GeV/c gives an average photon excess over background sources of 3.7% $\pm$ 1.0% (statistical) $\pm$4.1% (systematic) for peripheral collisions and 5.0%$\pm$0.8 (statistical) $\pm$5.8% (systematic) for central collisions. This is consistent (within 1σ) with the absence of photon excess in

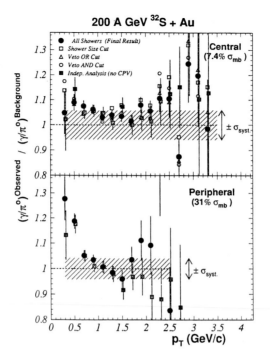

Figure 1: The ratio $(\gamma/\pi^0)^{obs}/(\gamma/\pi^0)^{bkgd}$ as a function of transverse momentum for peripheral and central collisions of 200·A GeV ^{32}S+Au. The errors on the data points (shown for the solid points only) indicate the statistical errors only. The shaded regions indicate the total estimated p_T-dependent systematic errors which bound the region corresponding to no photon excess.

both central and peripheral collisions. The difference between this final result and the preliminary results presented earlier[1] is due primarily to the difficulties associated with the determination of the π^0 yields.

To facilitate comparisons with earlier theoretical calculations, we have deduced from our data, as a function of p_T, upper limits at the 90% confidence level on the invariant yield of excess photons per central ^{32}S+Au collision. The level of these limits is similar to the excess photon yield reported in the preliminary analysis, and the limit results constitute, in themselves, an important finding of WA80. The upper limits are shown in Fig. 2 by means of arrows, together with the calculations of Srivastava and Sinha[7] and of Dumitru et al.[9] The comparisons with our preliminary data featured in Ref. 7 attracted a great deal of popular attention. For example, an article entitled "Smashed Atoms Recreate Quark-Gluon Soup of Big Bang" appeared in the October 29, 1994, issue of *The New Scientist*. Such claims are clearly exaggerated. The main point of Srivastava and Sinha was that if only a hadron gas is formed in S+Au central collisions, then the limited number of degrees of freedom leads to a very high initial temperature (about 400 MeV). This, in turn, results in a predicted photon yield that is far greater than that of our preliminary analysis. On the other hand, a good fit to the preliminary data is obtained on the assumption that the system is initially formed in the QGP phase (at an initial temperature of about 200 MeV) and that it then expands, cools, and undergoes a first-order phase transition to a hadron gas at a critical temperature of 160 MeV, followed by freeze-out at 100 MeV. It can be seen in Fig. 2 that this conclusion remains unchanged when the comparison is made to our upper limits on direct photon production. The dotted curve depicting the QGP plus hadron gas scenario is consistent with our results, while the solid curve of the hadron gas scenario with no phase transition far exceeds the experimental upper limits.

Of the published theoretical calculations, those of the Frankfurt group[9] are the most comprehensive. Addressing issues of dynamical evolution of the system, the authors used a three-fluid hydrodynamical model and were able to reproduce the rapidity distributions of negatively-charged hadrons at CERN/SPS energies as well as our transverse momentum distributions of neutral pions from central S+Au collisions at 200·A GeV. These additional comparisons with experimental results add credibility to their calculated thermal photon spectrum which is shown by the dashed line in Fig. 2 for the case of no phase transition. It is seen that this calculation also lies significantly above our upper limits. In contrast, the calculation of Ref. 9 in which a phase transition (possibly to a QGP) is assumed is consistent with our limits.

Some of the hadron-gas scenarios of Refs. 6–10 have been criticized as being unrealistic due to the restricted number of degrees of freedom that they consider. This and other related issues,were discussed recently at a meeting held at the Institute for Nuclear Theory at the University of Washington in Seattle, where our data were presented together with details of the analysis procedures used.[12] Several direct-photon production calculations were presented at the meeting in their preliminary stages, indicating continued interest in direct-photon measurements. Our upper-limit results provide an important benchmark for these efforts. Particularly, in a thermal scenario, they implicitly place an upper limit on the temperature that may have been attained in central collisions. Future comparisons of our photon-production limits to photon production estimates based on nonthermal models, such as cascade calculations, will be of great interest.

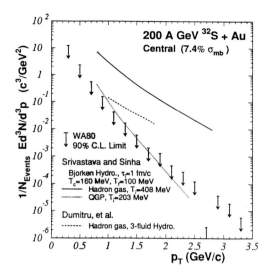

Figure 2: Upper limits at the 90% confidence level on the invariant excess photon yield per event for the 7.4% σ_{mb} most central collisions of 200·A GeV ^{32}S+Au. The solid curve is the calculated thermal photon production expected from a hot hadron gas taken from Ref. 7. The dashed curve is the result of a similar hadron gas calculation taken from Ref. 9. The dotted curve is the calculated thermal photon production expected in the case of a QGP formation also taken from Ref. 7.

Recently preliminary photon measurement results from Pb+Pb collisions at 158 GeV/nucleon have become available.[13] A large-acceptance photon spectrometer (LEDA), consisting of 10,080 independent lead-glass Cerenkov detectors, was used to make the measurements. LEDA covers approximately half of the full azimuth in the pseudorapidity range $2.3 \leq \eta < 3$. Application-specific integrated circuits developed at the Oak Ridge National Laboratory were used for the readout of the signals. A new high-voltage system using active photomultiplier bases with on-broad Greinacher voltage multipliers was also implemented, together with a new gain-monitoring system based on light-emission diodes and PIN-photodiodes. The high-stability and low-noise characteristics of the overall system resulted in an excellent energy resolution of $\sigma(E)/E = (5.5 \pm 0.6)\%/\sqrt{E(GeV)} + (0.8 \pm 0.2)\%$.

The preliminary invariant mass distribution of photon pairs is shown in Fig. 3 for a p_T range of 1.2 to 2.2 GeV. The π^0 peak is clearly visible, even in the raw data. The inset shows the background-subtracted data following the application of the standard event-mixing technique. Initial transverse momentum spectra of inclusive photons and of pions from a limited data sample are available, but are too preliminary to be shown here.

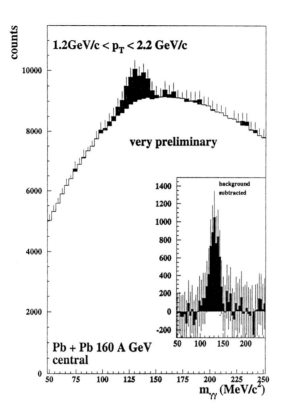

Figure 3: Raw invariant mass distribution of photon pairs. The inset depicts the background-subtracted distribution.

Acknowledgments

Discussions with T. C. Awes, P. W. Stankus, and G. R. Young are gratefully acknowledged. Shirley J. Ball's preparation of the manuscript is very much appreciated. Oak Ridge National Laboratory is managed by Lockheed Martin Energy Research Corp. for the U.S. Department of Energy under contract number DE-AC05-96OR22464.

References

1. R. Santo et al., *Nucl. Phys.* **A566**, 61c (1994).
2. R. Albrecht et al., *Phys. Rev. Lett.* **76**, 3506 (1996).
3. E. V. Shuryak, *Yad. Phys.* **28**, 796 (1978); [*Sov. J. Nucl. Phys.* **28**, 408 (1978)].
4. J. Kapusta, P. Lichard, and D. Seibert, *Phys. Rev. D* **44**, 2774 (1991).
5. L. Xiong, E. Shuryak, and G. E. Brown, *Phys. Rev. D* **46**, 3798 (1992).
6. E. V. Shuryak and L. Xiong, *Phys. Lett. B* **333**, 316 (1994).
7. D. K. Srivastava and B. Sinha, *Phys. Rev. Lett.* **73**, 2421 (1994).
8. J. J. Neumann, D. Siebert, and G. Fai, *Phys. Rev. C* **51**, 1460 (1995).
9. A. Dumitru, U. Katscher, J. A. Maruhn, H. Stöcker, W. Greiner, and D. H. Rischke, *Phys. Rev. C* **51**, 2166 (1995).
10. N. Arbex, U. Ornik, M. Plümer, A. Timmermann, and R. M. Weiner, *Phys. Lett. B* **354**, 307 (1995).
11. R. Albrecht et al., *Phys. Lett. B* **361**, 14 (1995).
12. P. W. Stankus, private communication.
13. Preliminary results of the WA98 Collaboration; T. K. Nayak, GSI preprint 96-07, February 1996.

SEARCH FOR STRANGELETS AND PRODUCTION OF ANTINUCLEI IN Pb-Pb COLLISIONS AT CERN

J. BERINGER, K. BORER, F. DITTUS, D. FREI, E. HUGENTOBLER,
R. KLINGENBERG, U. MOSER, T. PAL, K. PRETZL, J. SCHACHER,
F. STOFFEL

Lab. for High Energy Physics, University of Bern, Sidlerstr. 5,
CH-3012 Bern, Switzerland

K. ELSENER, K.D. LOHMANN

CERN, SL Division,
CH-1211 Geneva 23, Switzerland

C. BAGLIN, A. BUSSIERE, J.P. GUILLAUD

CNRS-IN2P3, LAPP Annecy,
F-74941 Annecy-le-Vieux, France

T. LINDEN, J. TUOMINIEMI

Dept. of Physics, University of Helsinki, PO Box 9,
FIN-00014 Helsinki, Finland

G. APPELQUIST, C. BOHM, B. SELLDEN, Q.P. ZHANG

Dept. of Physics, University of Stockholm, PO Box 6730,
S-11385 Stockholm, Sweden

PH. GORODETZKY

CNRS-IN2P3, CRN Strasbourg,
F-67037 Strasbourg, France

NEWMASS (NA52) COLLABORATION

The NA52 experiment searches for strange quark matter particles, so-called strangelets, and for antinuclei produced in Pb-Pb collisions at $158\,\mathrm{GeV}/c$ per nucleon at zero degree production angle. Upper limits on the production of strangelets covering a mass to charge ratio up to $120\,\mathrm{GeV}/c^2$ and lifetimes $> 1.2\,\mu\mathrm{s}$ are given. Differential production cross sections for antiprotons and antideuterons over a large rapidity range are presented. One antihelium-3 nucleus was observed. The data were taken during the 1994 lead beam period at CERN.

1 Introduction

The main goal of the NA52 experiment is the search for charged strangelets in relativistic heavy ion collisions. However, the detector is also ideally suited to study, amongst other things, the production of antinuclei. In the following the two aspects of the NA52 experiment are described and the latest results are presented.

It was pointed out by several authors [1,2] that strangelets could be produced in a baryon rich quark gluon plasma (QGP). Their production is due to a cooling process of the plasma, which results in an enhancement of strange quarks in the quark phase. The cooling mechanism of the plasma is started by the evaporation of pions, K^+ and K^0, which carry entropy and strange antiquarks away from the system. The remaining strange quarks in combination with other up and down quarks could form a strangelet. The formation of strangelets in heavy ion collisions has also been considered in coalescence [3] and thermodynamical models [4]. Strangelets are multiquark states, which consist of approximately equal numbers of u, d and s-quarks. On the basis of the Pauli exclusion principle such multiquark states become stable owing to the introduction of strangeness as an additional degree of freedom. Strangelets can exist in neutral or in charged form. By virtue of the large strange quark content, the charge to mass ratio of strangelets is expected to be small ($|Z|/A < 0.1$), which is used as a prominent experimental signature. Bag model calculations show that for sufficiently large masses ($m > 10\,\mathrm{GeV}/c^2$) strangelets could be absolutely stable or metastable with lifetimes sufficiently long to be detected in the NA52 experiment [5,6]. The discovery of stable strangelets would confirm the existence of a new, as yet unseen, groundstate of matter. It would not only have profound implications in nuclear and particle physics, but also support the astrophysical and cosmological hypotheses on the role of strange quark matter in the universe. For instance, strange quark matter was postulated to be a candidate for the dark matter in the universe [7], it may form the core of neutron stars [8] and explain the "Centauro" events observed in cosmic ray experiments [9]. Strangelets with lifetimes $> 10^{-7}$s have been previously searched for in BNL and CERN heavy ion experiments as well as in cosmic rays. A recent review is given in ref. [10]. The data of the NA52 experiment, shown here, represent the full statistics taken in 1994.

The NA52 spectrometer with its full particle identification capabilities allows to study also the production of antinuclei over a large rapidity range. Antinuclei produced in heavy ion collisions probe the space time evolution of the high density state which is formed in the interaction. Enhanced production of antinuclei is predicted by several models [11,12,13,14,15]. Most models assume, that antibaryons arise via coalescence of three antiquarks, and are therefore sensitive to the thermal antiquark distributions and the chemically equilibrated abundances in the QGP. It was conjectured that high antiquark densities in the QGP might lead to antinucleus abundances. However, baryon-antibaryon annihilations in the baryon dense hadron gas may considerably dilute those abundances. It is therefore of interest to measure and to compare antibaryon production in heavy ion collisions in the BNL and CERN energy domain, where different energy and baryon densities can be reached. Antiproton and antideuteron production in heavy ion collisions have already been studied in BNL [16] and CERN [17]. In this paper we report on antinuclei production in Pb-Pb collisions at zero production angle.

2 Experimental method

The experimental set up is shown in Fig. 1. The secondary beam line H6 in

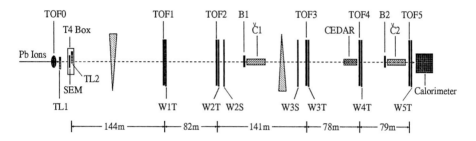

Figure 1: The NA52 setup.

the North Area of the SPS at CERN is used as a spectrometer. It accepts a momentum bite of 2.8 % and has a solid angle acceptance of 2.2 μsr. The incident lead ions are counted in a segmented, 0.4 mm thick quartz Čerenkov counter (TOF0). During the 1994 data taking the average lead beam intensity was $2.5 \cdot 10^7$ ions per spill. The mass of the particles produced at zero degree in a lead target is determined from the measurements of their rigidity (p/Z) in the spectrometer, of their velocity with the time of flight detectors (TOF0 to TOF5) and of their charge by their energy loss (dE/dx) in the TOF1-5 counters. The individual time resolutions of the TOF counters varied between $\sigma_t = 74$ ps and $\sigma_t = 105$ ps. A differential (CEDAR) and two threshold ($\check{C}_1$, $\check{C}_2$) Čerenkov counters are used for particle identification. They are especially useful at high rigidities, where the time of flight measurement is not sufficient to separate the particles. A segmented Scintillator/Uranium calorimeter adds further particle identification capabilities and redundancy to the measurements. Seven sets of multiwire proportional chambers (W1T-W5T, W2S, W3S) are used for particle tracking. A comprehensive description of the detector can be found in ref. [18].

3 Strangelet search

For the strangelet search the spectrometer was set at rigidities of $p/Z = \pm 100$ and $\pm 200\,\mathrm{GeV}/c$. At these rigidities, particles with a mass to charge ratio of 10 to $40\,\mathrm{GeV}/c^2$ are near midrapidity. For the $+200\,\mathrm{GeV}/c$ setting a 16 mm Pb target was used, while for $\pm 100\,\mathrm{GeV}/c$ and $-200\,\mathrm{GeV}/c$ a 40 mm Pb target was used. At each rigidity about $1.1 \cdot 10^{11}$ Pb-Pb interactions were recorded, with the exception of $+100\,\mathrm{GeV}/c$, where the obtained data are based on $2.1 \cdot 10^{11}$ interactions. Fig. 2 (b) shows $(m/Z)^2$ distributions obtained at $-100\,\mathrm{GeV}/c$. Heavy particles like $\bar{\mathrm{d}}$ and $\bar{\mathrm{t}}$ are identified by Čerenkov counter $\check{C}_1$. Antiprotons were tagged by the CEDAR counter. In the full data sample no heavy object with a mass to charge ratio $m/|Z|$ between 5 and $60\,\mathrm{GeV}/c^2$ at $p/Z = \pm 100\,\mathrm{GeV}/c$ and no event between 10 and $120\,\mathrm{GeV}/c^2$ at $p/Z = \pm 200\,\mathrm{GeV}/c$ has been observed.

Based on these results we calculate an upper limit for the production of strangelets with a lifetime of $t_{lab} > 1.2\,\mu s$. The sensitivity of the experiment is defined as $S(m) = 1/(N_{int} \cdot f(m))$, where N_{int} is the number of Pb-Pb interactions, and $f(m)$ is the detection probability for a strangelet within the acceptance of the spectrometer. In order to calculate $f(m)$, a strangelet production model [19] with a factorized phase space distribution was used:

$$\frac{\mathrm{d}^2 N}{\mathrm{d}y\,\mathrm{d}p_\perp} = \frac{4p_\perp}{\langle p_\perp \rangle^2} \exp\left(-\frac{2p_\perp}{\langle p_\perp \rangle}\right) \frac{1}{\sqrt{2\pi}\,\sigma_y} \exp\left(-\frac{(y - y_{\mathrm{cm}})^2}{2\sigma_y^2}\right) \quad (1)$$

Here, y is the rapidity of the strangelet, y_{cm} is the rapidity of the c.m.s. of the nucleons participating in the interaction ($y_{cm} = 2.9$ for Pb-Pb at $158 \cdot A\,\mathrm{GeV}/c$), σ_y is the width of the rapidity distribution which was taken to be $\sigma_y = 0.5$, and $\langle p_\perp \rangle$ is the mean transverse momentum of the strangelet. The probability $f(m)$ can then be calculated by integrating Eq. 1 over the acceptance in y and $p_\perp$ of the spectrometer. The obtained sensitivity curves are shown in Fig. 3. The two kinematic regions of $100\,\mathrm{GeV}/c$ and $200\,\mathrm{GeV}/c$ have been added by means of combining the individual sensitivities of each polarity to derive a common upper limit for the production probability of charged strangelets. The absolute value of the sensitivity strongly depends on the assumed model parameters, in particular on the mean $\langle p_\perp \rangle$ of the strangelets. Since the $\langle p_\perp \rangle$ of the strangelets is unknown several possibilities have been

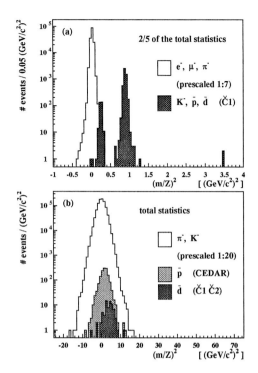

Figure 2: Mass distributions of particles reaching the TOF3 counter at a rigidity p/Z of $-10\,\mathrm{GeV}/c$ (a) and reaching the TOF5 counter at a rigidity of $-100\,\mathrm{GeV}/c$ (b) are shown.

considered. For example, the observed production of protons and lambdas in sulfur-nucleus collisions at $200 \cdot A\,\mathrm{GeV}/c$ favours $\langle p_\perp \rangle = 0.5\sqrt{m\mathrm{GeV}}$ [20]. However, as strangelets are possibly cooled remnants of a QGP it is reasonable to assume also much smaller values. The results are shown in Fig. 3 together with strangelet production probabilities a s estimated by Crawford et al. [21]. Assuming low values of $\langle p_\perp \rangle$ the experimental upper limits begin to overlap the predictions.

4 Production of antinuclei

The number of interactions recorded at the various rigidities are listed in Tab. 1. The data obtained at -100 and $-200\,\mathrm{GeV}/c$ were a by-product of the strange-

384

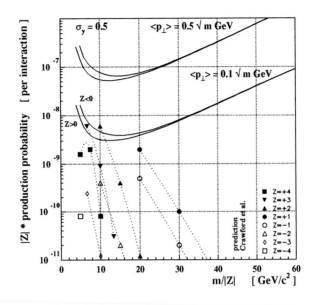

Figure 3: Experimental sensitivity for strangelets compared to predictions of ref.[21].

let search. It should be noted that at $-100\,\text{GeV}/c$ the antiprotons were tagged with the CEDAR only for a subsample of the total accumulated data. The data of low rigidities (-10 and $-20\,\text{GeV}/c$) were taken with a 4 mm thick Pb target. In order to evaluate background contributions to the particle yields, we also recorded data without a target at $-10\,\text{GeV}/c$. This background was mainly due to lead interactions in air and in materials in the T_4 box (see Fig. 1). In the analysis the low rigidity data (-10 and $-20\,\text{GeV}/c$) we required the particles to reach at least TOF3, while at the high rigidities (-100 and $-200\,\text{GeV}/c$) the particles were required to reach TOF5.

Fig. 2 (a) shows the mass distributions $(m/Z)^2$ for rigidities of -10 and $-100\,\text{GeV}/c$. The obtained mass-resolutions for antiprotons are $\sigma_m = 20\,\text{MeV}/c^2$ at $-10\,\text{GeV}/c$ and $\sigma_m = 800\,\text{MeV}/c^2$ at $-100\,\text{GeV}/c$. In order to demonstrate the particle identification capabilities with the TOF and the Čerenkov counters light particles, like pions and kaons are also shown in the Fig. 2. One antihelium-3 nucleus was observed at $-20\,\text{GeV}/c$ (see Tab. 1).

The invariant differential production cross section for particles with the

Table 1: The observed number of antinuclei and the obtained invariant differential production cross sections with the statistical and systematic errors are listed. The events were not selected for central collisions.

Rigidity [GeV/c]	Rapidity y	Cross section [barn GeV^{-2} c^3]	Observed number of particles	Number of Pb-Pb interactions
		$\overline{\mathrm{p}}$		
-10	3.06	$2.35 \pm 0.08 \pm 0.35$	12041	$1.7 \cdot 10^{10}$
-20	3.75	$1.87 \pm 0.07 \pm 0.28$	32947	$0.9 \cdot 10^{10}$
-100	5.36	$(0.79 \pm 0.02 \pm 0.12) \cdot 10^{-3}$	1415	$8.4 \cdot 10^{10}$
		$\overline{\mathrm{d}}$		
-10	2.38	$(0.65 \pm 0.38 \pm 0.10) \cdot 10^{-3}$	3	$1.7 \cdot 10^{10}$
-20	3.06	$(1.00 \pm 0.25 \pm 0.15) \cdot 10^{-3}$	16	$0.9 \cdot 10^{10}$
-100	4.67	$(29.0 \pm 3.4 \pm 4.4) \cdot 10^{-6}$	74	$1.2 \cdot 10^{11}$
-200	5.36	$(0.73 \pm 0.28 \pm 0.11) \cdot 10^{-6}$	7	$1.1 \cdot 10^{11}$
		$^3\overline{\mathrm{He}}$		
-20	3.35	$(16 \pm 16 \pm 2) \cdot 10^{-6}$	1	$0.9 \cdot 10^{10}$

energy E and the momentum p was evaluated from

$$E\frac{d^3\sigma}{dp^3} = \frac{N_S}{N_I} \cdot \frac{\eta}{\epsilon} \cdot \frac{1}{n \cdot \alpha} \cdot \frac{E}{p^3} \tag{2}$$

with N_S, the number of observed secondary particles, N_I, the number of incident lead ions, n, the number of target nuclei per unit area, and α, the spectrometer acceptance. The factor ϵ accounts for the absorption of the incident ions and the secondary particles in the production target. It also takes the absorption of secondary particles in the spectrometer as well as the trigger and reconstruction efficiencies into account. The spectrometer acceptance was calculated from a Monte Carlo simulation and the results were compared to the experimentally obtained beam profiles. From the empty target run a background correction factor η was applied. The obtained invariant production cross sections are summarized in Tab. 1 and shown in Fig. 4. From different contributions (acceptance, pile up, empty target background) we estimate a total systematic error of $\approx 15\%$ for all rigidities.

5 Summary

No evidence for the production of charged strangelets with $m/|Z|$ between 5 to 120 GeV/c and lifetimes $t_{lab} > 1.2\,\mu s$ was found in the 1994 data sample. The

386

quoted sensitivity is similar to the one previously obtained in S-W collisions at

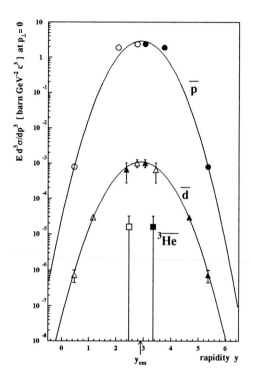

Figure 4: Invariant particle production cross sections (closed symbols) are shown as a function of rapidity. The open symbols are data points reflected at midrapidity ($y_{cm} = 2.9$). Only the statistical errors are drawn. The lines show Gaussian functions fitted through the measured data points while the mean value was fixed to midrapidity. The sigma of the Gaussian distribution is in both cases ≈ 0.6 units of rapidity. The events were not selected for central collisions.

$200 \, \text{GeV}/c$ per nucleon by this group[22]. It is planned for the future to increase the sensitivity by at least a factor ten. During the 1995 data taking with lead, the accumulated statistics was already increased by about a factor 20 for a spectrometer rigidity setting of $-200 \, \text{GeV}/c$. A similar sensitivity is planned to be reached for positive charged strangelets in a forthcoming run in the fall of 1996. Besides the strangelet search we have investigated antinuclei production at zero degree production angle over a wide range of rapidity. We observed one antihelium-3 nucleus. The $\overline{\text{d}}/\overline{\text{p}}$ ratio at mid rapidity turns out to be $4.2 \cdot 10^{-4}$.

Using a simple coalescence model we obtained a scaling factor $B_2 = (1.5 \pm 0.4) \cdot 10^{-3} \, \mathrm{GeV}^2/c^3$ for antideuterons and $B_3 = (8 \pm 8) \cdot 10^{-5} \, (\mathrm{GeV}^2/c^3)^2$ for antihelium-3. The scaling factor B_2 for antideuterons is roughly an order of magnitude lower than those for deuterons determined from $A + A$ collisions at Bevalac [23,24], indicating an increased antiproton source size at our energies. Further measurements of antinuclei production are planned for the future. The production of positively charged nuclei will be published soon in a forthcoming paper.

References

1. H.-C. Liu, G.L. Shaw, *Phys. Rev.* D **30**, 1137 (1984).
2. C. Greiner, P. Koch, H. Stöcker *Phys. Rev. Lett.* **58**, 2797 (1987).
3. A.J. Baltz et al., *Phys. Lett.* B **325**, 7 (1994).
4. P. Braun-Munzinger, these Proceedings.
5. E.P. Gilson, R.L. Jaffe, *Phys. Rev. Lett.* **73**, 332 (1993).
6. J. Madsen, *Phys. Rev.* D **47**, 5156 (1993).
7. E. Witten, *Phys. Rev.* D **30**, 272 (1984).
8. F. Weber, these Proceedings.
9. J.D. Bjorken, L.D. McLerran, *Phys. Rev.* D **20**, 2353 (1979).
10. B.S. Kumar, *Nucl. Phys.* A **590**, 29c (1995).
11. U. Heinz et al., *J. Phys.* G **12**, 1237 (1986).
12. J. Ellis et al., *Phys. Lett.* B **233**, 223 (1989).
13. S. Gavin et al., *Phys. Lett.* B **234**, 175 (1990).
14. J. Schaffner et al., *Z. Phys.* A **341**, 47 (1991).
15. H. Sorge et al., *Phys. Lett.* B **289**, 6 (1992).
16. M. Aoki et al., *Phys. Rev. Lett.* **69**, 2345 (1992).
17. NA44, NA35, NA52 in *Proceedings of QM95, Nucl. Phys.* A **590**, (1995).
18. K. Pretzl et al. in *Proceedings of the International Symposium on Strangeness and Quark Matter, Krete, Hellas, Sep. 1-5, 1994*, ed. G. Vassiliadis, A.D. Panagiotou, B.S. Kumar and J. Madsen (World Scientific, Singapore, 1995) pp. 230-244.
19. H. Crawford et al., *Phys. Rev.* D **45**, 857 (1992).
20. T. Alber et al., *Phys. Lett.* B **366**, 56 (1996).
21. H. Crawford et al., *Phys. Rev.* D **48**, 4474 (1993).
22. K. Borer et al., *Phys. Rev. Lett.* **72**, 1415 (1994).
23. S. Nagamiya et al., *Phys. Rev.* C **24**, 971 (1981).
24. R.L. Auble et al., *Phys. Rev.* C **28**, 1552 (1983).

THE PROSPECTS OF DENSE HADRONIC MATTER

Herbert Ströbele

Universität Frankfurt, Institut für Kernphysik,
August-Euler-Str. 6, 60486 Frankfurt

The evolution of particle production characteristics with the c.m. energy in nucleon-nucleon and nucleus-nucleus collisions is studied with the aim to provide the basis for extrapolations to RHIC and LHC energies. The pions are not very sensitive to the system size, if multiplicities and the mean transverse momentum are considered. However extrapolations to higher energies are problematic, because the onset of a changing behaviour may be present in the experimental data. The relative production probability of strange particles (to pions) increases strongly with energy up to AGS energies with a clear trend to level off beyond. Participant proton rapidity distributions for intermediate and heavy mass systems seem to scale with beam rapidity. Their energy loss is in the range 70% to 80% at SPS energies which is to be compared to 50% in p+p interactions.

1 Introduction

The obvious way to learn about the prospects of research with relativistic heavy ions beyond the turn of the millenium is an extrapolation from the past into the future. One of the outstanding goals of nuclear physics in the past has been the approach to the island of nuclear stability. After more than sixty years of research it is still a strong field and actually has, for the first time, demonstrated that the island of stability is within reach (see the corresponding contributions to the proceedings). In the nineteen-seventies matter densities far above the nuclear ground state density became accessible in laboratory experiments. This achievement stimulated renewed discussions on the equation of state of nuclear matter and its relevance for the understanding of neutron stars and supernovae explosions. Also this field is still very much alive over twenty years later in the contributions on SIS/BEVALAC physics to these proceedings. With the advent of even higher energy ion beams at the AGS and at CERN a new regime of energy density comes into reach. The justification of the involved experimental and theoretical efforts is the expectation that hadrons can no longer exist in hadronic matter at high energy densities. This implies that instead of a dense and hot blob of hadronic matter a fire-ball of quarks and gluons is formed in these ultrarelativistic heavy ion collisions. If the lifetime of this fire-ball is long enough it may evolve into a steady state system, a plasma of quarks and gluons. One of the important questions in this context concerns the initial conditions of the fireball. Is it more favourable for the approach to equilibrium to have the valence quarks of the nucleus in the incoming nuclei stay in the fireball or is the Bjorken picture of energy deposition in form of gluons and $q\bar{q}$ pairs a better choice. QCD lattice calculations cannot contribute (yet) to answering this question, because quark-chemical potentials are not included in the calculations. On the experimental side there is evidence that even at CERN SPS energies the projectile nucleons experience enough stopping to be found in the fireball. In ultrarelativistic heavy ion collisions at RHIC and LHC we expect the nucleons to retain a large part of their longitudinal energy and, thus, leave a net baryon number free fireball behind (at least at the LHC). Perhaps we have to wait for the first results from these experiments before we can judge in which energy regime the properties of the supposedly created Quark-Gluon-Plasma can be studied best. With the presently available results from heavy ion experiments we can, apart from pinning down the evidence for the formation of a Quark-Gluon-Plasma state, try to find possible systematic trends as function of decreasing baryon stopping which may allow to make reliable predictions for LHC energies.

In this contribution we will compare experimental results from nuclear collisions at high and low energies with small and large projectiles. The specific observables will be the multiplicity of light mesons (π), their average transverse momenta, ratios of strange to nonstrange particles, the slope parameters of transverse mass distributions of heavy particles, and the phase space distributions of protons which have participated in the collision.

2 Particle Yields

In general the number of produced particles per collision (system) increases with beam energy or equivalently with $\sqrt{s}$. This is true for hadron-hadron ($h + h$) collisions as well as for nucleus-nucleus ($A + A$) collisons. A comparison of $h + h$ with $A + A$ may reveal different systematic behaviour. In Fig. 1 particle production in $p + p$ interactions is summarized [1]. Shown are the mean multiplicities as function of s. For each particle species a steep rise near threshold and a saturation with increasing energy is observed. The threshold together with the curves are shifted with increasing particle mass to higher s except for the Λ. Its production threshold coincides with the one of the K^+. What is an appropriate way to compare $p + p$ data to the results from $A + A$ collisions? In a first step trivial differences between proton-proton and nucleon-nucleon ($N + N$) interactions have to be accounted for. Next one has to chose a common basis for the comparison of $A + A$ with $N + N$ collisions. The multiplicities of Fig. 1 are eveluted for inelastic $p + p$ interactions. An equivalent definition is difficult to find for $A + A$ collisions. One way to relate $A + A$ to $N + N$ collisions is the normalisation to the number of participating nucleons.

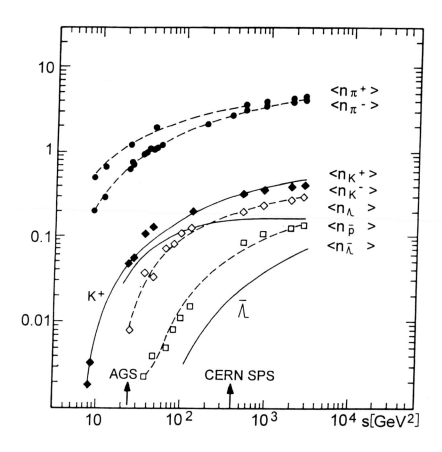

Figure 1: Mean multiplicity of produced particles in p+p interactions as function of the total
c.m.energy

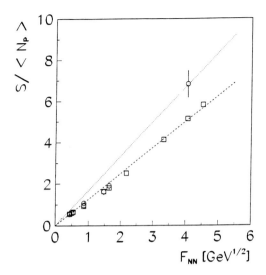

Figure 2: Entropy per participating nucleon as function of the Fermi variable F (approx. $s^{1/4}$) for nucleon-nucleon (squares) and central $A + A$ (circles) collisions (from reference[2])

The number of produced particles for a given species per participating nucleon as function of s can then be compared to Fig. 1, if the mean multiplicities are divided by two. This type of comparison has been done by Gazdzicki[2] for the pions. Instead of using the pion multiplicity directly he assumes a transfer of pions (or entropy) into the nucleons in order to account for the net absorption of pions in nuclear collisions. In Fig. 2 his results are shown as function of the "Fermi-variable" F which is essentially the square root of the free kinetic energy in the c.m. system. The $N + N$ data exhibit a linear increase up to SPS energies as already anticipated by Fermi. As to the $A + A$ data they seem to scale in the same way up to AGS energies $(F = 1.6~GEV^{1/2})$, but exhibit a pion excess of 30% at SPS energies in $S + S$ collisions. If this trend is confirmed or even accentuated by Pb data from the SPS extrapolations to RHIC and LHC energies will be impossible except if the beam energy gap between AGS and SPS is covered with several data points to pin down where and how the deviation from the $N + N$ scaling behaviour occurs. A similar analysis has been performed for strange particles[3]. Fig. 3a shows the ratio $E_s = (< \Lambda > + < K + \overline{K} >)/ < \pi >$ again as function of F for $p + p$ interactions. The shape of the correlation resembles those seen in Fig. 1 which means that the strange particle yields increase faster with F than the pion multiplicity. In Fig. 3b the corresponding plot for $A + A$ collisions is shown (note the different vertical scale and also a slightly different definition of F[2,3]). Obviously E_s in $A + A$ collisions is always greater than in $p + p$ interactions.

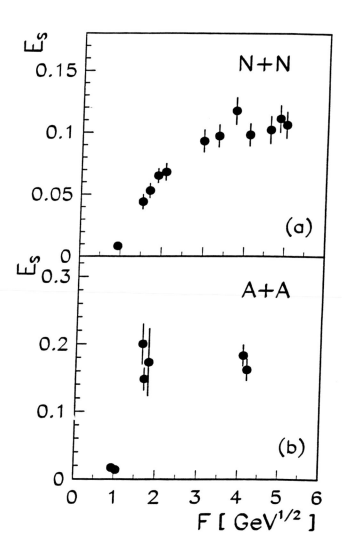

Figure 3: Ratio of the average number of strange particle $(\Lambda + K)$ to the average number of pions produced in nucleon-nucleon (a) and in central nucleus-nucleus (b) collisions as function of F (approx. $s^{1/4}$)

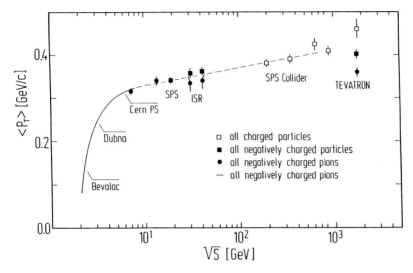

Figure 4: Mean transverse momentum of negatively charged particles as function of $s^{1/2}$ in $p + p$ and $p + pbar$ collisions

This strangeness enhancement is most pronounced at AGS energies and seems to saturate above yielding a factor of 1.5 for the ratio $E_s(A + A)/E_s(p + p)$ at SPS energy. The expected increase of E_s when going from AGS to SPS energies seems to have been compensated by the increase of the pion multiplicity observed already in Fig. 2.

The strong increase from Dubna $(F \sim 1)$ to AGS $(F \sim 2)$ energies in $A + A$ (a factor of 10 as compared to a factor of 5 in $p + p$) may be explained by the high density of baryon resonances observed in $A + A$ collisions at AGS.

3 Average Transverse Momenta

Measurements of the average transverse momenta in $p + p$ $(p + \overline{p})$ interactions as function of $\sqrt{s}$ show a similar behaviour as the mean particle multiplicities: a steep rise from low (BEVALAC/SIS) to medium (AGS) energies with a saturation beyond. In fact the mean transverse momentum of negatively charged pions remains constant to within some ten MeV/c above $\sqrt{s}$=10 GeV (see Fig. 4). The comparison with $A + A$ collision data is straight forward only at AGS energies for which the average transverse momenta of the pions have been determined in full phase space [4]. The value of 320 $\pm$5 MeV/c for

$< p_T >$ of all pions in Au+Au collisions at $\sqrt{s}$=5 GeV falls slightly above the universal curve in Fig. 4 which passes through 300 MeV/c at $\sqrt{s}$=5 GeV. At SPS energies the $< p_T >$ of identified pions has not yet been measured in full phase space. Results on negatively charged particles in central $S+S$ collisions[5] give $< p_T >= 355 \pm 5$ MeV/c again somewhat higher than the universal curve at $\sqrt{s}$=20 GeV (330 MeV/c). This difference is partly due to the admixture of K^- and $\bar{p}$. Here again we find very similar pion emission characteristics in $p+p$ and $A+A$ collisions. The pions seem to be intensitive to the environment they are emitted from. Only the detailed shapes of the transverse momentum distributions (not shown here) exhibit significant differences. The experimental results on heavy particles (K, p, Λ) look different. Their transverse momentum spectra are well described by a thermal distribution. This means that there is a 1-to-1 relation between slope parameter and average transverse momentum. The general observation is that $< p_T >$ of heavy particles increases significantly with system size. The NA44 collaboration reports an increase of the slope parameter (and equivalently of $< p_T >$) of K_s^o by 20% (50%) when going from $p + p$ to central $S + S$ $(Pb + Pb)$ collisions[6]. In the heaviest system at SPS energies $< p_T > (K_s^o)$ is as high as 600 MeV/c. This is higher than found in the most violent $p + \bar{p}$ collisions at $\sqrt{s}$=630 GeV as depicted in Fig. 5[7]. Going to masses as high as the one of the baryons (Λ and $\overline{\Lambda}$) the collider ($p+\bar{p}$) data vary between 600 and 1000 MeV/c. This is the same range as quoted for $< p_T > (\Lambda)$ in central $Pb + Pb$ collisions by NA49[8]. In summary, the heavy particles are a much more sensitive probe for the differences between $h+h$ and $A + A$ collisions than the light pions.

4   Phase Space Distributions of Participating Protons

The study of the phase space distribution of those particles which carry the baryon number of the incoming nucleons provide information about how much of the longitudinal energy of the impinging nucleons gets transferred to newly created particles and to the transverse degrees of freedom. Fig. 6 summarizes the data from light projectiles on heavy target at CERN energies. Shown are the rapidity distributions of the participant protons in various projectile-target combinations at 200 GeV/nucleon. For the data on $^{16}O + Au$ at 60 GeV/nucleon the rapidity was scaled up such that beam particles have the same rapidity. After proper weighing of the ordinate to account for the different projectiles we observe an almost perfect scaling, the data for different system fall on an universal line. This means that the relative rapidity loss of the projectile nucleons is independent of the size of the projectile. This scaling behaviour persists when comparing similar systems at different energies. In

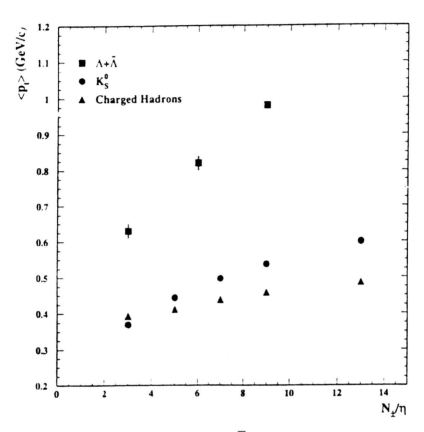

Figure 5: Average transverse momenta of $\Lambda + \overline{\Lambda}, K_s^o$ and charged hadrons as function of charged particle multiplicity per unit of rapidity in $p + \overline{p}$ collisions at $s^{1/2} = 630 GeV$

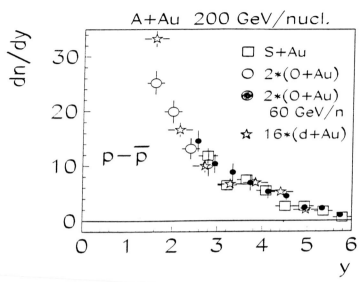

Figure 6: Rapidity distribution of participant protons (obtained from the difference between positive and negative particle distribution) in collisions of ^{32}S, ^{16}O, and d with Au

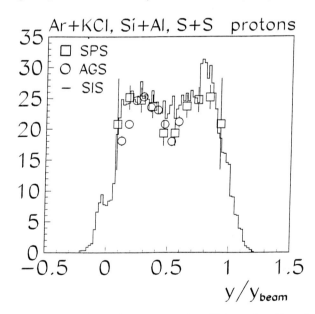

Figure 7: Rapidity distribution of participant protons in collisions of medium size nuclei at SIS/BEVALAC, AGS and SPS energies

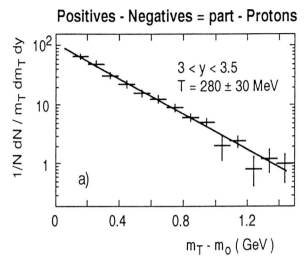

Figure 8: Transverse mass distribution of participant protons in central $Pb + Pb$ collisions at 158 GeV/nucleon

Fig. 7 Bevalac, AGS and SPS data on central collisions of Ar on KCl, Si on Al [9], and ^{32}S on ^{32}S [5] are compared. The rapidity distributions are rescaled to give a beam rapidity of 1. The agreement between the 3 distributions is excellent. A similar plot with $Pb + Pb$ $(Au + Au)$ data was shown recently at other conferences [10,11] with good agreement of the distributions at SIS/Bevalac and AGS energies. At CERN SPS energies the situation is not yet clear. A somewhat flatter distribution may indicate the onset of nuclear transparency in central $Pb + Pb$ collisions at SPS energies.

The total energy transferred to newly created particles can be determined, if the longitudinal <u>and</u> the <u>transverse</u> degrees of freedom of the participating baryons are taken into account. The transverse mass distribution of protons is shown in Fig. 8 [12]. The spectrum is characterized by an exponential falloff with a slope parameter of 280 MeV. The corresponding mean transverse momentum is 750 MeV/c. Combining the information in the rapidity and transverse mass distribution of participating protons in central $Pb + Pb$ collisions at SPS energies yields an energy loss of the incoming nucleons of 75 $\pm$10%, somewhat higher than observed in central $S + S$ collisions (70 $\pm$ 7%).

5 Summary

Extrapolation of the multiplicities of produced particles from AGS/SPS to RHIC and LHC energies is not straight forward, because there seems to be a change of its energy dependence in the currently accessible energy range. The average transverse momentum of pions in $A + A$ collisions is not expected to change with increasing $\sqrt{s}$ and system size. On the basis of high energy hadron collider data and results from nucleus-nucleus collsions an extrapolation to RHIC and perhaps also LHC $(A + A)$ physics should be possible. As to the rapidity distribution of those baryons which have resided in the impinging nuclei extrapolation from SPS to RHIC/LHC energies will always be problematic, because the details of the slowing down process is expected to vary significantly above presently available energies. The comparison of AGS and SPS data may reveal already now a different (relative) rapidity shift for the heaviest systems at the higher energy. More data from $Pb + Pb$ collisions are needed for a final evaluation.

1. A. M. Rossi et al., Nucl. Phys. B84 (75) 269
2. M. Gazdzicki, Z. Phys. C66 (95) 659
3. M. Gazdzicki and D. Rörich, Z. Phys. C, acc. for publication, June 96
4. Y. Akiba for the E866 Collaboration, contribution Quark Matter 96, Twelfth International Conference on Ultra-Relativistic Nucleus-Nucleus Collisions, May 1996
5. J. Bächler et al., NA35 Collaboration, PRL 72 (94) 1419
6. N. Xu for the NA44 Collaboration, contribution Quark Matter 96, Twelfth International Conference on Ultra-Relativistic Nucleus-Nucleus Collisions, May 1996
7. Bocquet et al., UA1 Collaboration, Phys. Lett. B366 (96) 441
8. P. Jones for the NA49 Collaboration, contribution Quark Matter 96, Twelfth International Conference on Ultra-Relativistic Nucleus-Nucleus Collisions, May 1996
9. T. Abbot et al., E802 Collaboration PRC 50 (94) 1024
10. T. Wienold, contribution Quark Matter 96, Twelfth International Conference on Ultra-Relativistic Nucleus-Nucleus Collisions, May 1996
11. J. Harris, Proceedings of the 12th Winter Workshop on Nuclear Dynamics, Editors W. Bauer and G. D. Westfall, Plen. Publ. Corp. N.Y. (1996)
12. R. Stock, "Hadron Spectra from Lead Collisions at 160 GeV per Nucleon"; proceedings of INPC, Peking, China (1995)

URQMD - a new molecular dynamics model from GANIL to CERN energies

S.A. Bass, M. Bleicher, M. Brandstetter, C. Ernst, L. Gerland, J. Konopka, S. Soff,
C. Spieles, H. Weber, L.A. Winckelmann and H. Stöcker
Institut für Theoretische Physik
Johann Wolfgang Goethe-Universität
Postfach 11 19 32
D-60054 Frankfurt am Main, Germany

C. Hartnack and J. Aichelin
SUBATECH, Ecole des Mines
4, rue Alfred Kastler
La Chantrerie
F-44070 Nantes, Cedex 03, France

N. Amelin
Joint Institute for Nuclear Research (JINR)
Dubna, 141980 Moscow region, Russia

The ultrarelativistic transport model, UQMD (Ultrarelativistic Quantum Molecular Dynamics), is used to analyze the physics of the excitation function of hadronic abundances, stopping and flow. This work bridges with one model consistently the entire available range of energies from 160 MeV/u to 160 GeV/u, even for the heaviest system Pb+Pb.

1 Introduction

The **FOPI, KaoS**, and **TAPS** experiments at GSI, the **EOS, E866** and **E877** experiments at Brookhaven and the **CERES, NA44, NA49** and **WA89** experiments at CERN are used to search for scaling violations as signatures of hot dense excited matter and the QCD phase transition[1,2,3,4]. The main emphasis of the analysis of the excitation function including its sensitivity to the QGP and chiral phase-transition[5] and the nuclear equation of state[6] has been directed towards the scaling of stopping with y_{proj} and its mass dependence from 160 MeV/n to 160 GeV/n and towards transverse degrees of freedom.

2 The URQMD model

URQMD is a hadronic transport model including strings[7]. Its collision term contains 49 different baryon species (including nucleon, delta and hyperon resonances with masses up to 2 GeV) and 25 different meson species (including strange meson resonances), which are supplemented by their corresponding antiparticle and all isospin-projected states. Full baryon/antibaryon symmetry is included. For excitations with higher masses a string picture is used. Elementary cross sections are fitted to available proton-proton data (see e.g. fig. 1).

The baryons, mesons and resonances which can be populated in URQMD are listed in table 1. All states listed can be produced in string decays, s-channel collisions or resonance decays.

The real part of the nucleon optical potential is modeled after the Skyrme ansatz, including Yukawa and Coulomb forces.

N	Δ	Λ	Σ	Ξ	Ω
938	1232	1116	1192	1317	1672
1440	1600	1405	1385	1530	
1520	1620	1520	1660	1690	
1535	1700	1600	1670	1820	
1650	1900	1670	1790	1950	
1675	1905	1690	1775		
1680	1910	1800	1915		
1700	1920	1810	1940		
1710	1930	1820	2030		
1720	1950	1830			
1990		2100			
		2110			

0^-	1^-	0^+	1^+	2^+
π	ρ	a_0	a_1	a_2
K	K^*	K_0^*	K_1^*	K_2^*
η	ω	f_0	f_1	f_2
η'	ϕ	σ	f_1'	f_2'

Table 1: List of Baryons, mesons, and resonances which are included in the URQMD model. In addition all charge conjugate states are treated on the same footing. For higher mass excitations meson- and (anti)baryon-strings are included.

The URQMD model allows for systematic studies of excitation functions over a wide range of energies in a unique way: the basic concepts and the physics input used in the calculation are the same for all energies. A relativistic cascade is applicable over the entire range of energies from 100 MeV/nucleon up to 200 GeV/nucleon. A preliminary molecular dynamics scheme using a hard Skyrme interaction is used between 100 MeV/nucleon and 4 GeV/nucleon.

One of the main goals of relativistic heavy ion collisions is the determination of the nuclear equation of state. At high energies, semiclassical cascade models in terms of scattering hadrons have proven to be rather accurate in explaining experimental data. Therefore it is of fundamental interest to extract the equation of state from such a microscopic model, i. e. to investigate the equilibrium limits and bulk properties, which are not an explicit input to the non-equilibrium transport approach with its complicated collision term (unlike

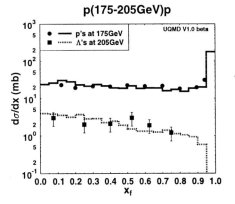

Figure 1: Feynman x distribution for protons and lambdas at 205 and 175 GeV/nucleon respectively. The URQMD calculation (histograms) is fitted to experimental data (points).

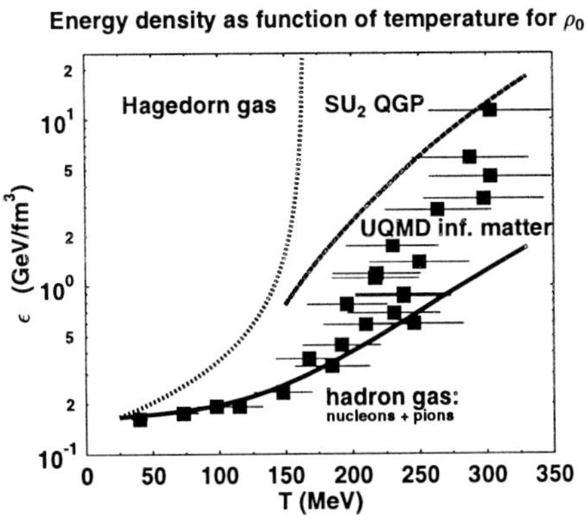

Figure 2: Energy density as a function of temperature (URQMD, cascade mode). The equations of state of a Hagedorn-gas, of a quark-gluon plasma, and of an ideal gas of nucleons and ultrarelativistic pions are also depicted.

e. g. in hydrodynamics). To simulate infinite hadronic systems we construct a box with periodic boundary and initialize nucleons at ground-state nuclear density randomly in phase space, varying the total energy density. After the system has equilibrated according to the URQMD simulation, the temperature is extracted by fitting the particles' momentum spectra. Alternatively, the temperature can be extracted from the relative abundances of different hadrons, e. g. the Δ/N ratio, which should yield the same temperature. This has been checked in the current simulation. The result of this procedure is plotted in Fig. 2 for a cascade calculation, i.e. taking only the imaginary part of the baryon–baryon, baryon–meson and meson–meson interactions into account. It appears that the energy density rises faster than T^4 at high temperatures of $T \approx 200 - 300$ MeV. This indicates an increase in the number of degrees of freedom. It may be interpreted as a consequence of the numerous high mass resonances and string excitations, which in a way release constituent quark degrees of freedom (but, of course, no free current quarks as in an ideal QGP). Investigations of equilibration times and relative particle and cluster abundances are in progress. Moreover, the admittedly poor statistics have to be improved.

3 Longitudinal degrees of freedom

Baryonic stopping is a necessary condition for the creation of hot and dense
nuclear matter. The key observable is the rapidity distribution of baryons
which is displayed in figure 3 for three presently used heavy ion accelerators.
In all cases a system as heavy as Au+Au or Pb+Pb exhibits a central pile-up
at midrapidity. However, the physical processes associated are different: The
average longitudinal momentum loss in the SIS energy regime is mainly due
to the creation of transverse momentum whereas at the AGS/SPS energies
abundant particle production eats up a considerable amount of the incident
beam energy. The form of the distributions change from a gaussian at SIS
energies to a plateau at AGS and finally to a slight two-bump structure at
CERN energies.

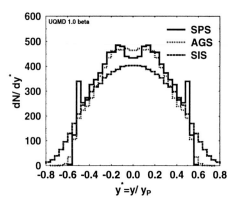 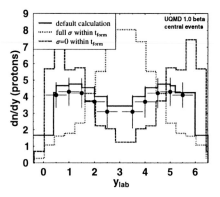

Figure 3: Rapidity distributions for Au+Au collisions at SIS (1 GeV/nucleon), AGS (10.6 GeV/nucleon) and CERN SPS energies (160 GeV/nucleon). All distributions have been normalized to the projectile rapidity in the center of mass frame.

Figure 4: Formation time dependence of stopping power. The cross sections of particles within their formation time introduce large uncertainties for their rapidity distribution.

At CERN/SPS energies baryon stopping is dominated not only by rescattering effects but also by the formation time after hard collisions in which
strings are excited. Within their formation time baryons originating from a
leading (di)quark interact with (2/3) 1/3 of their free cross section and mesons
with 1/2 of their free cross sections. The influence of the formation time is
shown in figure 4 for the system S+S at 200 GeV/nucleon. The default calculation (including formation time) reproduces the data fairly well whereas the

calculation with zero formation time (dotted line) exhibits total stopping. A calculation with zero cross section within the formation time, however, exhibits transparency.

In order to study this effect more closely the $\sqrt{s}$ distributions for Au+Au collisions at AGS and S+S collisions at SPS energies are analyzed. Figure 5 shows the respective distribution for Au+Au. The collision spectrum is dominated by BB collisons with full cross sections and exhibits a maximum at low energies. Approximately 20% of the collisions involve a diquark, i.e. a baryon originating from a string decay whose cross section is reduced to 2/3 of its full cross section.

In figure 6 the same analysis is performed for S+S at 200 GeV/nucleon. In contrast to the heavy system at AGS the collision spectrum exhibits two pronounced peaks dominated by full BB collisions, one in the beam energy range and one in the low (thermal) energy range. Approximately 50% of the collisions, most of them at intermediate $\sqrt{s}$ values, involve baryons stemming from string excitations whose cross sections are reduced by factors of 2/3 (referred to as *diquarks*) or 1/3 (referred to as *quarks*). The peak at high $\sqrt{s}$ values stems from the initial hard collisions whereas the peak at low energies is related to the late, thermal stages of the reaction. Since the reduction factor of 1/3 (2/3) does only take valence-quarks into account but neglects contributions of gluons and sea-quarks it is no undisputable quantity and contains a certain level of freedom – the same holds true for the formation time itself which cannot be derived from first principles.

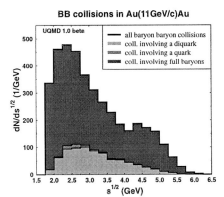

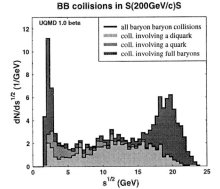

Figure 5: E_{CM}^{coll} distribution for baryon baryon collisions in a central Au+Au reaction at the AGS.

Figure 6: E_{CM}^{coll} distribution for baryon baryon collisions in a central S+S reaction at SPS energies.

4 Transverse degrees of freedom

The creation of transverse flow is strongly correlated with the underlying equation of state[6]. In particular it is believed that secondary minima as well as the quark-hadron phase transition lead to a weakening of the collective sideward flow[5]. The occurence of the phase transition should therefore be observable through abnormal behaviour (e.g. jumps) of the strength of collective motion of matter. Note that URQMD in its present form does not include any phase transition explicitly.

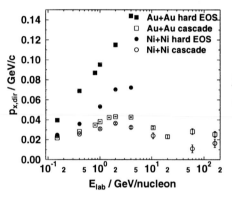 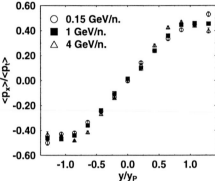

Figure 7: Excitation function of the total directed transverse momentum transfer p_x^{dir} for the Au+Au and Ni+Ni systems. URQMD including a hard equation of state (full symbols) are compared to the predictions of cascade calculations (open symbols).

Figure 8: Mean directed transverse momentum as a function of the scaled rapidity for Au+Au at b=4 fm. If the transverse flow is scaled with the mean transverse momentum, the sidewards does not depend on the bombarding energy.

For the purely hadronic scenario the averaged in plane transverse momenta for Ni+Ni and Au+Au in the 0.1 – 4 GeV/nucleon region are displayed in figure 7. Calculations employing a hard equation of state (full symbols) are compared to cascade simulations (open symbols). In the latter case only a slight mass dependence is observed. For the calculation with potentials the integrated directed transverse momentum push per baryon is more than twice as high for the heavier system which corroborates the importance of a non-trivial equation of state of hadronic matter.

For beam energies below 5 GeV/nucleon the amount of directed transverse momentum scales in the very same way as the total transverse momentum in the course of the reaction. Therefore the $\langle p_x \rangle$ versus rapidity divided by the

$\langle p_t \rangle$ is identical for all beam energies in the range of scaling as can be seen in figure 8.

5 Conclusion and outlook

A novel ultrarelativistic tranport model, dubbed URQMD (Ultrarelativistic Quantum Molecular Dynamics) is introduced. The code is written in a modular fashion and will be published[7]. The model is applied to analyze the physics of the excitation function of hadronic abundances, stopping and flow. This work bridges with one model consistently the entire available range of energies from 160 MeV/u to 160 GeV/u, even for the heaviest system Pb+Pb. Heavy systems exhibit a central pile-up at mid-rapidity over three orders of magnitude in beam energy. For SPS energies ambiguities arise due to the concept of formation time and the treatment of cross sections within that timeinterval. Nearly 50% of all baryon-baryon interactions in the course of the reaction are affected. A flow excitation function for Ni+Ni and Au+Au is presented. Strong effects of the real part of the nucleon nucleon optical potential are observed.

Acknowledgments

This work has been supported by the Bundesministerium für Bildung und Forschung (BMBF), the Deutsche Forschungsgemeinschaft (DFG), and the Gesellschaft für Schwerionenforschung mbH (GSI).

References

1. W. Scheid, R. Ligensa, and W. Greiner, Phys. Rev. Lett. **21**, 1479 (1968).
2. W. Scheid, H. Müller, and W. Greiner, Phys. Rev. Lett. **32**, 741 (1974).
3. H. Stöcker, J. A. Maruhn and W. Greiner, Phys. Rev Lett. **44**, 725 (1980).
4. for a comprehensive overview and the current status of the field, see e.g. Proc. of Quark Matter '95, Nucl. Phys. **A590** (1995).
5. D. H. Rischke, Y. Pursun, J. A. Maruhn, H. Stöcker and W. Greiner. Nucl. Phys. **A597** (1995), 701.
6. H. Stöcker and W. Greiner. Phys. Reports **137**, 277 (1986)
7. S. A. Bass, M. Bleicher, M. Brandstetter, C. Ernst, L. Gerland, C. Hartnack, J. Konopka, S. Soff, C. Spieles, H. Weber, L. A. Winckelmann, J. Aichelin, N. Amelin, H. Stöcker and W. Greiner
   URQMD source code and technical documentation, to be published.

COLORED CHAOS AND THE FORMATION OF THE QUARK-GLUON PLASMA

B. MÜLLER

Department of Physics, Duke University
Durham, NC 27708-0305, USA

Recent results revealing the universally chaotic nature of the Yang-Mills fields in the classical limit are discussed. The results enhance our understanding of the basis of statistical models for hadronic reactions and of the mechanisms of thermalization in non-Abelian gauge theories.

Multiparticle production and other phenomena of what today is called "soft" hadronic interactions have often been successfully described in the framework of statistical models. However, it has remained largely obscure why the assumption of a random statistical distribution of final states is warranted in high energy reactions that last not much longer than 1 fm/c (or 3×10^{-24}s), and why are the final states not dominated by coherent quantum states or collective excitations of a small subset of the available hadronic degrees of freedom?

Similar questions posed themselves in the context of Bohr's statistical model of compound nucleus reactions. In this case, the conceptual difficulties were eventually resolved by the insight that the highly excited compound nucleus is a chaotic quantum system [1,2] exhibiting rapid exchange of energy between the accessible degrees of freedom.

On the basis of Berry's conjecture [3]—that the amplitudes of energy eigenstates of chaotic quantum systems are stochastic variables—Srednicki recently showed [4] that the expectation values of physical operators in energy eigenstates of such systems coincide with their thermal expectation values. This result extends to the fluctuations of physical quantitites [5]. In other words, chaotic quantum systems can be considered as ergodic. Here I discuss recent numerical evidence that non-Abelian gauge theories, the basis of nuclear interactions, are strongly chaotic in the classical limit. This discovery, made more than a decade ago [6], has been explored in detail only recently. As I will argue, the results throw new light on the apparent success of thermal models for nuclear reactions and, in particular, allow for a novel and non-perturbative estimate of the formation time of a quark-gluon plasma.

1 Chaos and Ergodicity

Let us begin with some general remarks about the applicability of statistical concepts to a single system. A dynamical system exhibits *ergodic* behavior, if the time average of an observable A can be replaced by the phase space average

$$\lim_{T \to \infty} \frac{1}{T} \int_0^T dt \, A(t) = \langle A \rangle \equiv \frac{1}{Z_E} \int d\Gamma_E A(\Gamma_E). \tag{1}$$

$\langle A \rangle$ here denotes the *microcanonical* average, Z_E is the microcanonical partition function, and $d\Gamma_E$ is the phase space measure at constant energy. For systems with very many degrees of freedom it is equivalent to take the *canonical* average

$$\langle A \rangle = \frac{1}{Z(\beta)} \int d\Gamma A(\Gamma) \exp[-\beta E(\Gamma)], \tag{2}$$

where $Z(\beta) = \int d\Gamma \exp[-\beta E(\Gamma)]$ is the canonical partition function and the inverse temperature $\beta = 1/T$ is determined by the condition $E = -\partial(\ln Z)/\partial\beta$.

For practical applications it is crucial to know the time scale on which ergodicity is attained. It can be shown that this time scale is related to the rate h of exponential divergence of neighboring trajectories in phase space; this rate is called the (maximal) Lyapunov exponent [7]. The complete spectrum of Lyapunov exponents is defined as follows. Consider a given trajectory in phase space, $x_\alpha(t)$, where $\alpha = 1, \ldots \nu$ enumerates the degrees of freedom. $x_\alpha(t)$ is a solution of the classical equations of motion of the system. Now take a set of neighboring trajectories:

$$\tilde{x}_\alpha'^{(i)}(t) = x_\alpha(t) + \delta x_\alpha^{(i)}(t), \quad \alpha = 1, \ldots, \nu. \tag{3}$$

For infinitesimal δx_α they are solutions of a second-order linear differential equation of the form

$$D[x_\alpha(t)] \delta x_\alpha^{(i)}(t) = 0. \tag{4}$$

One can then obtain a complete orthogonal set of solutions of this equation; they define ν Lyapunov exponents λ_i according to

$$\lambda_i = \lim_{t \to \infty} \frac{1}{t} \ln \frac{\|\delta x_\alpha^{(i)}(t)\|}{\|\delta x_\alpha^{(i)}(0)\|}, \quad i = 1, \ldots, \nu. \tag{5}$$

In other words, for long times one has the norm of $\delta x_\alpha^{(i)}$ growing (or shrinking) as $\|\delta x_\alpha^{(i)}(t)\| \propto \exp(\lambda_i t)$. One usually assumes the Lyapunov exponents to the ordered in size:

$$\lambda_1 \equiv h \geq \lambda_2 \geq \ldots \geq \lambda_\nu. \tag{6}$$

For conservative (Hamiltonian) systems the Lyapunov exponents occur in pairs of equal magnitude, but opposite sign: $\lambda_i = -\lambda_{N+1-i}$. This is in accordance with Liouville's theorem which states that the volumes in phase space filled by an ensemble remains unchanged with time, implying that there must be a direction of contraction for every direction in which the phase space volume expands. For each conservation law there occur two vanishing Lyapunov exponents; the conservation of energy always ensures the existence of one such pair for a Hamiltonian system. Since the extent of the ensemble rapidly shrinks below any practially achievable resolution in the exponentially contracting directions, the *observable* volume in phase space grows as

$$\overline{\Gamma}(t) \propto \exp\left(t \sum_i \lambda_i \theta(\lambda_i)\right) \equiv \exp(\dot{S}_{KS} t) \tag{7}$$

where the sum only includes the positive Lyapunov exponents. The exponential growth rate of the observable phase space volume implies a *linear* rate of growth of the observable "coarse-grained" entropy associated with the ensemble. This rate, $\dot{S}_{KS} = \sum_i \lambda_i \theta(\lambda_i)$, is called the Kolmogorov-Sinai entropy, or short, KS-entropy. Dynamical systems that have a positive KS-entropy everywhere in phase space are called K-systems; they exhibit all the properties required for a statistical description on time scales that are long compared with the ratio between the equilibrium entropy S_{eq} and the KS-entropy, i.e. for times

$$t \gg \tau_s = S_{eq}/\dot{S}_{KS}. \tag{8}$$

An illustration of these properties is provided by the simple dynamical system

$$H(x, y; p_x p_y) = \tfrac{1}{2}\left(p_x^2 + p_y^2 + x^2 y^2\right), \tag{9}$$

which occurs as part of the extreme infrared limit of Yang-Mills fields[6]. The system described by the Hamiltonian (9) has a positive Lyapunov exponent $\lambda \approx 0.4$. Almost all its trajectories are unstable against small perturbations[8] and the analogous quantum system has been shown to exhibit a Wigner distribution of its level spacings[9]. The remarkable ability of this system is to randomize an initially localized phase space distribution. After a rather limited time the phase space distribution is indistinguishable from a microcanonical ensemble.

2   Chaos in Non-Abelian Gauge Theories

If we want to apply these concepts to non-Abelian gauge theories, we must consider these as classical Hamiltonian systems with many degrees of freedom, and we need a gauge invariant distance measure in the space of field configurations. The first part is easy; the Hamiltonian formulation of lattice gauge theory by Kogut and Susskind can form the basis for a study for the gauge group SU(n) of non-Abelian gauge theories as dynamical systems. The lattice Hamiltonian is expressed as (a denotes the lattice spacing)

$$H = \frac{g^2}{2a} \sum_\ell \text{tr}(E_\ell E_\ell^\dagger) + \frac{2}{g^2 a} \sum_p (n - \text{Re tr } U_p), \tag{10}$$

where electric field strength E_ℓ and the link variables U_ℓ are defined on the lattice links, and U_p denotes the ordered product of the U_ℓ around an elementary plaquette $p : U_p = U_4^\dagger U_3^\dagger U_2 U_1$. The U_ℓ are elements of the gauge group (SU(3) in the case of QCD) and the E_ℓ are elements of the associated Lie algebra. In the classical limit, the link variables U_ℓ are functions of time, and the electric field variables are given by

$$E_\ell = \frac{a}{ig^2} \dot{U}_\ell U_\ell^\dagger. \tag{11}$$

The Hamiltonian equations then provide a set of coupled equations for the time evolution of $U_\ell(t)$ and $E_\ell(t)$, which can be integrated numerically.

An appropriate measure for the distance between two field configurations is [10]

$$D[U_\ell, E_\ell; U_\ell', E_\ell'] = \frac{1}{2N_p} \left(a^2 \sum_\ell \left| \text{tr}(E_\ell E_\ell^\dagger) - \text{tr}(E_\ell' E_\ell'^\dagger) \right| + \sum_p \left| \text{tr } U_p - \text{tr } U_p' \right| \right). \tag{12}$$

It is gauge invariant, gives a vanishing distance between gauge equivalent field configurations, and goes over into

$$D[A, E; A', E'] \propto \frac{1}{2V} \int d^3x \left(|E^2 - E'^2| + |B^2 - B'^2| \right) \tag{13}$$

in the continuum limit, measuring the local differences in the electric and magnetic field energy. If one only wants to determine the largest Lyapunov exponent, it is sufficient to consider either the electric or the magnetic contribution to $D[U, U']$.

If one starts from two randomly chosen neighboring field configurations and integrates these in time, one finds that the distance quickly grows exponentially, until it saturates due to the compactness of the space of gauge fields. The growth rate h quickly reaches a constant limit as function of the lattice size N^3, if the energy density is kept fixed by choosing the same average energy per plaquette E_p in each case. It is easy to see [11] that the Hamiltonian

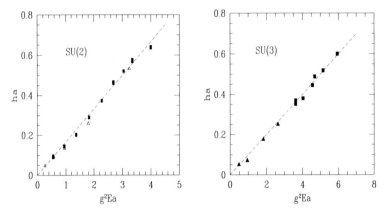

Figure 1: Maximal positive Lyapunov exponent as a function of the scaling parameter $(g^2 E_p a)$ for (a) SU(2), (b) SU(3) gauge theory.

(10) exhibits a scaling behavior such that the Lyapunov exponents, if they are universal functions of the average energy density, as expressed by E_p, can only depend on the dimensionless scaling variable $g^2 E_p a$. The nontrivial surprise is that, as shown in Figures 1a,b, the dependence is linear [10,12]

$$ha \equiv \lambda_1 a = b_n g^2 E_p a \quad \text{for SU}(n),\tag{14}$$

where $b_2 \approx \frac{1}{6}$ and $b_3 \approx \frac{1}{10}$. The linear relationship means that the lattice spacing drops out, yielding $h = b_n g^2 E_p$ independent of a. The maximal Lyapunov exponent hence has a well-defined continuum limit.

What about the other Lyapunov exponents? Their calculation for large lattices is prohibitively expensive, as there are in total $6(n^2 - 1)N^3$ degrees of freedom for SU(n) gauge theory on a N^3 lattice, but Gong[13] has evaluated the complete spectrum for SU(2) on lattices of size $N = 1, 2$, and 3. The result again is a surprise: when the Lyapunov exponents are scaled by the maximal one, and are plotted on the interval $[0, 1]$, the spectra for $N = 2$ and $N = 3$ are indistinguishable, and there is only a small difference between $N = 1$ and $N = 2$ (see Figure 2).

The Lyapunov spectrum for SU(2) shows three separate components: there are $(6N^3 - 1)$ positive and negative exponents each, and there are $(6N^3 + 2)$ exponents that converge to zero in the limit $t \to \infty$. Their vanishing reflects the existence of $(3N^3 + 1)$ conservation laws: Gauss' law at every lattice point and the overall energy conservation. Since the density of points on the line over the fixed interval $[0,1]$ grows as N^3, this implies that the sum over positive Lyapunov exponents increases like the volume of the lattice, yielding a constant

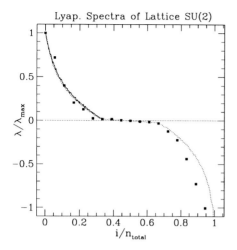

Figure 2: Spectrum of Lyapunov exponents for SU(2) lattice gauge theory. The black dots are for a 1^3, the dotted line is for a 2^3, and the solid line is for a 3^3 lattice. The $18N^3$ exponents are plotted on the fixed interval [0,1] to exhibit the scaling with N.

KS-entropy density $\dot{\sigma}_{KS}$ in the thermodynamic limit. Gong finds

$$\dot{S}_{KS} = \sum_i \lambda_i \theta(\lambda_i) = c_2 \lambda_1 N^3, \tag{15}$$

which together with (14) and the lattice volume $V = (Na)^3$ yields

$$\dot{\sigma}_{KS} = \frac{1}{V}\dot{S}_{KS} = \frac{1}{3}b_2 c_2 g^2 \varepsilon, \tag{16}$$

where $\varepsilon = 3E_p/a^3$ is the average energy density on the lattice. (Note that there are $3N^3$ plaquettes.) No one has yet calculated the complete Lyapunov spectrum for SU(3), but I expect a similar relationship as (15) to hold in that case, too. The coefficient c_2 is not completely independent of the scaling variable $g^2 E_p a$, but has a value around 2. We will return below to the question how the physically relevant value of $g^2 E_p a$ can be chosen.

3   Physics Perspectives

The instability of all degrees of freedom of the non-Abelian gauge field (in the classical limit) leads to a rapid "thermalization" of the energy density on the lattice. This is illustrated in Figure 3 showing the distribution of magnetic

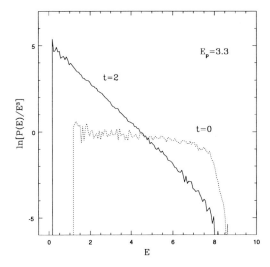

Figure 3: Logarithmic plot of the evolution of the magnetic energy density distribution on the lattice for SU(3) gauge theory. The distribution appears "thermalized" (exponential) after a time $t \approx 2$.

energy on the lattice plaquettes [14]. The initial state was chosen according to a random (not thermal) distribution of lattice link variables with vanishing electric field everywhere. Within two lattice units ($t/a = 2$) the energy has been equilibrated between electric and magnetic fields and, as the exponentially falling distribution shows, has assumed the form of a Gibbs distribution. The time scale for this "thermalization" is in good agreement with the time scale estimated from the inverse of the maximal Lyapunov exponent which is $h \approx 0.6$ in lattice units in SU(3) at this energy density.

The fact that the energy density thermalizes on a time scale much shorter than that required for the numerical determination of the Lyapunov exponents (typically $t/a = 1000$) allows us to relate the Lyapunov exponents to quantities in the presence of a thermal environment. First, we can replace the average energy per plaquette E_p in (14) by the "temperature" T, because

$$E_p = \frac{2}{3}(n^2 - 1)T \qquad (17)$$

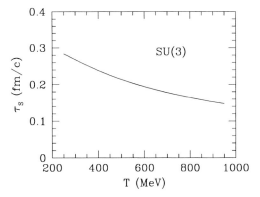

Figure 4: "Thermalization" time scale in SU(3) gauge theory as defined by the inverse of the maximal Lyapunov exponent (18), using $g(T)^2 = 16\pi^2/[11\ln(\pi T/\Lambda)^2]$ with $\Lambda = 200$ MeV.

for the classical, equilibrated SU(n) gauge field. This implies that

$$h \approx \begin{cases} \frac{1}{3}g^2T \approx 0.33g^2T & \text{for SU(2),} \\ \frac{8}{15}g^2T \approx 0.53g^2T & \text{for SU(3).} \end{cases} \tag{18}$$

We can use this result to obtain a model independent, nonperturbative estimate of the thermalization time solely due to gauge field dynamics in QCD. To compensate for the lack of "running" of the gauge coupling constant in the context of our classical gauge field calculation, we may use the one-loop result for

$$g(T)^2 = \frac{(4\pi)^2}{11\ln(\pi T/\Lambda)} \tag{19}$$

in (18) to evaluate $\tau_h = h^{-1}$, as shown in Figure 4. Clearly, this time is much smaller than 0.5 fm/c for all relevant temperatures, indicating a very rapid thermalization of the available energy. One should note that the Lyapunov exponents usually approach their asymptotic values from *above*, i.e. the dynamical instabilities are actually greater before the energy has been completely thermalized. This indicates that thermalization of field configurations far away from equilibrium may proceed even more rapidly.

It first appeared as a remarkable coincidence that the maximal Lyapunov exponents for SU(2) and SU(3) agree within numerical errors with the analytically calculated damping rate of a non-Abelian plasmon at rest [15]:

$$\Gamma = 2\gamma(0) = 6.635\frac{n}{24\pi}g^2T \approx \begin{cases} 0.35g^2T & \text{for SU(2),} \\ 0.53g^2T & \text{for SU(3).} \end{cases} \tag{20}$$

The reason for the factor 2 is that the plasmon pole is usually parametrized as $\omega_p(k) = \omega^*(k) - i\gamma(k)$, so that the energy density of the soft plasmon mode falls off as $\exp(-2\gamma(0)t)$. The coincidence becomes a little bit less surprising when one realizes that $\gamma(0)$ can be expressed as an integral of the form

$$\gamma(0) = -\frac{1}{2mg}\text{Im}\Pi(m_g - i\epsilon, k = 0) = Cg^2T \tag{21}$$

where the constant C involves an integral over the Bose distribution function of thermal gluons in the *classical* limit $(e^{\beta\omega} - 1)^{-1} \to (\beta\omega)^{-1}$. The plasmon damping rate is essentially determined by classical physics and independent of Planck's constant $\hbar$.

The observation that the Lyapunov exponent is numerically evaluated in the vicinity of a "thermalized" field configuration over time scales that are much longer than those of thermal fluctuations allows us to establish a connection between these two quantities. According to (5) the Lyapunov exponents are determined from the long time behavior of solutions of the linearized equation for the fluctuation $a_\mu(x,t)$ around an exact solution $A_\mu(x,t)$ of the Yang-Mills equations. In the continuum limit this equation reads:

$$\left(g_{\mu\nu}D(A)^2 - D_\mu(A)D_\nu(A)\right) a^\nu(x,t) - 2i[F_{\mu\nu}, a^\nu(x,t)] = 0 \tag{22}$$

where $D_\mu(A)$ denotes the gauge covariant derivative and $[F_{\mu\nu}, a^\nu]$ denotes the Lie algebra commutator. The initial value problem for (22) can be solved by means of the propagator in the background field:

$$a^\mu(x,t) = \int d^3x' \Delta^{\mu\nu}(x,t;x',0|A)a_\nu(x',0). \tag{23}$$

Recalling that the maximal Lyapunov exponent is obtained from the long time average of the growth rate of $a^\nu(x,t)$ and assuming ergodicity, we may replace the long-time average of the propagators by the canonical, i.e. thermal, average:

$$\overline{\Delta}^{\mu\nu} \to \Delta_T^{\mu\nu} \tag{24}$$

where $\Delta_T^{\mu\nu}$ denotes the *exact* finite temperature propagation function of the gauge field. Note that the causal (retarded) propagator describes only damped fluctuations, but here also exponentially growing perturbations appear because we consider the initial value problem (23), which corresponds to a different integration path in the complex frequency plane (see 14). The thermal average in (24) is calculated for a *classical* emsemble of gauge fields, whereas the usual perturbative approach is based on the quantum mechanical ensemble. However, this difference turns out to be irrelevant for the plasmon damping rate $\gamma(0)$, although it has a large effect on the effective plasmon mass m_g. This is not entirely fortuitous, because the damping rate is given by tree diagrams, such as Compton scattering and bremsstrahlung, that have an exact low-energy classical limit [14,15]. We can therefore identify the maximal Lyapunov exponent with (twice) the damping rate of the most unstable mode in a thermal non-Abelian plasma, which turns out to be a plasmon at rest [16].

4 Conclusions and Outlook

The short thermalization time scales of less than 1 fm/c found in our studies of the time evolution of classical non-Abelian gauge fields provides a theoretical foundation for the observation that final states in "soft" strong interaction physics are often populated statistically. The reason for the success of these classical studies is that the dynamical instabilities in thermal gauge theories are of order $g^2 T$, which is a classical inverse length or time scale that does not involve $\hbar$.

We are currently studying whether other quantities of order $g^2 T$, such as the thermal "spatial" string tension, can also be calculated in the framework of classical Yang-Mills theory. This consideration leads into the problem of deriving an effective quasi-classical theory for thermal Yang-Mills theories at the length scale $(g^2 T)^{-1}$ that consistently incorporates quantum effects from shorter distances in the form of transport coefficients. Presumably such an effective theory will contain a gauge invariant mass term of order $gT/\sqrt{\hbar}$ (as in the Taylor-Wong action [17]) and a Langevin noise term describing the fluctuations due to interactions with hard thermal modes [18].

Another interesting problem concerns the application of real-time evolution of gauge fields to processes far off equilibrium as they occur in the earliest stage of hadron-hadron or nucleus-nucleus interactions. We have recently studied the instability of the superposition of two counter-propagating plane waves, i.e. of a standing abelian plane wave, in SU(2) Yang-Mills theory [19]. Here one finds that the Lyapunov exponent is proportional to the *amplitude* of the wave, not to the energy as it is the case in random fields. Once the coherent wave is only slightly perturbed it decays rapidly, exciting modes of all wavelengths, and quickly generates a thermal energy spectrum. This phenomenon extends to the evolution of more realistic initial configurations, such as the interaction between non-Abelian wave packets [20].

416

Acknowledgements

I thank T. S. Biró, C. Gong, C. R. Hu, S. G. Matinyan and A. Trayanov for their enthusiastic help in unraveling the intricacies of chaotic dynamics in gauge theories. This work was supported in part by the U. S. Department of Energy (grant DE-FG02-96ER40945) and the North Carolina Supercomputing Program.

References

1. J. J. M. Verbaarschot, H. A. Weidenmüller, and M. R. Zirnbauer, *Phys. Rep.* **129** (1985) 367; see also: C. Mahaux and H. A. Weidenmüller, *Ann. Rev. Nucl. Part. Science* **29** (1979) 1.
2. G. E. Mitchell, E. G. Bilpuch, P. M. Endt, J. F. Shriner, and T. von Egidy, *Nucl. Instr. Meth.* **B46/47** (1991) 446.
3. M. V. Berry, *J. Phys.* **A10** (1977) 2083; P. Pechukas, *Phys. Rev. Lett.* **51** (1983) 943.
4. M. Srednicki, *Phys. Rev.* **E50** (1994) 888.
5. M. Srednicki, *J. Phys.* **A29** (1996) L75.
6. S. G. Matinyan, G. K. Savvidy, and N. G. Ter-Arutyunyan-Savvidy, *Zh. Eksp. Teor. Fiz.* **80** (1981) 830 [*Sov. Phys. JETP* **53** (1981) 421].
7. see e.g. A. J. Lichtenberg and M. A. Liebermann, *Regular and Stochastic Motion*, (Springer-Verlag, New York, 1983).
8. P. Dahlquist and G. Russberg, *Phys. Rev. Lett.* **65** (1990) 2837.
9. E. Haller, M. Köppel, and L. Cederbaum, *Phys. Rev. Lett.* **52** (1984) 1665.
10. B. Müller and A. Trayanov, *Phys. Rev. Lett.* **68** (1992) 3387.
11. T. S. Biró, C. Gong, B. Müller, and A. Trayanov, *Int. J. Mod. Phys.* **C5** (1994) 113.
12. C. Gong, *Phys. Lett.* **B298** (1993) 257.
13. C. Gong, *Phys. Rev.* **D49** (1994) 2642.
14. C. Gong, Dissertation, Duke University, 1994 (unpublished).
15. E. Braaten and R. D. Pisarski, *Phys. Rev.* **D42** (1990) 2156.
16. T. S. Biró, C. Gong, and B. Müller, *Phys. Rev.* **52** (1995) 1260.
17. J. C. Taylor and S. Wong, *Nucl. Phys.* **B346** (1990) 115.
18. C. Greiner and B. Müller, preprint DUKE-TH-96-99, ⟨hep-th/9605048⟩.
19. C. Gong, S. G. Matinyan, B. Müller, and A. Trayanov, *Phys. Rev.* **D49** (1994) R607.
20. C. R. Hu, et al., *Phys. Rev.* **D52** (1995) 2402; **D53** (1996) 3823.

LITTLE BANG AT BIG COLLIDERS: HEAVY ION PHYSICS AT LHC AND RHIC

J. Schukraft

CERN, Div. PPE, CH-1211 Geneva 23, Switzerland

Within the short time span of 20 years, the physics of ultra-relativistic heavy ion collisions will have evolved from light ion reactions at a laboratory energy of a few GeV/nucleon, first explored at the Brookhaven AGS starting in 1986, to using heavy projectiles at a centre of mass energy of several TeV/nucleon at the CERN LHC starting around 2005. After a brief description of the main machine parameters, this contribution will summarize the improvements and changes expected for the physics of heavy-ion reactions at these colliders as well as the experimental program planned at RHIC and LHC.

1 Introduction

The aim of high-energy heavy-ion physics is the study of strongly interacting matter at extreme energy densities. Statistical QCD predicts that, at sufficiently high density, there will be a transition from hadronic matter to a plasma of deconfined quarks and gluons — a transition which in the early universe took place in the inverse direction some 10^{-5} s after the Big Bang and which might play a role still today in the core of collapsing neutron stars.

According to T.D. Lee [1], high energy physics has traditionally focused on experiments in which higher and higher energies have been concentrated into a region with smaller and smaller spatial dimensions. This 'high energy frontier' has been extremely succesfull in establishing the basic ingredients of what we call now the Standard Model, and it will continue to be the prime means by which we can extend our knowledge towards higher mass scales and a more fundamental theory. However, some aspects and open problems in high energy physics might not be easy accesible by studying individual particles or two-body interactions only. In QCD, the mechanism of generating hadronic masses via the process of chiral symmetry breaking as well as the nature of confinememt might be linked to coherent, large scale properties of the quantum fields and the structure of the physical vacuum (we know this to be the case for a number of phenomena in solid state physics and QED such as supra-conductivity). In order to study the influence of the 'vaccum', or, in general, macroscopic aspects of QCD, we might have to investigate *bulk phenomena* by distributing *high energy* over a relatively *large volume.*

This approach, to distribute high energy over a large volume in order to study the macroscopic properties of the QCD vacuum — and maybe even to transform it from its low energy physical state to the high energy phase of a quark-gluon-plasma — is at the heart of ultra-relativistic heavy ion physics. We hope to explore and test QCD on a 'largerer', macroscopic scale and to address the fundamental questions of confinement and chiral-symmetry breaking. Moreover, the study of the QCD phase transition, which is the only one accessible to laboratory experiments, might well be of relevance in understanding in general the dynamical nature of phase transitions involving elementary quantum fields.

The study of ultra-relativistic heavy-ion collisions ($E/m >> 1$) is a rather new, but rapidly evolving field. After the pioneering experiments at the BEVALAC and in DUBNA with relativistic heavy ions ($E/m \approx 1$), the first experiments started in 1986 with light ions (A $\approx$ 30) almost simultaneously in Brookhaven with the AGS ($E_{beam} \approx 15$ GeV/nucleon) and at CERN with the SPS ($E_{beam} = 200$ GeV/nucleon). Really heavy ions (A $\approx$ 200) are available in the AGS since the end of 1992 and at the SPS since the end of 1994. When the colliders RHIC ($\sqrt{s} = 200$ GeV/n) and LHC ($\sqrt{s} = 5.5$ TeV/n) come into operation around the turn of the century, the available energy in the centre-of-mass will have increased by almost five orders of magnitude within 20 years. This unprecedented pace was made possible only by (re)using accelerators, and to some extent even detectors, built over a much longer time scale for use in high-energy physics. The following sections will summarize the physics and experiments to come in the latest (and possible last) heavy-ion machines, the colliders RHIC and LHC.

2   RHIC and LHC: history and machine parameters

RHIC: The idea to build a dedicated collider machine for ultra-relativistic heavy-ion collisions preceded the first fixed-target experiments at BNL and CERN by several years. In its 1983 long-range plan, the Nuclear Science Advisory Committee recommended RHIC as the highest priority future project for Nuclear Physics in the US, and the project was finally approved in 1990. Since 1992, the experimental program is essentially defined with two large detectors (STAR and PHENIX) and two smaller ones (PHOBOS and BRAHMS) in various stages of the approval process (originally there where 9 LoI's submitted after the first call for proposals in 1990).

RHIC, which uses the existing heavy-ion facilities at Brookhaven as injection chain, will be located in the existing 3.8 km tunnel of an earlier abandoned project (ISABELLE/CBA). Two independent rings with 3.45 T superconducting magnets will allow collisions between different nuclei (p–p, p–A, A–B), including protons, at a top energy of 250×(Z/A) GeV/n, i.e. up to a c.m.s energy of 500 GeV for p–p and 200 GeV/n for Au–Au. The current design luminosity is $2 \times 10^{26} [cm^{-2}s^{-1}]$ for Gold on Gold reactions and in excess of 10^{31} for p–p. For Au–Au, with a geometrical reaction cross section of ≈ 6 barn, the total interaction rate will be about 1000/s, out of which a few tens will be central collisions with the largest particle production and the highest energy densities. Luminosity upgrades by up to an order of magnitude might be feasible and are under consideration.

RHIC is expected to start operation in 1999. As a dedicated machine, it will run mostly with heavy ions (or protons for comparison data). There are six interaction regions available for experiments, out of which four have been assigned for the initial set of experiments mentioned above. In addition to heavy-ion physics, there are a number of experimental proposals to study p–p reactions with polarized beams at $\sqrt{s} = 500$ GeV.

LHC: The possibility to add a second machine, a hadron collider, on top of the LEP electron–positron collider at CERN, has been mentioned already in the mid 70's by Sir John Adams, then DG of CERN. When LEP was approved in 1981, the 27 km long tunnel was made sufficiently large to accommodate what is now known as the LHC. The idea to also use the LHC for heavy-ion collisions was first mentioned by C. Rubbia in the proceedings of the ECFA workshop on future colliders at La Thuile in 1987 [2]. It was however only during the preparation for the Aachen LHC workshop in 1990 [3], that the European heavy-ion community took up the challenge and discussed the potential physics program and detector concepts for nucleus–nucleus interactions at several TeV/nucleon.

Responding to a call for expressions of interest, three general-purpose pp detectors, three dedicated beauty experiments, and one heavy-ion detector were presented in 1992 [4]. Since then, the experimental program is taking shape, and two of the pp experiments (ATLAS, CMS) and the heavy-ion detector (ALICE) have been approved and submitted their technical proposals.

Also at CERN, the existing injector chain recently upgraded for the fixed-target Pb-program, will be used to inject into the LHC. In addition, the LEAR ring will accumulate and cool the beams to reach the highest luminosities. The 8.4 T superconducting magnets of the LHC will provide a maximum energy of $7 \times (Z/A)$ TeV, i.e. a c.m.s. energy of 14 TeV for p–p and 5.5 TeV/n for Pb–Pb. The design luminosity varies between $2 \times 10^{27} [cm^{-2}s^{-1}]$ for Pb–Pb (i.e. a factor of 10 above RHIC) and $> 10^{34}$ for p–p. A recently adopted change in the RF system of the LHC allows for both symmetric beams (i.e. p–p or A–A) as well as p–A or A–B reactions, however only at fixed magnetic rigidity because of the specific design of the LHC magnets.

The LHC project has been approved at the end of 1994 and, pending a final schedule to be established in 1997, is expected to start operation in 2005. As a dedicated pp machine, it will run most of the time with protons at top energy. Heavy ions will however be an integral part of the initial experimental program, with typically a few weeks per year foreseen for nuclear beams; i.e. a sharing similar to the current fixed-target operations at the SPS or AGS.

3 Physics: from fixed-target to colliders

3.1 Initial conditions in nucleus–nucleus collisions

The primary aim for studying nucleus–nucleus collisions at very high energies is to understand the phase diagram (or equation of state) of strongly interacting matter — from nucleons to hadrons to partons — and the search for the quark-gluon plasma (QGP). The predictions of lattice QCD are rather firm in that a transition to the QGP should exist in the vicinity of a critical temperature T_c of ≈ 200 MeV (whether the transition is of first order, second order, or only 'rapid' is however still a matter of debate). However, whether the QGP is actually *created* in heavy-ion collisions will depend on the dynamics of the reactions and in particular on the initial conditions of the system shortly after the collision. In order to reach the QGP, or even only to use macroscopic concepts (such as 'phase transition') and the language and variables of thermodynamics (such as 'temperature' or 'density'), the system has to be *extended* — i.e. its dimensions ought to be much larger than the typical scale of strong interactions — in (or near) equilibrium — i.e. its *lifetime* has to be larger than the relaxation times — and the *energy density* ϵ has to exceed the critical threshold for QGP formation which is predicted by QCD to be of the order of 1 - 3 GeV/fm^3, equivalent to a temperature T_c of 150 - 200 MeV or a baryon density ρ_c of 5 to 10 times normal nuclear matter density.

Present results from the ongoing fixed-target program indicate that already with the light ions available now the initial conditions could indeed be favourable for QGP formation. In head-on central collisions, hundreds of particles are produced per unit of rapidity, the system expands to a size of the order of 1000 fm^3, and initial energy densities are estimated to exceed 2 GeV/fm^3. There are indications of 'thermodynamics' in the hadronic sector, i.e. strong rescattering and approach to equilibrium (e.g. particle abundances, shape of p_t spectra) and there are also a number of 'puzzles' which defy straightforward explanation, at least up to now (e.g. anti-hyperons, continuum lepton pairs). However, the expansion is extremely fast, with an estimated lifetime of only a few fm/c, which in addition includes the time spent in the cooler hadronic phase towards the end of the evolution.

While these results show that we are already now certainly *close* to the requirements listed above for QGP formation, they are by no means measured to be *sufficient*. In particular the energy density estimates are inversely proportional to the assumed 'formation time', i.e. the time needed to reach thermal equilibrium, and might well be smaller by a factor of the order of two. Also the lifetime of the system seems marginal, and even if a QGP is formed it might simply not live long enough for its signals to clearly stand out from the background created in later, hadronic phases of the evolution.

In order to improve on the initial conditions, or, if the QGP should be found in the current experiments with heavy projectiles, in order to explore QCD thermodynamics in different regions of the phase diagram, we have to go to higher energies. While the physics *subject* and to some extent the *observables* stay essentially the *same* from the AGS all the way to the LHC, the *object* under study will be very *different*. There are a number of *qualitative* and *quantitative* changes which will be summarized in the next three sections.

3.2   Qualitative changes

Baryon density: High energy density can be reached in nuclear collisions essentially along two different routes, either via 'compression' of the nuclei, i.e. increase of the baryon density ρ as in a neutron star, or via 'heating', i.e. creation of a large number of new particles as in the early universe.

The highest baryon density will be reached at the energy where both target and projectile nucleons end up 'on top of each other' at midrapidity. This 'full stopping' regime could be in the energy range somewhere between AGS and SPS; its location should be known soon when baryon distributions with heavy projectiles will be available from both machines. RHIC might or might not have a baryon-free region at midrapidity, but LHC will most likely have zero net baryon density around $y_{cm} = 0$ and closely resemble the conditions prevailing in the early universe.

Energy density: The presence of a phase transition is seen in lattice QCD calculations by a rapid increase of the energy density at the critical temperature T_c. However, recent calculations indicate that the QGP deviates considerably from an ideal Stefan-Boltzmann behaviour around the transition temperature, as indicated by a sizeable difference in normalized energy density (ϵ/T^4) and pressure $(3P/T^4)$ [5]. It seem likely that an 'ideal' quark-gluon plasma will exist at an energy density which is reachable, if at all, only at the LHC $(T/T_c > 1.5, \epsilon/\epsilon_c > 5)$. The increase in energy density from SPS to LHC, about an order of magnitude (see below), would allow in an optimistic scenario to 'scan' across the interesting region above T_c and study the origin and nature of this effect (residual interactions, 'hadronic ghosts', ..), which is currently the subject of intense theoretical investigations.

Energy scale: In hadronic interactions at low energies ($\sqrt{s} < 50$ GeV), the average (minimum bias) particle production is dominated by soft, non-perturbative phenomena. The initial conditions and the dynamics in the pre-equilibrium phase, which should eventually lead to an equilibrated QGP, can therefore only be described phenomenologically and are subject to strong model dependence. At higher energies, semi-hard and hard QCD processes become more and more important ('minijets', $p_t > 2$, and jets, $p_t > 10$) [6]. These semi-hard and hard processes can, at least in principle, be calculated in the well-tested framework of perturbative QCD. Both initial conditions as well as the early dynamics, and therefore important information such as initial energy density and formation time, are accessible to *calculations*; a vast improvement compared to the hand-waving and dimensional arguments needed at lower energies. While substantial numerical uncertainties are still present to date in parton cascade models, related to the ill-defined boundary between 'soft' and 'hard', they are nevertheless expected to put the analysis and interpretation of nuclear collisions at the colliders on a much sounder basis.

3.3 Quantitative improvements

Besides being *different*, as mentioned in the previous paragraph, nuclear collisions at very high energies are also expected to provide a significantly *better* environment for QGP studies. Table 1 gives a summary (which includes some 'personal averaging') of some of the relevant parameters for central Pb–Pb collisions at the SPS, RHIC and LHC.

Table 1: Comparison between SPS, RHIC and LHC predictions for central Pb–Pb collisions: dN/dy = charged particle multiplicity; ϵ = energy density (assuming $\tau_0 = 1$ in all cases); V_f = final volume at freeze-out; τ_0 = formation (thermalization) time; τ_c = time spend in QGP phase. For a number of these parameters, the average *relative increase* compared to the SPS is shown in a separate row (i.e. the SPS numbers are one by definition).

	dN/dy	$\epsilon[GeV/fm^3]$	$V_f[fm^3]$	$\tau_0[fm/c]$	$\tau_c[fm/c]$	τ_c/τ_0
SPS	500 - 800	1.7 - 2.7	$\approx 10^3$	$\geq 1?$	$\leq 1?$	
		1	1		1	1
RHIC	700 - 2000	2.3 - 6.8	$\approx 7x10^3$	$\leq 1?$	1.5 - 4	
		2	7		3	5
LHC	3000 - 8000	8.0 - 27	$\approx 2x10^4$	$\leq 0.2?$	4 - 11	
		7	20		7	≥ 30

The charged particle multiplicity dN/dy will grow from about 500 at the SPS to perhaps up to 8000 at the LHC. The initial energy density, here calculated with a fixed formation time of $\tau_0 = 1$ fm/c (but see below), will grow proportionally by about an order of magnitude to values much larger than the critical density ($\epsilon_c \approx 1 - 3$ GeV/fm^3). The system expands to a final volume of more than 10.000 fm^3, certainly huge by the standards of QCD. Because of the large particle and energy densities, collisions are more frequent and relaxation times will decrease. In particular the initial thermalization time (or 'formation time') τ_0 has been estimated in parton cascade models to be as small as a fraction of a fm/c[6], down by a factor of about five from the 'dimensional' estimate of 1 fm/c used currently at low energies. And, probably most important, the time spent in the QGP phase during the evolution (τ_c) could increase at LHC to up to 10 fm/c. This might seem to be still rather short, but it is not so much the absolute lifetime that is relevant but rather the ratio of lifetime to relaxation times, e.g. the ratio τ_c/τ_0. With the lifetime growing and the relaxation times decreasing (the clock 'ticks faster' at high density) the net gain from SPS to LHC could be as much as a factor of 30. There will be more time available for the QGP to develop recognizable signals and less time for the hadronic phase to spoil them.

In summary, the system created in nuclear collisions at the colliders will be *hotter, bigger, and longer-lived.* The gain in all three variables is about an order of magnitude when going from SPS to LHC, and still a factor of 3 - 5 when comparing RHIC with LHC. The gain is all the more significant as in current experiments we are neither far below nor far above the phase transition, and therefore already a factor two could make a big difference.

3.4 Differences in selected signals

The high energy of the colliders will significantly increase the production cross-sections for a number of hard processes, which can serve as tools to probe the nature of the hadronic or partonic medium. The measurement of Υ production will give additional constraints on competing models describing the J/Ψ suppression, observed at current experiments, either in terms of deconfinement or hadronic interactions[8]. While Υ production rates at RHIC are marginal, they are sufficient at the LHC but will require a large acceptance experiment. The J/Ψ is abundantly produced at both machines.

Jets will be frequently produced at both colliders. The energy loss of the fast quarks or gluons is directly proportional to the Debye screening scale and could therefore be different in a dense hadronic system and a QGP[9]. This 'jet quenching' effect could be observed as a striking difference in the shape of the

p_t spectra for nucleus–nucleus and p–p collisions.

The high initial temperature associated with the high initial energy density will increase strongly the amount of thermal radiation (both in lepton pairs and direct photons) coming from a QGP. This radiation is difficult to discriminate from the background of hadronic decays, but at the colliders the signal-to-background ratio could be more favourable than at lower energies. If very high initial temperatures ($>$ 500 MeV) should indeed be reached, even charm quark production might increase and serve as a signal in a way similar to the strangeness enhancement predicted in a QGP at lower temperatures.

One of the most promising new features at very high particle multiplicity is the possibility of measuring a large number of observables *event-by-event*: dN/dy, particle ratios (π, K, p, γ), p_t-spectra, and even size and lifetime of each 'fireball' via particle interferometry. This will allow a detailed study of correlations and non-statistical fluctuations which would be washed out when averaging over many events. Such fluctuations are, in general, associated with *critical phenomena* in the vicinity of a phase transition, and lead to local or global differences in the events, even for similar initial conditions.

4 Experiments at RHIC and LHC

The following is a very short summary of the main features of the approved experiments; a more detailed description can be found in the proceedings of the 1993 Quark Matter conference [7].

4.1 RHIC

STAR: This experiment concentrates on hadronic probes, i.e. particle spectra and ratios, particle interferometry, and jets. Having a very large acceptance around midrapidity (2 units in y, full azimuth) and additional tracking chambers in the forward and backward hemispheres (y = 2 - 4), it will be able to do a detailed event-by-event analysis. It is housed in a large solenoid and employs a vertex detector, a large central TPC and additional forward/backward TPC's, an e.m. calorimeter and a large array of TOF counters.

PHENIX: PHENIX will measure both hadronic and e.m. signals, but concentrating on the later (i.e. dileptons, both electrons and muons, and photons). Its main components are an open axial field magnet, a two-arm lepton-photon spectrometer (with 1 steradian acceptance each) equipped with tracking chambers, a RICH and e.m. calorimeters, and two forward muon spectrometers. A part of one of the electron arms is in addition equipped with TOF counters for hadron identification.

PHOBOS: The physics goals of PHOBOS are to measure single particle spectra and correlations in particular at extremely low momenta ($15 < p_t < 600$ MeV/c). By the uncertainty principle, small momenta are associated with large length scales, and are therefore potentially very interesting in heavy-ion collisions. It has two very compact arms of silicon detectors for tracking, located in a 2 T dipole field on either side of the beam ($0 < \eta < 1.8$).

BRAHMS: This will be the only experiment which can access physics away from midrapidity. It will measure single inclusive hadron spectra both in the fragmentation region (up to y = 4) and around midrapidity with two movable spectrometer arms. In particular the forward arm, which has to cope with high momenta and particle densities, employs a large number of magnetic focusing elements, tracking chambers, and particle identification detectors (TOF, gas Cherencov, and RICH counters).

4.2   LHC

ALICE: Being the only dedicated heavy-ion detector foreseen at the LHC, ALICE will be a general-purpose experiment, aiming at a comprehensive study of hadrons, electrons, muons and photons produced in the collision of heavy nuclei, up to the highest particle multiplicities anticipated at the LHC.

The central part of ALICE, which covers $(90 \pm 45)°$ ($|\eta| < 0.9$) over the full azimuth, is embedded in a large magnet with a weak solenoidal field. The baseline design consists of a high-resolution inner tracking system, a cylindrical TPC, a particle identification array (TOF and RICH detectors), and a single-arm electromagnetic calorimeter. A forward muon spectrometer ($2° - 10°$, $\eta = 2.4 - 4$) has recently been added to study production and suppression of heavy quark resonances. ALICE will trigger on central collisions with a Zero Degree Calorimeter and measure multiplicity distributions over a large fraction of the available phase space.

5   Conclusions

The present exploratory program with light ions at CERN and Brookhaven has established the feasibility of high-energy ion–ion experiments with their abundant particle production. It has shown that high energy densities can indeed be obtained in these reactions and produced first hints for the onset of new, collective phenomena. The upcoming experiments with really heavy ions, both at BNL and CERN, should determine to what extent we can actually get into a regime of thermodynamic behaviour. They should lead to baryon densities close to or even exceeding the ones in the core of neutron stars, and

— if present estimates are correct — should have a chance to obtain more conclusive evidence for quark deconfinement.

The new generation of heavy-ion colliders with a centre-of-mass energy of 200 GeV/n (RHIC) or even 6 TeV/n (LHC) will bring us into the true high-energy heavy-ion regime; in particular the LHC will reach and even extend the energy range probed by cosmic-ray nucleus–nucleus collisions. Extrapolating from present results, all parameters relevant to the formation of the Quark–Gluon Plasma (QGP) will be more favourable: the energy density, the size and lifetime of the system, and the relaxation times should all improve by a large factor, at the LHC typically by an order of magnitude, compared to Pb-Pb collisions at the SPS. We should now get *average* energy densities well above the deconfinement threshold, and probe the QGP in its asymptotically free 'ideal gas' form. Unlike at lower energies, the central rapidity region will have nearly vanishing baryon number density, similar to the state of the early universe. It is possibly dominated in the early pre-equilibrium stage by a very dense system of 'mini-jets', which would lead to rapid thermalization and even higher initial temperatures than in the conventional scenario.

Heavy-ion collisions at these new machines are therefore expected to provide a *very different*, and *significantly better*, environment for the study of strongly interacting matter than existing accelerators. The experimental program is taking shape and the community is looking forward to exploring the 'wonderland' which might open up in the years to come.

References

1. T.D. Lee, Rev. Mod. Phys., Vol 47 (1975) 267 and Nucl. Phys. A533 (1993) 3c.
2. Workshops on Physics at Future Accelerators, La Thuile, Italy, January 1987, yellow report CERN 87-07
3. Proc. of the Large Hadron Collider Workshop, Aachen, Germany, October 1990, ECFA 90-133, CERN 90-10
4. Proc. of the General Meeting on LHC Physics & Detectors, Evian-les-Bains, France, March 1992, p. 95.
5. S. Gottlieb et al., Phys. Rev D35 (1987) 3972.
6. K. Geiger, Phys. Rev. D46 (1992) 4965 and 4986, X.N. Wang and M. Gyulassy, Phys. Rev. D44 (1992) 3501.
7. Proc. of the last Int. Conf. on Ultra-Relativistic Nucleus–Nucleus Collisions: Nucl. Phys. A525 (1991), A544 (1992), A566 (1994), A590 (1995).
8. F. Karsch and H. Satz, Z. Phys. C51 (1991) 209.
9. X.N. Wang and M. Gyulassy, Phys. Rev. Lett. 68 (1992) 1480.

TRANSVERSE SHOCKS IN THE TURBULENT GLUON PLASMA PRODUCED IN ULTRA-RELATIVISTIC A+A

M. GYULASSY[a], D. H. RISCHKE, and B. ZHANG

Physics Department, Pupin Physics Laboratories, Columbia University,
538 W 120th Street, New York, NY 10027, U.S.A.

Mini-jet production in ultra-relativistic nuclear collisions leads to initial conditions characterized by large fluctuations of the local energy density (hot spots) and of the collective flow field (turbulence). Assuming that local equilibrium is reached on a small time scale, ~ 0.5 fm/c, the transverse evolution of those initial conditions is computed using hydrodynamics. We find that a new class of collective flow phenomena (hadronic volcanoes) could arise under such conditions. This could be observable via enhanced azimuthal fluctuations of the transverse energy flow, $d^2 E_\perp / d\phi dy$.

1 Introduction

At energies $\sqrt{s} > 100$ AGeV, mini-jet production in central nuclear collisions is expected [1] to be the primary dynamical source of a plasma of gluons with an initial energy density an order of magnitude above the deconfinement and chiral symmetry scale, $\epsilon_c \sim 1$ GeV/fm^3. Many observable consequences of the formation of this new phase of matter have been predicted based on a variety of assumptions [2], and experiments are currently under construction to search for evidence for that so called quark-gluon plasma (QGP) at the Relativistic Heavy–Ion Collider (RHIC) at Brookhaven. Evidently, signatures depend sensitively on the assumed ensemble of initial conditions generated in such collisions. Most often it is assumed for simplicity that a homogeneous, cylindrically symmetric, and longitudinally boost-invariant quark-gluon plasma is created and thus that signatures can be computed ignoring fluctuations of the initial conditions themselves.

In this talk we point out, however, that the mini-jet formation mechanism leads to a rather inhomogeneous and turbulent ensemble of initial conditions that is characterized by a wide fluctuation spectrum of the local energy density (hot spots) and of the collective flow field (turbulence). In this case, some of the proposed signatures will be washed out while new ones will certainly arise. We show below that in fact a new type of collective "shock" phenomena may occur under these conditions that could be readily observed as unusual transverse energy fluctuations.

---

[a]Speaker

This topic is well suited to this symposium honoring the 60th birthday of Professor Greiner because he was the first to propose in 1974 the "shocking" idea that nuclear shock waves may be produced in nuclear collisions and serve as an ideal probe of the equation of state of dense matter [3]. Since the experimental discovery of nuclear shocks in the 1 AGeV range in 1984 and in the AGS energy range last year [b], much theoretical and experimental effort has been devoted to this topic over a large energy range: $E_{lab} = 100$ AMeV to 100 ATeV. The present work focuses on a new class of shock phenomena that could occur in quark-gluon plasmas near the top end of that energy scale.

To illustrate the novel type of collective flow patterns we have in mind, the hydrodynamic [4] evolution of initial conditions consisting of three static cylindrical "hot spots" are shown in Fig. 1.

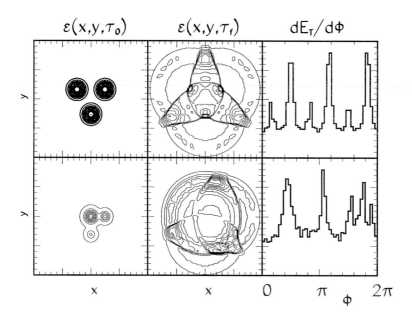

Figure 1: Hadronic volcanoes erupt due to shock formation in regions where expanding shells of matter evolved from inhomogeneous "hot spots" initial conditions (left panels) intersect. The signature of such volcanoes is the strong anisotropic azimuthal angle dependence of the transverse energy flow $dE_T/d\phi$ (right panels). Contours of constant energy density, $\epsilon(x, y, \tau)$, at the breakup time τ_f are shown in the middle panels. The top row shows the evolution from a symmetric initial configuration of hot spots. The bottom rows shows the evolution from a more realistic asymmetric configuration.

---

[b] see talks by J. Symons, J. Stachel, and P. Braun-Munzinger in these proceedings.

An ideal gas equation of state ($p = \epsilon/3$) is assumed. Shock waves are seen to be created where the three rapidly expanding shells of matter intersect at six points. The collective acceleration of matter in those shock zones can be clearly seen as strong peaks in the transverse energy flow. For the symmetric configuration (Fig. 1a), a "mercedes" pattern emerges (Fig. 1b), leading to three primary and three secondary shocks in Fig. 1c. For the asymmetric case in Figs. 1d,e the azimuthal angle dependence of the shock peaks is spread out and modified. The search for analogous enhanced anisotropic transverse energy fluctuations (or hadronic volcanoes[c]) as a function of rapidity and azimuth is the subject of the following discussion.

2   HIJING Initial Conditions

Mini-jet production is thought to be the dominant mechanism controlling the initial conditions in $A + A$ because the pQCD inclusive cross section for jets with moderate $p_\perp > p_0 \sim 1 - 2$ GeV rises to the value $\sigma_{jet}(p_0) > 10$ mb at RHIC energies. Compelling evidence for pQCD mini-jet dynamics has been observed in pp and $p\bar{p}$ reactions at collider energies[5].

Due to the nuclear geometry, the total number of mini-jet gluons produced in central $A + A$ collisions is expected to be $\sim A^{4/3} \sim 10^3$ times larger than in pp collisions. This simple geometric enhancement causes the rapidity density of mini-jet gluons to reach $dN_g/dy \sim 300 - 600$ in $Au + Au$ depending on the mini-jet transverse momentum cut-off scale $p_0 = 1 - 2$ GeV/c. The hadronization of this mini-jet gluon plasma via the string fragmentation mechanism in HIJING approximately doubles the final hadron transverse energy distribution due to the pedestal or "string" effect. In the pre-hadronization phase of the evolution we therefore include a "soft" component in terms of a background gas of soft gluons normalized such that together with the mini-jet component the final transverse energy distribution predicted by HIJING is reproduced.

The average initial energy density of the mini-jet gluon plasma can be estimated using the Bjorken formula: $\langle \epsilon(\tau) \rangle \approx (p_0/\tau_0 \pi R^2) \, dN_g/dy \, (\tau_0/\tau) \sim 40 \, (\tau_0/\tau) \text{ GeV/fm}^3$. This applies until the thermalization time is reached and work due to expansion must be considered.

To see that the initial conditions are in fact highly inhomogeneous and are dominated by a few "hot spots", we must calculate the *local* energy density $\epsilon(\tau, \mathbf{x}_\perp)$ instead of averaging over the transverse coordinate as in the Bjorken formula. The local energy density and transverse momentum density at $z = 0$

---

[c]T.D. Lee first used this term for anisotropy that could arise due to surface instabilities. In our case the "volcanoes" are dynamically generated from specific inhomogeneous initial conditions.

(the $y_{cm} = 0$ frame) can be computed from the HIJING list of produced gluons as

$$\left(\begin{array}{c} \epsilon(\tau, \mathbf{x}_\perp) \\ \vec{m}(\tau, \mathbf{x}_\perp) \end{array} \right) = \sum_\alpha \left(\begin{array}{c} 1 \\ \mathbf{v}_{\perp\alpha} \end{array} \right) \frac{\tau p_{\perp\alpha}^3}{1 + (\tau p_{\perp\alpha})^2} \, \delta^2(\mathbf{x}_\perp - \mathbf{x}_{\perp\alpha}(\tau)) \, \delta(y_\alpha) \ . \tag{1}$$

The sum is over the produced gluons with transverse and longitudinal momentum $(\mathbf{p}_{\perp\alpha}, p_{z\alpha} = p_{\perp\alpha} \sinh y_\alpha)$. The longitudinal and transverse coordinates of the production, $(z_\alpha = 0, \mathbf{x}_{\perp\alpha})$, are determined from the coordinates of the binary nucleon-nucleon collision from which the gluons originate. The above formula takes into account the free streaming of gluons not only along the z direction via the volume element $\tau \Delta y$ but also in the transverse direction via $\mathbf{x}_{\perp\alpha}(t) = \mathbf{x}_{\perp\alpha} + \mathbf{v}_{\perp\alpha}\tau$, where $\mathbf{v}_{\perp\alpha} = \mathbf{p}_{\perp\alpha}/p_\alpha$. The factor $1/(1 + (\hbar/\tau p_{\perp\alpha})^2)$ is the formation probability[6] of the gluon in the comoving frame. High-p_T gluons are produced first and gluons with lower p_T later according to the uncertainty principle leading to the so-called inside-outside picture of the dynamics. Before $\tau \sim \hbar/p_{\perp\alpha}$ that component of the radiation field is still part of the coherent Weizsäcker-Williams field of the passing nuclei[7].

The expression (1), when averaged over transverse coordinates, recovers the Bjorken expression $\langle \epsilon(\tau) \rangle \approx \epsilon_0(\tau_0/\tau)$ with vanishing average transverse momentum density, $\langle \vec{m}(\tau) \rangle = 0$.

To study the transverse coordinate dependence of the initial conditions, we must specify a transverse, $\Delta r_\perp$, and longitudinal, Δy, resolution scale. The densities coarse-grained on that resolution scale are obtained from (1) substituting

$$\delta^2(\mathbf{x}_\perp - \mathbf{x}_{\perp\alpha}(\tau)) \, \delta(y_\alpha) \rightarrow \frac{\Theta(\Delta y/2 - |y_\alpha|)}{\Delta r_\perp^2 \, \Delta y} \, \Theta(\Delta r_\perp/2 - |\mathbf{x}_{\perp\alpha}(\tau) - \mathbf{x}_\perp|) \ . \tag{2}$$

To determine the relevant resolution scale, we note that at time τ the local horizon for any gluon in the comoving ($y_g = 0$) frame has a radius $c\tau$. Thus at the thermalization time, τ_{th}, each gluon can be influenced by only a small neighborhood of radius $c\tau_{th} \approx 0.5$ fm of the mini-jet plasma. We take the transverse resolution scale to be the maximal causally connected diameter, $\Delta r_\perp = 2c\tau_{th} \approx 1$ fm and the rapidity width to be $\Delta y = 1$. Gluons moving with larger relative rapidity y are produced later, $\tau(y, p_\perp) \sim \cosh(y)/p_\perp$ due to time dilation at the boundary of the local horizon. The above choice of the resolution scale is the most optimistic from the point of view of *minimizing* fluctuations of the thermodynamic variables between the causally disconnected domains in the transverse plane. We note that the number, $N_d(\tau) = (R/\tau)^2$, of such causally disconnected domains in a nuclear area, πR^2, is initially very large. Even at later times, $\tau \sim 3$ fm/c, when the mean energy density falls below ϵ_c, several disconnected domains remain.

The soft component is modeled by ≈ 1700 low–$p_\perp$ gluons per unit rapidity with a Gaussian transverse momentum distribution with rms $p_\perp = 0.3$ GeV/c. This leads to about 500 GeV of transverse energy per unit rapidity as needed to reproduce the HIJING final dE_T/dy. However, these soft gluons only add a relatively small, approximately homogeneous contribution to the energy density, $\epsilon_s \approx 4\,\mathrm{GeV/fm}^3$, at $\tau_{th} = 0.5$ fm/c.

3 Hot Spots and Turbulence

Hot Spots and Turbulent Minijet Initial Conditions t=0.5 fm/c

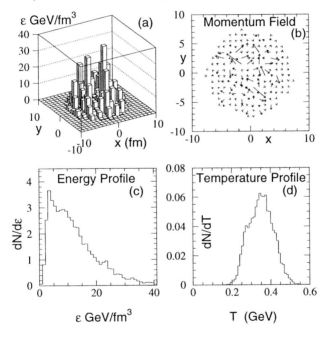

Figure 2: (a) Energy density distribution of hot spots in a typical HIJING event. Note the large fluctuations of the initial energy density across the transverse plane. (b) Momentum density fluctuations indicate considerable initial state turbulence. (c) Distribution of energy density on a $\Delta r_\perp = 1$ fm transverse coordinate resolution scale averaged over 200 HIJING events. Note that this resolution scale corresponds to the maximal allowed one by causality at proper time, $t = 0.5$ fm/c. (d) The thermal profile distribution on the same resolution scale has a rather large width. The mean and shape of course depend also on the equation of state (assumed here to be an ideal gas).

The inhomogeneous nature of the mini-jet plasma initial conditions in central $Au + Au$ collisions at RHIC is illustrated in Fig. 2. In Fig. 2a the energy density profile at an assumed thermalization time $\tau = 0.5$ fm/c of a typical HIJING ($b = 0$) event at $\sqrt{s} = 200$ AGeV is shown. The plasma in the central rapidity slice ($\Delta y_{cm} = 1$) exhibits large fluctuations (hot spots) because the average number of hard mini-jets produced per nucleon is only ~ 1 per unit rapidity at RHIC energies. The hot spots are also associated with a chaotic velocity field (turbulence) as seen in Fig. 2b. When averaged over many events the distribution of the proper energy density and temperature on a transverse resolution scale $\Delta r_\perp = 1$ fm are shown in Figs. 2c,d. Note that at that formation time the plasma cannot be characterized by a unique temperature. In fact the widths of those distributions which are controlled by the Glauber geometry of finite nuclei and the size of the mini-jet cross section obviously cannot be neglected. Signatures of plasma formation will differ considerably in this turbulent gluon scenario than in the conventional homogeneous hot-glue scenario. For example high–p_T direct photon production and heavy–quark production will be greatly enhanced in the hot spots [8]. In the next section we find that hadronic volcanoes similar to those in Fig. 1 also arise.

4   Hadronic Volcanoes and Transverse Energy Fluctuations

In Figure 3 the azimuthal angle dependence of the transverse energy flow, $dE_\perp/dyd\phi$, is shown for a typical HIJING initial condition. The left panels (Figs. 3a,c) correspond to an ideal gas equation of state, while the right panels (Figs. 3b,d) correspond to a first order Bag model equation of state [4]. The top two panels evolve the initial distribution in Fig. 2a with the momentum field in Fig. 2b artifically set to zero ($\vec{m}(\mathbf{x}_\perp, \tau_0) = 0$). This static inhomogeneous initial condition is most similar to the one in Fig. 1. The initial isotropic nature of the thermally folded transverse energy distribution is seen by the dashed lines in Figs. 3a,b. The lower panels (Figs. 3c,d) correspond to the evolution of the turbulent initial condition in Figs. 2a,b. The turbulence of the initial condition is revealed by the anisotropy of the dashed curves.

In all cases the magnitude of the transverse energy decreases with time due to the work done by the plasma undergoing longitudinal (Bjorken) expansion. More work is done by an ideal gas case than a plasma with first order transition because the latter equation of state is softer [4]. In the static hot spot case the production of hadronic volcanoes is indeed similar to Fig. 1. In the turbulent case, the effect is obscured somewhat by the large initial anisotropy. Nevertheless, in the turbulent case also several sharper peaks appear. Evidence for the collective origin of the hadronic volcanoes in the turbulent case must be looked for in higher order E_T correlation analysis, e.g., $\langle E_T(\phi)E_T(0)\rangle$. For a more detailed discussion of these results we refer to Ref.[8].

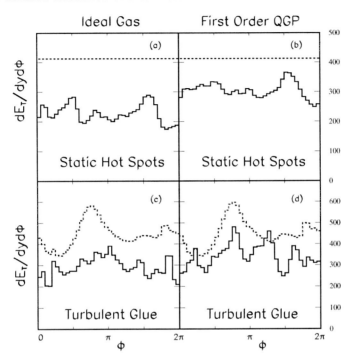

Figure 3: The evolution of transverse energy $dE_T/dyd\phi$ at $y = 0$ is shown for different inhomogeneous HIJING initial conditions and equations of state. In parts (a,b) the fluid velocity field shown in Fig. 2b is set to zero while in parts (c,d) the turbulent velocity field is included. The evolution in parts (a,c) assumes an ideal gas law $p = \epsilon/3$, while in parts (b,d) a first order Bag model transition is included. The dashed curve is the initial transverse energy distribution at the onset of hydrodynamic evolution. In the static hot spot cases the initial transverse energy is uniformly distributed in azimuth, while in the turbulent case large initial fluctuations are generated by the mini-jets. The solid curves show the final transverse energy distribution.

Acknowledgments

This work was supported by the Director, Office of Energy Research, Division of Nuclear Physics of the Office of High Energy and Nuclear Physics of the U.S. Department of Energy under Contract No. DE-FG-02-93ER-40764. D.H.R.

was partially supported by the Alexander von Humboldt–Stiftung under the Feodor–Lynen program.

References

1. K.J. Eskola and M. Gyulassy, Phys. Rev. C 47 (1993) 2329; K.J. Eskola, K. Kajantie, and J. Lindfors, Nucl. Phys. B 323 (1989) 37.
2. Quark Matter '93, eds. E. Stenlund, H.A. Gustavson, A. Oskarsson, and I. Otterlund, Nucl. Phys. A 566 (1994) 1; Quark Matter '95, eds. A. Poskanzer, J. Harris, L. Schroeder, Nucl. Phys. A 590 (1995) 1c.
3. W. Scheid, H. Müller, and W. Greiner, Phys. Rev. Lett. 32 (1974) 741; H. Stöcker and W. Greiner, Phys. Rep. 137 (1986) 277; W. Greiner, H. Stöcker, and A. Gallmann, eds., Hot and Dense Nuclear matter, NATO ASI Series B, Vol. 335 (Plenum, New York, 1993).
4. D.H. Rischke, S. Bernard, J.A. Maruhn, Nucl. Phys. A 595 (1995) 346; D.H. Rischke and M. Gyulassy, Nucl. Phys. A 597 (1996) 701.
5. X.N. Wang and M. Gyulassy, Phys. Rev. D 44 (1991) 3501; Phys. Rev. D 45 (1992) 844; Phys. Rev. Lett. 68 (1992) 1480; Phys. Lett. B 282 (1992) 466; Comp. Phys. Comm. 83 (1994) 307.
6. M. Gyulassy, X.N. Wang, Nucl. Phys. B 420 (1994) 583; X.N. Wang, M. Gyulassy, M. Plümer, Phys. Rev. D 51 (1995) 3436.
7. L. McLerran and R. Venugopalan, Phys. Rev. D 49 (94) 2233, 3352; Phys. Rev. D 50 (94) 2225.
8. M. Gyulassy, D.H. Rischke, and B. Zhang, in preparation.

SEARCH FOR EXOTIC FORMS OF STRANGE MATTER IN THE ALICE EXPERIMENT

J.P. COFFIN, C. KUHN, C. ROY, P. CROCHET, P. FINTZ, G. GUILLAUME,
F. JUNDT, A. MICHALON and F. RAMI

*Centre de Recherches Nucléaires, IN2P3-CNRS/Université Louis Pasteur B.P.28,
F-67037 STRASBOURG CEDEX 2, France*

Simulations have been performed with the aim to estimate the minimum yield within an event required for an exotic form of strange matter (Strangelets/MEMO's) to be seen with the central part of the ALICE detector. The calculations are restricted to specific assumptions on the *Strangelet* characteristics : Z=0-2, A$\simeq$ 2-15 GeV/c^2, and $\sim 10^{-20}$s and $\geq 10^{-7}$s for the charge, mass and lifetime, respectively.

1 Production of exotic forms of strange matter at the LHC

QGP formation should result in an enhanced production of strange quarks and it has been speculated that droplets of strange matter (i.e. strangelets) could be formed in a little big bang created in very high energy heavy-ion collisions[1]. Strange matter may also appear in the form of metastable exotic multi-hypernuclear objects (MEMO's) resulting from the coalescence of several strange baryons (e.g. Λ's). The formation of MEMO's does not necessarily require a QGP phase as it may result from strangelet decay, and vice-versa. Although those two forms of exotic matter would be nearly indistinguishable in their decay modes, the observation of either would clearly be of very high interest.

In order to assess the capability of an ultra relativistic heavy-ion experiment at LHC with the ALICE detector to recognize such unusual signals, we have investigated the sensitivity to strangelet/MEMO production[2].

In heavy-ion reactions strangelets and MEMO's might be found in the final state as objects with baryon number $A \approx 2 - 40$, Z/A ratio ranging from $\sim$ - 0.5 up to + 0.5, and fraction of strangeness within $f_s \sim 0.5 - 1.5$[1,2]. Strangelets should be created preferentially in a region with large net baryon density. The phase space covered by the ALICE detector ($-0.9 \leq \eta \leq 0.9$) is characterized by a low net baryon density and a chemical potential $\mu_B \approx 0$. These conditions are not favourable for strangelet formation (as opposed to strangelet production at large rapidities[3]). Recent theoretical developments[4] suggest, however, that strangelets could be produced also at the LHC as a result of local fluctuations in the net baryon number. As the average baryon and strangeness density at midrapidity is zero, strangelets and *anti*-strangelets would be produced in equal numbers.

Strangelets and MEMO's could be stable or metastable objects, and their stability, lifetime, and decay modes are strongly parameter dependent [2].

Strangelets (S) may be

i) unstable $(\tau < 10^{-20}s)$, in which case they decay via hyperon emission (Λ, Σ, Ξ) and meso-nucleonic strong interaction processes [1,2]:

$$S \to S' + N + \pi$$
$$S \to S' + N$$

ii) metastable $(\tau < 10^{-4}s)$, in which case they decay via weak interaction processes:

$$S \to S' + N$$
$$S \to S' + \pi$$
$$S \to S' + e + \nu$$

iii) stable $(\tau > 10^{-4}s)$.

In the following, for reason of simplicity, we will call strangelet any exotic form of strangeness regardless it is a strangelet or a MEMO.

2 Strangelet detection in ALICE

The search for strangelets with the ALICE detector [5,6] will depend on their lifetime and decay modes.

i) If these objects are stable or long-lived metastable $(> 10^{-7}s)$, they will travel across the complete detector (assuming $\beta \approx 0.1$) and should be seen as tracks with unusual charge-to-mass ratio.

ii) If they are metastable with a lifetime less than $10^{-7}s$, they may be unveiled, in principle, by identification of their decay products $(\pi, K, p, \Lambda,...)$ and by secondary vertex reconstruction.

iii) If they are unstable, thus of immediate decay $(< 10^{-20}s)$, their identification is very difficult. However, correlations of decay products like Λ, K, $\bar{K}$, π^- may still provide some information even in this case.

Here, we examine the sensitivity for a strangelet search in ALICE for cases (i) and (iii).

2.1 Stable or long-lived strangelets

In general, strangelets will have some non-integer value for their charge-to-mass ratio and may therefore be identified via dE/dx and/or time of flight (TOF) versus momentum per charge (p/Z). As an example, we consider strangelets with $Z = 1$ and $Z = 2$ and a mass between 6 and 15 GeV (i.e. $|Z/A| < 0.3$). This mass range is of particular interest as lower mass strangelets are less stable[1,2] while heavier objects have lower production cross-section. $S1$ refers to strangelets with Z=1 and a mass equal to 6 GeV/c^2.

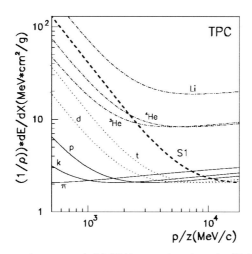

Figure 1: Mass stopping power $(1/\rho)dE/dx$ as a function of p/Z in the TPC for various particles and a strangelet $S1$ ($Z = 1$, mass= 6 GeV)

The mass stopping power $(1/\rho)dE/dx$, calculated analytically with the formula of Bethe-Bloch is shown in figure 1 for the TPC of the ALICE detector. Assuming a 7% energy resolution[5], one sees that strangelets may be resolved (with $> 3\sigma$ separation) from all other products over a large domain of p/Z. The additional use of TOF information would significantly improve the separation[2]; however, the TOF measurement has significant non-Gaussian tails which might limit its usefulness in the search for very rare events.

We have evaluated the probability for current stable particles to contaminate the strangelet energy-loss (E_S) distribution for $S1$ in the TPC. Figure 2 shows the probability (as a function of p/Z) of different particles having an energy loss in the domain $< E_S > \pm 3\sigma$ by assuming a purely Gaussian shape with $\sigma = 7\%$ for the dE/dx distribution. Below about 5 GeV/c, it is possible to discriminate strangelets reliably against protons and deuterons, as well as against π, kaons and tritons, below 4 GeV/c. However, He isotopes start to contaminate the distribution below 4 GeV/c.

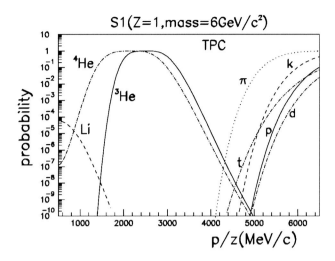

Figure 2: Probability as a function of p/Z, for different particles in the TPC to have an energy loss in the domain $< E_{S1} > \pm 3\sigma$ where E_{S1} is the strangelet $S1$ energy loss.

The total number of particles that can alter a strangelet signal is shown in Table 1 for $5 * 10^7$ central Pb-Pb events collected in ALICE in $10^6 s$.

Table 1: Yields of different particles for $5 * 10^7$ central Pb-Pb events.

π	K	p	d	t	^3He	^4He	^6Li
$5 * 10^{11}$	$1.25 * 10^{11}$	10^{11}	$5 * 10^8$	$5 * 10^6$	$5 * 10^6$	$5 * 10^4$	$5 * 10^2$

The event generator SHAKER [5] has been used to calculate the $\pi/K/p$ yields and the Quantum Statistical Model (QSM) [7] for light nuclei. All momentum distributions (including the one for strangelets) have been assumed to be thermal for a temperature ~ 200 MeV.

The maximum number (N_{max} with 0.99% confidence level) of contaminating particles, i.e. particles that fall within the energy-loss range of $S1$, is shown as a function of p/Z in figure 3 for $5 * 10^7$ analysed events. In order to identify strangelets, a minimum number $(N_{min S})$ has to be produced, such as $(N_{min S})$ is larger than the number $N_{max} = \sum_i nmax$ of contaminating i particles.

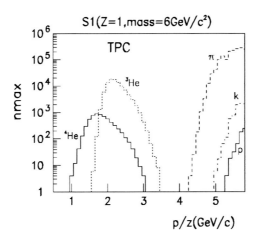

Figure 3: Maximum number $(nmax)$ of contaminating particles in the TPC as a function of p/Z, according to the probability values shown in Fig. 3 and the production rates given in table 1, assuming thermal distributions.

The p/Z domains where strangelet identification is possible in the TPC $(N_{min S} - N_{max} > 0)$ are shown in figure 4 for rates of strangelet production ranging from 10^{-1} up to 10^{-4} strangelet per event within the acceptance. Stable strangelets are identifiable for a multiplicity per event of the order of 10^{-4} or more.

The above limits depend on the assumption of a purely Gaussian shape of the dE/dx measurement, i.e. the analysis is dominated by statistical rather than systematic errors. The ALEPH dE/dx resolution has shown no significant departure from a Gaussian curve down to the lowest measured level of 10^{-3} ($\sim 3\sigma$)[8]. Below this limit, systematic errors in the dE/dx measurement are difficult to predict and the final sensitivity can only be assessed with data from the actual experiment. The above limits would not be affected if the distribution stays Gaussian down to a level of 10^{-6}.

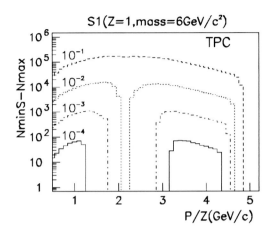

Figure 4: Difference between the minimum number (N_{minS}) of strangelets $S1$ and the maximum number of contaminating particles (N_{max}) as a function of p/Z for different assumptions on the production rate of $S1$ (10^{-1} to 10^{-4}) for $5 * 10^7$ central Pb-Pb events.

2.2 Unstable strangelets

We have investigated the possibility to recognize unstable strangelets for the simplest possible object, i.e. the H-dibaryon or dilambda which can be viewed as a *uuddss* strangelet or a $\Lambda\Lambda$ MEMO. We used for the dilambda an hypothetical mass and a width of m=2400 MeV and $\sigma = 40$ MeV, and a dominant decay mode into two Λ's. The efficiency necessary to reconstruct the $\Lambda \to p+\pi$ decay has been taken from the simulation described in reference 2. From the roughly 250 Λ's produced per event within the acceptance, about eight can be reconstructed in the $p + \pi$ decay channel.

The invariant mass distribution of all $\Lambda\Lambda$ pairs is shown in figure 5 for 10^5 events. The simulation includes the dilambda with a production multiplicity of 1.17 per event. The dilambda signal is shown in figure 6 after subtraction of the combinatorial background, whose shape can be estimated either by a fitting procedure [2] or by event mixing.

For a total statistics of $5 * 10^7$ events, the dilambda particle could be observed if it is produced at a rate of about 10^{-2} to 10^{-1} per event within the acceptance.

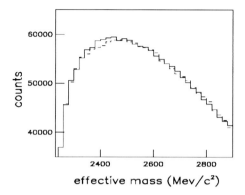

Figure 5: Simulated dilambda signal + combinatorial background $(S + B1)$ for 10^5 central Pb-Pb events (solid histogram). The dashed histogram represents the background (B2) obtained by fitting the $(S + B1)$ mass distribution in the region outside the dilambda peak.

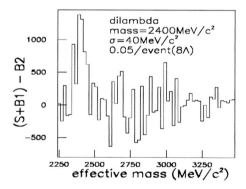

Figure 6: Background substracted dilambda invariant mass distribution. The peak observed around 2400 MeV would correspond to the effective mass of the dilambda.

3 Summary

The simulations presented here for strangelet and MEMO search are based on a straight use of the ALICE detector. If the sensitivity is rather modest, they show that a chance exists, providing LHC energies allow high energy density, high temperature and long plasma lifetime, to observe exotic forms of strange matter in an open geometry, over a large domain of lifetime and for different decay processes. This departs from most of the experiments devised so far to the search of these exotic objects.

References

1. For a review see: W. Greiner, Stöcker and A. Gallmann, 'The nuclear equation of state' NATO ASI series B: Physics, Vol. 335 (1994), Plenum Press New York.
2. J.P. Coffin et al., Internal Note ALICE 95-49
3. E. Gladysz-Dziadus and A. Panagiotou, Internal Note ALICE 95-18
4. C. Spieles, L. Gerland, H. Stöcker, C. Greiner, C. Kuhn and J.P. Coffin, Phys. Rev. Lett. 76, 1776 (1996)
5. Letter of Intent for A Large Ion Collider Experiment, CERN/LHCC/93-16, LHCC/I 4
6. ALICE Technical Proposal, CERN, LHCC 95-71, LHCC/P3
7. D. Hahn and H. Stöcker, Nucl. Phys. **A476**, 718 (1988); J. Konopka, private communication
8. D. Buskulic and the ALEPH collaboration Z. Phys. **C66**, 355 (1995)

FAST HADRONIZATION OF QUARK-GLUON PLASMA

L.P. Csernai[1], I.N. Mishustin[2,3], Á. Mócsy[1], D. Molnár[1] and Feng Zhong-han[1]

[1] *Section for Theoretical Physics, Department of Physics*
University of Bergen, Allegaten 55, 5007 Bergen, Norway,
[2] *The Niels Bohr Institute, Copenhagen University*
Blegdamsvej 17, DK- 2100 Copenhagen Ø, Denmark, and
[3] *The Kurchatov Institute, Russian Scientific Centre,*
Moscow, 123182 Russia

A hadronization scenario is presented where a central role is played by the chiral symmetry break-down in the expanding Quark-Gluon Plasma. This mechanism can become effective when thermal damping ceases after the thermal freeze-out of the quark system. We estimate time scales and spatial characteristics of chiral-symmetry breaking instabilities on the basis of an effective field-theoretical model.

1 Introduction

Let us discuss hadronization schenarios of QGP, which have important consequences and may explain several observations. The talk is based on earlier[1] and particularly on some recent works.[2,3,4]

The dynamics of the rehadronization of expanding and cooling QGP was discussed recently[1] and in the thermally overdamped limit the characteristic nucleation time was found to be about 50-100 fm/c. However, due to the occuring large supercooling other faster processes must play a role in the hadronization.[2,3,4]

Non-thermal processes are possible[2] if the hadronization happens together with, or after the thermal freeze-out of the plasma. In particular, it has been shown that in this case the release of the latent heat does not necessarily lead to an overheated final hadronic phase, as it was thought earlier[5,6].

This picture is in a qualitative agreement with recent experimental data. For instance, the enhanced production of strange antibaryons in central nucleus-nucleus collisions (by a factor 2÷3 compared to pp and pA reactions) was observed in several experiments[7]. Another interesting result is that data show almost equal yields of antilambdas and antiprotons at midrapidity compared to their ratio 0.2 in pp collisions. Both these observations can hardly be explained within the equilibrium hadronic scenario. As demonstrated, the data suggest high specific entropy of produced matter and zero chemical potential for strange particles. These findings are also consistent with the late hadronization of the QGP and no rescatterings of produced hadrons.

Recent initial studies of phase transition dynamics in effective field theoretical models, particularly in the linear sigma model,[8,9,10,12] show an alternative possibility for treating non-thermal dynamics. These models, however, are not connected directly to the preceeding thermal evolution of the system, but assume different initial conditions, like quenching, cold annealing or hot annealing [13].

In the present work our aim is to connect the final non-thermal chiral dynamics of the phase transition with the preceeding thermal- and fluid dynamical expansion and with the preceeding thermally dominated phase transition process via homogeneous nucleation [1,4]. As a result we will describe the whole reaction including thermal equilibrium and post freeze-out, non-thermal stages, thus being able to set the time-scale of the whole reaction and hadronization.

2 Chiral Dynamics

At late stages of the QGP evolution collisions between partons may have already ceased, and they interact only with the background fields, $\Phi = (\sigma, \vec{\pi})$. To describe such a system we use an effective field-theoretical model, namely the linear sigma model, where quarks are moving in the background chiral field. Several works have dealt recently with the similar problems [8,9,10], however, without introducing explicitly quark degrees of freedom.

2.1 The Lagrangian

The Lagrangian of the linear σ-model reads

$$\mathcal{L} = \bar{q} \left[i\gamma_\mu \partial^\mu - g(\sigma + i\gamma_5 \vec{\tau}\vec{\pi}) \right] q + \frac{1}{2} \left(\partial_\mu \sigma \, \partial^\mu \sigma + \partial_\mu \vec{\pi} \, \partial^\mu \vec{\pi} \right) - U(\sigma, \vec{\pi}), \quad (1)$$

where

$$U(\sigma, \vec{\pi}) = \frac{\lambda^2}{4} \left(\sigma^2 + \vec{\pi}^2 - v^2 \right)^2 - H\sigma + \left(m_\pi^2 f_\pi^2 - \frac{m_\pi^4}{4\lambda^2} \right) \quad (2)$$

is the so-called Mexican Hat potential, see Fig. 1. Here q stands for the light quark fields, (u, d), while σ and $\vec{\pi} = (\pi_1, \pi_2, \pi_3)$ are the scalar and pion fields which together form a 4-component chiral field $(\sigma, \vec{\pi})$. Without the $H\sigma$ term this Lagrangian is invariant with respect to the $SU_L(2) \otimes SU_R(2)$ chiral transformations.

The parameters in this Lagrangian are chosen in such a way that in *normal vacuum* chiral symmetry is spontaneously broken and expectation values of the meson fields are $\langle\sigma\rangle = f_\pi$, $\langle\vec{\pi}\rangle = 0$, where $f_\pi = 93$ MeV is the pion decay

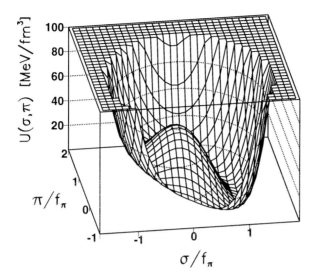

Figure 1: The 'Mexican Hat' potential, $U(\sigma, |\vec{\pi}|)$, of the Lagrangian with the parameters used in this work

constant. To have the correct pion mass in vacuum, $m_\pi = 138$ MeV, one should take

$$v^2 = f_\pi^2 - \frac{m_\pi^2}{\lambda^2}, \qquad H = f_\pi m_\pi^2. \tag{3}$$

and $m_\sigma^2 = 2\lambda^2 f_\pi^2 + m_\pi^2$, which can be chosen to be about 0.6 GeV (then $\lambda^2 \approx 20$). Sigmas represent stiff, "radial" excitations of the chiral field. The remaining coupling constant g can be fixed by requiringthat the effective quark mass, m, in broken vacuum, $m = g\langle\sigma\rangle = gf_\pi$, coincides with the constituent quark mass in hadrons, about $1/3$ of the nucleon mass $m = m_N/3$. This gives $g \approx (m_N/3)/f_\pi \approx 3.3$.

In addition to the normal vacuum state, $\langle\sigma\rangle = f_\pi$, $\langle\vec{\pi}\rangle = 0$, the Lagrangian (1) has one more stationary point on the top of the Mexican Hat potential (2). In the chiral limit, $m_\pi \to 0$, it corresponds to

$$\langle\sigma\rangle = 0, \quad \langle\vec{\pi}\rangle = 0. \tag{4}$$

This state becomes a true ground state of the matter at high density and/or temperature signaling the restoration of chiral symmetry, see Fig. 2. We assume that the system was initially in this chiral-symmetric phase. The difference between the energy densities of the symmetric and broken vacuum states, i.e. the bag constant, is $B \approx (\lambda^2/4) f_\pi^4 \sim m_\pi^4$ in this model.

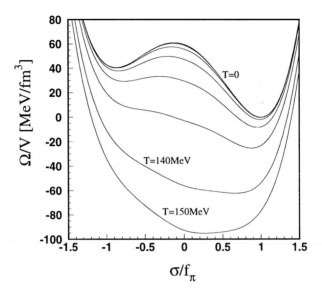

Figure 2: The $\Omega(T,\mu)/V$ thermodynamical potential including the potential energy of the σ-field, including the $H\sigma$-term, and the kinetic energy of the quarks and antiquarks, at $\mu = 0$ and for different temperatures from from $T = 0 - - - 200$ MeV in 20 MeV steps

In our further discussion we use the mean-field approximation, ignoring all loop contributions, considering σ and $\vec{\pi}$ as classical fields. The equation of motion

$$\partial_\mu \partial^\mu \sigma(x) + \lambda^2 \left[\sigma^2(x) + \vec{\pi}^2(x) - v^2\right] \sigma(x) - H = -g\rho_S(x),$$

$$\partial_\mu \partial^\mu \vec{\pi}(x) + \lambda^2 \left[\sigma^2(x) + \vec{\pi}^2(x) - v^2\right] \vec{\pi}(x) = -g\rho_P(x). \tag{5}$$

Here $\rho_S = \langle \bar{q}q \rangle$ and $\rho_P = i\langle \bar{q}\gamma_5 \vec{\tau}q \rangle$ are scalar and pseudoscalar quark densities, which should be determined self-consistently from the motion of q and $\bar{q}$ in background meson fields. The scalar and pseudoscalar densities can be represented as

$$\rho_S(x) = ga(x)\sigma(x), \quad \rho_P(x) = ga(x)\vec{\pi}(x), \tag{6}$$

where $a(x)$ is expressed in terms of the momentum distribution function, $f(x, p)$, as

$$a(x) = \nu_q \int \frac{d^4p}{(2\pi\hbar)^3} \, 2\delta(p^\mu p_\mu - m^2(x)) f(x, p)$$

$$\longrightarrow \frac{\nu_q}{(2\pi\hbar)^3} \int \frac{d^3p}{E(x,\mathbf{p})} [n_q(x,\mathbf{p}) + n_{\bar{q}}(x,\mathbf{p})]. \tag{7}$$

Here ν_q is the degeneracy factor of quarks, $E(x,\mathbf{p}) = \sqrt{m^2(x) + \mathbf{p}^2}$, $n_q(x,\mathbf{p})$ and $n_{\bar{q}}(x,\mathbf{p})$ are the occupation numbers of valence quarks and antiquarks.

2.2   Phase transition in the σ-model

The Mexican Hat potential without the $H\sigma$ term in our Lagrangian leads to a first-order phase transition. At $\mu = n_B = 0$ the critical temperature is $T_c = 135.1$ MeV. However, the transition is very weak first-order, with a small barrier in the Ω potential (taking into account quark and antiquark kinetic energies, $\Omega = E_q - TS_q + U(\sigma, \pi = 0)$).

The order of the phase transition in the sigma model is discussed in detail recently[11]. Quantum fluctuations modify the phase diagram of the transition and with some parametrizations the transition may become second-order or smoothen out. We do not take into account fluctuations in our estimates because we discuss a very fast process where fluctuations have no time to develop. If we keep the $H\sigma$ term in the potential the transition becomes a smooth transition, see Fig. 2. Thus, no phase transition in the strict sense takes place.

2.3   Relativistic kinetic equation

Using the Wigner function formalism [14,15,16] one can derive the relativistic transport equation for the σ-model Lagrangian (1). Disregarding spin polarization effects one can represent the Wigner matrix for quarks in the form

$$W(x,p) = f(x,p)\left[\sigma(x) - i\gamma_5\vec{\tau}\vec{\pi}(x) + p_\mu\gamma^\mu\right]. \tag{8}$$

In the quasiclassical approximation the scalar part of the Wigner function $f(x,p)$ obeys the relativistic Vlasov equation

$$\left[p^\mu\partial_\mu + \frac{1}{2}(\partial^\mu m^2(x))\frac{\partial}{\partial p^\mu}\right] f(x,p) = 0, \tag{9}$$

where the quark (antiquark) effective mass $m(x)$ is obtained self-consistently,

$$m^2(x) = g^2\left[\sigma^2(x) + \vec{\pi}^2(x)\right]. \tag{10}$$

This expression for m can be justified also by chiral symmetry arguments. The vanishing collision term on the right hand side of the transport equation (9) reflects the fact that we are describing the evolution of the system after freeze-out.

3 Post Freeze-out Expansion

First we study the boost-invariant expansion of the plasma in homogeneous background fields. We assume that the meson fields and the scalar density depend on the proper time τ only. We also assume one dimensional scaling flow.[17].

Under these assumptions one can prove that the Vlasov equation (9) is satisfied by any distribution function, $f(s, p)$, which besides $p_\perp$ depends on only one scaling variable,

$$s = \pm(\tau/\tau_0)\sqrt{(p^\mu u_\mu)^2 - m^2(\tau) - p_\perp^2}. \tag{11}$$

On the mass shell $s = (\tau/\tau_0)p_\parallel$, where $p_\parallel = \sqrt{m^2(\tau) + p_\perp^2}\sinh(y - \eta)$, is the longitudinal momentum in the local rest frame (moving with rapidity η).

It is natural to think that before and up to the freeze-out, at τ_0, the distribution functions of quarks and antiquarks are given by the Fermi-Dirac distributions. In Lorentz invariant form

$$n_q(x, \mathbf{p}) = \left[\exp\left(\frac{\sqrt{m^2(\tau_0) + p_\perp^2 + p_\parallel^2} - \mu_0}{T_0}\right) + 1\right]^{-1}, \tag{12}$$

and $n_{\bar{q}}(x, \mathbf{p}) = n_q(x, \mathbf{p}; \mu_0 \to -\mu_0)$. Here $\mu_0 = \mu(\tau_0)$ and $T_0 = T(\tau_0)$ are the chemical potential and temperature at the time of freeze-out. The scaling solution of the Vlasov equation at $\tau > \tau_0$ can be obtained now from eq. (12) by simply changing $p_\parallel$ to $p_\parallel \tau/\tau_0$. Thus the scaling solution of the Vlasov equation at $\tau \geq \tau_0$ can be obtained from eq. (12) as:

$$n_q(x, p)_{LR} = \left[\exp\left\{-\left(\sqrt{p_\parallel^2 + \left(\frac{\tau_0}{\tau}m_\perp(\tau_0)\right)^2} - \frac{\tau_0}{\tau}\mu_0\right)\middle/ \frac{\tau_0}{\tau}T_0\right\} + 1\right]^{-1}. \tag{13}$$

The post-freeze-out evolution of the momentum distribution of quarks and antiquarks is characterized by two features. First, it becomes narrower due to the growing effective mass $m(\tau)$ with increasing τ. Second, an anisotropy is developed in momentum space, because the $p_\parallel$ distribution additionally shrinks as τ_0/τ. It is energetically favorable for quarks and antiquarks to stay in the chiral-symmetric phase where their effective mass is close to zero (it vanishes in the chiral limit, $m_\pi \to 0$). The scaling solution for the case $m = 0$ was found and discussed in Ref. [18], and mentioned in Ref. [12].

4 Unstable Mode Analysis

Let us study the onset of instability associated with the chiral-symmetry-breaking transition along the lines presented in similar works[10,12]. In earlier works[2,3] we applied here a linear response method which was used earlier for analyzing spinodal instability in nuclear matter. Here we solve numerically the equation of motion (5) for Φ. Now instead of assuming $m = 0$ we evaluate the scalar and pseudoscalar densities numerically for arbitrary quark masses, thus enabling us to study the whole dynamics with the source terms not only the initial moments of instability. The source term was evaluated by using an 8 point Gaussian-Laguerre quadrature for the expression

$$a(x) = \frac{\nu_q (\hbar c)^2}{(2\pi\hbar)^3} \int \frac{d^3 p}{E(x, \mathbf{p})} [n_q(x, \mathbf{p}) + n_{\bar{q}}(x, \mathbf{p})] =$$

$$\frac{\nu_q T_0^2}{\pi^2 (\hbar c)^2} \frac{\alpha}{\sqrt{1-\alpha}} \int_0^\infty \frac{\arcsin \sqrt{(1 - \alpha^2)/\left(1 + \frac{m^2 c^2}{T_0^2 s^2}\right)}}{\exp(-s) + \exp\left(-s + \sqrt{s^2 + \frac{m_0^2 c^2}{T_0^2}}\right)} e^{-s} \, ds , \qquad (14)$$

where $\alpha = \tau_0/\tau$, $m_0 = m(\tau_0)$ is the quark mass at τ_0 and T_0 is the freeze-out temperature.

The possibility for instability in the symmetric model, $H = 0$, opens at some threshold time, τ_{tf}[2,3]. At $\tau = \tau_{tf}$ the instability just becomes possible for $k = 0$, and the corresponding growth time is infinitely long. When the system expands further Im ω grows rapidly. Quarks and antiquarks stabilize the system and make the transition less sharp.

If $H \neq 0$ the transition is gradual, but fast. The source term leads to a modification of the dynamics. Compared to the earlier dynamical studies with a quenched initial condition, however, the difference is not large and we get almost the same rapid rollover transition.

5 Discussion

The evolution of vacuum to the normal (spontaneously broken) ground state goes through the generation of the long wavelength oscillations of scalar and pion fields. These classical pion field configurations are characterized by (i) the long life time (> 20 fm/c), (ii) low momenta (< 40 MeV/c) of their decay products, pions, and (iii) large imbalance in isospin of produced pions.

The growth dynamics and the size of domains of the coherent chiral condensate were also studied in Refs. [8,9,10] in the quench approximation and in

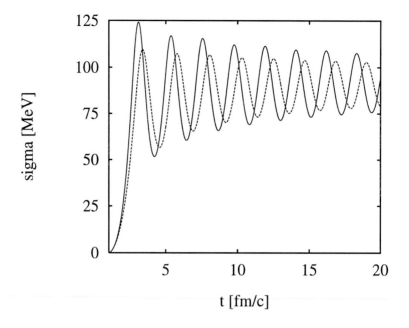

Figure 3: Time dependence of the σ field with (full line) and without (dashed line) the scalar and pseudoscalar quark density in the equation of motion, where the quark density was evaluated from the post freeze-out dynamics of the quarks. The calculation is started at the freeze-out time of $\tau_0 = 1$ fm/c from $\sigma(\tau_0) = \pi(\tau_0) = 0$ and $\dot{\sigma}(\tau_0) = 1$ MeV/fm, $\dot{\pi}(\tau_0) = 5$ MeV/fm. This initial condition is identical to the case presented in ref. [10].

Ref. [12]. A small domain size will reduce to large extent the signal of the coherent pion field. If the domains are large, of 3-7 fm size as predicted in Ref. [12], than in addition to the isospin alignment, the emitted pions should have very low relative momenta, say $p_t < 50$ Mev/c.

In conclusion we proposed a new scenario for rapid hadronization of the QGP at RHIC and LHC energies. The hadronization is associated with the transition from the initial, chirally symmetric state to the spontaneously broken final state. This process leads to the formation of $q - \bar{q}$ clusters surrounded by domains of excited vacuum with oscillating scalar and pion fields. The characteristic decomposition time is about few fm/c, i.e., much shorter than the time scale of homogeneous nucleation. The fast hadronization process should manifest itself by short emission times in HBT measurements. The specific signals of the chiral transition, i.e., large fluctuations in isospin and rapidity of produced pions, can be observed only on event-by-event basis. Other signals, such as excess of low p_t pions, enhanced strangeness production and vanishing in-medium effects should be seen even in event-averaged data.

Acknowledgments

We thank T. Biró, J. Bondorf, J. Kapusta, B. Müller, J. Németh, L.M. Satarov, and Ove Scavenius for stimulating discussions. This work was supported in part by NORDITA, EU-INTAS programme, International Science Foundation (Soros) and the Norwegian Research Council (NFR). One of us (I.N. M.) thanks the Niels Bohr Institute for kind hospitality and the Carlsberg Foundation for financial support.

References

1. L. P. Csernai and J. I. Kapusta, *Phys. Rev. Lett.* **69** (1992) 737; *Phys. Rev.* **D46** (1992) 1379.
2. L. P. Csernai and I.N. Mishustin, *Phys. Rev. Lett.* **74** (1995) 5005.
3. L. P. Csernai, I.N. Mishustin, and Á. Mócsy, *Heavy Ion Phys.* in press.
4. T. Csörgő and L.P. Csernai, *Phys. Lett.* **B333** (1994) 494.
5. J. Kapusta and A. Mekjian, *Phys. Rev.* **D33** (1986) 1304.
6. N. Bilić, et al., *Phys. Lett.* **B311** (1993) 266; J. Cleymans, et al., *Z. Phys.* **C58** (1993) 347.
7. S. Abatzis, et al., (WA85 Collaboration), *Phys. Lett.* **B316** (1993) 615.
8. K. Rajagopal and F. Wilczek, *Nucl. Phys.* **B379** (1993) 395; **B404** (1993) 577.
9. S. Gavin, A. Goksch and R.D. Pisarski, *Phys. Rev. Lett.* **72** (1993) 2143.
10. Z. Huang and X.N. Wang, *Phys. Rev.* **D49**, (1994) 4335.
11. J.I. Kapusta and A.M. Srivastava, *Phys. Rev.*
12. S. Gavin and B. Müller, *Phys. Lett.* **B329** (1994) 486.
13. M. Asakawa, Z. Huang, and X.N. Wong, *Phys. Rev. Lett* **74** (1995) 3126.
14. H.-Th. Elze, M. Gyulassy, D. Vasak, H. Heinz, H. Stöcker and W. Greiner, *Mod. Phys. Let.* **A2** (1987) 451.
15. W.-M. Zhang and L. Wilets, *Phys. Rev.* **D45** (1992) 1900.
16. A.B. Prozorkevich, S.A. Smoliansky, V.D. Toneev, (GSI-93-26, 1993) preprint.
17. J.D. Bjorken, *Phys. Rev.* **D27** (1983) 140.
18. J.I. Kapusta, L. McLerran and D.K. Srivastava, *Phys. Lett.* **B283** (1992) 145.

Nucleus-Nucleus Collisions at Highest Energies

M. Bleicher, N. Amelin, S. A. Bass, M. Brandstetter, A. Dumitru, C. Ernst,
L. Gerland, J. Konopka, S. Soff, C. Spieles, H. Weber, L. A. Winckelmann,
H. Stöcker

Institut für Theoretische Physik
Johann Wolfgang Goethe Universität
60054 Frankfurt am Main, Germany

The microscopic phasespace approach URQMD is used to investigate the stopping
power and particle production in heavy systems at SPS and RHIC energies. We
find no gap in the baryon rapidity distribution even at RHIC. For CERN energies
URQMD shows a pile up of baryons and a supression of multi-nucleon clusters at
midrapidity.

1 Motivation

One of the main aims of relativistic heavy ion collisions at collider energies is to
discover if the individual hadrons dissolve into a gas of free quarks and gluons
(quark-gluon-plasma, QGP) in the extremely compressed and heated hadronic
matter. This may happen in line with a transition into the chiral symmetric phase which modifies most hadron masses drastically. The achievable
energy deposition depends on the amount of stopping of the colliding nuclei.

2 The URQMD Model

The Ultrarelativistic Quantum Molecular Dynamics (URQMD)[1], is used to analyze the physics of the excitation function of hadronic abundances, stopping
and flow. This framework bridges with one model consistently the entire available range of energies from below SIS to CERN, even for the heaviest system
Pb+Pb. URQMD is a hadronic transport model including strings. Its collision
term contains 50 different baryon species (including nucleon, delta and hyperon
resonances with masses up to 2 GeV) and 25 different meson species (including strange meson resonances), which are supplemented by their corresponding
antiparticle and all isospin-projected states.

3 He-He Collisions at ISR

In general it appears to be an intricate problem to describe stopping behaviour
of baryons and pion production within one theoretical frame at very high
energies. If one wants to do LHC calculations it may be necessary to include

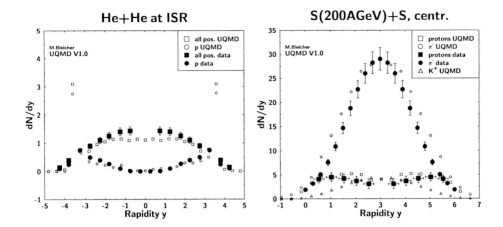

Figure 1: Left:Rapidity distribution of protons and positively charged particles for the reaction He+He at $\sqrt{s} = 31$ AGeV compared to the data [2]. Right: Rapidity distribution of protons and negatively charged pions for central S+S reactions at $E_{lab} = 200$ AGeV compared to the data [8].

multi-string-excitation as it is done in dual parton approach. To demonstrate the ability of URQMD to model a nucleus-nucleus collision even at the today highest available bombarding energies for heavy particles, we compare the calculated He+He collision at ISR with data [2] as shown in Fig.1 (left).

It is not surprising, that such a light system as helium is totally transparent. A baryon free area of 3 units in rapidity is produced. URQMD and the data agree well, the calculated produced particle yield may increase by 15% if one also simulates gluon jet events [3], which is not included yet.

4   SPS Energy Regime

The dominant reaction mechanism in the early stage of a reaction is the excitation of collision partners to resonances or strings[1]. Then secondary interactions, i. e. the annihilation of produced mesons on baryons, lead also to the formation of s channel resonances or strings, which may explain the strangeness enrichment [4] and (for masses larger than $3m_N$) allow for $\overline{N}N$ creation[5]. The escape probability for $\bar{p}$'s from the exploding nuclear matter enters via the free $N\overline{N}$ annihilation cross section. For central events of Pb+Pb at SPS approx. 85% of the produced anti-baryons are annihilated during the reaction.

These two counter–acting effects may be measured by the directed "anti-

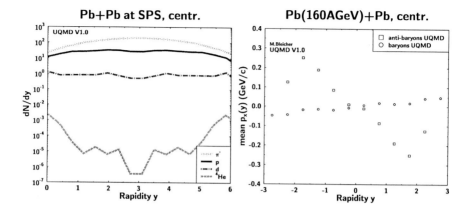

Figure 2: Left: Rapidity distribution of π^-, protons, deuterons and ^{4}He in central Pb+Pb reactions at $E_{lab} = 160$ AGeV. Right: Directed flow of baryons and antibaryons for the same system.

flow" of antimatter. The observable asymmetry for bouncing antimatter can be quantified by the mean p_x vs. rapidity (Fig.2, right). The anti–flow of antibaryons appears to be strongest for semicentral collisions, while for baryons the maximum p_x is at considerably smaller b–values. The latter is due to the pressure (i. e. the EOS) the former one due to absorption and geometry.

Comparisons of URQMD calculations to data from SIS to SPS is documented elsewhere[6]. Good agreement of baryon and meson production and dynamics has been achieved. An impression is given in Fig.1 (both) - protons as well as pions and kaons are shown.

Further insight into the collision geometry may be gained by looking at composite particle probes. Since URQMD does not include the production of light nuclei dynamically, cluster formation is added after strong freeze-out. (Freeze-out means after the last strong interaction of the particle.) We calculate the deuteron (helium) formation probability by projecting the nucleon pair phasespace on the deuteron wave function via the Wigner-function method as described in [7]. Especially for high bombarding energies or exotic clusters it is necessary to use this sophisticated method, since the complex phasespace distribution of the cluster ingredients has to be taken into account [9]. The Hulthén parametrization is used to describe the relative part of the deuteron

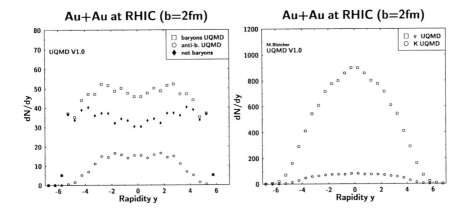

Figure 3: The system Au+Au at an energy of $\sqrt{s} = 200$AGeV (RHIC), central collisions selected. Left:Rapidity distribution of baryons, antibaryons and net-baryons. Right: Rapidity distribution of pions and kaons

wavefunction. The yield of deuterons is given by

$$\mathrm{d}N_d = \frac{1}{2}\frac{3}{4}\Big\langle \sum_{i,j} \rho_d^{\mathrm{w}}(\Delta\vec{R}, \Delta\vec{P})\Big\rangle \mathrm{d}^3(p_{i_p} + p_{j_n})\,.$$

The Wigner-transformed Hulthén wavefunction of the deuteron is denoted by ρ_d^{w}. The sum goes over all n and p pairs, whose relative distance $(\Delta\vec{R})$ and relative momentum $(\Delta\vec{P})$ are calculated in their rest frame at the earliest time after both nucleons have ceased to interact. The factors $\frac{1}{2}$ and $\frac{3}{4}$ account for the statistical spin and isospin projection on the deuteron state. The calculation of the high mass clusters is straight forward, e.g. by exchanging the Hulthén parametrization of the d wavefunction with a 4-body harmonic oscillator wavefunction[7] to describe the ^{4}He (See Fig. 2 (left), π^-, protons, as well as deuterons and ^{4}He are depicted). In contrast to the pile up of protons at midrapidity, cluster production is strongly supressed due to the high temperatures in the center of the collisions.

Calculations of H^0 ($\Lambda\Lambda$-clusters) for AGS and SPS energies have also be performed in this framework[10].

5 RHIC Estimates

As shown above URQMD seems to be well suited for an estimate of stopping power in Au+Au collisions at RHIC. However, we are well aware of the fact that for certain observables (e.g. high p_t components of particle spectra) the current framework is not sufficient, since it does not incorporate hard partonic scattering explicitely. Figure 3 (left) shows the results of our calculation for gold-gold at $\sqrt{s} = 200$ AGeV (RHIC). The nuclei suffer a mean rapidity shift of more than 2 units of rapidity. The mid-rapidity region is apparently not baryon free, in contrast to some earlier expectations. Our results are similiar to RQMD calculations[3].

We finally put our interest on the produced mesons. In Fig.3 (right) we show the rapidity distribution of kaons and pions. As mentioned above at this high energies modifications of the string fragmentation may be necessary. The inclusion of gluon jets as a first step will increase the meson multiplicities by about 15% [3].

6 Conclusion

At first some remarks on the validity of our calculation: It is believed to be a fact that a quark-gluon-plasma is created at RHIC. While our calculation does not assume a QGP, it certainly goes beyond the purely hadronic scenario - strings are excited and quarks and diquarks are subject to further interaction. From our calculation we infer that the interaction of leading quarks and diquarks dictate the stopping behaviour of heavy ion collisions even at SPS energies.

Acknowledgments

This work is supported by GSI, DFG and BMBF. M.B. wants to thank R. Mattiello for many inspiring ideas and fruitful discussions.

References

1. S. A. Bass, M. Bleicher, M. Brandstetter, A. Dumitru, C. Ernst, L. Gerland, J. Konopka, S. Soff, C. Spieles, H. Weber, L. A. Winckelmann, H. Stöcker and W. Greiner;
   source code and technical documentation, to be published in 1996
2. L. Otterlund, *Nucl. Phys.* **A418**(1983)98c

3. T. Schönfeld *et al.*, *Mod. Phys. Lett.* **A8**(1993)2631; T.Schönfeld, Ph.D. Thesis Universität Frankfurt 1993

4. R. Mattiello, *et al.*, *Phys. Rev. Lett.* **63**(1989)1459

5. A. Jahns et al., *Phys. Rev. Lett.* **72**(1994)3464 and refs. therein

6. S. A. Bass *et al.* contribution to this volume

7. R. Mattiello *et al.*, *Phys. Rev. Lett.* **74**(1995)2180; M. Gyulassy, K. Frankel, E.A. Remler, *Nucl. Phys.* **A402**(1983)596; E.A. Remler and A.P.Sathe, *Ann. Phys.* **91**(1975)295

8. H. Ströbele *et al.*, *Nucl. Phys.* **A525**(1991)59c

9. M. Bleicher *et al.*, *Phys. Lett.* **B361**(1995)10

10. C. Spieles *et al.* contribution to this volume

HYDRODYNAMICAL ANALYSIS OF SYMMETRIC NUCLEUS-NUCLEUS COLLISIONS AT SPS ENERGIES

B.R. SCHLEI, D. STROTTMAN

Theoretical Division, Los Alamos National Laboratory, Los Alamos, NM 87545

We present a theoretical study of ultrarelativistic heavy-ion data obtained at the CERN/SPS by the NA49 Collaboration using 3+1-dimensional relativistic hydrodynamics. We find excellent agreement with the rapidity spectra of negative hadrons and protons and with the correlation measurements for $Pb + Pb$ at 160 $AGeV$ (preliminary results). Within our model this implies that for $Pb + Pb$ a quark-gluon-plasma of initial volume 174 fm^3 with a lifetime 3.4 fm/c was formed.

1 Introduction

It is very appropriate to discuss relativistic heavy ions in a meeting honouring Walter Greiner. It is a field which he helped found, in which he has provided many of the seminal ideas, in particular the use of fluid dynamics and the equation of state of nuclear matter[1], and which he has continued to strongly influence[2,3,4].

A prediction of the theory of quantum chromodynamics is that at sufficiently high energy and/or baryon density, the quarks inside hadrons are deconfined and a quark-gluon plasma is created, albeit for only a very brief time. In the past several years there have been vigorous experimental programs in high energy heavy ion physics pursued at CERN and at the Brookhaven AGS. Ever higher energies and/or masses have been used to increase the lifetime of the hot matter by increasing the initial energy density, the size of the system, or both. The probability of preparing a strongly interacting system that shows thermodynamical behavior and therefore is treatable by well known thermodynamical or fluid dynamical methods increases with the size of the system.

2 Modeling the reaction using relativistic heavy ions

Fluid dynamics provides an intuitively simple description of heavy ion collisions: two nuclei smash into each other, are rapidly thermalised and compressed; the resulting fireball then expands and breaks up into bits of hadronic matter that ultimately reach the detectors. Because some of the underlying principle assumptions of fluid dynamics, *e.g.*, that of a zero mean free path and instant thermalisation, it has been long assumed that this model would not be valid at relativistic bombarding energies. The experimental observation at the AGS and CERN of complete or nearly complete stopping of matter has altered this perception. Many researchers over the past few years have successfully simulated heavy ion collisions using fluid dynamics and it continues to provide accurate estimates of experimental observables from relativistic heavy ion collisions.

A theory of relativistic fluid dynamics is not guaranteed to exist: it requires that a system of N particles (with $N \approx 1000$ or more) with $6N$ degrees of freedom be accurately represented by a system of 6 equations. Arguments to justify fluid dynamics in conventional physics have been made, $e.g.$, Bogoliubov[5], but their applicability in the relativistic regime is unclear. The apparent validity of fluid dynamics in describing hadronic reactions may be that five of the six Euler equations are kinematic constraints: conservation of mass and energy-momentum. The observables currently examined may simply be dominated by such constraints, although Bose-Einstein correlations may be a much more severe test. There are severe difficulties in relativistic fluid dynamics when one attempts to include dissipation[9,10]. There are ambiguities in the choice of the rest-frame[11,12,13]. All first-order theories are known to have difficulties such as being singular or to admit acausal solutions[14]. For a justification of hydrodynamics, the reader is referred to a recent text[8] or reviews of the subject[2,6,7].

There are several assumptions in the hydrodynamical approach: (i) the initial conditions, (ii) the equation of state (EOS), (iii) the magnitude of the viscosity and heat conduction, and (iv) models of the final stage of the reaction when hydrodynamics has ceased to be valid and one must worry about the emission processes, freeze-out, and evaporation.

We have undertaken a comprehensive study of data taken at CERN by the NA35 and the NA49 collaborations[15,16]. Results of Pb+Pb at 160 AGeV were obtained[17,18] using 3+1 dimensional relativistic fluid dynamics. There are several assumptions in the hydrodynamical approach: the initial conditions, the equation of state, and models of the final stage of the reaction when hydrodynamics has ceased to be valid, i.e., freeze-out. All the dynamics is contained in the equation, $P = P(\mu, T, n, ...)$, the equation of state. Initial conditions specified in terms of energy density and initial velocity profiles was chosen so as to reproduce the measured rapidity distribution of particles. An equation-of-state was assumed that contains a phase transition from the hadronic phase to a quark-gluon phase at a critical temperature $T_C = 200$ MeV. Because of the need for a relativistic framework, a covariant freeze-out scenario developed by Cooper and Frye[19] was used.

A reaction of two baryonic fluids leads to a deceleration of the projectile and target baryonic currents and thus to the spread of their width in momentum space. In our model we impose an initial rapidity field $y(z)$ on the fluid.

The kinetic energy of the two incoming baryonic fluids is converted into internal excitation (thermal energy) of a third fluid which is created in the central region. The relative fraction of the thermal energy inside the initial fireball volume fixes also the initial state of the formed fireball.

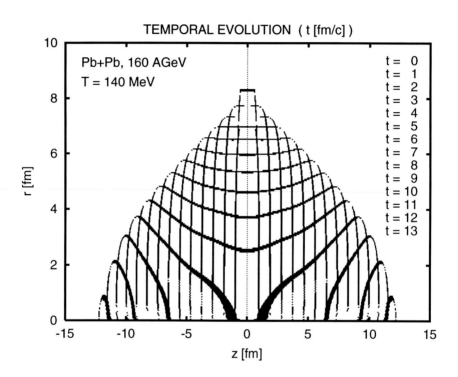

3 Discussion of Results

The numerical simulation reproduced simultaneously mesonic and baryonic rapidity and transverse momentum spectra[17,18]. The lifetime of the hot dense matter was calculated to be about 13 fm/c - the quark-gluon plasma persisted for 3.4 fm/c - and the collision time and baryon density increase is about 20-30% larger if a phase transition is present compared to the case in which there is no phase transition. The density achieved in the center of the hot fireball is calculated to be as much as 20 times that of normal nuclear matter. Under such extreme conditions as occurs in central collisions, a quark-gluon plasma is formed which in our calculations is 174 fm^3 which is an appreciable fraction of the total matter.

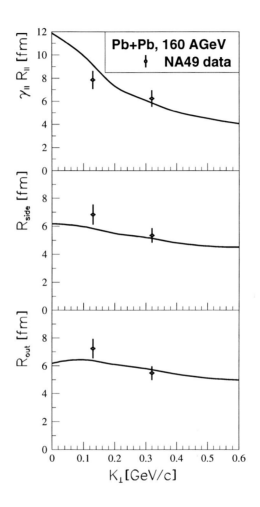

In the first figure we exhibit the space-time geometries of the hadron sources, *i.e.*, the time contour plots of the freeze-out hypersurface in the z-r plane. Each line represents the freeze-out hypersurface at a fixed time t. The numerical solution represented in the plot show an evolution of an initially disk-shaped fireball which emits hadrons from the very beginning. Due

462

to the effect of transverse, inwardly moving rarefaction waves, the transverse freeze-out positions move towards the center of the fireball. In the late stage of the hydrodynamical expansion the hadron-emitting fireball separates into two parts while cooling down until it ceases to emit.

Bose-Einstein correlation functions for pions have also been calculated. The preliminary NA49 data on interferometry are surprisingly well described as the second figure shows. In the case of pion interferometry a large fraction – 40% to 50% – of the pions originate from resonances which strongly influences the calculated radii. Bose-Einstein correlations are a very complicated observable defined through quantum statistics. The more resonances taken into account, the narrower the correlation function becomes because the resonance halo increases the effective source size. As seen in the first figure, the hadron source (the real fireball) is represented through a very complex freeze-out hypersurface. The longitudinal and transverse extensions of the fireball change dynamically as a function of time, rather than show up as static effective radii. This caveat should always be kept in mind when trying to interpret Bose-Einstein correlation data.

We have shown that data from heavy-ion experiments for single and double inclusive cross sections of mesons and baryons can be reproduced with a three-dimensional relativistic hydrodynamic description assuming an equation of state with a phase transition to a QGP. Our data analysis indicates an enhanced transverse flow in the case of $Pb + Pb$ collisions at CERN/SPS energies. The results of this work constitute further evidence that heavy-ion collisions in the SPS region show fluid dynamical behaviour and can be described by assuming an equation of state with a phase transition from a quark-gluon plasma to hadronic matter.

Acknowledgments

We wish to thank U. Ornik, M. Plümer and R.M. Weiner for many discussions during the course of this work. This work was supported by the Department of Energy and the Deutsche Forschungsgemeinschaft (DFG). B.R. Schlei acknowledges a DFG postdoctoral fellowship.

1. W. Scheid and W. Greiner, Z. Phys. **226** (1969) 365; W. Scheid, H. Müller and W. Greiner, Phys. Rev. Lett. **32** (1974) 741.
2. H. Stöcker and W. Greiner, Phys. Rep. **137** (1986) 277.
3. Proc. NATO Advanced Study Institute on the Nuclear Equation of State, ed. W. Greiner and H. Stöcker (Plenum Press, NY, NATO ASI series B 216, 1990).

4. Proc. of the NATO Advanced Study Institute on Hot and Dense Matter, Bodrum, Turkey, ed. W. Greiner, H. Stöcker and A. Gallmann (Plenum, NATO ASI series B 335, 1994)
5. N.N. Bogoliubov, in *Studies in Statistical Mechanics*, eds J. de Boer and G.E. Uhlenbeck (Interscience, N.Y., 1962).
6. P. Carruthers, Proc. N.Y. Acad. Sciences **229** (1974) 91.
7. J.A. Maruhn, in *Relativistic Heavy Ion Physics*, eds. L.P. Csernai and D.D. Strottman (World Scientific, Singapore 1991) Vol. 5 p. 82.
8. L.P. Csérnai: *Introduction to Relativistic Heavy Ion Collisions* (J. Wiley, 1993).
9. N.L. Balazs, Los Alamos preprint, LA-UR-92-3676.
10. D. Strottman, Nucl. Phys. **A566** (1994) 245c.
11. C. Eckart, Phys. Rev. **58** (1940) 919.
12. L.D. Landau, Izv. Akad. Nauk SSSR **17** (1953) 51.
13. B. Carter in *Relativistic Fluid Dynamics*, eds. A. Anile and Y. Choquet-Bruhat (Springer, Berlin, 1989).
14. W. A. Hiscock and L. Lindblom, Ann. Phys. **151** (1983) 466; W. A. Hiscock and L. Lindblom, Phys. Rev. **D 31** (1985) 725; W.A. Hiscock and L. Lindblom, Phys. Rev. **D35** (1987) 3723; T.S. Olson and W.A. Hiscock, Phys. Rev. **D41** (1990) 3687; W.A. Hiscock and L. Lindblom, Phys. Lett. **A131** (1988) 509; W.A. Hiscock and T.S. Olson, Phys. Lett. **A141** (1989) 125.
15. Th. Alber et al., Phys. Rev. Lett. **74** (1995) 1303; Th. Alber et al., Z. Phys. **C66** (1995) 77.
16. T. Alber for the Collaborations NA35 and NA49, Nucl. Phys. **A590** (1995) 453c; T. Alber et al., Phys. Rev. Lett. **75** (1995) 3814.
17. B.R. Schlei, U. Ornik, M. Plümer, D. Strottman, R.M. Weiner, "Hydrodynamical analysis of single inclusive spectra and Bose-Einstein correlations for $Pb + Pb$ at 160 $AGeV$", Los Alamos preprint LA-UR-95-3232 (1995), Phys. Lett. B, in press.
18. U. Ornik, M. Plümer, B.R. Schlei, D. Strottman, R.M. Weiner, "Hydrodynamical analysis of symmetric nucleus-nucleus collisions at CERN/SPS energies", Los Alamos preprint LA-UR-96-1298, submitted for publication to Phys. Rev. C.
19. F. Cooper and G. Frye, Phys. Rev D **10** (1974) 186.

RECENT PROGRESS UNDERSTANDING THE STRUCTURE FUNCTIONS AT SMALL x

LARRY MCLERRAN

Physics Department,
Univeristy of Minnesota,
Minneapolis, MN 55455, USA

I present a summary of recent work which attempts to understand the distribution functions of gluons at small x. The goal of this study is to compute both the structure functions for hadrons at very small x and gluon production. Due to the larger density of gluons in nuclei at equivalent values of x, large nuclei may be useful tools for studying such behaviour. This work uses the non-abelian generalization of Weizsacker-Williams fields generated by the valence quarks as the source of the gluon distribution functions. These fields produce, in a hadronic collision, time dependent fields which ultimately end up as produced gluons.

1 Introduction

One of the long standing goals of hadronic physics is to predict the structure functions of various constituents of hadrons. This could be accomplished by computing the hadronic wavefunction. Then the distribution functions would in a physical gauge, such as light cone gauge, be simply the Fock space contributions of various components of the hadronic wavefunction. As a side benefit, one would compute fluctuations and correlations in various quantities by considering expectation values of operators of higher dimension.

Another long standing problem in hadronic physics is the nature of the pomeron. This elusive phantom of hadronic physics is the presumed cause of the growing cross section for high energy hadronic scattering and the source of inelastically produced particles. At small enough x, it is believed that the density of gluons which compose the pomeron becomes very large. Moreover, it is expected that there is universal behaviour the growth with decreasing x for sufficiently small x for various types of hadron multiplicity distributions.

Another outstanding problem is to describe the initial conditions for the matter produced in heavy ion collisions, or for that matter for hadronic matter produced in hadron collisions in general. These issues arise naturally in the Geiger-Mueller parton cascade model of nuclear collisions where one takes the parton distribution functions as an initial distribution of on mass shell gluons which are the source of the scattering process.[1]

In the Geiger-Mueller parton cascade picture, the distribution functions which arise from the infinite momentum frame nuclear wavefunction are described both by their momentum space distribution and their coordinate space worldlines. The gluons are on mass shell. They scatter incoherently so that the initial state distribution changes and tends towards a local equilibrium distribution.

The problems with this description are

1. The cascade description simultaneously specifies both the coordinates and momenta of the partons. This is inconsistent with the uncertainty principle $\Delta p \Delta x \geq 1$. The uncertainty principle should be important at early times as the partons arise from the infinite momentum frame nuclear wavefunction.

2. The scattering is incoherent so that charge coherence effects are ignored. At high densities corresponding to the earliest times, the parton wavelength is long compared to the typical parton separation, so that these effects should be large.

In this talk, I shall attempt to describe recent work which both attempts to compute the gluon and quark distribution functions and scattering for hadrons at very small x and for very heavy nuclei.[2][3] The work is similar in spirit to work done by A. Mueller to compute structure functions for heavy quark systems.[4]. In the description below, a very heavy nucleus is used as the idealized hadron, since we would expect that one can use the techniques we have developed at larger values of x than for a hadron. These techniques may be largely generalized to the case of a hadron without difficulty in principle.

2   The Approximations

We shall use several kinematic approximations to simplify the problem. The first is that the valence quarks are always traveling close to light velocity collinear with the beam axis. The emission of gluons of low x or finite p_T will not change the velocity of the valence quarks by a significant amount. The valence quarks are therefore recoilless sources of charge moving at near light velocity along the beam axis.

The second set of approximations set the geometry for the valence quarks. Since the typical transverse momentum of the gluons we consider will be large compared to the inverse nuclear curvature, $p_T \gg 1/R$, we can approximate the nucleus as infinite in transverse extent. For gluons with $x \ll 1Fm/R$, the nucleus will appear to that parton as a Lorentz contracted disk of longitudinal extent much less than a Fermi. We see therefore that the valence quarks appear as a thin sheet of charge of infinite transverse extent propagating along the light cone.

For large nuclei, the density of gluons per unit rapidity per unit transverse area is large

$$\Lambda^2 = \frac{1}{\pi R^2} \frac{dN}{dy} >> \Lambda_{QCD}^2 \tag{1}$$

This justifies a weak coupling analysis.

We shall compute the gluon density dN/dxd^2p_T only for transverse momentum $p_T >> 1/Fm$. On this resolution scale, the valence quarks inside any box of transverse resolution scale $\Delta x << 1Fm$ will come from different nucleons. On the other hand, if we require that $\Delta x >> A^{-1/6}$, then the number of valence quarks within our resolution in transverse area is $\rho(\Delta x)^2 >> 1$. In this case, the color charge associated with these quarks is very large, and may therefore be treated classically.

Finally, for a nucleus traveling at the speed of light, the transverse coordinates of the quarks are frozen. They cannot move during any measurement process which takes a finite time. On the other hand, the positions of the valence quarks must vary from measurement to measurement as a consequence of the initial conditions. Since on the transverse resolution scale of interest, the valence quarks come from different nucleons, they are uncorrelated. The color charge density is therefore uncorrelated in transverse extent, and Gaussian distributed.

The mathematical problem we must solve is therefore a fixed charge density on a thin sheet of infinite transverse extent traveling at the speed of light. To compute average quantities, one must average over all sources of color charge with a weight

$$\int [d\rho] exp\{-\frac{1}{2\mu^2} \int d^2x_T \rho^2(x_T)\} \tag{2}$$

In this equation, a parameter μ has appeared. The quantity μ^2 has the dimensions of a charge squared per unit area, and is essentially $4/3$ times the number valence quarks per unit area. It goes as $A^{1/3}$ for large A. Since it is the only scale which characterizes the theory, for large values of this scale the coupling constant must be small, and weak coupling methods are justified.

The analysis described above includes the contribution to the charge density associated with the valence quarks. These valence quarks produce a gluon field and this gluon field carries charge. Since this induced charge is small per unit rapidity, since it muse be proportional to α_s which is assumed to be small, most of the charge induced by the gluons, as measured by a field at some value of x must come from a source at even smaller values of x. Therefore even for the induced charge due to gluon radiation, it too is effectively on a sheet of infinitesmal thickness. The effective action for such fields including the gluon radiation can be written as

$$\int [d\rho] exp\{- \int dy d^2 x_T \frac{1}{2\mu^2(y)} \rho^2(y, x_T)\} \tag{3}$$

Here the charge density is now what per unit rapidity per unit transverse area. The rapidity variable is taken to be

$$y = y^0 - ln\{x^{0-}/x^-\} \tag{4}$$

where $x^- = \frac{1}{\sqrt{2}}(t - z)$ is the light cone coordinate conjugate to the light cone momentum p^+. The rapidity y^0 is the beam rapidity and x^{0-} is the minimal allowed $x^- \sim 1/P^+$ where P^+ is the beam momentum.. The charge squared per unit area seen on the two dimensional sheet is

$$\chi(x_T) = \int_y^{y^0} dy \mu^2(y, x_T) \tag{5}$$

For hadrons, the object of the analysis will be to determine μ^2 and to justify the form of the above effective action.

3   Solution for the 1 Nucleus Problem

The solution for the problem of 1 nucleus is very simple.[2] On either side of the sheet of valence charge, there is no source. We therefore expect that the fields will be a gauge transformation of vacuum fields. Across the sheet there can be a discontinuity. The magnitude of the discontinuity is determined by the value of the valence charge distribution. We therefore have as a solution

$$\begin{aligned} A^\pm &= 0 \\ A^i &= \theta(x^-)\alpha^i(x_T) \end{aligned} \tag{6}$$

where

$$\alpha^i(x_T) = \frac{-1}{ig} U(x_T) \nabla_T^i U^\dagger(x_T) \tag{7}$$

where $x^\pm = (x^0 \pm x^3)/\sqrt{2}$ and where U is an element of the gauge group. The field is singular at $x^- = 0$. It must be regularized and smeared over the longitudinal extent of the source. This smearing is most conveniently done in terms of the space-time rapidity variable y.

The field α is determined in terms of ρ by the equation condition

$$\vec{D}(A) \cdot d/dy \ \vec{\alpha} = g\rho(y, x_T) \tag{8}$$

The solution outside the source has the property that the longitudinal electric and magnetic fields vanish. The transverse magnetic is everywhere orthogonal to the transverse electric. The fields are delta functions and only exist in the sheet of the nucleus. These fields are therefore manifestly unstringy.

In general when one includes the full rapidity dependence of the field, the problem can be solved. In fact the full solution can be averaged over a random source and the full transverse momentum dependence of the intrinsic p_T distributions of the gluons can be determined[5] The result is that the distribution function goes as $1/p_T^2$ at large p_T and softens to a slowly varying power of a logarithm at small p_T

One can also include the effect on the charge density of the induced gluon radiation. This leads to a renormalization group equation for the charge density which gives it dependence upon rapidity. In addition, the induced charge is singular at short distances and must be regularized. This regularization scale is the Q^2 at which the gluon distribution is measured. This leads to a set of renormalization group equations which determine the charge squared per unit rapidity. These corrections appear to generate the Lipatov enhancement at small x.[6] A derivation of these results will be presented later.[5]

For $p_T >> \mu$, our approximations break down. Nevertheless for some range of $p_T >> \mu$, one can argue that our naive analysis treating the valence quarks as classical sources is valid. This will be true until bremsstrahlung corrections to our classical analysis make the transverse momentum distribution fall more rapidly than a power of p_T. In the range $\alpha_S\mu << p_T << \mu$, the spectrum of gluons is typical of a Weizsacker-Williams distribution and $dN/d^2 p_T \sim 1/p_T^2$. At smaller values $\Lambda_{QCD} << p_T << \alpha_S\mu$, the distribution is almost flat. Therefore the dominant contribution to the gluon distribution function comes from $p_T \geq \alpha_S\mu$. This contribution gives $dN/dy \sim A$ up to logarithms of A. The contribution from $p_T \sim \Lambda_{QCD}$ is small and proportional to $A^{2/3}$. The non-perturbative component is therefore screened and feels only the surface of the nucleus. The perturbative component feels the entire nucleus.

4 Nucleus-Nucleus Collisions

For the two nucleus problem, we begin before the collision with a linear superposition of two one nucleus solutions.[3] Since the electric and magnetic fields only exist in the sheet of the valence quark distribution, nothing happens until the two nuclei pass through one another. The most general solution after the collision must be boost covariant, and cannot depend upon the space time rapidity

$$y = \frac{1}{2} ln\{\frac{t+z}{t-z}\} \tag{9}$$

except for factors which are required by covariance. The proper time

$$\tau = \sqrt{t^2 - z^2} \tag{10}$$

will generate the nontrivial time dependence.

The most general boost covariant solution is

$$\begin{aligned} A^\pm &= x^\pm \alpha^\pm(\tau, x_T) \\ A^i &= \alpha^i(\tau, x_T) \end{aligned} \tag{11}$$

This solution has $dN/dy \sim const$, and is the analog of the Bjorken hydrodynamic solution.[7]

The fields above solve the equations of motion plus boundary conditions. The evolution is non-linear for $p_T \leq \alpha_S \mu$. This turns out to be the region from which most gluons are emitted which will participate in the Geiger-Mueller cascade.

As time evolves, the fields decrease in magnitude. After some time $\tau \sim 1/\mu$, one can show that the fields become weak and the equations of motion linearize. At this time, the fields describe far separated gluons which can serve as the initial conditions for a cascade computation.

One can explicitly compute the gluon production in the region where $p_T >> \alpha_S \mu$. This semi-hard region is not the dominant region for gluon production, but it should provide some qualitative understanding. The dominant Feynman diagram which gives particle production is one where the two valence quarks emit gluons and they fuse to form a single on mass shell gluon. This diagram is different than the hard gluon-gluon scattering diagram where the gluon which is produced by fusion would produce two on mass shell gluons or where the two gluons produced by the valence quark would scatter to make two on mass shell gluons.

The mechanism relevant for our equations gives $dN/d^2 p_T \sim \alpha_S^3/p_T^4$ which is one power of α_S less than the mechanism of the standard parton model. The reason why we get a contribution from our process is because the gluons which are part of the distribution functions which scatter to produce the on mass shell gluon are off mass shell and carry intrinsic p_T. Of course at a momentum where the gluon distribution begins to fall more rapidly than $1/p_T^2$, the standard parton model hard scattering process begins to dominate.

The A dependence of the gluon production can be estimated by scaling arguments. The total gluon production should scale as A up to logarithms of A. The high p_T tail where $p_T >> A^{1/6}$ scales as $A^{4/3}$.

5   Summary

The problem of the gluon distribution functions and their implications for nucleus-nucleus scattering is potentially solvable in QCD for thick enough nuclei. The gluon distribution may also be determined for hadrons at sufficiently small x. This involves constructing the hadron wavefunction at small x. To do this construction one needs classical fields, and renormalization group equations which determine the induced charge due to gluon radiation.

For the two nucleus problem, there are a variety of issues related to those described above which must be integrated into the collision problem. For the classical problem, it is not really clear how to solve the evolution equations for the classical fields. This is presumably a quite complicated numerical problem.

Of course the question remains for nuclei of $A \sim 200$ and realistic collider energies, is A large enough so that we can make our weak coupling approximations? Although our methods might provide some insight into the dynamic of nuclear collisions and structure functions, our ultimate goal is to describe realistic nuclei. Assuming a multiplicty of 3000 particles per unit rapidity for gold-gold collisions at RHIC, we estimate a typical p_T for gluons of order 1-2 GeV. This is marginally consistent with weak coupling analysis. On the other hand at LHC, the typical p_T might be as high as 3-5 GeV, and weak coupling methods should be fairly reliable.

Acknowledgments

I gratefully acknowledge the contributions of my coworkers on this project: Alejandro Ayala-Mercado, Jamal Jalilian-Marian, Alex Kovner, Raju Venugopalan and Heribert Weigert. This work was supported under DOE contracts DOE High Energy DE-AC02-83ER40105 and DOE-Nuclear DE-FG02-87ER-40328.

References

1. K. Geiger and B. Mueller, *Nucl. Phys.* **B369**, 600 (1992); **CERN-TH 7313/94** (1994).
2. L. McLerran and R. Venugopalan, *Phys. Rev.* **D49**, 2233 (1994); **D49**, 3352 (1994); **D50**, 2225 (1995);
3. A. Kovner, L. McLerran and H. Weigert, *Phys. Rev.* **D52**, 6231 (1995).
4. A. H. Mueller, *Nucl. Phys,* **B307**, 34 (1988); **B317**, 573 (1989); **B335**, 115 (1990).
5. A. Kovner, L. McLerran and H. Weigert, work in progress.

6. E. A. Kuraev, L. N. Lipatov and V. S. Fadin, *Sov. Phys. JETP* **45**, 2 (1977).
7. J. D. Bjorken, *Phys. Rev.* **D27**, 140 (1983).
8. A. Ayala-Mercado, J. Jalilian-Marian, L. McLerran and R. Venugopalan, *Phys. Rev.* **D52**, 3809 (1985).
9. A. Ayala-Mercado, J. Jalilian-Marian, L. McLerran and R. Venugopalan, *Phys. Rev.* **D53**, 458 (1996).

THERMALIZATION OF SOFT FIELDS

C. Greiner

Institut für Theoretische Physik, Universität Giessen,
D-35392 Giessen, Germany

B. Müller

Department of Physics, Duke University, Durham,
North Carolina, 27708-0305, USA

We discuss the semi-classical limit of the equation of motion for the long-distance ("soft") modes in a scalar quantum field theory when the hard modes of the field are in thermal equilibrium, separated by a momentum cut-off $|\mathbf{p}| \leq k_c \ll T$. In particular we adress the question of how thermalization is achieved microscopically. The answer will lead to a stochastic interpretation for the effective action, resulting in dissipation and decoherence for the evolution of the soft modes.

1 Introduction

Solutions of the classical field equations in Minkowski space have been widely used in recent years to describe long-distance properties of quantum fields that require a nonperturbative analysis. These applications include: the diffusion rate of the topological charge in the electroweak gauge theory [1], the thermalization rate of nonabelian gauge fields [2], as well as a wide range of cosmological problems, such as thermalization, decoherence, and structure formation [3,4,5]. Many of these studies are concerned with dynamical properties of quantum fields near thermal equilibrium. Classical treatments of the long-distance dynamics of bosonic quantum fields at high temperature are based on the observation that the average thermal amplitude of low-momentum modes is large. For a weakly coupled quantum field the occupation number of a mode with wave vector $\mathbf{p}$ and frequency $\omega_{\mathbf{p}}$ is given by the Bose distribution

$$n(\omega_{\mathbf{p}}) = \left(e^{\hbar \omega_{\mathbf{p}}/T} - 1 \right)^{-1}. \qquad (1)$$

For temperatures T much higher than the (dynamical) mass scale m^* of the quantum field, the occupation number becomes large and approaches the classical equipartition limit

$$n(\omega_{\mathbf{p}}) \xrightarrow{|\mathbf{p}| \to 0} \frac{T}{m^*} \gg 1. \qquad (2)$$

At a closer look, however, the cogency of this heuristic argument suffers considerably. The thermodynamics of a classical field is only defined if an ultraviolet cut-off k_c is imposed on the momentum $\mathbf{p}$ such as a finite lattice spacing a. As an example consider the dynamical mass gap

$$m^{*2} \sim \frac{g^2}{2} \int \frac{d^3p}{(2\pi)^3} \frac{1}{\omega_{\mathbf{p}}} n(\omega_{\mathbf{p}}) \sim \frac{g^2}{2} T\hbar k_c \ , \qquad (3)$$

in the high-temperature limit (and for $k_c \gg m^*$), which contrasts with the results $m^{*2} \sim g^2 T^2$ in the full quantum theory. Many, if not most, thermodynamical properties of the classical field depend strongly on the value of the cut-off parameter k_c and diverge in the continuum limit ($k_c \to \infty$). In a correct semi-classical treatment of the soft modes the hard modes thus cannot be neclgected, but it should incorporate their influence in a consistent way. The aim of our work is to derive a correct effective action for the soft modes when integrating out the hard modes assumed to stay in thermal equilibrium. Other questions also arise: What coupling constant g should be used in the classical calculation? How does the temperature enter into the classical equation? Which mechanism ensures that the attained temperature equals that of the underlying thermal field theory, if the classically described modes thermalize due to their interaction? On the other hand, if the low-momentum field modes are governed by truly nonperturbative physics, such as in the case of non-abelian gauge theories at high temperature, we have to rely on a numerical treatment of the long-distance dynamics on the basis of an effective action $S_{\text{eff}}^{k_c}[\phi]$. The major advantage then is that all the short-distance dynamics that can be treated perturbatively is already contained in the effective action, thus the numerical evaluation can concentrate on the nonperturbative sector alone where quantal corrections among the soft modes are of minor importance while classical (thermal) effects dominate.

Our approach leads us to the formulation of an effective real-time dynamics for the soft modes of the quantum field in terms of a stochastic, dissipative action. The resulting stochastic forces describe the exchange of energy and other quantum numbers with the perturbatively treated hard modes, whereas the additional dissipative terms describe the eventual approach to thermal equilibrium of the soft modes. In the first Section we summarize the main conceptual ideas of obtaining an effective action (or influence functional) for a quantum mechanical system X interacting with a heat bath Q. Generalizing these ideas we then report the outcome of our investigation for a self-coupled scalar quantum field. Most important for a clear understanding of the arising new terms is then how to obtain the proper Markovian approximation of the corresponding nonlocal terms and how dissipation emerges in the semi-classical equation of motion for the soft modes. We conclude with some speculations about the generalization to gauge theories.

2   The influence functional technique

A quantum mechanical system X (described by the variable x) interacts with a bath Q (described by the variable q). At some specified initial time $t = t_i$ the combined system $(X \cup Q)$ is described by the full density matrix

$$\rho_{X \cup Q}(t_i) = \rho_{X \cup Q}(x_i, q_i, x_i', q_i'; t_i). \tag{4}$$

By formally integrating out the bath degrees of freedom q one obtains an effective interaction $S_{\mathrm{eff}}[x(s), x'(s)]$ describing the evolution of the system degrees of freedom $x(s)$. For this one assumes that the initial density matrix is uncorrelated in its variables x and q, i.e. $\rho_{X \cup Q}(t_i) = \rho_X(t_i) \otimes \rho_Q(t_i)$. Introducing the reduced density matrix ρ_r as

$$\rho_r(x, x'; t) = \mathrm{Tr}_{(q)} \{\rho(x, q; x', q'; t)\}, \tag{5}$$

one finds that its evolution in time can be put in the general form

$$\rho_r(t) = \int dx_i \int dx_i' \int_{x_i}^{x} \mathcal{D}x \int_{x_i'}^{x'} \mathcal{D}x' \, e^{\frac{i}{\hbar}(S_x[x] - S_x[x'])} \, e^{\frac{i}{\hbar} S_{IF}[x,x']} \rho_r(t_i) \tag{6}$$

where the influence functional $S_{IF}[x, x']$ [6] is given as

$$e^{\frac{i}{\hbar} S_{IF}[x,x']} = \int dq \int_{q_i}^{q} \mathcal{D}q \int_{q_i'}^{q} \mathcal{D}q' \, e^{\frac{i}{\hbar}(S_Q[q] - S_Q[q'] + S_{\mathrm{int}}[x,q] - S_{\mathrm{int}}[x',q'])} \rho_Q(t_i). \tag{7}$$

Thus one is led to say that ρ_r evolves in time according to the effective interaction

$$S_{\mathrm{eff}}[x, x'] = S_x[x] - S_x[x'] + S_{IF}[x, x']. \tag{8}$$

Expanding S_{IF} up to second order in x and x', the general structure of S_{IF} is given by [6]

$$\begin{aligned}
S_{IF}[x, x'] &\approx \int_{t_i}^{t} ds \, F(s)(x(s) - x'(s)) \\
&+ \frac{1}{2} \int_{t_i}^{t} ds_1 ds_2 \, \big(x(s_1) - x'(s_1)\big) R(s_1, s_2) \big(x(s_2) + x'(s_2)\big) \\
&+ \frac{i}{2} \int_{t_i}^{t} ds_1 ds_2 \, \big(x(s_1) - x'(s_1)\big) I(s_1, s_2) \big(x(s_2) - x'(s_2)\big). \tag{9}
\end{aligned}$$

If $\mathcal{L}_{\mathrm{int}}$ has the form $\mathcal{L}_{\mathrm{int}}[x, q] = x(t) \, \Xi(q(t))$, the "force" F, the response function R, and I are obtained as

$$F(s) = \langle \Xi(q(s)) \rangle_{\rho_Q}$$

$$R(t_1, t_2) = \frac{i}{\hbar} \langle [\Xi(q(t_1)), \Xi(q(t_2))] \rangle \, \theta(t_1 - t_2)$$

$$I(t_1, t_2) = \frac{1}{2\hbar} \Big(\langle \{\Xi(q(t)1)), \Xi(q(t_2))\} \rangle - 2 \langle \Xi(q(t_1)) \rangle \langle \Xi(q(t_2)) \rangle \Big) \ . \quad (10)$$

If the system behaves quasi-classically, the density matrix ρ_r becomes nearly diagonal. Then from (6), the major paths contributing to the evolution of the reduced density matrix are obtained by extremizing the effective action $S_{\text{eff}}[x, x']$. Introducing the variables $\bar{x} = (x + x')/2$, $\Delta = x - x'$, from (9) one has by expansion up to second order

$$\frac{i}{\hbar} S_{\text{eff}}(\bar{x}, \Delta) = \frac{i}{\hbar} \Delta \otimes \left(\frac{\delta S_X[\bar{x}]}{\delta \bar{x}(s)} + F(s) + R \otimes \bar{x} \right) - \frac{1}{2\hbar} \Delta \otimes I \otimes \Delta \ . \quad (11)$$

Thus the *average* quasi-classical equation of motion becomes

$$-\frac{\delta S_X[\bar{x}]}{\delta \bar{x}(s)} - F(s) - \int_{t_i}^{s} ds' R(s, s') \bar{x}(s') = 0 \ . \quad (12)$$

This equation, in return, has to be interpreted as an average over random, fluctuating forces. The last contribution in (11) leads to decoherence because any path contributing with sizeable $|\Delta| > 0$ over past time becomes *exponentially* suppressed—hence, the imaginary part I drives the system to quasi-classical behavior. For *short* periods in time, however, fluctuations in Δ can appear stochastically. These act as random "kicks" on the actual trajectory. To see this in more detail, one defines the real stochastic influence action

$$\tilde{S}_{\text{IF}}[x, x', \xi] = \text{Re}(S_{\text{IF}}[x, x']) + \int_{t_i}^{t} ds \, \xi(s) \, (x(s) - x'(s)) \, , \quad (13)$$

where $\xi(s)$ is interpreted as an external force [7], randomly distributed by a Gaussian distribution with zero mean:

$$P[\xi(s)] = \frac{1}{N} \exp \left(-\frac{1}{2\hbar} \int_{t_i}^{t} ds_1 ds_2 \xi(s_1) I^{-1}(s_1, s_2) \xi(s_2) \right) . \quad (14)$$

The influence functional $S_{\text{IF}}[x, x']$ is regained as the characteristic functional over the average of the random forces. The imaginary part I also defines the correlation of the random forces. Accordingly, from (13) follows

$$-\frac{\delta S_X[\bar{x}]}{\delta \bar{x}(s)} - F(s) - \int_{t_i}^{s} ds' R(s, s') \bar{x}(s') = \xi(s) \quad (15)$$

the effective Langevin like equation dictating the dynamics of the system variable in the quasi-classical regime. We do expect an essentially semiclassical evolution when the excitation energy of the system (divided by $\hbar$) is large so that the oscillatory phases of paths neighboring the classical path interferes destructively. In addition, all paths where the time averaged Δ is sufficiently large will be exponentially suppressed by the decoherence factor. Hence, the system will be driven by decoherence into the quasi-classical regime, depending crucially on the magnitude $\sqrt{\langle \xi^2 \rangle} \sim \sqrt{I}$ of the random forces.

3 Effective Equation of Motion for Classical Soft Modes

Our goal is the derivation of an effective action for the (classical) dynamics in Minkowski space of the low-momentum modes of a scalar quantum field Φ near thermal equilibrium. We thus divide the Fourier components of the field Φ into low-momentum ('soft') modes with $|\mathbf{p}| \leq k_c$ and high-momentum ('hard') modes with $|\mathbf{p}| > k_c$ by defining

$$\Phi(\mathbf{p}, t) = \Phi(\mathbf{p}, t)\theta(k_c - |\mathbf{p}|) + \Phi(\mathbf{p}, t)\theta(|\mathbf{p}| - k_c) \equiv \phi(\mathbf{p}, t) + \varphi(\mathbf{p}, t). \quad (16)$$

It is clear from the introduction that the cutoff scale k_c should be much smaller than the temperature T, i.e. $k_c \ll T$. We assume that all hard modes φ are, and remain, thermally occupied by a Bose distribution n.

The influence action $S_{\mathrm{IF}}[\phi, \phi'; t_f]$ is readily generalized from the previous section as

$$e^{iS_{\mathrm{IF}}[\phi,\phi';t=t_f]} = \int d\varphi_f d\varphi_i d\varphi_i' \int_{\varphi_i}^{\varphi_f} \mathcal{D}\varphi \int_{\varphi_i'}^{\varphi_f} \mathcal{D}\varphi'$$
$$\cdot e^{i(S[\varphi]+S_{\mathrm{int}}[\phi,\varphi]-S[\varphi']-S_{\mathrm{int}}[\phi',\varphi'])} \rho_{\mathrm{h}}(\varphi_i, \varphi_i'; t_i) \quad (17)$$

where $\rho_h(\varphi_i, \varphi_i')$ is the density matrix of initial, thermal configurations of the hard field modes and

$$S_{\mathrm{int}}[\phi, \varphi] = -\int_{t_i}^{t} d^4x \left[\tfrac{1}{6}g^2 \left(\phi^3\varphi + \tfrac{3}{2}\phi^2\varphi^2 + \phi\varphi^3 \right) + \tfrac{1}{4!}g^2\varphi^4 \right]. \quad (18)$$

describes the interaction of the soft and hard fields. It turns out that one must include all diagrams up to order g^4 in the perturbative expansion, even if $g^2 \ll 1$. All non-vanishing diagrams contributing to S_{IF} up to second order in S_{int} are shown in Figure 1. The inclusion of the diagrams of order g^4 is crucial, because this is the lowest order at which the real-time effective action develops an imaginary part which gives rise to stochastic terms as well as the corresponding real part will give rise to dissipative terms.

As we argued in the introduction, the dynamics of the soft modes under the stochastic action is governed by (quasi-)classical physics, if we assume that

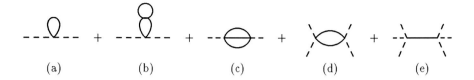

Figure 1: Feynman diagrams contributing to the influence action up to order g^4. Full lines denote hard modes and dashed lines correspond to soft modes.

the soft modes are sufficiently populated. This assumption will be warranted consistently by the full dynamics. Variation with respect to $\phi_\Delta = (\phi - \phi')$ yields the semi-classical corrections to the equation of motion. Before doing this, one has to deal with the imaginary part of the influence action. As outlined in the previous section, it has to be interpreted as a stochastic "external" force driving the soft modes[8]. One obtains the following equation of motion for the soft modes:

$$\Box\phi(x) + (\tilde{m}^2 + m^{*2})\phi(x) + \frac{\tilde{g}^2}{6}\phi^3(x) \tag{19}$$

$$+ \sum_{N=1}^{3} \frac{1}{(2N-1)!}\phi(x)^{N-1} \int_{t_i}^{t} d^4y \, \mathrm{Re}\left[\tilde{\Gamma}_{2N}^{k_c}(x-y)\right] \phi(y)^N = \sum_{N=1}^{3} \phi(x)^{N-1}\xi_N(x) \; .$$

$\Gamma_N^{k_c}$ labels the amputated $2N$-point vertex function, restricted to contributions solely from the hard modes with $|\mathbf{p}| > k_c$. Since the ultraviolet divergences of the vertex function $\Gamma_{2N}^{k_c}$ are independent of the low-momentum cut-off k_c and of the temperature, they can be absorbed in the standard counter terms for mass and coupling constant renormalization. Hence, $\tilde{m}$ and $\tilde{g}$ denote the renormalized mass and coupling constant, respectively, and $\tilde{\Gamma}_{2N}^{k_c}$ are the renormalized vertex functions. m^* describes the dynamically generated mass by means of the tadpole contributions (a) and (b) of Figure 1. Equation (19) is graphically represented in Figure 2. The terms on the right-hand side represent stochastic source, mass, etc., terms for the soft field modes. In particular, the term $\xi_1(x)$ indicates that the soft modes do not obey a source-free equation, rather, they are in constant contact with randomly fluctuating sources.

The equation of motion (19) resembles a Langevin equation analogous to the situation of quantum Brownian motion[7]. The contributions of $\mathrm{Re}\left(S_{\mathrm{IF}}^{(c,d,e)}\right)$ will be responsible for dissipation. However, as it stands, it is clearly nonlocal in time, and one has to identify the appropriate Markovian limit of this equation. To further evaluate the dissipative terms one has to insert an approximation

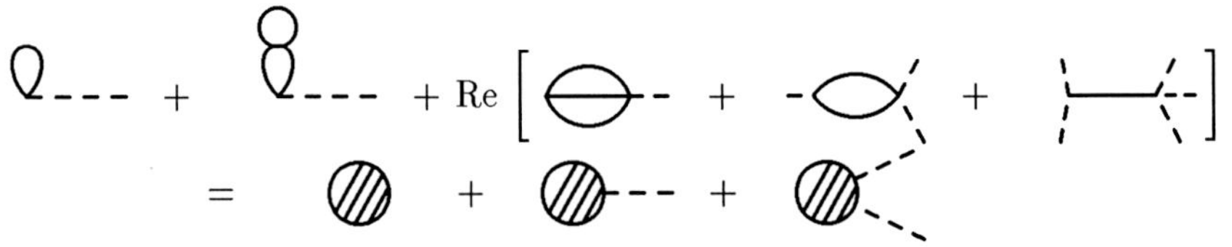

FIG.2. Graphical representation of the classical equation (19) for the soft field modes. The noise terms $\phi^{N-1}\xi_N$ are shown as blobs with $(N-1)$ external legs.

for the time dependence of the soft modes $\phi(\mathbf{y} = \mathbf{x} - \tilde{\mathbf{x}}, t_y = t - \tau)$ under the integrals. The soft modes will oscillate with a certain spectrum of frequencies. This is true also for the spatial Fourier modes $\phi(\mathbf{k}, t)$. In the weak coupling regime, when the soft modes do propagate alsmost freely, however, one expects, that they oscillate with some pronounced frequency

$$E_k \approx \sqrt{\tilde{m}^2 + m^{*2} + \mathbf{k}^2} \ . \tag{20}$$

We thus can approximately express the soft modes at earlier times $(t - \tau)$ by their value at the time t in the linear harmonic approximation

$$\phi(\mathbf{k}, t - \tau) \approx \phi(\mathbf{k}, t) \cos E_k \tau - \dot{\phi}(\mathbf{k}, t) \frac{1}{E_k} \sin E_k \tau \ . \tag{21}$$

The $\dot{\phi}(\mathbf{k}, t)$ contribution turns out to be *essential* and cannot be neglected. Our strategy is completely equivalent to the quasi-particle approximation used in the more familiar kinetic theories when evaluating the collision term in transport processes[9]: A certain mode is assumed to propagate with a specific frequency, the quasi-particle energy. This approximation is valid if the spectral function of the quasi-particle is sharply peaked at this frequency. Moreover, pursuing the analogy with standard kinetic theory, the equation of motion (19) becomes instantaneous as all soft quasi-particle modes $\phi(\mathbf{k}, t)$ are evaluated at the same time. We thus obtain the Markovian limit of eq. (19).

In an intermediate step[8] one defines the *memory* kernels, which coincides up to some factor with the amputated 2N-point vertex function $\mathrm{Re}\Gamma_N^{k_c}(\mathbf{k}, \omega)$:

$$\mathcal{M}^{(c)}(\mathbf{k}, \omega) = -\frac{\pi}{24} \ \tilde{g}^4 \ (i) \int_{k_c} \frac{d^3 q_1 d^3 q_2}{(2\pi)^6} \frac{1}{\omega_1} \frac{1}{\omega_2} \frac{1}{\omega_3} \theta(|\mathbf{k} - \mathbf{q}_1 - \mathbf{q}_2| - k_c) \tag{22}$$

$$\times \{ \ [(1 + n_1)(1 + n_2)(1 + n_3) - n_1 n_2 n_3] \delta(\omega - \omega_1 - \omega_2 - \omega_3)$$

$$+ \ [(1 + n_1) n_2 (1 + n_3) - n_1 (1 + n_2) n_3] \delta(\omega - \omega_1 + \omega_2 - \omega_3)$$

$$+ \ [n_1 (1 + n_2)(1 + n_3) - (1 + n_1) n_2 n_3] \delta(\omega + \omega_1 - \omega_2 - \omega_3)$$

$$+ \quad [n_1 n_2(1 + n_3) - (1 + n_1)(1 + n_2)n_3]\,\delta(\omega + \omega_1 + \omega_2 - \omega_3)$$
$$+ \quad [(1 + n_1)(1 + n_2)n_3 - n_1 n_2(1 + n_3)]\,\delta(\omega - \omega_1 - \omega_2 + \omega_3)$$
$$+ \quad [(1 + n_1)n_2 n_3 - n_1(1 + n_2)(1 + n_3)]\,\delta(\omega - \omega_1 + \omega_2 + \omega_3)$$
$$+ \quad [n_1(1 + n_2)n_3 - (1 + n_1)n_2(1 + n_3)]\,\delta(\omega + \omega_1 - \omega_2 + \omega_3)$$
$$+ \quad [n_1 n_2 n_3 - (1 + n_1)(1 + n_2)(1 + n_3)]\,\delta(\omega + \omega_1 + \omega_2 + \omega_3)\}$$

where $\omega_1 = \omega_{\mathbf{q}_1}, \omega_2 = \omega_{\mathbf{q}_2}, \omega_3 = \omega_{\mathbf{k} - \mathbf{q}_1 - \mathbf{q}_2}$ and $n_\alpha = n(\omega_{\mathbf{q}_\alpha})$, and

$$\mathcal{M}^{(d)}(\mathbf{k}, \omega) = -\frac{\pi}{8}\tilde{g}^4\,(i)\,\int_{k_c} \frac{d^3 q}{(2\pi)^3}\frac{1}{\omega_1}\frac{1}{\omega_2}\theta(|\mathbf{k} - \mathbf{q}| - k_c) \tag{23}$$
$$\times\{\quad [(1 + n_1)(1 + n_2) - n_1 n_2]\,\delta(\omega - \omega_1 - \omega_2)$$
$$+ \quad [n_1(1 + n_2) - (1 + n_1)n_2]\,\delta(\omega + \omega_1 - \omega_2)$$
$$+ \quad [(1 + n_1)n_2 - n_1(1 + n_2)]\,\delta(\omega - \omega_1 + \omega_2)$$
$$+ \quad [n_1 n_2 - (1 + n_1)(1 + n_2)]\,\delta(\omega + \omega_1 + \omega_2)\}$$

where $\omega_1 = \omega_{\mathbf{q}}, \, \omega_2 = \omega_{\mathbf{k} - \mathbf{q}}$, and

$$\mathcal{M}^{(e)}(\mathbf{k}, \omega) = -\frac{\pi}{12}\tilde{g}^4\,(i)\frac{1}{\omega_{\mathbf{k}}}\,\theta(|\mathbf{k}| - k_c)\cdot\{[(1 + n(\omega_{\mathbf{k}})) - n(\omega_{\mathbf{k}})]\,\delta(\omega - \omega_{\mathbf{k}})$$
$$+ [n(\omega_{\mathbf{k}}) - (1 + n(\omega_{\mathbf{k}}))]\,\delta(\omega + \omega_{\mathbf{k}})\}\,. \tag{24}$$

The memory kernels $\mathcal{M}^{(i)}(\mathbf{k}, \omega)$ are directly related to the discontinuities of the respective self-energy insertions of the hard modes. Each kernel is composed of two contributions: the direct part $\Gamma_d(\mathbf{k}, \omega)$, or the "loss" term, and the inverse part $\Gamma_i(\mathbf{k}, \omega)$, or "gain" term:

$$i\mathcal{M}(\mathbf{k}, \omega) \propto \Gamma_d(\mathbf{k}, \omega) - \Gamma_i(\mathbf{k}, \omega)\,, \tag{25}$$

where Γ_d and Γ_i, respectively, are positive and can be identified with the decay rate of a quasi-particle with momentum $\mathbf{k}$ and energy ω into the allowed open channels or the production rate due to the inverse processes.

When inserting the approximation (21) for the soft quasi-particle modes into the semi-infinite time integrals, one notices that the contributions with an odd power in ϕ contain the on-shell reaction rates (22-24). Intuitively we expect them to describe the dissipative aspects of the equation of motion for the soft modes as energy is transferred by scattering (or production) processes between the soft and hard modes. Since this interpretation is most obvious for the first term (diagram (c)), we discuss this term in some more detail. Fourier transforming (19) for the mode $\mathbf{k}$, one ends up with

$$\ddot{\phi}(\mathbf{k}, t) + \mathbf{k}^2\phi(\mathbf{k}, t) + \tilde{m}^2\phi(\mathbf{k}, t) + \frac{\tilde{g}^2}{6}\otimes\phi^3 \tag{26}$$
$$+\phi(\mathbf{k}, t)P\int\frac{d\omega}{2\pi}\frac{i\mathcal{M}^{(c)}(\mathbf{k}, \omega)}{E_k - \omega} + \dot{\phi}(\mathbf{k}, t)\frac{i\mathcal{M}^{(c)}(\mathbf{k}, E_k)}{2E_k} + \ldots = \xi_1(\mathbf{k}, t) + \ldots\,.$$

The first additional term can be readily interpreted as momentum dependent contribution to the mass term. The second term has the familiar structure of a velocity dependent damping term, i.e. $\eta\dot{\phi}(\mathbf{k}, t)$, where the friction coefficient

$$\eta^{(c)}(\mathbf{k}) = \frac{i\mathcal{M}^{(c)}(\mathbf{k}, E_k)}{2E_k} \geq 0 \tag{27}$$

is in fact positive. It is exactly related to the rate $(\Gamma_d - \Gamma_i)$ with which a small disturbance relaxes towards equilibrium by coupling to the various open channels. The damping rate for the soft modes in this particular channel follows as $\gamma^{(c)}(\mathbf{k}) = \frac{1}{2}\eta^{(c)}(\mathbf{k})$. In the case $k_c \to 0$, the dynamical mass $(m^*)^2$ to order $\tilde{g}^2$ is given by the tadpole diagram (a) in Figure 1: $(m^*)^2 = \tilde{g}^2 T^2/24$, which is the dominant contribution to the mass in the high temperatue limit. Inserting this for the dispersion relation of the energy, i.e. $E_0 = m^*$, the damping coefficient $\eta^{(c)}$ is exactly related to twice the damping rate of plasmons in Φ^4-theory [10]:

$$\eta^{(c)} = 2\gamma(m^*, 0) = \frac{\tilde{g}^3 T}{32\sqrt{24}\pi} . \tag{28}$$

Following the above observation a similar interpretation of the two other dissipative terms holds: (1) The principal value contributions give rise to a (generally momentum dependent) vertex renormalization and to a new (again momentum dependent) six-point vertex; (2) the on-shell contributions can be associated with damping processes involving two hard particles and two soft modes or involving one hard particle interacting with three soft modes, respectively. The friction coefficients can be approximately stated as [8]

$$\eta^{(d)}(\mathbf{k}, t) \approx \frac{1}{4E_k} \int^{k_c} \frac{d^3k_1}{(2\pi)^3} |\phi(\mathbf{k}_1, t)|^2 \tag{29}$$
$$\times \left(i\mathcal{M}^{(d)}(\mathbf{k} + \mathbf{k}_1, E_k + E_{k_1}) + i\mathcal{M}^{(d)}(\mathbf{k} + \mathbf{k}_1, E_k - E_{k_1}) \right) ,$$

$$\eta^{(e)}(\mathbf{k}, t) \approx \frac{3}{4E_k} \int^{k_c} \frac{d^3k_1 d^3k_2}{(2\pi)^6} |\phi(\mathbf{k}_1, t)|^2 |\phi(\mathbf{k}_2, t)|^2 \tag{30}$$
$$\times \left(i\mathcal{M}^{(e)}(\mathbf{k} + \mathbf{k}_1 + \mathbf{k}_2, E_k + E_{k_1} + E_{k_2}) \right.$$
$$+ i\mathcal{M}^{(e)}(\mathbf{k} + \mathbf{k}_1 + \mathbf{k}_2, E_k - E_{k_1} - E_{k_2})$$
$$\left. + 2i\mathcal{M}^{(e)}(\mathbf{k} + \mathbf{k}_1 + \mathbf{k}_2, E_k + E_{k_1} - E_{k_2}) \right) .$$

Dissipation thus corresponds directly to the imaginary parts of the self energy insertions $\mathrm{Re}\Gamma_{2N}^{k_c}(\mathbf{k}, \omega)$ in the response term of eq. (19), as one would have expected from standard finite temperature field theory.

The fluctuation-dissipation-theorem ensures that the soft modes approach thermal equilibrium precisely at the temperature T of the hard modes, when they evolve under the equation of motion (19). In particular the noise terms will continuously 'heat' the system whereas the dissipative part of the response function counteracts. Equilibrium is achieved when the system has thermalized to the temperature dictated by the bath. This aspect guarantees that the soft modes will become populated thermally, i.e. their amplitudes become large, which justifies our basic assumption.

Summarizing, we restate the quasi-classical equation of motion for the soft modes $\phi(\mathbf{k}, t)$ in their approximate instantaneous form:

$$\frac{\partial^2 \phi}{\partial t^2} + \left(\mathbf{k}^2 + \tilde{m}^2 + \sum_{i=a,b,c} \mu_{1,k_c}^i\right) \phi + \left(\frac{\tilde{g}^2}{6} + \mu_{2,k_c}^{(d)}\right) \otimes \phi^3 +$$

$$\mu_{3,k_c}^{(e)} \otimes \phi^5 + \sum_{i=c,d,e} \eta_{k_c}^{(i)} \dot{\phi} \approx \sum_{N=1}^{3} \xi_n \otimes \phi^{N-1} , \tag{31}$$

where $\otimes$ denotes a convolution in momentum space. The coefficients μ_i depend on $\mathbf{k}$ and t, as well as on the momentum cut-off k_c. In a massless theory an explicit calculation shows that the leading cut-off dependence of μ_1^k derives from the thermal one-loop contribution [11]:

$$\mu_{1,k_c}^{(a)} \longrightarrow \frac{\tilde{g}^2 T^2}{24} - \frac{\tilde{g}^2 k_c T}{4\pi} \qquad \text{for } \tilde{m} = 0 . \tag{32}$$

As discussed by Bödeker et al [11], the cut-off dependence of μ_1 is exactly balanced by the dynamically generated mass due to the self-interaction among the soft modes, if these are also in thermal equilibrium. The same should also hold for the other k_c-dependent constants in (31).

4 Conclusions and Outlook

In summary, we have derived a consistent set of temporally local transport equations for the long-distance modes of a self-coupled scalar field. Our approach is based on the analysis of time scales showing that the soft field modes, at high temperature and weak coupling, oscillate on a time scale shorter than the characteristic damping time. A temporally local (Markovian) equation can only be derived for the Fourier components of the soft field modes, i.e. the occupation amplitudes, but not for the field itself. The dominant damping mechanism for the infrared field mode ($k_c = 0$) is generally given by the two-loop "sunset" diagram, corresponding to emission or absorption of the soft mode on hard thermal quanta.

Non-Abelian gauge fields at finite temperature would present an interesting application of the methods discussed so far, because their infrared sector remains non-perturbative even at high temperature due to the absence of perturbative screening of magnetostatic fields. This phenomenon has been widely studied in the Euclidean formalism, and its resolution by a sequence of effective actions has been proposed. The emergence of a finite correlation lengths in the magnetostatic sector is essentially a classical phenomenon which occurs at the momentum scale $g^2 T$ [12]. It should be fully contained in the classical effective action obtained by integrating out hard modes with momenta $|\mathbf{p}| > k_c > g^2 T$. The problem in gauge theories is that one cannot employ a low-momentum cut-off k_c because it is not gauge invariant. An infrared cut-off scheme that violates neither gauge nor Lorentz invariance is therefore preferable.

References

1. D.Yu. Grigoriev and V. A. Rubakov, *Nucl. Phys.* **B299**, 6719 (1988); D.Yu. Grigoriev, V. A. Rubakov, and M. E. Shaposhnikov, *Nucl. Phys.* **B326**, 737 (1989).
2. B. Müller and A. Trayanov, *Phys. Rev. Lett.* **68**, 3387 (1992); T. S. Biró, C. Gong, and B. Müller, *Phys. Rev.* **D52**, 1260 (1995).
3. B. L. Hu, J. P. Paz, and Y. Zhang, *Phys. Rev.* **D45**, 2843 (1993); *Phys. Rev.* **D47**, 1576 (1993).
4. R. H. Brandenberger and A. C. Davis, *Phys. Lett.* **B332**, 305 (1994).
5. N. Turok, *Phys. Rev. Lett.* **63**, 2625 (1989); D. N. Spergel, N. Turok, W. H. Press, and B. S. Ryden, *Phys. Rev.* **D43**, 1038 (1991).
6. R. Feynman and F. Vernon, *Ann. Phys. (N.Y.)* **24**, 118 (1963); R.P. Feynman and A.R. Hibbs, 'Quantum Mechanics and Path Integrals', McGraw-Hill Inc. (1965).
7. A. Schmid, *J. Low Temp.* **49**, 609 (1982); A.O. Caldeira and A.J. Leggett, *Physica* **121A**, 587 (1983).
8. C. Greiner and B. Müller, 'Classical Fields near Thermal Equilibrium', Giessen preprint (1996), subm. to . *Phys. Rev.* **D**.
9. S.R. DeGroot, W.A. van Leeuwen and Ch.G. van Weert, 'Relativistic Kinetic Theory', North-Holland-Publishing-Company (1980).
10. R. P. Parwani, *Phys. Rev.* **D45**, 4695 (1992).
11. D. Bödeker, L. McLerran, and A. Smilga, *Phys. Rev.* **D52**, 4675 (1995).
12. A. D. Linde, *Phys. Lett.* **B96**, 239 (1980); D. J. Gross, R. D. Pisarski, and L. G. Yaffe, *Rev. Mod. Phys.* **53**, 43 (1981).

ELECTROMAGNETIC SIGNALS FROM DECONFINED MATTER RESULTING FROM ULTRARELATIVISTIC HEAVY-ION COLLISIONS[a]

B. KÄMPFER[1,2], A. PESHIER[2], M. HENTSCHEL[1], G. SOFF[1]

[1]*Institut für Theoretische Physik, TU Dresden,*
01062 Dresden, Germany

[2]*Forschungszentrum Rossendorf, Institut für Kern- und Hadronenphysik,*
PF 510119, 01314 Dresden, Germany

O.P. PAVLENKO
Institute for Theoretical Physics,
252143 Kiev - 143, Ukraine

Electromagnetic radiation off strongly interacting matter resulting from ultrarelativistic heavy-ion collisions is estimated. We consider virtual photons (dileptons) and real photons produced in the course of cooling thermalized matter. The chemical evolution of initially undersaturated parton matter coupled to the longitudinal and transverse expansion is followed. Due to the strong transverse expansion the so-called $M_\perp$ scaling is restored. Some estimates of contributions of the pre-equilibrium stage are presented.

1 Introduction

Electromagnetic signals (leptons and photons) can leave nearly undisturbed systems of strongly interacting matter. Therefore, despite of the low production cross sections, the lepton pairs and photons are considered as useful probes of the dynamics of heavy-ion collisions.[1,2] In particular, it is expected that the electromagnetic radiation off highly excited matter, produced in high-energy nuclear collisions, can signalize whether deconfined matter (i.e., the quark-gluon plasma) is transiently created. Unfortunately, there are many sources of dileptons and photons:

(i) Hard first-chance parton collisions produce direct radiation, namely the Drell-Yan yield; in addition, at high beam energies, a copious charm and beauty production is expected,[3,4] which causes via semileptonic decays a strong feeding of the lepton channel.

(ii) The pre-equilibrium contribution is still a matter of debate, but there are some hints that this component might not be negligible (see below).

(iii) What we are mostly interested in is the radiation from thermalized parton (or deconfined) matter.

---

[a]This contribution is dedicated to Walter Greiner on the occasion of his 60*th* birthday.

(iv) The hadronizing parton matter will also radiate electromagnetic signals. Theoretical estimates are still hampered by poor knowledge of the nature of the confinement transition.

(v) Hadronic reactions cause a strong radiation component. A not yet convincingly resolved issue concerns the relation of the yield from this stage to the deconfined matter yield.

(vi) Hadronic decays also contribute substantially to the full spectrum.

This causes problems in unfolding the space-time integrated spectra and in identifying doubtless a possible contribution from deconfined matter. One has therefore to employ sufficiently realistic models of the reaction dynamics in order to estimate the various contributions in different phase space regions.

In the present note we report mainly results of calculations relying on a hydrodynamical evolution scenario for thermalized matter which includes also chemical equilibration processes on the parton level. A few remarks on pre-equilibrium spectra are added.

2 Spectra of dileptons

Details of our dynamical model for thermalized matter are described in reference.[5] In a nut shell: Transverse expansion is superimposed on scale-invariant longitudinal expansion. The transverse expansion is dealt with our global hydrodynamics.[6] Since deconfined matter is predicted to be created in a chemical off-equilibrium state[3,7,8] we follow the evolution of thermalized matter by means of local rate equations. We assume such a fast equilibration that at confinement temperature a chemical equilibrium is reached, and we utilize the standard procedure to follow the confinement transition of hadronizing matter. In the hadronic phase we include the recently proposed effective form factor[9] for describing the dilepton rates. It is assumed that the contributions from the hadronic decays after freeze-out can be subtracted from experimental spectra, so this component is not included in our model. We utilize here a bag model equation of state in the deconfined phase, but we reduce the latent heat by scaling up the effective number of degrees in the hadron phase (see reference[5] for details).

We now present the results of our calculations. The initial conditions for the thermalized era are characterized by the time $\tau_0 = 0.32$ fm/c, temperature $T_0 = 550$ MeV, chemical potential $\mu = 0$, gluon fugacity $\lambda_g^0 = 0.5$ or 0.25, quark fugacity $\lambda_q^0 = \frac{1}{5}\lambda_g^0$ (unless $\lambda_{q,g} = 1$), transverse radius $R_\perp^0 = 7$ fm, transverse velocity $v_\perp^0 = 0$. Fugacity means here phase space occupancy.

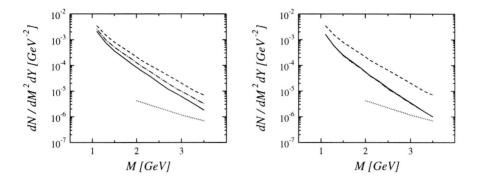

Figure 1: Invariant mass spectra of dileptons for initial temperatures $\hat{T}_0 = T_0(\lambda_g^0)^{-1/3}$ (left panel, corresponding to constant entropy density s_0) and $\hat{T}_0 = T_0(\lambda_g^0)^{-1/4}$ (right panel, corresponding to constant energy density e_0), and $\lambda_q^0 = \frac{1}{5}\lambda_g^0$ when ever $\lambda_g^0 < 1$. The dotted lines depict the Drell Yan background, while the dashed ($\lambda_g^0 = 1$), full ($\lambda_g^0 = 0.5$) and dot-dashed lines ($\lambda_g^0 = 0.25$) display the total yields from the thermal era (parton gas plus mixed phase plus hadron gas).

2.1   Invariant mass spectra

Figure 1 displays the invariant mass spectrum for all thermal contributions summed and the Drell Yan yield. The latter one is for RHIC conditions ($\sqrt{s} = 200$ GeV) and scaled for central collisions of nuclei with mass number $A = 200$ by $A^{4/3}$; Duke-Owens structure functions, set 1.1 with K factor 2, are used. Figure 1 should be compared with figure 3 in reference,[5] where T_0 is kept fixed. Here we keep the initial entropy or energy density fixed, therefore the actual initial temperatures are $\hat{T}_0 = T_0(\lambda_g^0)^{-1/3}$ or $\hat{T}_0 = T_0(\lambda_g^0)^{-1/4}$. One can observe in figure 1 that indeed the smaller initial fugacity is nearly compensated by the higher temperature. We display here the spectra only in the relevant region up to or slightly above the J/ψ. At much higher invariant mass, say $M > 10$ GeV, the various curves cross and, despite the lower initial fugacity, the yield related to highest initial temperature (here for $\lambda_g^0 = 0.25$) is largest. So the expectation,[10] that a smaller fugacity is compensated by higher temperature, can be considered as a rough approximation. It should be noticed that the transverse expansion for the given initial temperatures is essential to reduce the life time of the hadron phase, so that the yield from the deconfined stage shines out.

2.2 Transverse momentum spectra

The featureless continuum spectrum does not contain too much information and it seems difficult to read off the underlying dominating radiation source (as demonstrated in reference, [5] a dominating hadron gas source could show structures related to hadronic resonances). Therefore, one should look for other observables. One possibility is to explore peculiarities of the transverse spectra at fixed rapidity Y and fixed transverse mass $M_\perp = \sqrt{M^2 + q_\perp^2}$ as a function of the transverse pair momentum $q_\perp$.

In an initially strongly quark-undersaturated plasma one might expect that the electromagnetic basic process for dilepton production, $q\bar{q} \to \gamma^* \to \mu^+\mu^-$, is not the dominating reaction, but the Compton like reactions $qg \to q\gamma^* \to q\mu^+\mu^-$ and $\bar{q}g \to \bar{q}\gamma^* \to \bar{q}\mu^+\mu^-$ dominate. In figure 2 these yields are displayed together with the other lowest-order α_s annihilation process $q\bar{q} \to g\gamma^* \to g\mu^+\mu^-$. Here the chemical equilibration process is assumed to be fast or slow. Despite the quark undersaturation the electromagnetic annihilation still dominates unless near the kinematical boundary, where the Compton process becomes important. Figure 2 indicates that, for rough estimates, the electromagnetic annihilation within accuracy of factor 2 is sufficient to calculate the rates. Within this accuracy also the $M_\perp$ scaling (i.e., the $q_\perp$ independence of the spectra with $M_\perp = const$ and $Y = const$) is restored due to the transverse motion, see figure 3. As well known, both the transverse expansion and hadronic form factors usually destroy the $M_\perp$ scaling. However, for the given initial conditions the hadronic era is shortened, so that the dominating deconfinement phase displays as expected the scaling property approximately (cf. however reference [11]).

Notice that regularization procedures of the α_s processes for dileptons radiated off chemical non-equilibrium parton matter need further investigations. [12]

2.3 Rapidity dependence

While the above presented results are for the midrapidity region $Y \approx 0$ and for vanishing chemical potential, one can also explore the rapidity dependence of dileptons from baryon-rich matter. [13] In this respect we refer the interested reader to references. [14,15]

3 Photons

Along the above lines the photon spectra have been calculated. [6] In contrast to the dileptons, these photons, not stemming from hadronic decays,

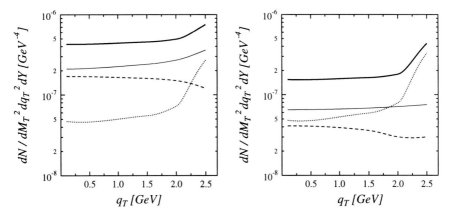

Figure 2: Transverse momentum dilepton spectra for a evolving parton gas with initial conditions $T_0 = 550$ MeV and $\lambda_g^0 = 0.25$; $M_\perp = 2.6$ GeV. The strong coupling parameter is $\alpha_s = 0.3$. Left (right) panel: fast (slow) chemical equilibration is assumed. Thin (dashed, dotted) lines: contributions from the electromagnetic annihilation (QCD annihilation, QCD Compton like) process; heavy full lines: sum of these contributions. This figure should be compared with figure 6 in reference. [5]

are still searched for. The recent experimental attempts resulted only in upper bounds. [16] One possible way to separate a possible thermal component is to analyze diphotons from the lowest-order process $q\bar{q} \rightarrow \gamma\gamma$ and to exclude the hadronic decay products by kinematical cuts. Results of our implementation of the diphoton rates [17] in the above evolution scenario will be presented elsewhere. It should be noticed that the previous WA80 experiment and the present WA93 configuration are so highly segmented that such measurements should be easily accessible.

4 Pre-equilibrium dileptons

The recent studies [18,19] point to rather strong contributions from the pre-equilibrium stage, i.e., the time when first-chance collisions had happened and the main part of entropy is produced and the initial coherence of the nuclear parton wave function is destroyed, but the parton distributions does not yet look like a thermal one.

To pin down the relation of the thermal yield and the pre-equilibrium rate we employ here a schematic model. First we display in figure 4 the time evolution of the dilepton yield from a thermally equilibrating parton gas. [18] It is seen that the relaxation time scale influences strongly the yield. The heavy dots indicate the time when roughly 75% of the total yield is produced;

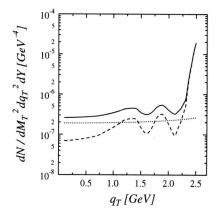
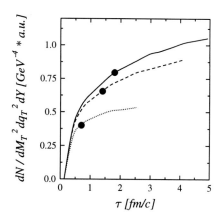

Figure 3: Transverse momentum spectra of dileptons for initial conditions $T_0 = 550$ MeV and $\lambda_g^0 = 0.25$ for $M_\perp = 2.6$ GeV. Dashed (dotted) line: contribution from hadron (parton) gas; full lines: total yields from thermalized matter; fast equilibration is assumed. This figure should be compared with figure 5 in reference.[5]

Figure 4: Time evolution of the scaled dilepton spectrum (arbitrary units) from a thermally equilibrating parton gas for invariant mass $M = 2.6$ GeV and transverse momentum $q_\perp = 200$ MeV. The relaxation times are $\tau_{rel} = 10$ fm/c (full line), 1 fm/c (dashed line), 0.13 fm/c (dotted line).

obviously this happens within about 25% of the life time of the parton gas.

The solution of the Boltzmann equation in relaxation time approximation has two components

$$f(\tau) = \exp\left\{\frac{\tau_0 - \tau}{\tau_{rel}}\right\} f_0 + \int_{\tau_0}^{\tau} d\tau' \tau_{rel}^{-1} \exp\left\{\frac{\tau_0 - \tau}{\tau_{rel}}\right\} f_{eq}(T(\tau')), \qquad (1)$$

where f is the parton distribution function (assumed to be scale-invariant), τ_{rel} denotes the relaxation time, and f_{eq} stands for the thermal equilibrium distribution function (with temperature parameter T determined by an integral equation, cf. reference[18] for details). The dilepton rate has within the kinetic theory framework the structure

$$\frac{dN}{d^4x\, d^4Q} = \int \frac{d^3p_1}{2E_1\,(2\pi)^3} \frac{d^3p_2}{2E_2\,(2\pi)^3} |\mathcal{M}|^2 (2\pi)^4 \delta^4(p_1 + p_2 - Q), \qquad (2)$$

with $\mathcal{M}$ for the matrix element of the fusion reaction $q\bar{q} \to \mu^+\mu^-$. Due to the structure of the equation (2) the rate takes the form

$$\frac{dN}{d^4x\, d^4Q} = \int \{\cdots f_0\, f_0 + \cdots f_0\, f_{eq} + \cdots f_{eq}\, f_{eq}\}, \qquad (3)$$

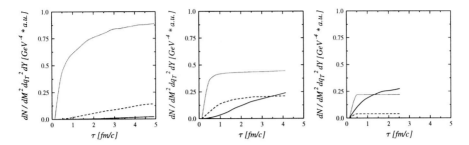

Figure 5: The same as in figure 4, but for the individual contributions: the components N_1, N_2 and N_{12} are depicted by dotted, full, and dashed lines, respectively. The left (middle, right) panel is for the relaxation time $\tau_{rel} = 10$ (1, 0.13) fm/c.

i.e., the dileptons are produced by collisions among the non-equilibrated partons ($dN_1 \propto f_0 f_0$), and among the equilibrated ($dN_2 \propto f_{eq} f_{eq}$), and by collisions of non-equilibrated with equilibrated partons ($dN_{12} \propto f_0 f_{eq}$). These three components are displayed in figure 5. Here, as in figure 4, the initial conditions are so that the energy density corresponds to the energy density of an equilibrium system at $T = 500$ MeV. The initial distribution function f_0 at initial time $\tau_0 = 0.13$ fm/ is chosen as $f_0 \propto \delta(Y - y) \exp\{-p_\perp^2/K\}$ with y as space-time rapidity and $K = 0.6$ GeV^2, cf. [18] One can observe that even for the comparatively short standard equilibration time scale of $\tau_{rel} = 1$fm/c the non-equilibrium component still dominates; only for extremely rapid equilibration the equilibrium component represents the main part in the rate. From this one can conclude that probably the pre-equilibrium dileptons are an important and non-negligible part of the spectrum, despite the short relaxation found in the parton transport simulations.[3,7,8] This holds in particular for the harder dileptons.

5 Summary

In summary we report here mainly estimates of the dilepton spectra to be expected under RHIC conditions, where a substantial part of the transverse energy is produced by perturbative hard or semi-hard QCD processes. Single photon spectra have been calculated within the same dynamical framework.

Acknowledgments

Valuable discussions with K. Redlich are gratefully acknowledged. This work is supported by BMBF (grant 06DR666), DFG and GSI.

References

1. I. Tserruya, *Nucl. Phys.* A **590**, 127c (1995).
2. H.S. Matis and the DLS collaboration, *Nucl. Phys.* A **583**, 617 (1995).
3. P. Levai, B. Müller, X.N. Wang, *Phys. Rev.* C **51**, 3326 (1995).
4. P.L. McGaughey, E. Quack, P.V. Ruuskanen, R. Vogt, X.N. Wang, *Int. J. Mod. Phys.* A **10**, 2999 (1995);
   R. Vogt, preprint LBL-37105 (1996);
   Z. Lin, M. Gyulassy, *Phys. Rev.* C **51**, 2177 (1995).
5. B. Kämpfer, O.P. Pavlenko, A. Peshier, G. Soff, *Phys. Rev.* C **52**, 2704 (1995).
6. B. Kämpfer, O.P. Pavlenko, *Z. Phys.* C **42**, 491 (1994).
7. K. Geiger, *Phys. Rev.* D **46**, 4965 (1992), *Phys. Rev.* D **46**, 4986 (1992), *Phys. Rev.* D **47**, 133 (1992), *Phys. Rev.* D **47**, 4986 (1992);
   K. Geiger, J.I. Kapusta, *Phys. Rev.* D **47**, 4905 (1993).
8. T.S. Biro, E. van Doorn, B. Müller, M.H. Thoma, X.N. Wang, *Phys. Rev.* C **48**, 1275 (1993).
9. D.K. Srivastava, J. Pan, V. Emel'yanov, C. Gale, *Phys. Lett.* B **329**, 157 (1994).
10. E.V. Shuryak, *Phys. Rev. Lett.* **68**, 3270 (1992).
11. B. Kämpfer, O.P. Pavlenko, *Phys. Rev.* C **49**, 2716 (1994).
12. J. Cleymans, I. Dadič, *Phys. Rev.* D **47**, 160 (1993).
13. C. Spieles, M. Bleicher, A. Dumitru, C. Greiner, M. Hofmann, A. Jahns, U. Katscher, R. Matiello, J. Schaffner, H. Sorge, L. Winckelmann, J. Maruhn, H. Stöcker, W. Greiner, *Nucl. Phys.* A **590**, 271c (1995).
14. A. Dumitru, D.H. Rischke, Th. Schönfeld, L. Winckelmann, H. Stöcker, W. Greiner, *Phys. Rev. Lett.* **70**, 2860 (1993);
    A. Dumitru, U. Katscher, J.A. Maruhn, H. Stöcker, W. Greiner, D.H. Rischke, *Z. Phys.* A **353**, 187 (1995), *Phys. Rev.* C **51**, 2166 (1995).
15. B. Kämpfer, O.P. Pavlenko, M.I. Gorenstein, A. Peshier, G. Soff, *Z. Phys.* A **353**, 71 (1995).
16. WA80 collaboration, preprint submitted to *Phys. Rev. Lett.* , (1996).
    T.C. Awes, *Nucl. Phys.* A **590**, 81c (1995).
17. R. Yohida, T. Miyazaki, M. Kadoya, *Phys. Rev.* D **35**, 388 (1987);
    K. Redlich, *Phys. Rev.* D **36**, 3378 (1987);
    S. Hirasawa, M. Kadoya, T. Miyazaki, *Phys. Lett.* B **218**, 263 (1989).
18. B. Kämpfer, O.P. Pavlenko, *Phys. Lett.* B **62**, 127 (1992).
19. K. Geiger, J.I. Kapusta, *Phys. Rev. Lett.* **70**, 1920 (1993);
    K. Geiger, *Phys. Rev. Lett.* **71**, 3075 (1993).

Dynamics of Strangeness Production and Strange Matter Formation [*]

C. Spieles, M. Bleicher, L. Gerland, H. Stöcker
Institut für Theoretische Physik, J. W. Goethe-Universität,
D-60054 Frankfurt am Main, Germany

C. Greiner
Institut für Theoretische Physik, J. Liebig-Universität,
D-35392 Giessen, Germany

1 Introduction

We want to draw the attention to the dynamics of a (finite) hadronizing quark matter drop. Strange and antistrange quarks do not hadronize at the same time for a baryon-rich system[1]. Both the hadronic and the quark matter phases enter the strange sector $f_s \neq 0$ of the phase diagram almost immediately, which has up to now been neglected in almost all calculations of the time evolution of the system. Therefore it seems questionable, whether final particle yields reflect the actual thermodynamic properties of the system at a certain stage of the evolution. We put special interest on the possible formation of exotic states, namely strangelets (multistrange quark clusters). They may exist as (meta-)stable exotic isomers of nuclear matter [2]. It was speculated that strange matter might exist also as metastable exotic multi-strange (baryonic) objects (MEMO's [3]). The possible creation — in heavy ion collisions — of long-lived remnants of the quark-gluon-plasma, cooled and charged up with strangeness by the emission of pions and kaons, was proposed in [1,4,5]. Strangelets can serve as signatures for the creation of a quark gluon plasma. Currently, both at the BNL-AGS and at the CERN-SPS experiments are carried out to search for MEMO's and strangelets, e. g. by the E864, E878 and the NA52 collaborations[9,10].

2 The model

We adopt a model [5] for the hadronization and space-time evolution of quark matter droplet. We assume a first order phase transition of the QGP to hadron gas. The expansion of the QGP droplet is described in a hybrid-like model, which takes into account equilibrium as well as nonequilibrium features of the

---

[*]Supported by GSI, BMBF, DFG

process by the following two crucial, yet oversimplifying (and to some extent controversial) assumptions: (1) the plasma sphere is permantently surrounded by a thin layer of hadron gas, with which it stays in perfect equilibrium (Gibbs conditions) during the whole evolution; in particular the strangeness degree of freedom stays in chemical equilibrium because the complete hadronic particle production is driven by the plasma phase. (2) The nonequilibrium radiation is incorporated by a time dependent freeze-out of hadrons from the outer layers of the hadron phase surrounding the QGP droplet. During the expansion, the volume increase of the system thus competes with the decrease due to the freeze–out. The global properties like (decreasing) S/A and (increasing) f_s of the remaining two-phase system then change in time according to the following differential equations for the baryon number, the entropy, and the net strangeness number of the total system:

$$\frac{d}{dt} A^{tot} = -\Gamma A^{HG}$$

$$\frac{d}{dt} S^{tot} = -\Gamma S^{HG} \qquad (1)$$

$$\frac{d}{dt} (N_s - N_{\bar{s}})^{tot} = -\Gamma (N_s - N_{\bar{s}})^{HG} ,$$

where $\Gamma = \frac{1}{A^{HG}} \left(\frac{\Delta A^{HG}}{\Delta t} \right)_{ev}$ is the effective ('universal') rate of particles (of converted hadron gas volume) evaporated from the hadron phase. The equation of state consists of the bag model for the quark gluon plasma and a mixture of relativistic Bose–Einstein and Fermi–Dirac gases of well established strange and non–strange hadrons up to 2 GeV in Hagedorn's eigenvolume correction for the hadron matter[1]. Thus, one solves simultaneously the equations of motion (1) and the Gibbs phase equilibrium conditions for the intrinsic variables, i.e. the chemical potentials and the temperature, as functions of time.

3   Strangelet distillation at low μ/T

In[6] it was shown that large local net-baryon and net-strangeness fluctuations as well as a small but finite amount of stopping can occur at RHIC and LHC. This can provide suitable initial conditions for the possible creation of strange matter in colliders. A phase transition (e. g. a chiral one) can further increase the strange matter formation probability. In[6] it was further demonstrated with the present model that the high initial entropies per baryon do not hinder the distillation of strangelets, however, they require more time for the evaporation and cooling process.

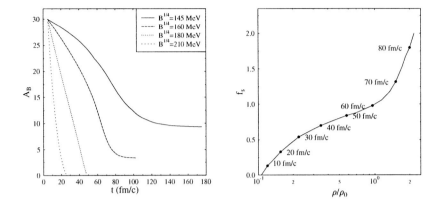

Figure 1: Time evolution of the baryon number for a QGP droplet with $A_{\mathrm{B}}^{\mathrm{init}} = 30$, $S/A^{\mathrm{init}} = 200$, $f_s^{\mathrm{init}} = 0.7$ and different bag constants (left). Evolution of a QGP droplet with baryon number $A_{\mathrm{B}}^{\mathrm{init}} = 30$ for $S/A^{\mathrm{init}} = 200$ and $f_s^{\mathrm{init}} = 0$. The bag constant is $B^{1/4} = 160$ MeV. Shown is the baryon density and the corresponding strangeness fraction (right).

Fig. 1 (left) shows the time evolution of the baryon number for $S/A^{\mathrm{init}} = 200$ and $f_s^{\mathrm{init}} = 0.7$ for various bag constants. For $B^{1/4} < 180$ MeV a cold strangelet emerges from the expansion and evaporation process, while the droplet completely hadronizes for bag constants $B^{1/4} \geq 180$ MeV (for $B^{1/4} = 210$ MeV hadronization proceeds without any significant cooling of the quark phase, although the specific entropy S/A decreases by a factor of 2 from 200 to only 100). The strangeness separation works also in these cases, and leads to large final values of the net strangeness content, $f_s \stackrel{>}{\sim} 1.5 - 2$. However, then the volume of the drop becomes small, it decays and the strange quarks hadronize into Λ-particles and other strange hadrons. For even higher bag constants $B^{1/4} \approx 250$ MeV neither the baryon concentration effect nor strangeness distillery occurs (Fig. 4).

Fig. 1 (right) shows the evolution of the two-phase system for $S/A^{\mathrm{init}} = 200$, $f_s^{\mathrm{init}} = 0$ and for a bag constant $B^{1/4} = 160$ MeV in the plane of the strangeness fraction vs. the baryon density. The baryon density increases by more than one order of magnitude! Correspondingly, the chemical potential rises as drastically during the evolution, namely from $\mu^i = 16$ MeV to $\mu^f > 200$ MeV. The strangeness separation mechanism drives the chemical potential of the strange quarks from $\mu_s^i = 0$ up to $\mu_s^f \approx 400$ MeV. Thus, the thermodynamical and chemical properties during the time evolution are quite different from the initial conditions of the system.

Fig. 1 illustrates the increase of the baryon density in the plasma droplet

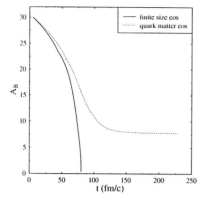 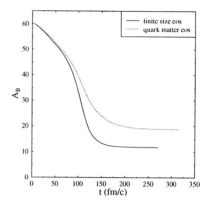

Figure 2: Time evolution of the net baryon number of a QGP droplet, calculated with (full line) and without (dashed line) finite size corrections to the quark matter equation of state. The initial conditions are $f_s^{\text{init}} = 0$ and $S/A^{\text{init}} = 200$. The bag constant is $B^{1/4} = 145$ MeV.

as an inherent feature of the dynamics of the phase transition (cf. [7]). The origin of this result lies in the fact that the baryon number in the quark–gluon phase is carried by quarks with $m_q \ll T_C$, while the baryon density in the hadron phase is suppressed by a Boltzmann factor $\exp(-m_{\text{baryon}}/T_C)$ with $m_{\text{baryon}} \gg T_C$. Mainly mesons (pions and kaons) are created in the hadronic phase. More relative entropy S/A than baryon number is carried away in the hadronization and evaporation process[5], i.e. $(S/A)^{HG} \gg (S/A)^{QGP}$. Ultimately, whether $(S/A)^{HG}$ is larger or smaller than $(S/A)^{QGP}$ at finite, nonvanishing chemical potentials might theoretically only be proven rigorously by lattice gauge calculations in the future. However, model equations of state do suggest such a behaviour, which would open such intriguing possibilities as baryon inhomogenities in ultrarelativistic heavy ion collisions as well as in the early universe.

4   Finite size effects

The bag model equation of state for infinite quark matter is certainly a very rough approximation. Regarding finite size effects the leading correction to the quark matter equation of state is the curvature term. For massless quarks the volume term of the gandcanonical potential suffers the following modification (including the gluon contribution)[8]:

$$\Omega_C = (\frac{1}{8\pi^2}\mu_q^2 + \frac{11}{72}T^2)C \tag{2}$$

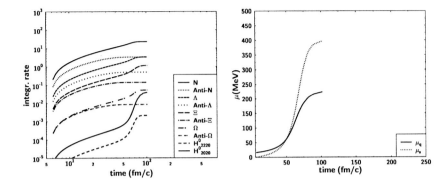

Figure 3: Integrated rates of particles, evaporated out of a hadronizing QGP droplet as functions of time (left) and the corresponding (strange) quark chemical potential (right). The initial conditions are $f_s^{\mathrm{init}} = 0$, $S/A^{\mathrm{init}} = 200$ and $A_B^{\mathrm{init}} = 30$. The bag constant is $B^{1/4} = 160$ MeV, the mass of the H^0 is varied between $m = 2020$ MeV and $m = 2220$ MeV.

with $C = 8\pi R$ being the curvature of the spherical bag surface. From this one can easily derive all thermodynamic quantities and study the evolution of the two-phase system QGP/hadron gas in the above described model. It shows that even in the case of a favourable bag constant $B^{1/4} = 145$ MeV a quark blob with an initial net baryon number of $A_B^{\mathrm{init}} = 30$ will completely hadronize — in contrast to the calculation with the unmodified equation of state (Fig. 2). Of course, the difference between the dynamics according to the two equations of state is reduced for larger systems. Still, it can be speculated that shell effects may allow for the formation of rather small strangelets which are stable. Moreover, the introduction of a more realistic hadronic equation of state (e. g. with the help of a relativistic mean field theory including adequate interactions for strange hadrons[3]) might modify this pessimistic picture again.

5 Particle rates from the hadronizing plasma

Enhanced production of strange particles in relativistic nuclear collisions has received much attention recently[9,10]. In particular thermal models have been developed and applied[11,12] to explain (strange) particle yields and to extract the characteristic thermodynamic properties of the system (a few macroscopic parameters) from them. In our model the picture of a sudden hadronization which is supposed in these studies is only one possible outcome. Under more general assumptions the observed particle rates have to be put in relation to the whole time evolution of the system.

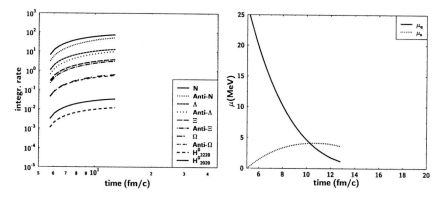

Figure 4: Integrated rates of particles, evaporated out of a hadronizing QGP droplet as functions of time (left) and the corresponding (strange) quark chemical potential (right). The initial conditions are $f_s^{init} = 0$, $S/A^{init} = 200$ and $A_B^{init} = 30$. The bag constant is $B^{1/4} = 250$ MeV, the mass of the H^0 is varied between $m = 2020$ MeV and $m = 2220$ MeV.

The integrated particle rates and the quark chemical potentials as functions of time have been calculated for two different scenarios: In Fig. 3 the results are plotted for a bag constant of $B^{1/4} = 160$ MeV which is favorable for the strangeness distillation. In Fig. 4 a very high bag constant of $B^{1/4} = 250$ MeV is used. This results in a very rapid (and complete) hadronization without significant cooling. Obviously, in the first case the particle rates reflect the massive changes of the chemical potentials during the evolution (which is the result of the strangeness distillation process). Note that e. g. the Λ's are emitted mostly at the late stage, whereas the $\bar{\Lambda}$'s stem almost exclusively from the early stage. The $\bar{\Lambda}/\Lambda$ ratio is therefore not a meaningful quantity (if one takes it naively), since the two yields represent different sources! For the other choice of the bag constant the present model renders more or less the picture which is claimed by 'static' thermal models: the plasma fireball decomposes very fast into hadrons (watch the different time scales of Figs. 3 and 4) and the quark chemical potentials stay low compared to the temperature. Time dependent rates of the hypothetic H^0 Dibaryon are also shown in Figs. 3 and 4. This particle is introduced to the hadronic resonance gas with its appropriate quantum numbers and two different assumed masses. It appears that the distillation mechanism gives rise to H^0 yields of the same order as the $\bar{\Omega}$'s (Fig. 3) if the mass is $m_{H^0} \approx 2020$ MeV. For the high bag constant the H^0 yields are much more suppressed as compared to the strange (anti-)baryons. The absolute yields of the H^0 do not change much, since the system emits the particles at significantly higher temperature (due to the high bag constant).

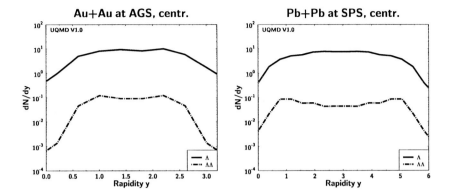

Figure 5: Rapidity distributions of hyperons and $\Lambda-\Lambda$-clusters calculated with URQMD1.0β plus a clustering procedure according to the Wigner-function method. Shown are the spectra for central collisions of Au+Au at 10.7 GeV (left) and Pb+Pb at 160 GeV (right).

6 Hyper–cluster formation in a microscopic model

We now apply the Ultrarelativistic Quantum Molecular Dynamics 1.0β[13,14], a semiclassical transport model, to calculate the abundances of strange baryon-clusters in relativistic heavy ion collisions. The model is based on classical propagation of hadrons and stochastic scattering (s channel excitation of baryonic and mesonic resonances/strings, t channel excitation, deexcitation and decay). In order to extract hyper-cluster formation probabilities the Λ pair phase space after strong freeze-out is projected on the assumed dilambda wave function (harmonic oscillator) via the Wigner-function method as described in [15]. According to the weak coupling between Λ's in mean-field calculations[3] we assume the same coupling for $\Lambda\Lambda$-cluster as for deuterons[15].

In Fig. 5 the calculated rapidity distributions of hyperons and $\Lambda\Lambda$-clusters are shown for central reactions of heavy systems at AGS and SPS energies. The multiplicities of Λ's plus Σ^0's in inelastic p+p reactions are 0.088 ± 0.003 at 14.6 GeV/c and 0.234 ± 0.005 at 200 GeV/c with the present version of the model. These numbers are given to assess the absolute yields in A+A collisions. The hyperon rapidity density stays almost constant when going from AGS to SPS energies, the dN/dy of the hyper-clusters even drops slightly at midrapidity. This is due to the higher temperature which gives rise to higher relative momenta and therefore a reduced cluster probability. The $\Lambda/\Lambda\Lambda$ ratio is approximately 100, which can be compared to the Λ/H^0 ratios which result from the expanding quark gluon plasma (see last section).

References

1. C. Greiner, P. Koch and H. Stöcker, *Phys. Rev. Lett.* **58**, 1825 (1987); C. Greiner, D. H. Rischke, H. Stöcker and P. Koch, *Phys. Rev.* D **38**, 2797 (1988).

2. A. R. Bodmer, *Phys. Rev.* D **4**, 1601 (1971); S. A. Chin and A. K. Kerman, *Phys. Rev. Lett.* **43**, 1292 (1979); J. D. Bjorken and L. D. McLerran, *Phys. Rev.* D **20**, 2353 (1979).

3. J. Schaffner, C. Greiner, C. B. Dover, A. Gal, H. Stöcker, Phys. Rev. Lett. **71**, 1328 (1993)

4. H.-C. Liu and G.L. Shaw, *Phys. Rev.* D **30**, 1137 (1984).

5. C. Greiner and H. Stöcker, *Phys. Rev.* D **44**, 3517 (1992).

6. C. Spieles, L. Gerland, H. Stöcker, C. Greiner, C. Kuhn, J.P. Coffin, *Phys. Rev. Lett.* **76**, 1776 (1996),

7. E. Witten, *Phys. Rev.* D **30**, 272 (1984); E. Farhi and R. L. Jaffe, *Phys. Rev.* D **30**, 2379 (1984).

8. J. Madsen, *Phys. Rev. Lett.* **70**, 391 (1993), and contribution to International Symposium on Strangeness and Quark Matter (Sept. 1-5, 1994) Crete (Greece) World Scientific, 1995.

9. Proc. of the Int. Conf. on Ultrarelativistic Nucleus-Nucleus Collisions, Quark Matter '95, Monterey, CA, USA, Nucl. Phys. **A590** (1995).

10. Proc. of the Int. Conf. on Strangeness in Hadronic Matter, S'95, Tucson, AZ, USA, AIP Press, Woodbury, NY (1995).

11. J. Letessier, A. Tounsi, U. Heinz, J. Sollfrank, J. Rafelski, *Phys. Rev.* D **51**, 3408 (1995).

12. P. Braun-Munzinger, J. Stachel, J.P. Wessels, N. Xu *Phys. Lett.* B **365**, 1 (1996); P. Braun-Munzinger, J. Stachel, *J. Phys.* G **21**, L17 (1995).

13. S. A. Bass, M. Bleicher, M. Brandstetter, A. Dumitru, C. Ernst, L. Gerland, J. Konopka, S. Soff, C. Spieles, H. Weber, L. A. Winckelmann, N. Amelin H. Stöcker and W. Greiner; source code and technical documentation, to be published; S. A. Bass et al., contribution to this volume.

14. M. Bleicher et al., contribution to this volume.

15. R. Mattiello et al., contribution to this volume, and private communication.

PROBING THE BOTTOM OF DIRAC SEA

I.N. MISHUSTIN

The Niels Bohr Institute, University of Copenhagen
Blegdamsvei 17, DK-2100 Copenhagen Ø, Denmark
and
Department of General and Nuclear Physics, the Kurchatov Institute,
Moscow 123182, Russia

Saturation properties of nuclear matter are studied within the chiral Nambu–Jona-Lasinio model formulated on the level of constituent nucleons. In the mean-field approximation the model supports only one bound state which can be placed at the nuclear saturation point by proper choice of model parameters. This solution is characterized by a small nucleon effective mass which goes to zero in the chiral limit. The cut-off momentum Λ, which determines the density of active states in the Dirac sea, is quite small, $\Lambda \approx 300$ MeV. Comparison with the Walecka model is made.

1 Introduction

The Dirac sea was invented more than 60 years ago and since that time it attracts imagination of physicists. The Dirac sea is associated with the filled negative-energy states of fermions. It manifests itself in the production of antiparticles, which are interpreted as holes in the Dirac sea. In strong enough external fields the transitions between negative and positive-energy states become possible. Very famous examples of this kind are the pair production in a homogeneous electric field [1] and the spontaneous production of electron-positron pairs in supercritical Coulomb field [2].

In a field-theoretical description of nuclei and nuclear matter, like the Walecka model [3], where nucleons are described by the Dirac equation, the existence of the Dirac sea is often ignored. Sometimes, in approaches like QHD-I [4], it is renormalized out by introducing polynomial self-interactions of meson fields. The mean scalar and vector potentials predicted by such models are typically large, say, $U_S = -360$ MeV and $U_V = 300$ MeV. In the mean potential acting on nucleons in ordinary nuclei these two large contributions almost cancel each other. The resulting potential $U_N = U_S + U_V = -60$ MeV is quite normal in the nuclear scale. As follows from G-parity transformation, for antinucleons both contributions have the same sign. Therefore, the mean potential acting on antinucleons $U_{\overline{N}} = U_S - U_V = -660$ MeV is extremely deep in the nuclear scale. It is already comparable with the free nucleon mass $m_N = 938$ MeV. This observation was first made by Dürr and Teller in 1956 [5]! It is remarkable that the nuclear potentials extracted recently on the basis of the QCD sum rules [6,7] have the same magnitude.

The mean potentials certainly grow with baryon density. If one would extrapolate the picture of Dirac nucleons interacting with mean meson fields to high densities, one would find [8,9] that $U_{\overline{N}} = -2m_N$ at about $(3 \div 7)$ times normal nuclear density $\rho_0 = 0.16$ fm^{-3} . Such high densities can be reached in relativistic heavy-ion collisions as well as in neutron stars. This would mean that a piece of nuclear matter at such a density is unstable with respect to the spontaneous production of baryon-antibaryon pairs on its surface [8,9]. The negative-energy states may be excited also by strong time-dependent meson fields generated in relativistic nuclear collisions [10].

In these simple considerations the negative-energy states were treated on the level of the single particle Dirac equation and their influence on the energy of the many-body system was ignored. Therefore, the question arises, is it at all legitimate to use such a simple picture? To answer this question one should go to a deeper level and consider field-theoretical models which include the Dirac sea explicitly. The Nambu–Jona-Lasinio (NJL) model [11] is a most famous model of this kind. It was proposed originally as a tool providing the dynamical generation of nucleon mass. One of the most important properties of this model is that it obeys explicit chiral symmetry. In recent years the NJL model is widely used for describing hadron properties in terms of constituent quarks [12,13]. For nuclear matter this kind of models was first applied in ref. [14].

The paper is organized as follows. In the next Section the NJL model with $SU(2)_L \otimes SU(2)_R$ chiral symmetry is formulated in terms of nucleonic degrees of freedom. In Section 3 the gap equation and constraints on the model parameters are discussed. In Section 4 the model is applied for studying saturation properties of nuclear matter within the mean-field approximation. The comparison with the mean-field Walecka model is given in Section 5. Conclusions are summarized in Section 6.

2   Formulation of the NJL model for nuclear matter

Many nonperturbative aspects of QCD can be studied on the basis of effective models obeying the symmetry properties of QCD such as chiral symmetry and scale invariance. One of the most successful models of this kind is the Nambu–Jona-Lasinio (NJL) model formulated on the level of constituent quarks (for review see [12,13]). However, constituent quarks are not adequate degrees of freedom for nuclear matter at normal density. From nuclear phenomenology we know that nucleons and mesons are effective degrees of freedom at least at densities around the normal nuclear density $\rho \sim \rho_0$. Therefore, below we formulate the NJL model for interacting nucleons. In such approach we also avoid the question of confinement. At the same time one should expect that due to the composite structure of the nucleon this approach can not be extrapolated too far in density.

The NJL Lagrangian for a spinor nucleon field N, as introduced by Nambu–Jona-Lasinio in 1961[11], is written as

$$\mathcal{L} = \mathcal{L}_0 + \mathcal{L}_{int}, \tag{1}$$

where

$$\mathcal{L}_0 = \overline{N} \left(i\gamma^\mu \partial_\mu - m_0 \right) N \tag{2}$$

is the free Dirac Lagrangian for nucleons with a "bare" mass m_0, and $\mathcal{L}_{int}$ is the 4-fermion interaction Lagrangian. In general form it can be represented as

$$\mathcal{L}_{int} = \frac{G_S}{2} \left[\left(\overline{N}N \right)^2 + \left(i\overline{N}\gamma_5\tau N \right)^2 \right] - \frac{G_V}{2} \left[\left(\overline{N}\gamma^\mu N \right)^2 + \left(i\overline{N}\gamma_5\gamma^\mu\tau N \right)^2 \right]. \tag{3}$$

Here $\tau = (\tau_1, \tau_2, \tau_3)$ are isospin Pauli matrices, γ^μ are Dirac matrices with $\gamma_5 = i\gamma^0\gamma^1\gamma^2\gamma^3$. At vanishing m_0 this Lagrangian is invariant under chiral transformations from $SU(2)_L \otimes SU(2)_R$ group. The coupling constants G_S and G_V have the dimension (energy)$^{-2}$, that is why the theory is non-renormalizable. It is quite natural for such an effective theory as the NJL model. To avoid divergences one should introduce a suitable regularization scheme. Below we adopt the 3-dimensional regularization scheme, $p < \Lambda$, with the cut-off momentum Λ treated as an adjustable parameter. The structure of the vacuum of the NJL model is illustrated in Fig. 1 (upper part).

A remarkable property of the NJL model is the dynamical generation of the fermion mass if the coupling strength G_S is sufficiently large. In this case the vacuum ground state hosts a pair condensate $\langle \overline{N}N \rangle$ and the pion emerges as a Goldstown boson, associated with the spontaneous breaking of chiral symmetry. Then PCAC and soft pion theorems follow in a quite natural way.

Below we adopt the mean-field (Hartree) approximation replacing $(\overline{N}\Gamma N)^2$ by $2\overline{N}\Gamma N \langle \overline{N}\Gamma N \rangle - \langle \overline{N}\Gamma N \rangle^2$, where $\langle \overline{N}\Gamma N \rangle$ is the ground state expectation value of a corresponding bilinear form and Γ is any combination of matrices appearing in $\mathcal{L}_{int}$.

3 Gap equation and constraints on the model parameters

Here we consider the iso-symmetric nuclear matter at zero temperature. In this case all terms containing τ matrices vanish and the NJL Lagrangian is

502

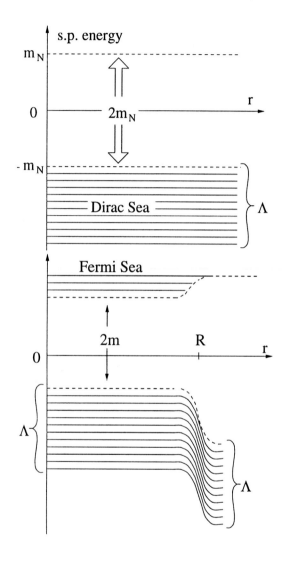

Figure 1: Schematic view of single-particle states in the NJL model. Upper part: in vacuum only negative-energy states up to cut-off momentum Λ are occupied (Dirac sea). Lower part: in medium also positive-energy states up to p_F are occupied (Fermi sea). The gap between positive- and negative-energy states is reduced from $2m_N$ in vacuum to $2m$ in medium.

written as

$$\mathcal{L} = \overline{N} \left(i\gamma^\mu \partial_\mu - m_0 \right) N + \frac{G_S}{2} \left(\overline{N} N \right)^2 - \frac{G_V}{2} \left(\overline{N} \gamma^\mu N \right)^2 . \qquad (4)$$

In the mean-field approximation this Lagrangian leads to the Dirac equation

$$\left(i\gamma^\mu D_\mu - m \right) N = 0, \qquad (5)$$

where the covariant derivative is defined as

$$D_\mu = \partial_\mu + iG_V \langle \overline{N} \gamma_\mu N \rangle \qquad (6)$$

and the effective (constituent) nucleon mass is

$$m = m_0 - G_S \langle \overline{N} N \rangle. \qquad (7)$$

The second term in equation (6) provides a density-dependent shift in energy of all single particle states. The gap equation (7) is a self-consistency relation for the nucleon mass. The scalar density $\langle \overline{N} N \rangle$ contains contributions from both the low-continuum states (vacuum contribution) and the occupied upper-continuum states (matter contribution). In medium the gap between these two families of states is reduced compared to the vacuum, i.e. $2m < 2m_N$. The combined effect of the energy shift and the gap reduction on the single-particle states is shown schematically in Fig. 1 (lower part).

In general the scalar density is expressed as [12]

$$\rho_S = \langle \overline{N} N \rangle = \nu_N \int \frac{d^3 p}{(2\pi\hbar)^3} \frac{m}{\sqrt{p^2 + m^2}} \left[-1 + n_N(p) + n_{\overline{N}}(p) \right], \qquad (8)$$

where $\nu_N = 4$ is degeneracy factor for nucleons, $n_N(p)$ and $n_{\overline{N}}(p)$ are the Fermi-Dirac occupation numbers of nucleons and antinucleons. The baryon density is

$$\rho_B = \langle \overline{N} \gamma^0 N \rangle = \nu_N \int \frac{d^3 p}{(2\pi\hbar)^3} \left[n_N(p) - n_{\overline{N}}(p) \right]. \qquad (9)$$

The non-renormalizability of the NJL model manifests itself in the appearance of divergent integrals as e.g. the vacuum contribution to the scalar density (term with -1 in eq. (8)). As mentioned above such integrals are regularized by introducing the cut-off momentum Λ. At zero temperature and finite baryon density $n_{\overline{N}}(p) = 0$ and $n_N(p) = \theta(p_F - p)$ where p_F is the Fermi momentum related to the baryon density as

$$\rho_B = \frac{\nu_N p_F^3}{6\pi^2} . \qquad (10)$$

The scalar density can be easily calculated now,

$$\rho_S = -\nu_N \int_{p_F}^{\Lambda} \frac{d^3 p}{(2\pi\hbar)^3} \frac{m}{\sqrt{p^2 + m^2}} = -\frac{\nu_N m}{4\pi^2} \left[\Lambda^2 \Phi\left(\frac{m}{\Lambda}\right) - p_F^2 \Phi\left(\frac{m}{p_F}\right) \right], \quad (11)$$

where

$$\Phi(x) = \sqrt{1 + x^2} - \frac{x^2}{2} \ln \frac{\sqrt{1 + x^2} + 1}{\sqrt{1 + x^2} - 1}. \quad (12)$$

This function has limits: $\Phi(x \to 0) = 1$ and $\Phi(x \to \infty) = 2/(3x)$.

There are two important constraints on the parameters of the model under consideration. First is quite obvious: the nucleon mass in vacuum should be equal to the physical mass $m_N = 938$ MeV. Therefore, the gap equation in vacuum ($p_F = 0$) should be fulfilled with $m = m_N$,

$$m_N = m_0 + \nu_N G_S \frac{m\Lambda^2}{4\pi^2} \Phi\left(\frac{m_N}{\Lambda}\right). \quad (13)$$

Another constraint comes from the soft pion phenomenology. Indeed, the strength of the explicit chiral-symmetry-breaking term $-m_0 \overline{N} N$ is controlled by the PCAC relation. The easy way to see this is to use the bosonization procedure by introducing the "meson" fields

$$\sigma = -\frac{G_S}{g} \overline{N} N, \quad \pi = -\frac{G_S}{g} \overline{N} i\gamma_5 \tau N$$

and integrating out the fermionic degrees of freedom. Here g is the meson-nucleon coupling constant which is related to the nucleon mass and pion decay constant $f_\pi = 93$ MeV as $g f_\pi = m_N$. The symmetry-breaking term is written now as $\frac{g m_0}{G_S} \sigma$. But it should be equal to $m_\pi^2 f_\pi \sigma$ to fulfil the correct PCAC equation. Therefore, one arrives at the relation

$$\frac{m_N m_0}{G_S} = m_\pi^2 f_\pi^2, \quad (14)$$

which after substituting G_S from the gap equation (7) acquires a form of the famous GOR relation [16].

Therefore, the model has 4 adjustable parameters: G_S, G_V, m_0 and Λ and two constraints, eqs. (13) and (14). In addition, one should bear in mind that the "non-chiral" part of the nucleon mass should not be too large, say $m_0 < 160$ MeV, according to ref. [15]. Now we are going to check whether enough freedom remains to reproduce correctly the nuclear ground state.

4 Ground state of nuclear matter

In nuclear matter, in addition to the filled negative-energy states, also the positive-energy states are filled up to the Fermi energy. In the mean field approximation the energy density of infinite nuclear matter is

$$\mathcal{E}(\rho_B) = \nu_N \int \frac{d^3p}{(2\pi)^3} \sqrt{p^2 + m^2} \left[-1 + n_N(p) + n_{\overline{N}}(p) \right] + \frac{G_S}{2} \rho_S^2 + \frac{G_V}{2} \rho_B^2. \quad (15)$$

This expression contains the density-independent contribution, $\mathcal{E}(\rho_B = 0)$, i.e. the vacuum energy, which should be subtracted. Then the energy per baryon in infinite nuclear matter is written as

$$\frac{E}{N} = \frac{\mathcal{E}(\rho_B) - \mathcal{E}(\rho_B = 0)}{\rho_B} \quad (16)$$

In practical calculations it is convenient to eliminate the scalar density from eq. (15) by using the gap equation (7), $\rho_S = (m_0 - m_N)/G_S$.

The energy per baryon $E/N(m, \rho_B)$ as a function of constituent nucleon mass m and baryon density ρ_B has a quite remarkable behaviour. At given ρ_B it has a local minimum determining the constituent mass $m(\rho_B)$ at this density. Local minima of E/N form a valley which starts at $m = m_N$ at zero density and goes to the point where $m \approx 0$ at some critical density ρ_c. This signals the restoration of chiral symmetry at higher densities. The critical density is easily calculated in the chiral limit, $m_0 = 0$. One can see that $m = 0$ becomes the only solution of the gap equation at

$$p_F^c = \Lambda \sqrt{1 - \frac{4\pi^2}{\nu_N G_S \Lambda^2}}. \quad (17)$$

Along the valley the energy per baryon is rather flat due to the approximate cancellation of different contributions. But at $\rho_B > \rho_c$ the vector interaction dominates and the energy grows rapidly. Due to this flatness, in the case $m_0=0$ the absolute minimum of E/N appears either at $m = m_N$, $\rho_B=0$ or at $m = 0$, $\rho_B = \rho_c$. So there is no normal nuclear bound state, only the abnormal state with $m = 0$ appears! At $m_0 \neq 0$ the situation is not much different. The energy per baryon has only one bound state which goes into the abnormal state at $m_0 \to 0$. One can try to identify this bound state with the nuclear ground state. So one can choose two free parameters of the model so that the absolute minimum of E/N would appear at $\rho_B = \rho_0 = 0.16$ fm^{-3} and $E/N - m_N$=-16 MeV, i.e. at the correct saturation point of nuclear matter. Since no freedom remains any more, all other characteristics of this bound state, e.g. the nucleon effective mass at ρ_0, are fixed by this choice. Such set of parameters is

$$\Lambda = 300\,\text{MeV}, \quad m_0 = 86\,\text{MeV}, \quad G_S m_N^2 = 423, \quad G_V m_N^2 = 380. \tag{18}$$

The behaviour of E/N as a function of ρ_B for this set of parameters is shown in Fig. 2. One can see that equation of state is rather soft at $\rho_B < \rho_0$ and quite stiff at $\rho_B > \rho_0$. It is interesting to see how this curve would look like if chiral symmetry would be perfect i.e. if $m_0=0$. The results of calculation with $m_0=0$ and all other parameters as in eq. (18) are also shown in Fig. 2. One sees that the saturation density is increased very little but the binding energy is changed from 16 to 60 MeV. The nucleon effective mass as a function of density is shown in Fig. 3. It decreases almost linearly down to small values where deviations from linear behaviour start. The value predicted at normal density is quite small, $m/m_N=0.32$, for $m_0=86$ MeV. As expected, the nucleon effective mass at ρ_0 goes to zero in the case $m_0=0$. According to eq. (17) the critical density in this case practically coincides with ρ_0.

It is not clear whether one can take seriously the above solution or not. The small value of cut-off momentum $\Lambda < m_N$ makes doubtful the validity of the mean-field approximation. The applicability of the Dirac equation to the composite particles like nucleons is also problematic. On the other hand the range of the attractive interaction $\Lambda^{-1} \approx 0.7$ fm is in agreement with the phenomenology of nucleon-nucleon interaction. Note that in ref. [14], where from the beginning m_0 was put to zero, the saturation of nuclear matter was not achieved unless an additional term, scalar-vector coupling, was introduced in the Lagrangian.

An alternative way is to formulate the NJL model on the constituent quark level and use it as an effective bag model. This can be done by simply changing ν_N to $\nu_q = 2N_c N_f$, where $N_c = 3$ is the number of colors and $N_f = 2$ is the number of flavours (u and d quarks only). Now the parameter m_0 is fixed by the current quark masses ($m_0 \approx 4$ MeV) while the constituent quark mass in vacuum m_q can be varied in some limits ($300 \div 500$ MeV). For instance, in ref. [17] the following parameters have been found to get correct properties of pseudoscalar and vector mesons:

$$\Lambda = 1050\,\text{MeV}, \quad m_0 = 3.3\,\text{MeV}, \quad G_S \Lambda^2 = 10.1, \quad G_V \Lambda^2 = 14.4, \tag{19}$$

which gives $m_q=463$ MeV. These parameters are in a striking disagreement with those in eq. (18). They do not give any bound state in the vicinity of ρ_0 and the critical density for the chiral transition is pushed to high baryon densities. Thus, the parameters which give correct meson properties are incompatible with the saturation properties of nuclear matter and vise versa.

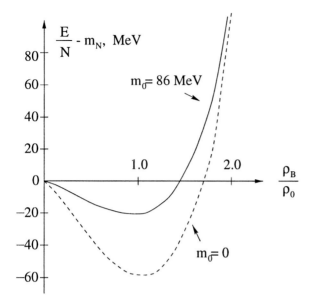

Figure 2: Energy per nucleon E/N as a function of baryon density in units of normal nuclear density $\rho_0 = 0.16 fm^{-3}$. Solid line labeled by $m_0 = 86$ MeV shows the calculation with model parameters from eq. (18). Dashed line shows the result for the same parameters but $m_0 = 0$.

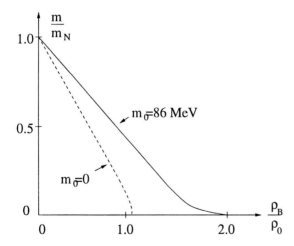

Figure 3: Nucleon effective mass (in units of free nucleon mass) as a function of baryon density. Notations are the same as in Fig. 2.

5 Comparison with Walecka models

It is instructive to make comparison of the NJL model with the Walecka model [3] represented by the Lagrangian

$$\mathcal{L} = \overline{N}\left(i\gamma^\mu\partial_\mu - m_N\right)N + g_S\overline{N}N\varphi - g_V\overline{N}\gamma^\mu N\omega_\mu - \frac{1}{2}m_S^2\varphi^2 + \frac{1}{2}m_V^2\omega^\mu\omega_\mu, \quad (20)$$

where φ and ω^μ are respectively the scalar and vector fields with masses m_S and m_V, coupled to nucleons with coupling constants g_S and g_V. We have dropped out terms containing derivatives of the meson fields which vanish in homogeneous matter at rest. By using the variational equations one can express the meson fields in terms of nucleon scalar and vector densities

$$\varphi = \frac{g_S}{m_S^2}\langle\overline{N}N\rangle, \quad \omega^0 = \frac{g_V}{m_V^2}\langle\overline{N}\gamma^0 N\rangle. \quad (21)$$

The saturation density and binding energy of nuclear matter can be reproduced with the model parameters [18]

$$C_S^2 = \frac{g_S^2}{m_S^2}m_N^2 = 358, \quad C_V^2 = \frac{g_V^2}{m_V^2}m_N^2 = 274. \quad (22)$$

Then the nucleon effective mass at ρ_0 is $m/m_N \approx 0.54$.

Inserting expressions (21) into the Lagrangian (20) one can see that the nonlinear terms look very similar to those in the NJL Lagrangian (4). This fact was used in ref. [19] to "prove" that the two models are equivalent. But in fact there remains a principal difference between these two models in the nature of the nucleon mass. In the Walecka model the Dirac Lagrangian contains the (large) free nucleon mass that makes this model non-chiral symmetric. In contrast, in the chiral-symmetric NJL model the (small) bare mass appears in the Dirac Lagrangian and the physical nucleon mass is generated by the mechanism of the spontaneous symmetry breaking. Strictly speaking the Walecka model corresponds to the "no Dirac sea" limit of the NJL model with $m_0 = m_N$. In other words, it can be obtained by the low-density expansion of the NJL Lagrangian around the vacuum state $\rho_B = 0$, $m = m_N$. But this expansion can not be continued to the saturation density of nuclear matter. One can conclude that the nuclear saturation density $\rho_0 = 0.16$ fm^{-3} is not small enough to apply the chiral perturbation theory. This is clearly demonstrated by the different nature of the bound state predicted by the two models. Also coupling strengths corresponding to the bound state are very different (compare C_S^2 and C_V^2 from eq. (22), respectively, with $G_S m_N^2$ and $G_V m_N^2$ from eq. (18)).

Nonlinear versions of the Walecka model are very successful in describing properties of finite nuclei. They include self-interaction terms of the scalar field in the form of the Boguta-Bodmer potential [20]

$$V(\varphi) = \frac{1}{2}m_S^2\varphi^2 + \frac{1}{3}b\varphi^3 + \frac{1}{4}c\varphi^4, \tag{23}$$

where b and c are additional adjustable parameters which are usually expressed in terms of $C_3 = b/(m_N g_S^3)$ and $C_4 = c/g_S^4$. Best fits require negative values of the parameter C_4 (see for instance ref. [21]).

It is interesting to note that negative c values are easy to understand from the view point of the chiral symmetry. Indeed, in this case the behaviour of the potential (23) becomes more similar to the behaviour of the effective scalar potential of the NJL model. To see this one should eliminate φ from eq. (23) in terms of constituent mass $m = m_N - g_S\varphi$ and compare it with corresponding scalar potential of the NJL model. In the exactly chiral-symmetric theory the potential should be symmetric with respect to the change $m \to -m$ and have zero slope at the point $m = 0$. In the general case the condition on the slope at $m = 0$ can be written as

$$\overline{C} \equiv C_S^{-2} + C_3 + C_4 = \frac{m_\pi^2 f_\pi^2}{m_N^4} \approx 2 \cdot 10^{-4} \tag{24}$$

In the original Walecka model ($C_3 = C_4 = 0$) one obtains $\overline{C} = C_S^{-2} \approx 0.0028$. In the best nonlinear fits, like NL-Z [21], $\overline{C} \approx 0.0016$. One sees the tendency towards chiral symmetry. Nevertheless these models are still far away from the chiral behaviour.

6   Conclusions

It is demonstrated that in the Nambu–Jona-Lasinio model there exists a possibility for only one bound state of matter. In the case of exact chiral symmetry (no symmetry breaking terms in the Lagrangian, $m_0 = 0$) this state would correspond to the phase with restored chiral symmetry and massless nucleons. When explicit symmetry-breaking terms are included ($m_0 \neq 0$), one can still find a bound state corresponding to the saturation point of nuclear matter but with small nuclear effective mass, $m/m_N \approx 0.3$. One may think that higher order effects associated with short-range nucleon-nucleon correlations and corrections to the mean-field approximation will lead to a more satisfactory description of this ground state. If this picture is qualitatively correct one can make the conclusions: 1) in the world with perfect chiral symmetry nuclei would be just droplets of the chiral liquid with massless constituents,

2) on the level of nucleonic degrees of freedom the cut-off momentum is quite small, $\Lambda \approx 300$ MeV i.e. the Dirac sea is quite shallow, 3) a small ($\sim \Lambda^3$) density of active negative-energy states should reduce significantly the yield od baryon-antibaryon pairs produced by strong meson fields. An alternative possibility is that the NJL model is meaningful only on the level of constituent quarks and, thus, it is incompatible with the saturation properties of nuclear matter.

Acknowledgments

This work is devoted to Walter Greiner who really loves exotic seas and most of all, the Dirac sea. The author thanks M. Gyulassy, L. McLerran, U. Mosel, M. Rho, L.M. Satarov, J. Schaffner, O. Scavenius, H. Stöcker, W. Weise and W. Greiner for fruitful discussions. This work was supported in part by EU–INTAS Grant No. 94–3405. The author also thanks the Niels Bohr Institute for kind hospitality and Carlsberg Foundation for financial support.

References

1. J. Schwinger, *Phys. Rev.* **82**, 664 (1951).
2. W. Pieper and W. Greiner, *Z. Phys.* **A218**, 327 (1969)
3. B.D. Serot and J.D. Walecka, *Adv. in Nucl. Phys.* **16**, 1 (1985).
4. B.D. Serot, Rep. Prog. Phys. **55**, 1855 (1992).
5. H.P. Düerr and E. Teller, *Phys. Rev.* **101**, 494 (1956).
6. T.D. Cohen, R.J. Furnstahl and D.K. Griegel, *Phys. Rev.* **C45**, 1881 (1992).
7. M. Lutz, S. Klimt and W. Weise, *Nucl. Phys.* **A542**, 521 (1992).
8. I.N. Mishustin, *Yad. Fiz. [Sov. J. Nucl. Phys.]* **52**, 1135 (1990).
9. I.N. Mishustin, L.M. Satarov, J. Schaffner, H. Stöcker and W. Greiner, *J. Phys. G: Nucl. Part. Phys.* **19**, 1303 (1993).
10. I.N. Mishustin, L.M. Satarov, H. Stöcker and W. Greiner, *Phys. Rev.* **C52**, 3315 (1995).
11. Y. Nambu and G. Jona-Lasinio, *Phys. Rev.* **122**, 345 (1961); **124**, 246 (1961).
12. U. Vogl and W. Weise, *Prog. Part. Nucl. Phys.* **27**, 195 (1991).
13. S. Klevansky, *Rev. Mod. Phys.* **64**, 649 (1992).
14. V. Koch, T.S. Biro, J. Kunz and U. Mosel, *Phys. Lett.* **B185**, 1 (1987); V. Koch, B. Blättel and U. Mosel, *Phys. Lett.* **B194**, 331 (1987).
15. X. Ji, *Phys. Rev. Lett.* **74**, 1071 (1995).
16. M. Gell-Mann, R.J. Oakes and B. Renner, *Phys. Rev.* **175**, 2195 (1968).

17. A. Polleri, R.A. Broglia, P.M. Pizzochero and N.N. Scoccola, Preprint NTGMI-95-4 (1995).

18. C.J. Horowitz and B.D. Serot, *Nucl. Phys.* **A368**, 303 (1981).

19. G. Gelmini and B. Ritzi, *Phys. Lett.* **B357**, 431 (1995).

20. J. Boguta and A.R. Bodmer, *Nucl. Phys.* **A292**, 414 (1977).

21. M. Rufa, P.-G. Reinhardt, J.A. Maruhn, W. Greiner and M.R. Strayer, *Phys. Rev.* **C38**, 390 (1988).

THE NUCLEON SPECTRAL FUNCTION AT FINITE TEMPERATURE IN NUCLEAR MATTER

G. RÖPKE

MPG Arbeitsgruppe "Theoretische Vielteilchenphysik"
Universität Rostock, D-18051 Rostock, Germany
and
Bogoliubov Laboratory of Theoretical Physics,
Joint Institute for Nuclear Research, 141980, Dubna, Russia

The nucleon self-energy and spectral function are calculated for nuclear matter at finite temperatures and different densities. In particular, the formation of bound states and the behaviour of the scattering phase shifts in a dense medium as well as the transition to superfluidity are discussed. Medium modifications of microscopic quantities such as cross sections, reaction rates and bound state wave functions are of importance for the abundances and energy spectra of clusters in hot dense matter.

1 Nuclear matter at intermediate excitation energy

Nuclear matter at intermediate excitation energies is an interesting object to investigate the medium modification of single-particle properties, two-particle properties, etc., in dense quantum systems. Some results obtained by our Rostock group will be reported. In particular, we are interested in densities $\rho \leq \rho_0 = 0.17$ fm^{-3} and temperatures $T \leq 50$ MeV, see Fig. 1.

The interaction between the nucleons has been given by various (Paris[2], Argonne[3], Bonn[4], Nijmegen[5]) potentials, especially by separable ones [6] which are convenient to find an explicit solution of the T-matrix in ladder approximation.

2 Self-consistent approximation for the two-particle propagator

The two-particle propagator is related to the T-matrix which is obtained in ladder approximation from the Bethe-Goldstone equation

$$T(12, 1'2', z) = V(12, 1'2') + \sum V(12, 34)\, G_2^0(34, z)\, T(34, 1'2', z). \quad (1)$$

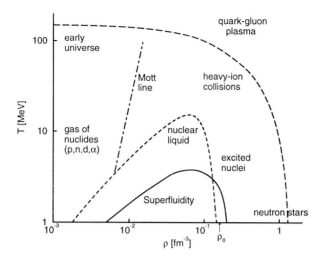

Figure 1: Temperature-density plane of nuclear matter. Due to the interaction, correlations arise in nuclear matter such as bound states, continuum correlations and condensates. The modification of single-particle properties, the dissolution of bound states near the Mott line, the occurrence of a first order phase transition analogous to the liquid-gas phase transition (dashed) and the occurrence of superfluidity are of importance to astrophysics and laboratory heavy ion reactions [1].

The solution can be given in form of a continuous fraction

$$T(12, 1'2', z) = \cfrac{V(12, 1'2')}{1 - \sum_{34} \cfrac{V(12,34)}{V(12,1'2')} \cfrac{G_2^0(34,z)V(34,1'2')}{1 - \sum_{56}\left[\frac{V(34,56)}{V(34,1'2')} - \frac{V(12,56)}{V(12,1'2')}\right]\frac{G_2^0(56,z)V(56,1'2')}{1 - \cdots}}} \quad (2)$$

which becomes particularly simple for separable interactions, where the continuous fraction ends after the first term. The uncorrelated propagation of two particles is given by

$$G_2^0(1, 2, z) = \int \frac{d\omega_1}{2\pi} \int \frac{d\omega_2}{2\pi} \frac{1 - f(\omega_1) - f(\omega_2)}{\omega_1 + \omega_2 - z} A(1, \omega_1) A(2, \omega_2) \quad (3)$$

containing the spectral function

$$A(1, \omega_1) = \frac{2 \operatorname{Im} \Sigma(1, \omega_1 - i0)}{[\omega_1 - \epsilon_1 - \operatorname{Re} \Sigma(1, \omega_1)]^2 + [\operatorname{Im} \Sigma(1, \omega_1 - i0)]^2} \quad (4)$$

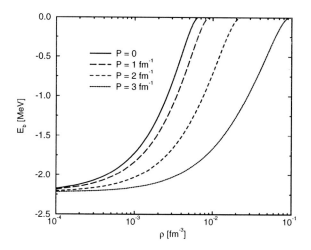

Figure 2: In-medium deuteron binding energies [9] as a function of the nuclear matter density for temperature $T = 10$ MeV and for several values of the total momentum P. The strongest effect of the Pauli blocking occurs for zero momentum deuterons. Their binding energies vanish at the so-called Mott density [10], cf. Fig. 1.

which, in turn, is given by the self-energy. The self-energy is related to the T-matrix according to

$$\text{Im}\Sigma(1,\omega_1 - i0) = \sum_2 \int \frac{d\omega_2}{2\pi} A(2,\omega_2) \left[f(\omega_2) + g(\omega_1 + \omega_2) \right] \text{Im} T_{\text{ex}}(12,12,\omega_1 + \omega_2 - i0)$$

(5)

and the real part follows from the Kramers-Kronig relation

$$\text{Re}\,\Sigma(1,\omega_1) = \Sigma^{\text{HF}}(1) + \mathcal{P} \int \frac{d\omega_2}{\pi} \frac{\text{Im}\,\Sigma(1,\omega_1 - i0)}{\omega_1 - \omega_2}.$$

(6)

Therefore, we have to solve a coupled system of equations in a self-consistent way. Furthermore, the results for the self-energy and the spectral function are improved if, in addition to the ladder approximation, further diagrams are included in calculating the T-matrix.

3 Two-particle states in hot dense matter

Using the angle averaged quasiparticle approximation for G_2^0, Eq. (3), as a first step of an iteration process, the solution of the T-matrix can immediately

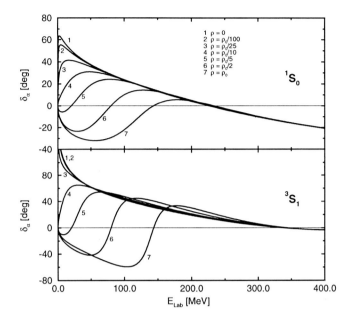

Figure 3: In-medium scattering phase shifts as function of lab. energy for nuclear matter at different densities, temperature $T = 10$ MeV and zero total momentum[9]. In the triplet channel, the phase shift at zero energy performs a jump of magnitude π at the Mott density in accordance with the Levinson theorem .

be found[7]. In particular, the binding energies of bound states (see Fig. 2) as well as the scattering phase shifts given by $\tan \delta_\alpha = \mathrm{Im}\, T_\alpha / \mathrm{Re}\, T_\alpha$ (see Fig. 3) are depending on density, temperature and total momentum.

Having the T-matrix at our disposal, the differential cross section as well as the total nucleon-nucleon cross section are evaluated[8], e. g.

$$\sigma(p, P) = [2\, \mathrm{Im}\, G_2^0(p, P)]^2 \sum_{J, L, L'} \frac{(2J+1)}{(2s_1+1)(2s_2+1)} \frac{4\pi}{p^2} |T^{LL'}(p, p, P)|^2. \quad (7)$$

As discussed below, the low-temperature enhancement[7] near $E_{\mathrm{lab}} = 4\mu$ of the in-medium nucleon-nucleon cross section shown in Fig. 4 can be interpreted as critical scattering near the superfluidity transition.

4 Single-particle spectral function

To go beyond the quasiparticle approximation, we consider the first iteration of the spectral function, Eq. (4), where two-particle states are included in the self-energy. Results for different densities are shown in Figs. 5, 6, 7.

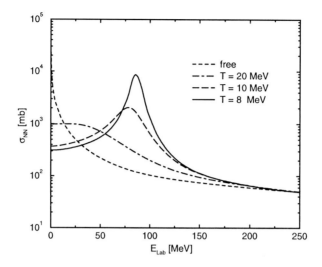

Figure 4: In-medium nucleon-nucleon cross section as a function of the lab. energy for nuclear matter at different temperatures, density $\rho_0/2$ and zero total momentum [9].

Evidently, the quasiparticle approximation breaks down when the inclusion of correlation leads to a more complicate line profile of the spectral function. A fully self-consistent calculation would smoothen sharp structures such as the contribution of bound states, since bound states are of finite width. Furthermore, the approximation for the self-energy is improved by including further diagrams.

5 Generalized Beth-Uhlenbeck equation of state

Having the nucleon spectral function at our disposal, many important physical properties can be evaluated. For instance, due to nucleon-nucleon correlations the occupation of states with high momenta is enhanced [12]. This modification of the nucleon momentum distribution is accessible by quasi-elastic electron scattering experiments.

Thermodynamic properties of hot dense matter are given by equations of state, e.g. the density as a function of the temperature and of the chemical potential. The pressure as a thermodynamic potential is obtained from the

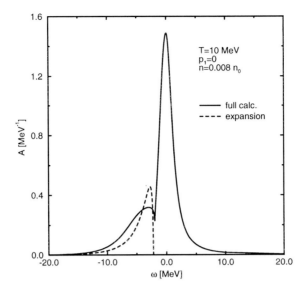

Figure 5: Imaginary part of the nucleon self-energy at density $\rho_0/100$ for temperatures $T=10$ MeV and zero momentum [11]. In addition to the broadened quasiparticle peak a satellite stucture due to the bound state contributions is seen at densities below the Mott density.

density by an integration procedure. From the spectral function we find

$$\rho(\beta,\mu) = \frac{1}{\Omega} \sum_1 \int \frac{d\omega_1}{2\pi} \frac{1}{e^{\beta(\omega_1-\mu)}+1} A(1,\omega_1). \tag{8}$$

With the expression for the spectral function given above, the total density is decomposed into three parts, the quasiparticle contribution, the bound state contribution, and the contribution of continuum correlations [7]. The composition of nuclear matter at finite temperature is shown in Fig. 8. Evidently an approach where few-nucleon states are treated in an uncorrelated medium is no longer valid in regions where the main contribution to the density is due to correlated nucleons.

6   Superfluidity transition

Singularities of the solution of the T-matrix $T(\omega)$ at ω_b are related to bound states. If furthermore $\omega_b = 2\mu$, the singularity of the Bose distribution func-

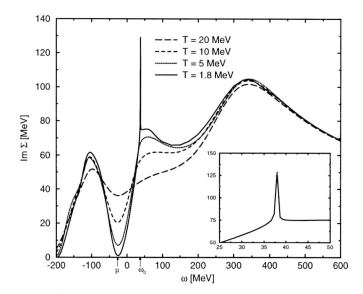

Figure 6: Imaginary part of the nucleon self-energy at the saturation density for four different temperatures and zero momentum [9,12]. With decreasing temperatures, Im Σ goes to zero at $\omega = \mu$. The sharp peak at $\omega_0 = 2\mu - \epsilon_1$ is considered as a precursor effect of the superfluidity phase transition.

tion leads to a divergency which is related to the transition to superfluidity (Thouless criterion).

In symmetric nuclear matter, the pairing instability with highest temperature occurs in the triplet channel first considered by Alm et al.[13] At low densities, the composition of nuclear matter is dominated by bound states (deuterons), and Bose-Einstein condensation (BEC) will occur at low temperatures. Above the Mott density, the transition to superfluidity is described by Cooper pairing. The crossover from BEC to BCS is an interesting phenomenon in nuclear matter. To give an adequate description of the superfluidity transition near the Mott line, see Fig. 8, correlations in the medium have to be included.

7   Three-particle reactions in a medium

The treatment of two-particle states can be extended to the treatment of few-body systems in a hot dense medium. Using the Green function method, Faddeev-type equations are derived containing also mean field corrections and

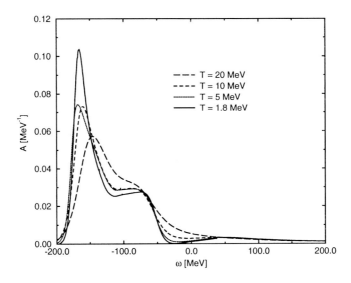

Figure 7: Nucleon spectral function as a function of the energy argument at the saturation density for four different temperatures and zero momentum [9,12].

Pauli blocking terms. In particular, the formation and break-up cross sections for deuterons are modified due to the shift of the energy threshold for the reaction and due to phase space occupation [14].

8   Medium effects in Heavy Ion Reactions

Despite the necessity of a theory for finite systems in nonequilibrium, concepts of thermal equilibrium or transport models have been applied to describe the results of heavy ion collisions. However, the concept of a free mixture of different components in chemical equilibrium is not applicable to nuclear matter at densities near ρ_0. If one prefers to use results from equilibrium hot dense matter to give an interpretation of HIC experiments, one has to take into account the medium modifications of the few-body properties.

The medium modifications of the nucleon-nucleon cross section, see Fig. 4, should be considered as an input to BUU simulations. Such corrections are of relevance to hydrodynamic calculations and, in particular, to the evaluation of the balance energy [15]. From the in-medium T-matrix, medium modifications of different processes such as the photon emission or the neutrino [16] (Urca) process have been evaluated. Transport coefficients and linear response theory

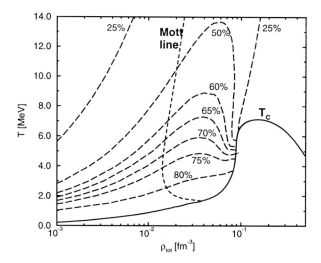

Figure 8: Temperature-density plane of symmetric nuclear matter showing lines of equal concentration of correlated nucleons, the Mott line, and the critical temperature of the onset of superfluidity.

can be formulated including medium modifications in hot dense matter.

An important phenomenon is cluster production in hot dense matter. Within a coalescence model, usually the properties of isolated bound states (energy, wave function etc.) are considered which, however, have to be replaced by in-medium properties. Depletions of the cluster abundances and modifications of the energy spectra of clusters in hot dense matter are consequences of medium corrections [17]. Neglecting these effects, contradictory conclusions about the properties of hot nuclear matter can arise.

9 Outlook

i) A self-consistent treatment of few-body properties in a correlated medium should be achieved. For instance, the single-particle spectral function or the pair amplitude for a condensate has to be evaluated beyond the quasiparticle picture.

ii) To make contact with experiments such as collisions with heavy ions, electrons, etc., one has to consider finite systems (cf. local density approximations) and nonequilibrium processes (hydrodynamic/kinetic equations including correlations in nonequilibrium) instead of equilibrium nuclear matter.

iii) A consequent quantum statistical approach is demanded to describe the formation of clusters and their energy distribution in hot dense matter.

iv) At higher densities and/or higher temperatures, relativistic effects, formation of new particles and the possible transition from hadronic matter to quark-gluon matter have to be considered.

Acknowledgments

I thank my coworkers T. Alm, M. Beyer, D. Blaschke, K. Morawetz, A. Schnell, A. Sedrakian and H. Stein which contributed to the results reviewed here.

References

1. H. Stöcker and W. Greiner, Phys. Reports **137**, 277 (1986).
2. M. Lacombe, B. Loiseau, J.M. Richard, R. Vinh Mau, J. Pires, and R. Tourreil, *Phys. Rev.* C **21**, 861 (1980).
3. R.B. Wiringa, R.A. Smith, and T.L. Ainsworth, *Phys. Rev.* C **29**, 1207 (1984).
4. R. Machleidt, K. Holinde, and Ch. Elster, Phys. Rep. **149**, 1 (1987).
5. V.G.J. Stoks, R.A.M. Klomp, C.P.F. Terheggen, and J.J. de Swart, *Phys. Lett.* B **267**, 159 (1991).
6. J. Haidenbauer and W. Plessas, *Phys. Rev.* C **30**, 1822 (1984).
7. M. Schmidt, G. Röpke, and H. Schulz, Ann. Phys. (N.Y.) **202**, 57 (1990).
8. T. Alm, G. Röpke, and M. Schmidt, *Phys. Rev.* C **50**, 31 (1994).
9. A. Schnell, PhD Thesis, Rostock 1996.
10. G. Röpke, L. Münchow, and H. Schulz, *Nucl. Phys.* A **379**, 536 (1982).
11. T. Alm, G. Röpke, A. Schnell, and H. Stein, *Phys. Lett.* B **346**, 233 (1995).
12. T. Alm, G. Röpke, A. Schnell, N.H. Kwong, and H.S. Köhler, *Phys. Rev.* C **53**, 2181 (1996).
13. T. Alm, G. Röpke, and M. Schmidt, *Z. Phys.* A **337**, 487 (1990).
14. M. Beyer, G. Röpke, and A.D. Sedrakian, *Phys. Lett.* B, in print.
15. T. Alm, G. Röpke, W. Bauer, F. Daffin, and M. Schmidt, *Nucl. Phys.* A **587**, 815 (1995).
16. D. Blaschke, G. Röpke, A.D. Sedrakian, and D.V. Voskresensky, Mon. Not. R. Astron. Soc. **273**, 596 (1995).
17. G. Röpke, H. Schulz, L.N. Andronenko, A.A. Kotov, W. Neubert, and E.N. Volnin, *Phys. Rev.* C **31**, 1556 (1985).

PARTICLE PRODUCTION BY TIME–DEPENDENT MESON FIELDS IN RELATIVISTIC HEAVY–ION COLLISIONS

I.N. MISHUSTIN[a,b], L.M. SATAROV[a]

[a] *The Kurchatov Institute, 123182 Moscow, Russia*
[b] *The Niels Bohr Institute, DK–2100 Copenhagen Ø, Denmark*

H. STÖCKER

Institut für Theoretische Physik, J.W. Goethe Universität,
D–60054 Frankfurt am Main, Germany

We study bremsstrahlung of real and virtual ω–mesons due to the collective deceleration of nuclei in the course of ultrarelativistic heavy–ion collisions. It is shown that decays of these mesons give a noticeable contribution to the observed yields of the baryon–antibaryon pairs, dileptons and pions. Excitation functions and rapidity distributions of particles produced by this mechanism are calculated under some simplifying assumptions about the space–time variation of baryon currents in a nuclear collision process. The calculated multiplicities of coherently produced particles grow fast with the bombarding energy, reaching a saturation above the RHIC energy. In the case of central Au+Au collisions the bremsstrahlung mechanism becomes comparable with particle production in incoherent hadron–hadron collisions above the AGS energies. The rapidity spectra of antibaryons and pions exhibit a characteristic two–hump structure which is a consequence of incomplete projectile–target stopping at the initial stage of the reaction. The predicted distribution of e^+e^- pairs has a strong peak at small e^+e^- invariant masses.

1   Introduction

It is widely accepted now that strong meson fields exist in ordinary nuclei, which are successfully described by the relativistic mean–field models [1]. According to this picture, strong time–dependent meson fields should be generated in the course of relativistic heavy–ion collisions. Within this framework several new collective phenomena were predicted: the filamentation instability of interpenetrating nuclei [2] and the spontaneous creation of baryon–antibaryon $(B\overline{B})$ pairs in a superdense baryon–rich matter [3]. Using the approach developed earlier for the pion [4] and photon [5] bremsstrahlung, we suggested recently [6] a new mechanism of the $B\overline{B}$ pair production by the collective bremsstrahlung and decay of meson fields in relativistic heavy–ion collisions. Such pairs may be produced at sufficiently high bombarding energies when characteristic Fourier frequencies of meson fields exceed the energy gap between the positive and negative energy states of baryons.

In the lowest order approximation the production of the $B\overline{B}$ pair may be considered as a two–step process $A_p A_t \to \omega_* \to B\overline{B}$. Here $A_p(A_t)$ denotes the projectile (target) nucleus and ω_*, the off–mass–shell vector meson [a]. The first step in the above reaction is a bremsstrahlung process leading to the production of virtual ω mesons with masses $M > 2m_B$ (m_B is the baryon mass). The second step is the conversion of these mesons into $B\overline{B}$ pairs. Obviously, this process is forbidden for a real ω meson which has the mass $m_\omega \simeq 0.783$ GeV and the relatively small width $\Gamma_\omega \simeq 8.4$ MeV. It is clear that an analogous bremsstrahlung mechanism may produce also pions (by decays of quasireal mesons $\omega \to \pi^+\pi^0\pi^-$) and low mass dileptons ($M_{l+l^-} \lesssim m_\omega$).

In this work we study the bremsstrahlung of vector meson fields originated from the collective deceleration of the projectile and target nuclei at the initial stage of a heavy–ion collision. The various channels of the bremsstrahlung conversion, including the production of the $N\overline{N}$ pairs, pions and dileptons are considered with emphasize to their observable signals.

2 Particle production by bremsstrahlung of nuclear meson fields

By the analogy to the Walecka model we introduce the vector meson field $\omega^\mu(x)$ coupled to the 4–current $J^\mu(x)$ of nucleons participating in a heavy–ion collision at a given impact parameter. The equation of motion defining the space–time behavior of $\omega^\mu(x)$ may be written as ($c = \hbar = 1$)

$$(\partial^\nu \partial_\nu + m_\omega^2)\,\omega^\mu(x) = g_V J^\mu(x)\,, \tag{1}$$

where g_V is the ωN coupling constant. In the mean–field approximation the quantum fluctuation of J^μ are disregarded and the vector meson field is purely classical. From Eq. (1) one can see that excitation of propagating waves in a vacuum (this is equivalent to the bremsstrahlung process) is possible if the Fourier transformed baryonic current

$$J^\mu(p) = \int \mathrm{d}^4 x\, J^\mu(x) e^{ipx} \tag{2}$$

is nonzero in the time–like region $p^2 = m_\omega^2$.

In the following we study the bremsstrahlung process in the lowest order approximation neglecting the back reaction and reabsorption of the emitted vector mesons, i.e. treating J^μ as an external current. From Eq. (1) one can calculate the energy flux of the vector field at a large distance from the collision

---

[a] As discussed in Ref. [6], at relativistic bombarding energies bremsstrahlung of the scalar meson field is small as compared to the vector meson field. Due to this reason we disregard here the contribution of the scalar meson bremsstrahlung.

region [4]. This leads to the following formulae for the momentum distribution of real ω–mesons emitted in a heavy–ion collision

$$E_\omega \frac{\mathrm{d}^3 N_\omega}{\mathrm{d}^3 p} = S(E_\omega, \boldsymbol{p}), \tag{3}$$

where $E_\omega = \sqrt{m_\omega^2 + \boldsymbol{p}^2}$ and

$$S(p) = \frac{g_V^2}{16\pi^3} |J_\mu^*(p) J^\mu(p)| \tag{4}$$

is a source function [2]. In our model the latter is fully determined by the collective motion of the projectile and target nucleons.

To take into account the off–mass–shell effects we characterize virtual ω mesons by their mass M and total width Γ_{ω_*}. The spectral function for the off–shell ω mesons may be written as

$$\rho(M) = \frac{2}{\pi} \frac{M \Gamma_{\omega_*}}{(M^2 - m_\omega^2)^2 + m_\omega^2 \Gamma_{\omega_*}^2}. \tag{5}$$

To calculate the distribution of virtual mesons in their 4–momentum p we use the formulae [7]

$$\frac{\mathrm{d}^4 N_{\omega_*}}{\mathrm{d}^4 p} = \rho(M) S(p), \tag{6}$$

where $M \equiv \sqrt{p^2}$. In the limit $\Gamma_{\omega_*} \to 0$ one can replace $\rho(M)$ by $2\delta(M^2 - m_\omega^2)$. In this case Eq. (6) becomes equivalent to the formulae (3) for the spectrum of the on–mass–shell vector mesons.

Below we consider the most important channels of the virtual ω decay: $i = 3\pi, N\overline{N}, e^+ e^-, \mu^+ \mu^-$. The total width Γ_{ω_*} can be decomposed into the sum over the partial decay widths $\Gamma(\omega_* \to i)$:

$$\Gamma_{\omega_*} = \sum_i \Gamma(\omega_* \to i). \tag{7}$$

Then the distribution over the total 4–momentum of particles in a given decay channel i may be written as

$$\frac{\mathrm{d}^4 N_{\omega_* \to i}}{\mathrm{d}^4 p} = B(\omega_* \to i) \frac{\mathrm{d}^4 N_{\omega_*}}{\mathrm{d}^4 p}, \tag{8}$$

where $B(\omega_* \to i) \equiv \Gamma(\omega_* \to i)/\Gamma_{\omega_*}$ is the branching ratio of the i-th decay channel. The latter is a function of the invariant mass of decay particles M.

To calculate the baryon 4–current $J^\mu(p)$ determining the source function $S(p)$ we use the simple picture of a high–energy heavy–ion collision suggested in Ref.[6] Below we consider collisions of identical nuclei $(A_p = A_t = A)$ at zero impact parameter. In the equal velocity frame the projectile and target nuclei initially move towards each other with the velocities $\pm v_0$ where $v_0 = \tanh y_0 = (1 - 4m_N^2/s)^{1/2}$ is determined by the c.m. bombarding energy per nucleon $\sqrt{s}$. In the "frozen density" approximation[6] the internal compression and transverse motion of nuclear matter are disregarded at the early (interpenetration) stage of the reaction. Within this approximation the colliding nuclei move as a whole along the beam axis with instantaneous velocities $\dot{\xi}_p = -\dot{\xi}_t \equiv \xi(t)$. The projectile velocity $\dot{\xi}(t)$ is a decreasing function of time, which we choose in the form[4]

$$\dot{\xi}(t) = v_f + \frac{v_0 - v_f}{1 + e^{t/\tau}}, \tag{9}$$

where τ is the effective deceleration time and v_f is the final velocity of nuclei (at $t \to +\infty$).

In our approximation the Fourier transforms $J^\mu(p)$ are totally determined by the projectile trajectory $\xi(t)$[6]:

$$J^0(p) = \frac{p_\parallel}{p_0} J^3(p) = 2A \int\limits_{-\infty}^{\infty} dt\, e^{ip_0 t} \cos[p_\parallel \xi(t)]\, F\left(\sqrt{p_T^2 + p_\parallel^2 \cdot [1 - \dot{\xi}^2(t)]}\right), \tag{10}$$

where $p_\parallel$ and $\boldsymbol{p}_T$ are, respectively, the longitudinal and transverse components of the three–momentum $\boldsymbol{p}$, $F(q)$ is the density form factor of the initial nuclei

$$F(q) \equiv \frac{1}{A} \int d^3 r \rho(r) e^{-i\boldsymbol{q}\,\boldsymbol{r}}. \tag{11}$$

The time integrals in Eq. (10) were calculated numerically assuming the Woods–Saxon distribution of the nuclear density $\rho(r)$. According to Eqs. (4), (10) the source function $S(p)$ vanishes at $p_\parallel = 0$. As a result, at high bombarding energies single particle distributions have a dip in a central rapidity region (see Figs. 2–3 and Ref.[6]). The two–hump structure of the rapidity spectra is a consequence of the incomplete mutual stopping of nuclei at the initial stage of a heavy–ion collision. On the other hand, the conventional mechanism of particle production in incoherent hadron–hadron collisions results in rapidity distributions of pions and antiprotons with a single central maximum even at high bombarding energies[8,9].

In this work we use the same choice of the coupling constant g_V and the stopping parameters τ, v_f as in Ref.[6] In particular, it is assumed that τ equals a half of the nuclear passage time

$$\tau = R/\sinh y_0 \,, \tag{12}$$

where R is the geometrical radius of initial nuclei. Instead of v_f we introduce the c.m. rapidity loss δy:

$$v_f = \tanh\left(y_0 - \delta y\right). \tag{13}$$

In the case of a central Au+Au collision we take the energy–independent value[8] $\delta y = 2.4$ for $\sqrt{s} > 10$ GeV, and assume full stopping ($\delta y = y_0$) for lower bombarding energies.

The partial width of the 3π decay channel is calculated assuming that $\Gamma(\omega_* \to 3\pi)$ is proportional to the three–body phase space volume [2]. The normalization constant is determined from the condition that $B(\omega_* \to 3\pi)$ equals the observed value $B(\omega \to 3\pi) = 0.89$ at $M = m_\omega$. Calculation of the $\omega_* \to N\overline{N}$ matrix element in the lowest order approximation in g_V gives the result

$$\Gamma(\omega_* \to N\overline{N}) = \frac{g_V^2}{6\pi}\sqrt{M^2 - 4m_N^2}\left(1 + \frac{2m_N^2}{M^2}\right)\Theta(M - 2m_N), \tag{14}$$

where $\Theta(x) = \frac{1}{2}(1 + \mathrm{sign}\,x)$. After substituting (14) into Eq. (6) and omitting the second term in the denominator of $\rho(M)$ one arrives at the distribution over the pair 4–momentum obtained earlier in Ref. [6]

The dilepton production is studied by calculating the matrix elements of the process $\omega_* \to \gamma_* \to l^+l^-$ where γ_* is a virtual photon. This calculation leads to the result ($l = \mathrm{e}, \mu$) [10]

$$\frac{\Gamma(\omega_* \to l^+l^-)}{\Gamma(\omega \to l^+l^-)} = \left(\frac{m_\omega}{M}\right)^6 \frac{M^2 + 2m_l^2}{m_\omega^2 + 2m_l^2}\sqrt{\frac{M^2 - 4m_l^2}{m_\omega^2 - 4m_l^2}}\,\Theta(M - 2m_l), \tag{15}$$

where m_l is the lepton mass.

3 Results

Below we present the results of numerical calculations obtained within the model described in the preceding section. Some of the model predictions concerning the $B\overline{B}$ production have been already published in Ref. [6] Fig. 1 shows the π^- and $\overline{p}$ multiplicities as functions of the bombarding energy in the case of central Au+Au collisions. For comparison, we show experimental data on the π^- multiplicities in the 11.6 AGeV/c Au+Au (circle) [11] and 160 AGeV Pb+Pb (triangle) [12] central collisions. One can see that the multiplicity of pions produced by bremsstrahlung exhibits a rapid growth between the AGS ($\sqrt{s} \simeq 5$ AGeV) and the SPS ($\sqrt{s} \simeq 20$ AGeV) energies and saturates above the RHIC ($\sqrt{s} \simeq 200$ AGeV) energy region. It is interesting that the bremsstrahlung component of pion yield becomes comparable with pion production in incoherent hadron–hadron collisions [9] already at the SPS bombarding energies. Note, however, that actual pion and antiproton yields, especially for heavy combinations of nuclei, may be reduced due to the absorption and annihilation neglected in the present model.

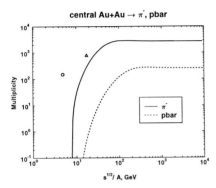

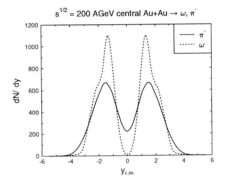

Figure 1: Excitation functions of π^- mesons (solid line) and antiprotons (dashed line) produced by bremsstrahlung in central Au+Au collision. Circle and triangle are experimental data on π^- multiplicity (see text).

Figure 2: Rapidity distributions of ω (solid line) and π^- (dashed line) mesons produced by bremsstrahlung in central Au+Au collision at RHIC bombarding energy.

The results on the π^- rapidity spectra are represented in Figs. 2–3. The spectra are calculated in the approximation $\Gamma_{\omega_*} = 0$. Here we use the kinematic formulae connecting the pion spectrum and the "primordial" distribution of vector mesons, Eq. (3), suggested in Ref. [2] Similarly to the case of antiprotons [6], the pion rapidity spectrum has a pronounced dip at $y_{c.m.} \simeq 0$. The two–hump structure of the π and $\bar{p}$ spectra may serve as a signature of the bremsstrahlung mechanism. According to Fig. 3 this structure can be seen only at high enough bombarding energies.

Fig. 4 shows the distributions over invariant masses of particles produced in different channels of bremsstrahlung conversion in the case of a central Au+Au collision at the SPS energy. Note that the mass spectrum of e^+e^- pairs created by the bremsstrahlung mechanism has a strong peak at invariant masses below the ω meson mass. On the other hand, attempts to explain the low mass dilepton yield by the conventional incoherent mechanisms (e.g. due to the $\pi\pi \to \rho \to e^+e^-$ processes) strongly underestimate the experimental data [13].

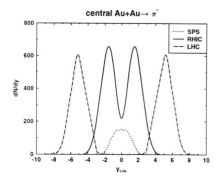

Figure 3: Rapidity spectra of π^- mesons in central Au+Au collisions at different bombarding energies.

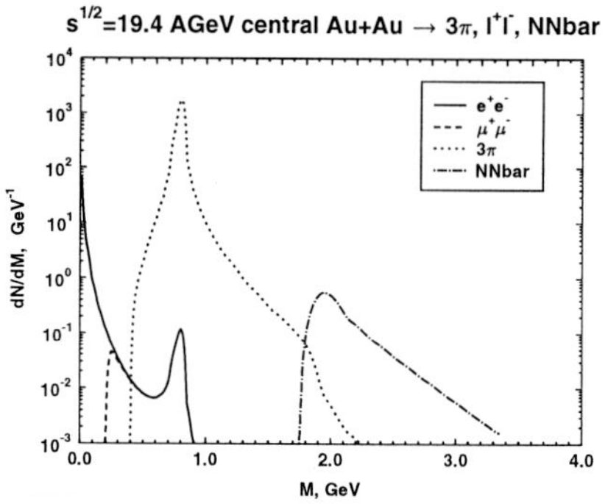

Figure 4: Distributions over invariant masses of particles in different decay channels of virtual ω mesons produced in central Au+Au collision at SPS energy.

4 Conclusions

In this work we have shown that the coherent bremsstrahlung of the vector meson field may be an important source of particle production already at the SPS bombarding energies. The observable signals of this mechanism may be the two–hump structure of pion and antibaryon rapidity spectra as well as the enhanced yield of low mass dileptons. The sharp energy and A–dependence of pion, dilepton and antibaryon excitation functions can be also a signature of the considered mechanism. The latter may be responsible, at least partly, for a rapid increase of the pion multiplicity observed in transition from the AGS to SPS energy [14].

Acknowledgments

This work is dedicated to Walter Greiner who initiated studies of the collective bremsstrahlung in nuclear collisions. The authors thank Yu.B. Ivanov,

S. Schramm and L.A. Winckelmann for valuable discussions. This work has been supported in part by the EU–INTAS Grant No. 94–3405. We acknowledge also the financial support from GSI, BMFT and DFG. One of us (I.N.M.) thanks the Niels Bohr Institute for the kind hospitality and the Carlsberg Foundation for the financial support.

References

1. B.D. Serot and J.D. Walecka, *Adv. in Nucl. Phys.* **16**, 1 (1985).
2. Yu.B. Ivanov, *Nucl. Phys.* A **495**, 633 (1989).
3. I.N. Mishustin, *Yad. Fiz. [Sov. J. Nucl. Phys.]* **52**, 1135 (1990);
   I.N. Mishustin, L.M. Satarov, J. Schaffner, H. Stöcker and W. Greiner, *J. Phys. G: Nucl. Part. Phys.* **19**, 1303 (1993).
4. D. Vasak, H. Stöcker, B. Müller and W. Greiner, *Phys. Lett.* **93B**, 243 (1980);
   D. Vasak, B. Müller and W. Greiner, *Phys. Scr.* **22**, 25 (1980).
5. T. Lippert, U. Becker, N. Grün, W. Scheid and G. Soff, *Phys. Lett.* B **207**, 366 (1988).
6. I.N. Mishustin, L.M. Satarov, H. Stöcker and W. Greiner, *Phys. Rev.* C **52**, 3323 (1995).
7. J. Knoll and D.N. Voskresensky, Report GSI-95-63, (Darmstadt, GSI, 1995); *Ann. Phys.* (N.Y.), in print.
8. Th. Schönfeld, H. Sorge, H. Stöcker and W. Greiner, Mod. Phys. Lett. A **8**, 2631 (1993).
9. N.S. Amelin, H. Stöcker, W. Greiner, N. Armesto, M.A. Braun and C. Pajares, *Phys. Rev.* C **52**, 362 (1995); N.S. Amelin, private communication.
10. P. Koch, *Z. Phys.* C **57**, 283 (1993).
11. M. Gonin, *Nucl. Phys.* A **533**, 799c (1993).
12. S. Margetis et al., Preprint LBL–36883, (Berkeley, LBL, 1995).
13. G. Agakishiev et al. (CERES Collab.), *Phys. Rev. Lett.* **75**, 1272 (1995).
14. M. Gaździcki, D. Röhrich, *Z. Phys.* C **65**, 215 (1995).

SHOCK WAVES IN HEAVY ION COLLISIONS

M.I. GORENSTEIN

Bogolyubov Institute for Theoretical Physics,
252143 Kiev-143, Ukraine

Shock waves in relativistic nuclear collisions are considered for proceses involving the deconfinement-confinement phase transitions. A general scheme of relativistic shock phenomena is formulated. The compression shock signature of quark-gluon plasma formation in nucleus-nucleus collisions is presented. We consider also the time-like shock hadronization of a supercooled quark-gluon plasma.

1 Introduction

Walter Greiner proposed nuclear shock wave mechanism almost 30 years ago [1]. Compression shocks were studied to extract the nuclear matter equation of state (EOS) from experimental data on heavy ion collisions (see, for example, [2]). A new step has been done to consider shock-like phase transitions between hadronic matter (HM) and quark-gluon plasma (QGP). Different scenarios involving shocks as a possible mechanism for both the deconfinement transition and for QGP hadronization have been suggested. The dynamical basis of these model considerations is the energy-momentum and baryonic number conservation across the discontinuity transition front. The entropy growth condition (thermodynamical stability) and mechanical stability conditions [3] should be additionally checked to guarantee shock existence.

In Section 2 a general picture of relativistic shock waves is considered. EOS for strongly interacting matter is discussed in Section 3. In Section 4 the compression shock model of the deconfinement transition is presented and in Section 5 we study the shock wave hadronization of the QGP.

2 Relativistic Shock Waves

The system evolution in relativistic hydrodynamics is governed by the energy-momentum tensor $T^{\mu\nu} = (\epsilon + p)u^\mu u^\nu - pg^{\mu\nu}$ and conserved charge currents (in our applications to heavy ion collisions it is the baryonic current nu^μ). They consist of local thermodynamical fluid quantities (energy density, ϵ, pressure, p, baryonic number density n) and the collective four-velocity $u^\mu = (1 - \mathbf{v}^2)^{-1/2}(1, \mathbf{v})$. The continuous flows are the solutions of the hydrodynamical equations $\partial T^{\mu\nu}/\partial x^\mu = 0$, $\partial nu^\mu/\partial x^\mu = 0$ with specified initial and boundary conditions. These equations are nothing more than the differential form of the energy-momentum and baryonic number conservation laws. Along with these continuous flows, the conservation laws can also be realized in the form of discontinuous hydrodynamical flows which are called shock waves and satisfy the following equations:

$$T_o^{\mu\nu}\Lambda_\nu = T^{\mu\nu}\Lambda_\nu \ , \qquad n_o u_o^\mu \Lambda_\mu = n u^\mu \Lambda_\mu \ , \tag{1}$$

where Λ^μ is the unit 4-vector normal to the discontinuity hypersurface. In Eq. (1) the zero index corresponds to the initial state ahead of the shock front and quantities without an index are the final state values behind it. A general derivation of the shock equations (valid for both space-like and time-like normal vectors Λ^μ) was given in Ref. [4].

The important constraint on the transitions (1) (thermodynamical stability condition) is the requirement of non-decreasing entropy: $su^\mu\Lambda_\mu \geq s_o u_o^\mu\Lambda_\mu$.

To simplify our consideration we consider only one-dimensional hydrodynamical motion in what follows. To study the s.l. shock transitions (i.e., with space-like Λ^μ) one can always choose the Lorentz frame where the shock front is at rest. Then the hypersurface of shock discontinuity is $x_{s.l.sh} = const$ and $\Lambda^\mu = (0, 1)$. The shock equations (1) in this (standard) case are:

$$T_o^{01} = T^{01}, \ \ T_o^{11} = T^{11} \ , \quad n_o u_o^1 = n u^1 \ . \tag{2}$$

From Eq. (2) one obtains

$$v_o^2 = \frac{(p - p_o)(\epsilon + p_o)}{(\epsilon - \epsilon_o)(\epsilon_o + p)} \ , \quad v^2 = \frac{(p - p_o)(\epsilon_o + p)}{(\epsilon - \epsilon_o)(\epsilon + p_o)} \ . \tag{3}$$

Substituting (3) into the last equation in (2) we obtain the well known Taub adiabat equation (TA) [5]

$$n^2 X^2 - n_o^2 X_o^2 - (p - p_o)(X + X_o) = 0 \ , \tag{4}$$

in which $X \equiv (\epsilon + p)/n^2$, and it therefore contains only thermodynamical variables.

For discontinuities on a hypersurface with a time-like normal vector Λ^μ (t.l. shocks) one can always choose another convenient Lorentz frame ("simultaneous system") where the hypersurface of the discontinuity is $t_{t.l.sh} = const$ and $\Lambda^\nu = (1, 0)$. Equations (1) are then

$$T_o^{00} = T^{00}, \ \ T_o^{10} = T^{10} \ , \quad n_o u_o^0 = n u^0 \ . \tag{5}$$

From Eq. (5) we find

$$\tilde{v}_o^2 = \frac{(\epsilon - \epsilon_o)(\epsilon_o + p)}{(p - p_o)(\epsilon + p_o)}, \quad \tilde{v}^2 = \frac{(\epsilon - \epsilon_o)(\epsilon + p_o)}{(p - p_o)(\epsilon_o + p)} \ , \tag{6}$$

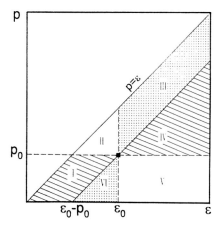

Figure 1: Possible final states in the $\epsilon - p$ plane for shock transitions from the intial state (ϵ_o, p_o). I and IV are the physical regions for s.l. shocks, III and VI for t.l. shocks. II and V are unphysical regions for both types of shocks. Note, that only states with $p \leq \epsilon$ are possible for any physical equation of state in relativistic theory.

where we use " $\sim$ " sign to distinguish the t.l. shock case (6) from the standard s.l. shocks (3). Substituting (6) into the last equation in (5) one finds the equation for t.l. shocks which is identical to the TA of Eq. (4). We stress that the intermediate steps (Eqs. (6) and (3)) are, however, quite different. Note, that the two solutions, Eqs. (6) and (3), are connected to each other by simple relations,

$$\tilde{v}_o^2 = \frac{1}{v_o^2}, \quad \tilde{v}^2 = \frac{1}{v^2} \; , \tag{7}$$

between velocities for s.l. shocks and t.l. shocks. These relations show that only one type of transitions can be realized for a given initial state and final state. Physical regions $[0,1)$ for v_o^2, v^2 (3) and for $\tilde{v}_o^2, \tilde{v}^2$ (6) can be easily found in $(\epsilon\text{-}p)$-plane and they are shown in Fig. 1.

   If one takes as the initial state only states which are thermodynamically equilibrated, a TA passes then through the point (ϵ_o, p_o) (which is called the centre of TA) and lies as a whole in the regions I and IV in Fig. 1. Phase transitions via the compressional s.l. shocks into region IV and rarefaction s.l. shocks into region I are called detonation and deflagration respectively. For supercooled initial QGP states the TA no longer passes through the point (ϵ_o, p_o) and new possibilities of t.l. shock hadronization (i.e., (5,6) shock transitions to regions III and VI in Fig. 1) appear.

3   Strongly Interacting Matter EOS

To study any hydrodynamical problem we need an EOS which expresses local thermodynamical parameters in terms of two independent variables. The most covenient way is to present EOS in the form $p = p(T, \mu)$ with temperature T and baryonic chemical potential μ as independent variables. Thermodynamical functions can be found then from the thermodynamical identeties: $s = (\partial p/\partial T)_\mu$, $n = (\partial p/\partial \mu)_T$, $\epsilon = Ts + \mu n - p$.

There is no uniqe simple pattern of a HM EOS and a whole variety of different hadronic EOS models can be found in the literature. In the mean field approximation HM propeties are described within the Walecka model[6] and its generalizations (see [7]). Another possibility is a thermodynamically consistent "excluded volume" model[8] where the hadronic pressure can be expressed in terms of ideal gas pressures as (v_i are the proper volume parameters for a hadron of type "i"):

$$p_h = \sum_i p_i^{id}(T, \tilde{\mu}_i) , \quad \tilde{\mu}_i = \mu_i - v_i p_h . \tag{8}$$

The most popular model to describe the QGP phase is the bag model EOS (see, for esample,[9]): $\epsilon_q(T, \mu) = \epsilon_{id}(T, \mu) + B$, $p_q(T, \mu) = p_{id}(T, \mu) - B$, where ϵ_{id} and p_{id} are the ideal gas expressions for a quark-gluon gas and B is the non-perturbative vacuum pressure. The bag model EOS is however not supported by the Monte Carlo lattice calculations in $SU(3)$-gluodynamics. The MC data above the deconfinement transition temperature T_c show approximate ideal gas behaviour for the energy density and much stronger deviation from the ideal gluon gas for the pressure. A simple formula for the gluon pressure, which is thermodynamically consistent with an ideal gas energy density, has been suggested[10]: $\epsilon(T) = \sigma T^4$, $p(T) = \frac{1}{3}\sigma T^4 - AT$, ($A = const > 0$) . To reproduce the properties of the gluon plasma the "cut-off" model[11] was proposed. It omits the low-momentum contribution ($k \leq K_c$) of the gluon spectrum. In contrast to the bag model EOS for the QGP, the cut-off model reproduces the lattice (no dynamical quarks) results reasonably well (see [12]). Recently phenomenological models with temperature dependent gluon effective mass have been discussed (see [13]).

If HM and QGP pressure functions are known, the whole phase diagram can be constructed according to the ususal Gibbs procedure for the systems with a first order phase transition: the equation $p_h(T, \mu) = p_q(T, \mu)$. is defined a phase transition line in the $T - \mu$ plane.

534

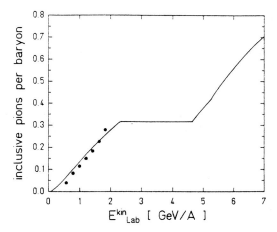

Figure 2: The π-meson multiplicity as a function of E_{Lab}^{kin}/A. Dots are experimental data.

4 Compression Shock Deconfinement Transition

In the compressional shock model of heavy ion collisions the discontinuity in the thermodynamical parameters of strongly interacting matter appears due to the initial hydrodynamical (velocity) discontinuity in the system of colliding nuclei. The physical picture of (one-dimensional) compression shocks in heavy ion collisions is quite transparent. The stability problem was studied in Ref.[14].

In Ref.[15] the compression shock adiabat was calculated with the mean-field model for HM and bag model for QGP. A point on the shock adiabat (4) is fixed by the incident energy as (m_N is the nucleon mass)

$$E_{lab}^{kin}/A = 2m_N \left[\left(\frac{\epsilon/n}{\epsilon_o/n_o} \right)^2 - 1 \right] , \qquad (9)$$

where the normal nuclear density $n_o = 0.16 \ fm^{-3}$, energy density $\epsilon_o = m_N n_o - 16 \ MeV$ and $p_o = 0$ define the initial state (the centre of the Taub adiabat in Eq. (4)). Constructing the compression shock adiabat we find that there is a region of unstable single-shock transitions. The stable physical solution in this case is the so-called generalized shock adiabat (see[14] for details). The calculations[15] shown in Fig. 2, gives a plateau-like structure of the pion multiplicity per nucleon in some interval of the bombarding energy. It originates due to the existence of the mixed phase. The qualitative features are independent of the precise values of the numerical parameters in the EOS and may serve as a signature of the deconfinement phase transition.

5 Shock Wave QGP Hadronization

The status of the QGP shock-like hadronization, suggested in Ref.[16], is rather different from that of the compression shock deconfinement transition. Firstly, the origin of a hadronization discontinuity is quite different. It can result from smooth initial conditions in the course of the QGP expansion if a transition between HM and QGP is a 1-st order phase transition. In this case one can also expect the appearance of supercooled QGP states ahead of the discontinuity front and superheated HM states behind it. Secondly, the space-time properties of a discontinuity hypersurface as well as the physical meaning of the collective velocities have not yet been studied in this case. At last, both types of the discontinuity-like hadronization – deflagration (rarefaction shock) and detonation (compression shock) – are possible. A recent consideration of a detonation QGP hadronization can be found in Ref.[17].

We discuss the discontinuity-like hadronization model constructed in Ref.[18]. By admitting the existence of supercooled QGP and superheated HM we have obtained the picture of discontinuity-like transitions which is essentially richer than only standard compression and rarefaction shocks. In our consideration we have restricted ourselves to the EOS (8) for a pion-nucleon gas with a common proper volume hadron parameter v and have chosen the "cut-off" model for the QGP.

If we admit supercooled QGP states for the initial states in shock transitions, the TA in the $(\epsilon–p)$-plane does not pass through the initial (non-equilibrium) point (ϵ_o, p_o). The TA (4) for a supercooled initial QGP state lies in the following regions of the $(\epsilon–p)$-plane (see Fig. 1): I (s.l. deflagration), II (unphysical region both for (3) and (6)), III (t.l. detonation) and IV (s.l. detonation). A shok hypersurface for QGP hadronization is most probably the hypersurface with time-like normal vector. It leads to a t.l. detonation of supercooled QGP. Final HM states in this case belong to the region III in Fig. 1.

Summary

The summary of my talk can be seen in Figs. 1 and 2.

A complete simple picture of all shock wave transitions is presented in Fig. 1. For any initial state (ϵ_o, p_o) there are 4 different types of shock transitions: compression and rarefaction space-like shocks (regions IV, I in Fig. 1), and compression and rarefaction time-like shocks (regions III, VI in Fig. 1). Reality of the transition is additionally specified by a space-time geometry of the discontinuity front [4] and by the shock stability conditions [3, 14].

The compression shock model of heavy ion collisions leads to the prediction of a deconfinement phase transition at rather small bombarding energies. A signature of this transition is a plateau-like structure of the pion multiplicity shown in Fig. 2. The specific dependence of pion multiplicity on collision energy is still a "prediction" as up to now there are no experimental data on heavy ion collisions in this energy region.

Conlucions

Relativistic shock phenomena deserve further studies. " What we need is hot and dense nuclear matter, and this achieved through the shock wave mechanism in its different forms" [19].

Acknowledgments

I am very much indebted to Walter Greiner for his permanent support of my research efforts. In Sections 4 and 5 a brief presentation of papers [15] and [18] was done. I am thankful to my coauthors K.A. Bugaev, D.H. Rischke, H.G. Miller, R.M. Quick and R.A. Ritchie for the fruitful collaboration. I am also thankful to L.P. Csernai for useful discussions of time-like shocks. I acknowledge the financial support of the INTAS Grant No 94-3405.

References

1. W. Scheid and W. Greiner, Ann. Phys. **48**, 493 (1968); Z. Phys. **226**, 364 (1969).
2. R. Stock, Phys. Rep. **135**, 259 (1986);
   H. Stöcker and W. Greiner, Phys. Rep. **137**, 277 (1986);
   R.B. Clare and D.D. Strottman, Phys. Rep. **141**, 177 (1986).
3. M.I. Gorenstein and V.I. Zhdanov, Z. Phys. C **34**, 79 (1987);
   K.A. Bugaev and M.I. Gorenstein, J. Phys. G **13**, 1231 (1987);
   K.A. Bugaev, M.I. Gorenstein and V.I. Zhdanov, Z. Phys. C **39**, 365 (1988).
4. L. Csernai, Zh. Eksp. Teor. Fiz. (Russ.) **92**, 379 (1987); Sov. Phys. JETP **65**, 216 (1987).
5. A.H. Taub, Phys. Rev. **74**, 328 (1948).
6. B.D. Serot and J.D. Walecka, Adv. Nucl. Phys. **16**, 1 (1986).
7. K.A. Bugaev and M.I. Gorenstein, Z. Phys. C **43**, 261 (1989);
   M.I. Gorenstein et al., J. Phys. G **19**, L69 (1993).

8. D.H. Rischke, M.I. Gorenstein, H. Stöcker and W. Greiner, Z. Phys. C **51**, 485 (1991); J. Cleymans, M.I. Gorenstein, J. Stalnacke and E. Suhonen, Phys. Scripta **48**, 277 (1993).

9. E.V. Shuryak, Phys. Rep. **61**, 71 (1980);
J. Cleymans, R.V. Gavai and E. Suhonen, Phys. Rep. **130**, 192 (1986).

10. M.I. Gorenstein and O.A. Mogilevsky, Z. Phys. C **38**, 161 (1988).

11. F. Karsch, Z. Phys. C **38**, 147 (1988).

12. D.H. Rischke et al., Phys. Lett. B **278**, 19 (1992); Z. Phys. C **56**, 325 (1992); M.I. Gorenstein, W. Greiner and St. Mrówczyński, Phys. Lett. B **286**, 365 (1992).

13. M. I. Gorenstein and Shin Nan Yang, Phys. Rev. D **9** (1995) 5206.

14. K.A. Bugaev, M.I. Gorenstein, B. Kämpfer and V.I. Zhdanov, Phys. Rev. D **40**, 2903 (1989).

15. K.A. Bugaev, M.I. Gorenstein, D.H. Rischke, Phys. Lett. B **255**, 18 (1991); JETP Pis'ma (Russ.) **52**, 1121 (1990).

16. L. Van Hove, Z. Phys. C **21**, 93 (1983); C **27**, 15 (1985).

17. N. Bilić, J. Cleymans, E. Suhonen and D.W. von Oertzen, Phys. Lett. B **311**, 266 (1993); N. Bilić, J. Cleymans, K. Redlich and E. Suhonen, Z. Phys. C **38**, 525 (1994).

18. M.I. Gorenstein, H.G. Miller, R.M. Quick, R.A. Ritchie, Phys. Lett. B **340**, 109 (1994).

19. W. Greiner, Preface to "Hot and Dense Nuclear Matter", NATO ASI, 1993

COLLECTIVE FLOW IN AU+AU COLLISIONS AT 10.7 GEV/A

H. C. BRITT, M. N. NAMBOODIRI, T. C. SANGSTER
LAWRENCE LIVERMORE NATIONAL LABORATORY, LIVERMORE, CA 94550

FOR EXPERIMENT E866 (THE E802 COLLABORATION)
ANL, BNL, UC BERKELEY, UC RIVERSIDE, COLUMBIA, INS(TOKYO), KYOTO,
LLNL, MARYLAND, MIT, NYU, TOKYO, TSUKUBA

ABSTRACT

A forward projectile scintillator hodoscope observing projectile spectators is used in conjunction with a PHOSWICH scintillator array at target rapidities to measure the mean transverse momentum imparted to protons, deuterons, and tritons emitted in the target pseudo rapidity region (-1.0<eta<0.5). Results are presented as a function of pseudo rapidity and energy measured in a zero degree calorimeter (proportional to impact parameter). Near eta=0 for particles within the PHOSWICH energy acceptance (30 MeV < E < 200 MeV), the measured P_x values scale with the mass of the detected particle. The results are uncorrected for the dispersion in the reaction plane determination.

INTRODUCTION

A major goal in the study of relativistic heavy ion collisions is the determination of the equation of state, EOS, of nuclear matter under conditions of temperature and density far removed from those normally encountered in the studies of nuclei near their ground states. An experimental tool sensitive to the EOS in hot dense systems is the study of nuclear flow[1]. For heavy ions at energies up to 1 GeV/nucleon, observed at Bevalac and GSI, the collision can traverse an initial compression up to roughly twice nuclear ground state densities and a subsequent expansion culminating with the final emission of nucleons and fragments at a freeze out density of the order of 1/4 the density for stable nuclei. In these collisions compressional bounce off of spectator nucleons and fragments in the reaction plane, squeeze out of participant nucleons and pions and collective radial expansion have been predicted [1-3] and observed.[4]

At the higher energies, 10-200 A GeV, available at AGS and CERN similar flow effects have been predicted,[5-7] It is expected that in this energy regime it will be possible to study reactions in which initial compression can reach up to 3 times normal nuclear densities and in this case the resulting flow characteristics may become sensitive to many new characteristics of the EOS such as the behavior of collision cross sections for nucleons and excited baryons and mesons at high relative momentum.[8-10] At the AGS compressional bounce off has been reported from experiments E866[11], E877[12], and E875.[13] In addition, it has been calculated that radial flow may measurably effect the spectral shapes for composite fragments, (d, t, He), emitted at mid rapidities.[7] In this paper we report on measurements of directional flow, "bounce off", for protons, deuterons and tritons from the reaction Au+Au at a bombarding energy of 10.7 A GeV.

APPARATUS

The results reported are a part of the experiment E866 which represents modifications of the earlier E802 experiment[14] in order to study the Au+Au system. As part of experiment E866 a forward angle scintillator hodoscope was installed to determine on an event-by-event basis the orientation of the reaction plane from an observation of the mean position of the projectile spectator fragments. An improved array of PHOSWICH scintillator telescopes was also installed to measure protons, deuterons and tritons emitted from the target spectator.

The projectile spectator hodoscope, HODO, was installed in front of the E802 zero degree calorimeter, ZCAL, at a distance of 11.5 meters from the target. It consists of a 40 element array of 1cm x 8mm x 40cm scintillator slats oriented in the X direction followed by an equivalent array oriented in the Y direction. The scintillator slats were read out at both ends by phototubes. The pulse height spectra from individual slats showed distinct peaks for particles of Z=1,2,3 and various combinations of these particles. After calibration the pulse heights in all of the slats hit were used to determine the mean position in X and Y for the projectile spectator matter for a given event. The mean position of an undeflected beam particle was determined by direct measurement by mixing a fraction of the beam events into the trigger. Because of the change of the beam position with time during a spill from the AGS, the calibration of the beam position was done as a function of time during the spill and was repeated for each data run. Details of the procedure will be reported in a detailed instrumentation paper.[15] For each event the position of the spectator center of gravity could be connected by a line to the estimated position of the undeflected beam to give an orientation for the reaction plane.

The PHOS array was an upgrade of the array used previously[16] in experiment E859 to measure proton and deuteron spectra at target rapidities.[17] It consisted of 100 fast/slow scintillator modules arranged in the cylindrical geometry shown in Fig. 1. 60 newly constructed modules were positioned at a distance of 1 meter in the angular region 40 to 90 degrees and 40 modules from experiment E859 were used in the angular region 90 to 130 degrees. All modules provided separation between p, d, and t and energy acceptance in the range 30 to 200 MeV.

METHOD

The trigger used in accumulating this data set was a minimum bias plus a requirement that at least one PHOS module had an event. Then for each identified p, d, or t the azimuthal angle of the appropriate module relative to the identified reaction plane was determined. Then distributions of the two dimensional momentum projections, P_x vs P_y , were developed for each module and particle. These distributions could also be developed as a function of the centrality of the collision as determined from the energy in ZCAL.

There is a significant dispersion in the reaction plane determination due to several factors and this has the effect of decreasing the measured flow signal below the value characteristic of a precise reaction plane determination. The major sources of this dispersion are the fluctuations in the charged particle emissions relative to the reaction plane and the uncertainty in the position of the beam spot. The reaction plane determination contains large uncertainties for very central collisions and, therefore, the present experiment is most useful for measurements involving hard peripheral collisions ($E_{ZCAL} > 1000$) which is also where one expects the directed flow signal to be maximal. The effective dispersion in the measured reaction plane orientations was estimated using a GEANT Monte Carlo routine which reproduced conditions of the experimental geometry, target thickness and various absorbers. A set of model calculations from RQMD (version1.07)[5] was used as input to the Monte Carlo. Results for the phi distribution relative to the true reaction plane are shown in Fig. 2 summed over all events. A gaussian fit to this distribution gives a standard deviation of approximately 40 degrees. Distributions of the predicted flow signal,$<P_x>$, are shown in Fig. 3 with and without this dispersion included and it is seen that a decrease of approximately 25% in the apparent signal results from this dispersion. These results are dominated by the hard peripheral collisions and the dispersion effect for more central collisions can be expected to be somewhat greater.

RESULTS

A composite of results obtained for protons, deuterons and tritons is shown in Fig. 4. The 2D distribution P_x vs P_y shows that the mean of the overall distribution of events is symmetric in P_y and shifted to negative values for P_x. (The deflection of the projectile spectator is defined to be in the +X direction.) The results can be seen more quantitatively as a function of ZCAL energy and pseudo rapidity in other panels of Fig. 4. The results for $E_{ZCAL}>$ 1000 channels and target rapidities -0.5<eta<0.5 show an "apparent" flow, equal within experimental uncertainties, of approximately 60 MeV/c/A for p, d and t. When combined with the average reaction plane dispersion from Fig. 3 these results suggest a corrected flow value of about 80 MeV/c/A. However, the dispersion corrections are still preliminary and all effects due to the energy acceptance of the PHOS modules have not been quantitatively incorporated.

These preliminary results suggest that the flow velocity in the spectator region does not increase with increasing particle mass as has been predicted for complex particles emitted in the participant region in recent RQMD calculations.[7] This result appears consistent with the calculations[7] and may illustrate a qualitative difference between the collective radial expansion of the participant region and the compressional "bounce off" that is measured in the spectator region.

For more central collisions (E_{ZCAL}<1000) the dispersion corrections are expected to change rapidly and so it is not clear whether the apparent decrease in $<P_x>$ is due to this or is a real decrease in $<P_x>$. Similar dispersion effects may be important at far backward angles (-1.0<eta<-0.5). Because of the difficulty of unfolding this dispersion from the multidimensional data set, we plan in the future to compare to predictions of various theoretical models which have been filtered through the acceptance criteria of the experiment.

542

We are very pleased that this contribution is part of the festschrift in honor of Walter Greiner's 60th birthday. His personal guidance and the theoretical support from the Institute for Theoretical Physics, Frankfurt, has been critical in the development of our relativistic heavy ion program at Livermore. It is particularly appropriate that our current work is in the field of collective flow where he and his coworkers have been responsible for the initial theoretical insights and many of the subsequent advances. We are pleased to acknowledge the help of G. Peilert in performing some of the RQMD calculations. Experiment E866 is supported by the U.S. Department of Energy (ANL, BNL, UC-Berkeley, UC-Riverside, Columbia, LLNL, and MIT), by NASA (UC-Berkeley), and by the US-Japan High Energy Physics Collaboration Treaty.

REFERENCES

1. H. Stöcker, J. A. Maruhn, and W. Greiner, Phys. Rev. Lett 44, 725 (1982).
2. H. Stöcker, et al., Phys. Rev. C 25, 1873 (1982).
3. H. Stöcker and W. Greiner, Phys. Rep. 137, 277 (1986).
4. H. H. Gutbrod, A. M. Pozkznzer, and H. G. Ritter, Rep. Prog. Phys. 52 1267 (1989).
5. H. Sorge, H. Stöcker and W. Greiner, Ann. Phys. (NY) 192, 266 (1989); Nucl. Phys. A498, 567c (1989).
6. S. H. Kahana, Y. Pang and T. J. Schlagel, Nucl. Phys. A566, 465c (1993); Y. Pang, D. E. Kahana, S. H. Kahana and T. J. Schlagel, Nucl Phys. A590, 565c (1995); D. E. Kahana, D. Keane, Y. Pang, T. Schlagel and S. Wang, preprint 1995.
7. R. Mattiello, A. Jahns, H. Sorge, H. Stöcker, and W. Greiner, Phys. Rev. Lett. 74 2180 (1995); R. Mattiello, contribution to this conference.
8. J. Aichelin, A. Rosenhauer, G. Peilert, H. Stöcker and W. Greiner, Phys. Rev. Lett. 58, 1926 (1987).
9. F. deJong and R. Malfliet, Phys. Rev. C 46, 2567 (1992).
10. J. Ellis, J. Kapusta and K. Olive, Phys. Lett. B273, 122 (1991).
11. T. C. Sangster, Presentation at Quark Matter 95, Monterey, CA (1995).
12. J. Barrette, et al., Phys. Rev. Lett 73 2532 (1994), J Stachel, contribution to this conference.
13. G. Singh and P. L. Jain, Phys. Rev. C 49, 3320 (1994); P. L. Jain, G. Singh and A. Mukhodhyay, Phys. Rev. Lett. 74, 1534 (1995).
14. T. Abbott, et al., Nucl. Instr. Meth. A290, 41 (1990).
15. T. C. Sangster, et al., In preparation
16. J. B. Costales, et al., Nucl Instr. Meth. A330, 183 (1993).
17 L. Ahle et al, submitted to Phys. Rev. C.

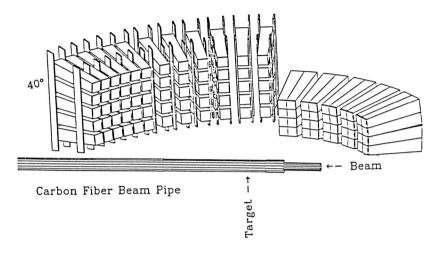

Fig 1a. Schematic Drawing of Phoswich Array

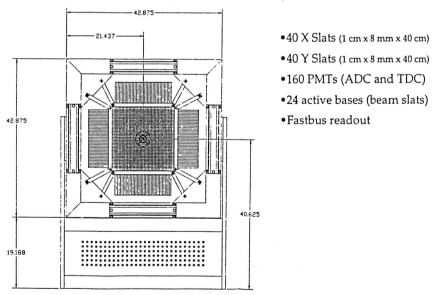

• 40 X Slats (1 cm x 8 mm x 40 cm)
• 40 Y Slats (1 cm x 8 mm x 40 cm)
• 160 PMTs (ADC and TDC)
• 24 active bases (beam slats)
• Fastbus readout

Fig 1b. Schematic Drawing of Projectile Hodoscope

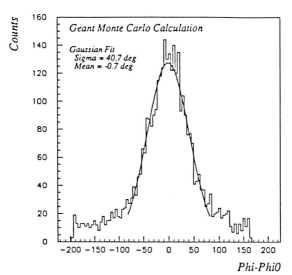

Figure 2. Monte Carlo GEANT calculations of the distribution of phi angles determined in the experiment relative to the true reaction plane using events from ROMD.[5,7]

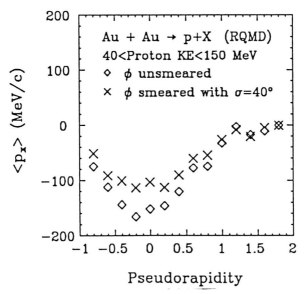

Figure 3. Distribution of predicted flow signal, <P$_x$>, from RQMD with no acceptance corrections and with a dispersion (sigma) in the reaction plane determination of 40 degrees.

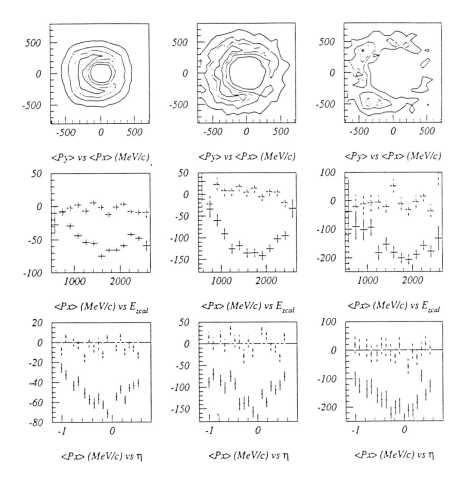

Figure 4. Experimental results for protons, deuterons and tritons observed in the PHOSWICH array. Two dimensional mean transverse momenta, P_x Vs P_y, top. Mean $<P_x>$ versus energy deposited in ZCAL, middle. Mean $<P_x>$ versus pseudorapidity, bottom. Proton data is shown in right panels, deuteron data in middle panels, triton data in left panels.

SCALING OF NUCLEAR STOPPING OBSERVED
IN CENTRAL COLLISIONS OF HEAVY NUCLEI

JOHN W. HARRIS [a]

Lawrence Berkeley National Laboratory
University of California, Berkeley, CA, 94720

The observation of a simple scaling dependence of the proton rapidity distributions resulting from central collisions of heavy nuclei over the incident energy range E_{lab} = 0.25 to 160 A-GeV is reported. The proton rapidity distributions for heavy systems (mass $\sim$ 200 + 200), unlike those for lighter systems (mass $\sim$ 30 + 30), are observed to scale as the rapidity gap between the incident nuclei over this entire incident energy range. These results for the heavy systems are consistent with two very different theoretical approaches - full stopping and longitudinal expansion, and a multiple collision model approach.

1 Introduction

It is of considerable interest to determine and understand the stopping power of nuclei at relativistic energies. The nuclear stopping power is a measure of the degree to which the energy of the relative motion of two colliding nuclei is transformed into other degrees of freedom. The amount of nuclear stopping governs parameters such as the energy and volume of the interaction region (and therefore energy density), the reaction dynamics and the extent to which conditions are favorable for formation of a high density or deconfined phase of matter. Increased thermalization of the incident energy, higher energy densities during the collision and a combination of increased particle production and collective flow are expected to accompany increased nuclear stopping. Thus, an understanding of nuclear stopping measurements at the presently available collision energies should provide insight into the energy and baryon densities that will be reached in the future in nuclear collisions at higher energies at the Relativistic Heavy Ion Collider (RHIC) at Brookhaven National Laboratory and the Large Hadron Collider (LHC) at CERN.

With the advent of heavy nuclear beams at relativistic energies, Au at the Brookhaven AGS and Pb at the CERN SPS, and sophisticated detectors at the somewhat lower energies of the LBL-Berkeley Bevalac and GSI-Darmstadt SIS, new information on the stopping power of nuclei has recently been obtained. This paper reports the observation of a simple scaling dependence of the proton rapidity distributions resulting from central collisions of heavy nuclei over the

---

[a] Present address: Physics Department, Yale University, New Haven CT 06520-8124

incident energy range, $E_{lab} = 0.25$ to 160 A-GeV, available from these relativistic heavy ion accelerators. The proton rapidity distributions for heavy systems are observed to scale as the rapidity gap between the incident nuclei over this entire incident energy range. This behavior is a result of a large amount of nuclear stopping in the collisions and is suggestive of a common mechanism (or mechanisms) acting over this energy range. It also suggests that considerably higher energy and baryon densities should be expected in future experiments with heavy nuclei at the higher energies of the relativistic heavy ion colliders. This scaling dependence of the proton rapidity distributions in central collisions of heavy nuclei will be presented and discussed below.

2 Experimental Results on Nuclear Stopping

Experiments determine the stopping power of colliding nuclei by measuring: i) final-state proton rapidity distributions, which represent the redistribution during the collision of valence quarks away from the initial distributions in the beam and target nuclei; ii) the energy remaining in forward-going baryons after the collision to compare to the incident energy carried into the reaction by the baryons; and iii) the transverse energy distributions which represent the energy transformed into produced particles and their kinetic motion. All of these measurements will provide information on the amount of energy that is transformed from the incident energy of relative motion into other degrees of freedom, and thus on the stopping power. This paper will concentrate on proton final-state rapidity [b] distributions to study the stopping power. The stopping power has typically been measured in this way as the average rapidity loss of the projectile nucleon or nucleons. [1]

Many systems have been studied at the AGS and the SPS to determine the nuclear stopping power. In order to gain a better understanding of the stopping and the reaction dynamics for these two energy regimes, it is instructive to investigate measurements in symmetric or quasi-symmetric nuclear collisions. The average rapidity loss Δy for protons measured in minimum bias proton-proton (pp) interactions is 0.71 to 0.76 at the AGS (for $p_{beam} = 12$ - 14.6 GeV/c, respectively) [2] and 0.9 at the SPS (for $p_{beam} = 200$ GeV/c), [3] as compiled recently by Ref. [4]. The Δy for protons is 0.79 for peripheral collisions of Si + Al at the AGS (for $p_{beam} = 14.6$ GeV/c) [5] and for $p - \bar{p}$ is 0.8 for peripheral S + S at the SPS (for $p_{beam} = 200$ GeV/c). [6] Thus, the rapidity distributions for protons from pp interactions [2] and peripheral nucleus-nucleus

---

[b] Note for reference that the rapidity variable is defined as $y = 1/2 \ln((E + p_\parallel)/(E - p_\parallel))$, where E is the total energy of the particle and $p_\parallel$ is its momentum component along the beam direction.

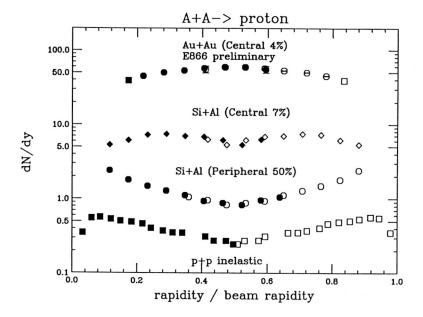

Figure 1: Distributions of proton dn/dy as a function of rapidity (normalized to beam rapidity) from interactions of proton + proton, peripheral and central Si + Al, and central Au + Au at the AGS plotted from bottom to top, respectively. Solid symbols are measured data and open symbols are data reflected about mid-rapidity. (Figure from Ref.8)

interactions at the AGS [5,7] and the SPS [6] are similar and peaked forward and backward in the c.m. frame, near the projectile and target rapidities. This can be seen in the lower part of Fig. 1 where, for example, the proton distributions from pp interactions [2] and Si + Al peripheral collisions [5,7] at the AGS are presented. [8] On the average the final-state protons exhibit only a small amount of stopping, represented by the displacement away from the incident proton projectile and target rapidity, in pp and *peripheral* AA interactions.

For *central* collisions of intermediate mass nuclei ($A \sim 30$) at the AGS the proton rapidity distribution spreads over the entire rapidity space. This can be seen in Fig. 1, where the proton rapidity distribution for 14.6 A-GeV Si + Al central collisions is displayed. The distribution for central collisions is rather flat and appears to exhibit two broad peaks located approximately half-way between the target/projectile and the c.m. rapidities. This is to be compared to the distribution for *peripheral* collisions, which peaks near the projectile and target rapidities. There is significantly more stopping in central Si + Al collisions, where the final-state protons are displaced farther from the initial beam and target rapidities, than in peripheral ones. The proton Δy measured for central Si + Al at the AGS (for $p_{beam} = 14.6$ GeV/c)[5] is 0.97 and for central S + S at the SPS (for $p_{beam} = 200$ GeV/c)[6] is 1.7. Also displayed in Fig. 1 is the rapidity distribution of protons for central collisions of 10.6 A-GeV Au + Au from the AGS. For this heavy system the proton rapidity distribution exhibits a peak at midrapidity ($y/y_{beam} = 0.5$). This corresponds to $\Delta y = 1.02$. The protons in central collisions of the heavier system are shifted farther away from the projectile and target rapidities than those in the lighter systems. Thus, at the AGS energies the heavier Au + Au system is more efficient at stopping the incoming matter than the lighter Si + Al system.

For a better understanding of nuclear stopping and to be able to project to the higher energies of the future heavy ion colliders, a comparison of results for central collisions of various systems at several incident energies is necessary. Displayed in Fig. 2 are scaled rapidity distributions $dn/d(y/y_{beam})$ as a function of normalized rapidity y/y_{beam}. The ordinate, $dn/d(y/y_{beam})$, is the number of "protons" per unit fraction of beam rapidity. This allows an absolute comparison of different mass systems on a normalized scale. The abscissa, y/y_{beam}, allows the comparison of systems with different incident energies (beam rapidities) on the same scale. In Fig. 2, the term "protons" refers to measured and identified protons at all energies except for the SPS. The data from NA35 and NA49[6,9] at the SPS are net charge distributions, which after all corrections, correspond to protons minus anti-protons. This is a measure of the distribution of valence quarks after the interaction, since produced quarks are subtracted out in the experimental determination of net charge. (See refs. [6,9] for more details.) Furthermore, since the $p\bar{p}$ production rate at the AGS and lower energies is small compared to the number of incoming protons (and therefore the measured outgoing protons), the proton distributions as measured at the lower energies and the net charge distributions $(p - \bar{p})$ measured at the SPS can be compared, as in Fig. 2.

The distributions displayed in Fig. 2 naturally fall into two categories. A lower group (with lower net charge and mass) in Fig. 2 results from central collisions of intermediate mass systems: 0.4 A-GeV Ca + Ca $\rightarrow$ p from the Plastic Ball[10] at the LBL-Bevalac, 14.6 A-GeV Si + Al $\rightarrow$ p from E802[5] at the BNL-AGS and 200 A-GeV S + S $\rightarrow p - \bar{p}$ from NA35[6] at the CERN-SPS. The normalized rapidity distributions for central collisions in the mass 30 + 30 system at 14.6 A-GeV (Si + Al) and 200 A-GeV (S + S) are very similar, flat with rather broad peaks centered between projectile (target) and mid-rapidity.

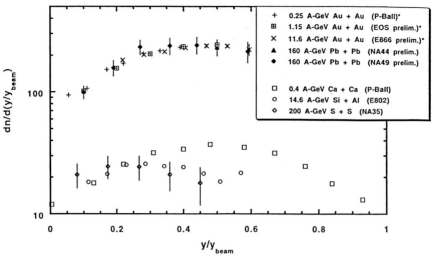

Figure 2: Scaled proton rapidity distributions, dn/d(y/y$_{beam}$) as a function of y/y$_{beam}$, for central collisions of various nuclear systems as listed in the legend. The grouping at the top of the figure corresponds to heavy systems (Au + Au and Pb + Pb), whereas the lower grouping represents the intermediate mass systems (Si + Al, S + S, Ca + Ca). The 1.15 A-GeV Au + Au and 160 A-GeV Pb + Pb data are preliminary.

For comparison the lower energy 0.4 A-GeV Ca + Ca data are quite different from the mass 30 + 30 data, with a peak at mid-rapidity. In central collisions of intermediate mass systems, it appears that there is a significant change that occurs between the energy of the Bevalac and the higher energies of the AGS and SPS (assuming negligible difference in the stopping of mass 30 + 30 systems and that of mass 40 + 40 systems). However, between the AGS and SPS energies there appears to be little change for these systems.

Also displayed in Fig. 2 are rapidity distributions for mass ~ 200 + 200 systems, appearing as the group of measurements in the upper half of the figure. Displayed are data for 0.25 A-GeV Au + Au → p from the Plastic Ball [10] at the LBL-Bevalac, preliminary 1.15 A-GeV Au + Au → p from the EOS-TPC [11] at the LBL-Bevalac, 11.6 A-GeV Au + Au → p from E866 [8] at the BNL-AGS, [c] preliminary 160 A-GeV Au + Au → $p - \bar{p}$ from NA44 [12] at

---

[c] the proton dn/dy data of Refs. [8,10,11] denoted by * in Fig. 2 have been renormalized to represent the total proton number in the system.

the CERN-SPS, [d] and preliminary 160 A-GeV Au + Au → $p - \bar{p}$ from NA49 [9] at the CERN-SPS. The mass ∼ 200 + 200 distributions peak and exhibit a significant "pile-up" of matter at midrapidity (y/ybeam = 0.5). The heavier systems appear to be more efficient at stopping the incoming matter than the intermediate mass systems. Thus, higher energy densities should be reached when colliding the heavier systems.

The shapes of the normalized rapidity distributions for protons from central collisions of 0.25, 1.15, and 10.6 A-GeV Au + Au and the $p - \bar{p}$ normalized rapidity distributions from central collisions of 160 A-GeV Pb + Pb are identical, within errors. This similarity over such a large range of incident energies is intriguing. By comparison, the mass 30 + 30 systems are more transparent to the incident energy than the heavier mass 200 + 200 systems. An exception is the mass 40 + 40 system at the lowest energy, where the shape of the distribution is identical to that of the mass 200 + 200 system at all energies (see Fig. 2). This, in itself, suggests a similarity in the dynamics of stopping in 0.4 A-GeV Ca + Ca central collisions with that of the Au + Au and Pb + Pb systems at all energies studied (0.25 - 160 A-GeV).

It is clear that there is presently insufficient energy available with the present-day accelerators to really measure the stopping in heavy systems, since the rapidity gap between projectile and target is too small to see any change in the proton rapidity distribution such as that observed in going from 0.4 A-GeV Ca + Ca to 14.6 A-GeV Si + Al and 200 A-GeV S + S. To be able to project the stopping power to higher energies, than presently available, one must study the systematic differences in stopping as a function of both mass and energy and understand in a quantitative model the energy dependence of the differences between the mass 30 + 30 systems and the mass 200 + 200 systems. This could lead to an understanding of the stopping as a function of nuclear thickness and incident energy, which is necessary to quantitatively predict stopping and energy densities for higher energy collisions.

3 Theoretical Approaches

Since the energy available in the collision increases as the rapidity when the collision energy is increased, the observed rapidity scaling of the proton rapidity distributions in the heavy systems is consistent with full stopping and longitudinal expansion of the system, as in hydrodynamics. [13] On the other hand, it is also completely consistent with a multiple collision approach, such as the multiple-chain model [1] which predicts a proton pile-up at midrapidity in collisions of heavy nuclei at energies up to and including the SPS energy.

---

[d] derived from the measured proton yield [12] and subtraction of a $\bar{p}$ yield of 8 %.

For intermediate mass nuclei, multiple collision models predict proton rapidity distributions which have a dip at midrapidity as observed in Figs. 1 - 2, with average rapidity losses for protons similar to those observed. It is desireable to be able to distinguish experimentally between these two very different mechanisms (Landau hydrodynamics and multiple collisions). Perhaps, use of the transverse momentum distributions in addition to the rapidity distributions for various particle species can make the distinction. Furthermore, it is interesting to note that calculations using the approach of Lorentz invariant relativistic quantum molecular dynamics (RQMD) predicted a peak at mid-rapidity for protons from the heavy systems at the energies of the AGS[14] and the SPS,[15] prior to the availability of such beams for experiments. A version of RQMD, which incorporates multi-hadronic intermediate states (color ropes), [16] has also predicted a similar peak at mid-rapidity for heavy systems at a RHIC c.m. energy ten times higher than the highest energy in the present study. In various calculations as the incident energy increases, there appear to exist novel reaction mechanisms for the transfer of energy and baryon number, such as color ropes in RQMD,[16] or multi-quark clusters in VENUS[17] which are a result of the high density. Furthermore, a nonlinear energy loss mechanism[18,19] would result in significantly increased stopping. It appears that the resolution of these stopping and energy loss issues as well as the understanding of such novel "non-hadronic" reaction mechanisms may only be resolvable by experiments at the higher energies of future heavy ion colliders.

4   Summary and Conclusions

A simple scaling dependence of the proton rapidity distributions resulting from central collisions of heavy nuclei over the incident energy range, $E_{lab} = 0.25$ to 160 A-GeV has been observed. The proton rapidity distributions for heavy systems peak at mid-rapidity and scale as the rapidity gap between the incident nuclei over the entire incident energy range of this study. This behavior is a result of a large amount of nuclear stopping in these collisions and suggests that considerably higher energy and baryon densities should be expected in future experiments with heavy nuclei at higher energies.

   In contrast, the protons from pp and peripheral AA interactions are peaked near the beam and target rapidities and exhibit only a small amount of stopping. Those from central collisions of intermediate mass nuclei (A $\sim$ 30) at the AGS and SPS energies are spread out over the entire rapidity space, are peaked away from midrapidity, and exhibit more stopping than for peripheral collisions but less than that of the heavy systems. An interesting result is that the shape of the proton rapidity distribution for central collisions of intermediate mass nuclei (A ~ 40) at 0.4 A-GeV is identical to the proton distributions for central collisions of heavy nuclei (A ~ 200) at all energies. This suggests a similarity between the dynamics of stopping in 0.4 A-GeV Ca + Ca central collisions and the Au + Au and Pb + Pb systems at all energies studied (0.25 - 160 A-GeV). This leads to the hypothesis that there is insufficient energy available with present-day accelerators to measure the stopping in the heavy systems. The rapidity gap between projectile and target is too small to see a change in the proton rapidity distribution for the heavy systems. Such a change is observed for the intermediate mass systems when the energy is increased from 0.4 A-GeV (Ca + Ca) to 14.6 A-GeV (Si + Al) and 200 A-GeV (S + S). A measurement of the stopping systematics in intermediate mass systems in the energy regime between 0.4 A-GeV and 14.6 A-GeV would be extremely useful. This could lead to an understanding of the stopping as a function of nuclear thickness and incident energy, which is necessary to be able to predict quantitatively stopping and energy densities for heavier systems at higher collision energies.

The results for the heavy systems are consistent with two very different theoretical approaches, which will require differentiation by experiment: one is full stopping and longitudinal expansion, and the other is a multiple collision approach of nucleons. In order to make predictions for collisions at higher energies it is necessary to consider interactions at the partonic level. Even here, new theoretical ground is being broken and the resolution of these stopping and energy loss issues, as well as the understanding of novel "non-hadronic" reaction mechanisms incorporated in models, may only be resolvable by future heavy ion collider experiments.

Acknowledgments

I wish to thank R. Stock, B. Müller, T. Wienold, C. Chasman and K. Kinder-Geiger for interesting discussions and comments on the subject of stopping. I thank my colleagues in NA49, and those in NA44 and EOS-TPC for permission to use preliminary data. I am grateful for the support of the Alexander von Humboldt Foundation and the hospitality of the Institut für Kernphysik of the Universität Frankfurt during part of this work. This work was supported in part by the Director, Office of Energy Research, Division of Nuclear Physics of the U.S. Department of Energy under Contract DE-AC03-76SF00098.

References

1. S. Date, M. Gyulassy and H. Sumiyoshi, Phys. Rev. D32, 619 (1985).
2. V. Blobel et al., Nucl. Phys. B69, 454 (1974).
3. K. Jaeger et al., Phys. Rev. D11, 2405 (1975).
4. F. Videbaek and O. Hansen, Phys. Rev. C52, 2684 (1995).
5. T. Abbott et al., Phys. Rev. C50, 1024 (1994).
6. J. Baechler et al., Phys. Rev. Lett. 72, 1419 (1994).
7. J. Barrette, et al. Phys. Rev. C50, 3047 (1994).
8. F. Videbaek et al., Nucl. Phys. A590, 249c (1995).
9. P. Seyboth, et al. Proc. of the XXV International Symposium on Multiparticle Dynamics, Stara Lesnia, Slovakia (1995).
10. H.H. Gutbrod et al., Z. Phys. A337, 57 (1990).
11. T. Wienold and the EOS-TPC Collaboration, Annual Report of the Nuclear Science Division, Lawrence Berkeley Laboratory Report (1996).
12. J. Dodd, et al., Proc. of the XXV International Symposium on Multiparticle Dynamics, Stara Lesnia, Slovakia (1995).
13. E.L. Feinberg, Z. Phys. C38, 229 (1988).
14. H. Sorge, et al., Phys. Lett. B243, 7 (1990).
15. A. von Keitz, et al., Phys. Lett. B263, 353 (1991).
16. H. Sorge, et al., Phys. Lett. B289, 6 (1992).
17. J. Aichelin and K. Werner, Phys. Lett. B300, 158 (1993).
18. M. Gyulassy and X.N. Wang, Nucl. Phys. B420, 583 (1994).
19. R. Baier, et al., Phys. Lett. B345, 277 (1995).

HYDRODYNAMICS AND COLLECTIVE BEHAVIOUR IN RELATIVISTIC NUCLEAR COLLISIONS

D.H. RISCHKE, M. GYULASSY

Physics Department, Pupin Physics Laboratories, Columbia University,
538 W 120th Street, New York, NY 10027, U.S.A.

Y. PÜRSÜN, H. STÖCKER, J.A. MARUHN

Institut für Theoretische Physik, Johann Wolfgang Goethe–Universität Frankfurt/M.
Robert–Mayer–Str. 8–10, D–60054 Frankfurt/M., Germany

Hydrodynamics is applied to describe the dynamics of relativistic heavy-ion collisions. The focus of the present study is the influence of a possible (phase) transition to the quark–gluon plasma in the nuclear matter equation of state on collective observables, such as the lifetime of the system and the transverse directed flow of matter. It is shown that such a transition leads to a softening of the equation of state, and consequently to a time-delayed expansion which is in principle observable via two–particle correlation functions. Moreover, the delayed expansion leads to a local minimum in the excitation function of transverse directed flow around AGS energies.

1 Introduction and Conclusions

Hydrodynamics has found widespread application in studying the dynamical evolution of heavy-ion collisions [1]. It was found that compressional shock waves, first predicted by Scheid and Greiner to occur in such collisions [2], lead to collective flow phenomena like sideward deflection of matter in the reaction plane ("side-splash" and "bounce-off") as well as azimuthal deflection out of the reaction plane ("squeeze-out"). The confirmation of these collective flow effects by BEVALAC experiments [3] was one of the main successes of the fluid-dynamical picture.

One of the primary goals of present relativistic heavy-ion physics is the creation and experimental observation of the so-called quark–gluon–plasma (QGP) phase of matter, predicted by lattice calculations of quantum chromodynamics (QCD) [4]. A class of interesting signals for the QGP, which are directly related to the QCD equation of state (EoS) as measured on the lattice, emerge from the influence of that EoS on the *collective* dynamical evolution of the system. Relativistic hydrodynamics is the most suitable approach to study these signals, since it is the only dynamical model which provides a *direct* link between collective observables and the EoS.

It was shown [5,6,7,8] that the transition to the QGP *softens* the EoS in the transition region, and thus reduces the tendency of matter to expand on account of its internal pressure. This, in turn, delays the expansion and considerably prolongs the lifetime of the system. It was moreover shown [8] that this prolongation of the lifetime (as compared to the expansion of an ideal gas without transition) is in principle observable via an enhancement of the ratio of inverse widths, R_{out}/R_{side}, of the two–particle correlation function in out– and side–direction. (This signal was originally proposed by Pratt and Bertsch [9].) Another aspect [10,11] of the delayed expansion is the reduction of the transverse directed flow in semi-peripheral collisions that can be readily tested experimentally at fixed target energies [12].

In this paper we summarize the essential physics and observable consequences of the softening of the EoS in the transition region, namely, the time-delayed expansion and the subsequent enhancement of R_{out}/R_{side}, and the disappearance of the transverse directed flow. Natural units $\hbar = c = k_B = 1$ are used throughout this paper.

2 The QCD Phase Transition and Softening of the Equation of State

Available lattice data for the entropy density in full QCD can be approximated by the simple parametrization [7,8,13]

$$\frac{s}{s_c}(T) = \left[\frac{T}{T_c}\right]^3 \left(1 + \frac{d_Q - d_H}{d_Q + d_H} \tanh\left[\frac{T - T_c}{\Delta T}\right]\right) , \tag{1}$$

where $s_c = const. \times (d_Q + d_H) T_c^3$ is the entropy density at T_c. Pressure p and energy density ϵ follow then from thermodynamical relationships. For $\Delta T = 0$, the EoS (1) reduces to the MIT bag EoS [14] with a strong first order phase transition between QGP and hadronic phase. In that case, the ratio d_Q/d_H determines the latent heat (density) of the transition, $\epsilon_Q - \epsilon_H \equiv 4B$.

In Fig. 1 we show (a) the entropy density and (b) the energy density as functions of temperature, and (c) the pressure and (d) the velocity of sound squared $c_S^2 \equiv dp/d\epsilon$ as functions of energy density for $\Delta T = 0$, $0.1\,T_c$, and an ideal gas with d_H degrees of freedom for $d_Q/d_H = 37/3$. Figs. 1 (a,b) present the thermodynamic functions in a form to facilitate comparison with lattice data. Present lattice data for full QCD can be approximated with a choice of ΔT in the range $0 \leq \Delta T < 0.1\,T_c$. In the hydrodynamical context, however, Figs. 1 (c,d) are more relevant. As can be seen in (c), for $\Delta T = 0$ the pressure stays constant in the mixed phase $\epsilon_H \leq \epsilon \leq \epsilon_Q$. Hydrodynamical expansion is, however, driven by pressure *gradients*. It is therefore (the square of) the velocity of sound $c_S^2 = dp/d\epsilon$, Fig. 1 (d), that is the most relevant measure of the system's tendency to expand. For $\Delta T = 0$, the velocity of sound vanishes

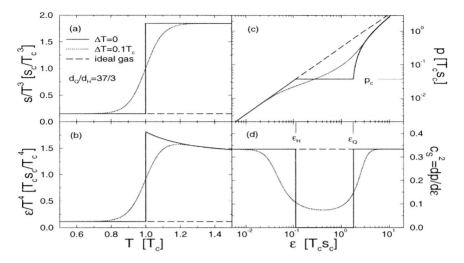

Figure 1: Equation of state.

in the mixed phase, i.e., mixed phase matter does not expand at all on its own account, even if there are strong gradients in the energy density. For finite ΔT, pressure gradients are finite, but still smaller than for an ideal gas EoS, and therefore the system's tendency to expand is also reduced, cf. Fig. 1 (d).

The reduction of c_S^2 in the transition region is commonly referred to as "softening" of the EoS, the respective region of energy densities is called "soft region"[5,6,7,8]. For matter passing through that region during the expansion phase, the flow will temporarily slow down or even possibly stall under suitable conditions and consequently lead to a time delay in the expansion of the system.

3   Hydrodynamics

Hydrodynamics is defined by local energy–momentum conservation,

$$\partial_\mu T^{\mu\nu} = 0 \; . \tag{2}$$

Under the assumption of local thermodynamical equilibrium (the so-called "ideal fluid" approximation) the energy–momentum tensor $T^{\mu\nu}$ assumes the particularly simple form [15]

$$T^{\mu\nu} = (\epsilon + p)\, u^\mu u^\nu - p\, g^{\mu\nu} \; , \tag{3}$$

where $u^\mu = \gamma\,(1, \mathbf{v})$ is the 4–velocity of the fluid ($\mathbf{v}$ is the 3–velocity, $\gamma \equiv (1 - \mathbf{v}^2)^{-1/2}$, $u_\mu u^\mu = 1$), and $g^{\mu\nu} = \mathrm{diag}(+, -, -, -)$ is the metric tensor. The

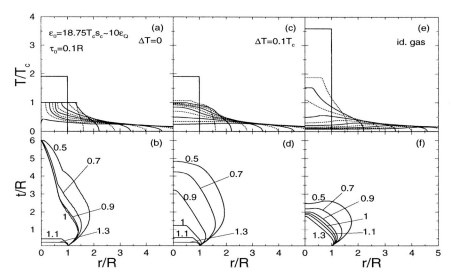

Figure 2: Transverse expansion in the Bjorken model.

system of equations (2) is closed by choosing an EoS in the form $p = p(\epsilon)$, i.e., as depicted in Fig. 1 (c). In the ideal fluid approximation, the (equilibrium) EoS is the *only* input to the hydrodynamical equations of motion (2) that relates to properties of the matter under consideration and is thus able to influence the dynamical evolution of the system. The final results are uniquely determined once a particular initial condition and a decoupling ("freeze-out") hypersurface are specified.

For finite baryon density, one has to also take into account local conservation of baryon number,

$$\partial_\mu N^\mu = 0 , \tag{4}$$

where $N^\mu = n u^\mu$ is the baryon 4–current (in the ideal fluid approximation), n is the baryon density in the local rest frame of a fluid element. In this case, the EoS has in general to be provided in the form $p = p(\epsilon, n)$ (see Fig. 1 of Ref. [11] for an explicit example).

4 Delayed Expansion and Two-Particle Correlations

In this section we discuss the delayed expansion and observable consequences in the so-called Bjorken model [16] and for the EoS (1). The main assumption of Bjorken's model is longitudinal boost invariance which implies that

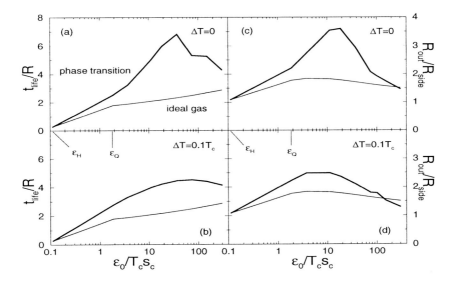

Figure 3: Lifetimes and inverse widths in the Bjorken model.

the longitudinal flow velocity of matter is always given by $v^z \equiv z/t$ [16]. The initial conditions are specified at constant proper time $\tau \equiv \sqrt{t^2 - z^2}$. We fix $\tau_0 = 0.1\,R$, motivated by the fact that for Au+Au collisions at RHIC energies equilibration is expected [17] to occur after 0.5 fm, while the initial radius R of the hot zone is on the order of 5 fm. Fig. 2 shows hydrodynamic solutions for the (cylindrically symmetric) transverse expansion of a "Bjorken cylinder" (at $z = 0$), for an initial (homogeneously distributed) energy density $\epsilon_0 = 18.75\,T_c\,s_c \sim 10\,\epsilon_Q \simeq 14$ GeV fm^{-3}. This value is expected to be reached through mini-jet production at RHIC energies [18]. As one observes, the system spends considerable time in the "soft region" of the EoS (corresponding to temperatures around T_c), and therefore the expansion is delayed, Figs. 2 (a–d), in comparison to the ideal gas case, Figs. 2 (e,f).

Figs. 3 (a,b) show the lifetime[a] of the system (assuming it "freezes out" at a temperature $T = 0.7\,T_c$) as a function of initial energy density ϵ_0 for the Bjorken expansion. One observes a distinguished maximum in the lifetime associated with the transition to the QGP. Note that this maximum of the lifetime occurs not exactly at energy densities corresponding to the "soft region" of the EoS. This is due to the fact that the strong dilution on account of the (ever present) longitudinal velocity field has to be compensated so that

---

[a] Here defined as the intercept of an isotherm in Figs. 2 (b,d,f) with the t–axis.

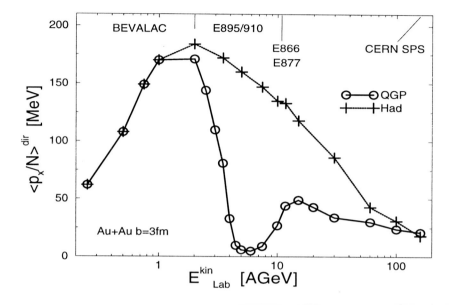

Figure 4: The transverse directed flow as calculated from 3+1 dimensional hydrodynamics.

the systems stays long enough in the "soft" transition region. If the system were initially at rest, the maximum would (as naively expected) occur around $\epsilon_0 \sim \epsilon_Q$ [7,8].

Figs. 3 (c,d) show the ratio of inverse widths of the two–pion correlation function in out– and side– directions, R_{out}/R_{side}, as a function of ϵ_0 for the Bjorken expansion [b]. Note that this ratio mirrors closely the dependence of the lifetime on initial conditions in Figs. 3 (a,b). The effect is maximized around initial energy densities expected to be reached at the RHIC collider [18]. The enhancement over the ideal gas case is of the order of 40–100% (for $\Delta T = 0.1\,T_c$ to $\Delta T = 0$) and should therefore be in principle experimentally observable.

5 Dis- and Reappearance of Transverse Directed Flow

The softening of the EoS and the delay in the expansion has an interesting consequence for semi-peripheral heavy-ion collisions at AGS energies. If the hot, compressed (baryon-rich) matter in the central zone undergoes a transition

---

[b]Details on how to calculate this quantity can be found in Ref. [8].

to the QGP, its tendency to expand is reduced, similarly as discussed above. This prevents the deflection of spectator matter, as it would occur for a stiff EoS with a stronger tendency to expand, for instance a purely hadronic EoS without phase transition [11]. As shown in Fig. 4, this effect is observable in the excitation function of the transverse directed flow per baryon,

$$\langle p_x/N \rangle^{dir} = \frac{1}{N} \int_{-y_{CM}}^{y_{CM}} dy \ \langle p_x/N \rangle(y) \ \frac{dN}{dy} \ \text{sgn}(y) \ . \tag{5}$$

The overall decrease of this quantity above $E_{\text{Lab}}^{\text{kin}} \sim 2$ AGeV observed for both EoS's is simply due to the fact that faster spectators are less easily deflected by the hot, expanding participant matter. One clearly observes a dramatic *drop* between BEVALAC and AGS beam energies and an *increase* beyond ~ 10 AGeV for the EoS with phase transition as compared to the calculation with the pure hadronic EoS. Thus, *there is a local minimum in the excitation function of the directed transverse (in-reaction-plane) collective flow around* ~ 6 AGeV, which is again related to the phase transition to the QGP and the existence of a "soft region" in the nuclear matter EoS. Note that the position of the minimum strongly depends on the EoS. It may easily shift to higher beam energies, if more resonances are included in the hadronic part of the EoS. Also, absolute values for the directed flow cannot yet be compared to experimentally measured ones, since at this stage freeze-out has not been performed. Moreover, viscosity effects are neglected in the ideal hydrodynamic picture, which are known to have a strong influence on flow [19]. The main point is, however, that irrespective of these quantitative uncertainties, the *minimum* is a generic *qualitative* signal for a transition from hadron to quark and gluon degrees of freedom in the nuclear matter EoS.

Acknowledgments

This work was supported by the Director, Office of Energy Research, Division of Nuclear Physics of the Office of High Energy and Nuclear Physics of the U.S. Department of Energy under Contract No. DE-FG-02-93ER-40764. D.H.R. gratefully acknowledges support by the University Frankfurt/M. and the University of Tennessee.

References

1. For a review, see:
   H. Stöcker and W. Greiner, Phys. Rep. 137 (1986) 277,
   R.B. Clare and D.D. Strottman, Phys. Rep. 141 (1986) 177.

2. W. Scheid, H. Müller, W. Greiner, Phys. Rev. Lett. 32 (1974) 741.
3. H.A. Gustaffson et al., Phys. Rev. Lett. 52 (1984) 1590,
   H.H. Gutbrod et al., Phys. Lett. B 216 (1989) 267,
   H.G. Ritter and the EOS collaboration, Nucl. Phys. A 583 (1995) 491c,
   M.D. Partlan and the EOS collaboration, Phys. Rev. Lett. 75 (1995) 2100.
4. see, for instance: E. Laermann, Proc. of "Quark Matter '96", May 20 – 24, 1996, Heidelberg, Germany (to appear in Nuclear Physics A).
5. C.M. Hung and E.V. Shuryak, Phys. Rev. Lett. 75 (1995) 4003.
6. D.H. Rischke, S. Bernard, J.A. Maruhn, Nucl. Phys. A 595 (1995) 346.
7. D.H. Rischke and M. Gyulassy, Nucl. Phys. A 597 (1996) 701.
8. D.H. Rischke and M. Gyulassy, Columbia University preprint CU–TP–756 (submitted to Nucl. Phys. A).
9. S. Pratt, Phys. Rev. C 49 (1994) 2722, Phys. Rev. D 33 (1986) 1314,
   G. Bertsch, Nucl. Phys. A 498 (1989) 173c.
10. L.V. Bravina, N.S. Amelin, L.P. Csernai, P. Levai, D. Strottman, Nucl. Phys. A 566 (1994) 461c.
11. D.H. Rischke, Y. Pürsün, J.A. Maruhn, H. Stöcker, W. Greiner, preprint CU–TP–695, nucl–th/9505014 (to be published in Heavy Ion Phys.).
12. J. Barrette et al. (E877 collaboration), Phys. Rev. Lett. 73 (1994) 2532,
    Nucl. Phys. A 590 (1995) 259c,
    Y. Zhang and J.P. Wessels (E877 collaboration), Nucl. Phys. A 590 (1995) 557c,
    G. Rai and the E895 collaboration, LBL PUB–5399 (1993).
13. J.P. Blaizot and J.Y. Ollitrault, Phys. Rev. D 36 (1987) 916.
14. A. Chodos, R.L. Jaffe, K. Johnson, C.B. Thorn, V.F. Weisskopf, Phys. Rev. D 9 (1974) 3471.
15. L.D. Landau and E.M. Lifshitz, "Fluid mechanics" (Pergamon, New York, 1959).
16. J.D. Bjorken, Phys. Rev. D 27 (1983) 140,
    G. Baym, B.L. Friman, J.P. Blaizot, M. Soyeur, W. Czyż, Nucl. Phys. A 407 (1983) 541.
17. X.-N. Wang, M. Gyulassy, M. Plümer, Phys. Rev. D 51 (1995) 3436,
    R. Baier, Yu.L. Dokshitzer, S. Peigné, D. Schiff, Phys. Lett. B 345 (1995) 277.
18. E. Shuryak, Phys. Rev. Lett. 68 (1992) 3270,
    K.J. Eskola and M. Gyulassy, Phys. Rev. C 47 (1993) 2329.
19. W. Schmidt, U. Katscher, B. Waldhauser, J.A. Maruhn, H. Stöcker, W. Greiner, Phys. Rev. C 47 (1993) 2782.

NUCLEAR CLUSTERS AS A PROBE FOR EXPANSION, FLOW AND BARYONIC MEAN FIELDS IN ULTRARELATIVISTIC HEAVY ION REACTIONS

R. Mattiello [1,2]
H. Sorge [2]
and H. Stöcker [3],

[1] *Brookhaven National Laboratory, Upton, New York 11973*
[2] *New York State University of Stony Brook, Stony Brook, NY 11974*
[3] *Institut für Theoretische Physik, J.W. Goethe Universität*
D-60054 Frankfurt am Main, Germany

A phase space coalescence description for cluster formation in relativistic nucleus-nucleus collisions is presented. The momentum distributions of nuclear clusters d,t and He are predicted for central Au(11.6AGeV)Au reactions in the framework of the RQMD transport approach. Transverse expansion leads to a strong shoulder-arm shape and different inverse slope parameters in the transverse spectra of nuclear clusters deviating markedly from thermal distributions. A clear "bounce-off" event shape is seen: the averaged transverse flow velocities in the reaction plane are for clusters markedly larger than for protons. The cluster yields –particularly at low p_t at midrapidities– and the in-plane (anti)flow of clusters and pions change markedly if baryon potential-interaction is included. This allows to study the transient mean fields at high density via the event shape analysis of nucleons, nucleon clusters and pions.

1 Introduction

One of the challenges of modern heavy ion physics is the extraction of the equation of state for extremely excited nuclear matter. In particular the creation and study of matter at high net baryon density has received much attention recently. A rapid restoration of chiral symmetry with increasing baryon density is predicted by all approaches which embody this fundamental aspect of QCD [1,2]. Beam energies between 10 to 15 AGeV – as studied experimentally at the BNL-AGS [3] – seem to be well suited to stop two heavy ingoing nuclei and to create the desired high baryon densities. Mean fields [4] may give important contributions to the compression-induced pressure and could – via the flow effect – be accessible to experimental observation just as in the 1GeV region [5]. The bounce-off for protons has been observed at 10 GeV/n [6] as well as azimuthally asymmetric particle correlations in the projectile hemisphere [7]. These experimental discoveries encourage us to investigate the formation of nuclear clusters – as compared to light hadrons – for which flow can even dominate the momentum spectra [9].

2 Cluster Coalescence

We extend a phase space coalescence approach which was successfully applied to deuteron production at bombarding energies around 1AGeV [10,11] and 10-15AGeV [12,13] in order to calculate yields for light nuclear clusters (A≤4). The formation probability of light baryon clusters d, t and He is calculated by projecting a final nucleon phasespace distribution - here generated by the RQMD model (1.07) - on cluster wave-functions via the Wigner-function method [14]. The "source function" for the nucleons is defined by the "freeze-out" positions x_i^μ and momenta p_i^μ of nucleons after their last scattering or decay. The basic assumption is that the single particle phase space density given by the transport model is a good approximation to the density operator in the Wigner-representation. Under these conditions an approximation to the density operator for M-particle states can be constructed by the assumption of uncorrelated emission:

$$\frac{\binom{N}{M}}{N^M} \sum_{Z=0}^{M} \binom{M}{Z} \left[\prod_{i=1}^{Z} (2\pi)^3 f_p(\vec{x}_i, \vec{p}_i) \left(\left| \frac{1}{2}\frac{1}{2} \right> < \frac{1}{2}\frac{1}{2} \right| \right) \frac{1}{2} \left(\left| \frac{1}{2} s_i^3 \right> < \frac{1}{2} s_i^3 \right| \right) \right]$$

$$\times \left[\prod_{i=Z+1}^{M} (2\pi)^3 f_n(\vec{x}_i, \vec{p}_i) \left(\left| \frac{1}{2} - \frac{1}{2} \right> < \frac{1}{2} - \frac{1}{2} \right| \right) \frac{1}{2} \left(\left| \frac{1}{2} s_i^3 \right> < \frac{1}{2} s_i^3 \right| \right) \right] \quad (1)$$

Z denotes the charge number of the M-nucleon combination and $f_n(\vec{x}, \vec{p})$, $f_p(\vec{x}, \vec{p})$ are the neutron and proton phase space density, respectively.

The cluster wave function is assumed to separate in collective and relative part. The Wigner-density of the relative wave function is given by

$$\hat{\rho}_C^W := (|TT_3><TT_3|)(|SS_3><SS_3|)\rho_C^W(t_1, q_1; ...; t_{M-1}, q_{M-1}) \quad (2)$$

The $\vec{t}_i(\vec{x}_1, ..., \vec{x}_M)$ and $\vec{q}_i(\vec{p}_1, ..., \vec{p}_M)$ for (i=1,...,M-1) are the $M-1$ relative coordinates of the nucleons within the cluster. S, S_3, T, T_3 are the spin and isospin quantum numbers of the cluster state.

The formation of cluster states is defined by the trace over the statistical operator $\hat{\rho}$ and the projector on the individual cluster wave function $|\Psi_C><\Psi_C|$ in the Wigner-representation. In a Monte Carlo formulation – appropriate for the application to microscopic transport calculations – this formation rate can be expressed by [15]

$$dN_M = g \left\langle \sum_{\substack{i_1,...,i_M \\ i_1<...<i_M}} \rho_C^W(\vec{t}_{i_1}, \vec{q}_{i_1}; ...; \vec{t}_{i_{M-1}}, \vec{q}_{i_{M-1}})) \right\rangle$$

$$\times d^3 t_{i_1} d^3 q_{i_1} ... d^3 t_{i_{M-1}} d^3 q_{i_{M-1}} \quad . \quad (3)$$

$< ... >$ denotes event averaging. The sum runs for each event over all M-nucleon combinations with coordinates in position and momentum space taken at equal time in the M-nucleon rest frame (cms) immediately after all cluster nucleons have frozen out. The calculated numbers contain higher mass fragments by construction. The number of A> 4 clusters is small, however, for rapidity values $|y - y_{mid}| < 1$. The factor g contains both isospin projection and spin averaging. For the deuteron the spin and isospin degrees of freedom lead to the correction $g(d) = 3/8$. Note that the factorization of the phase space density and internal quantum numbers in Eq. 1 leads to an isospin projection factor $1/2$ in addition to the usually applied spin correction $3/4$. The reason for this additional correction is the lack of many body correlations, i.e. the phase space part of $\hat{\rho}$ contains the densities of all states without respect to symmetry concerning particle exchange. Therefore, we use in our approch an equal (statistical) weight for all possible states in the internal degrees of freedom in order to reduce to the correct number of states that are allowed by the Pauli principle. For ^{3}H and He states the statistical corrections are $g(^3\text{He}) = g(^3\text{H}) = \frac{1}{12}$ and $g(^4\text{He}) = 1/96$.

For the deuteron we assume a Hulthén-wave function derived from a Yukawa-type potential interaction [16,17]. For triton, ^{3}He and ^{4}He states we use 3- and 4-particle harmonic oscillators with different coupling strength. The coupling constants are adjusted to the mean square charge radii of the diverse cluster states (see e.g. [18,19,20])

At nuclear ground state density the nuclear mean field may be decomposed into two large pieces: an attractive scalar field provided by the quark condensate and/or correlated two-pion exchange (the σ field), and a repulsive vector potential (the ω field) [21]. It is expected that the momentum dependence [11,22], the excitation into resonances [23] and the transition to quark matter [24] will play a crucial role fields in highly excited and dense matter. In the following we demonstrate the sensitivity of flow observables to mean fields by comparing two schematic cases: In the first case the potentials are switched off (i.e. the *cascade mode* is used). The second scenario uses potential type interactions which define effective baryon masses in a medium [25]

$$p_i^2 - m_i^2 - V_i = 0$$

and thus simulate the effect of mean fields. Here

$$(2m_N)^{-1} V_i = +\frac{1}{2} \sum_{j, j \neq i} \alpha_{ij} \left(\frac{\rho_{ij}}{\rho_0} \right) + \frac{\beta}{\gamma + 1} \left[\sum_{j, j \neq i} \left(\frac{\rho_{ij}}{\rho_0} \right) \right]^{\gamma} \qquad (4)$$

with ρ_{ij} a Gaussian of the CMS distance vector normalized to one, ρ_0 ground state matter density and $\alpha = -0.4356\text{GeV}, \beta = 0.385\text{GeV}, \gamma = 7/6$ parameters

Deuteron Formation

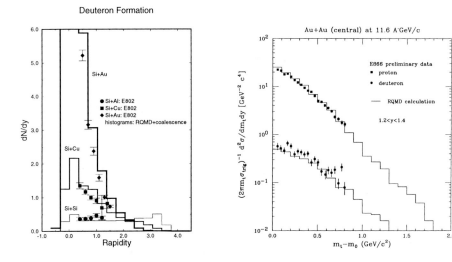

Figure 1: Left: Rapidity distributions of deuterons in Si+Si (b< 1fm), Si+Cu (b< 1.5fm) and Si+Au (b< 3fm) reactions at 14.6AGeV calculated from RQMD simulations including potential interactions for baryons (solid histograms). The symbols show E802-data from Ref. [28] for central Si+Al, Cu and Au reactions. Right: Transverse mass spectra for protons and deuterons in central Au(11.6AGeV)Au reactions at $y_{LAB} = 1.3$. RQMD-simulations including potential interactions for baryons (histograms) are compared with preliminary E866-data (symbols) from Ref. [31].

which are adjusted to the saturation properties of nuclear matter (binding energy and compressibility). As has been stated in[26], the experimental data for nucleus-nucleus reactions at 10-15AGeV seem to indicate more repulsion than just given by a pure density dependence as in Eq. (3). This additional repulsion is probably caused by the internal excitations of the dense matter produced, i.e. a momentum dependence of nuclear forces. we use the same approach as in [26], i.e. we harden the density dependence of the potentials in order to get agreement with proton singles spectra. (This is achieved by switching off the attractive 2-body force in the $\Delta\Delta$ and NB* channel – $\alpha_{\Delta\Delta} = \alpha_{NB*} = 0$ – thus explaining the index pair (ij) in Eq. (3).)

3   Transverse Expansion and Cluster Flow

The production of clusters has been measured and analyzed recently for central and peripheral reactions p+A, Si+A and Au+Au at AGS-energies [27,28,3,29] and S+A at 200AGeV [30]. Comparisons between coalescence calculations for deuterons at $p_t = 0$ and measurements for pA reactions have been discussed

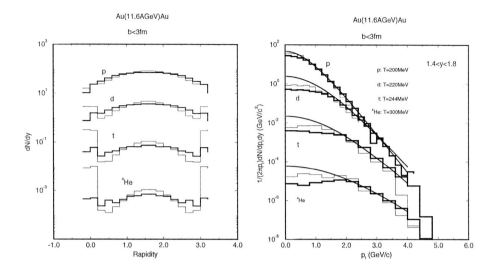

Figure 2: Rapidity distributions (l.h.s) and transverse momentum spectra (r.h.s) of p,t and He in Au(11.6AGeV)Au ($b < 3fm$) calculated with RQMD+coalescence. Potential interactions for baryons (bold solid histograms) are compared to cascade calculations (solid histograms). The solid lines show Boltzmann-parametrizations for the high-momentum tail (see text).

already in [27,13]. In Fig. 1 we compare results for deuterons in central Si+A reactions at 14.6AGeV [12] with E802 data [28]. Furthermore, calculations for transverse mass spectra of protons and deuterons are shown for the reaction Au(11.6AGeV)Au (b<3fm) compared to preliminary E866-data [31]. On the level of the systematic errors in the measurements ($\approx 15\%$,[32,28]) we find good agreement – even for the strong temperature splitting between protons and deuterons in massive reactions.

The most prominent observables for transverse expansion and collective flow are the characteristic shoulder-arm shape and different apparent temperatures in the transverse mass spectra of particles with different mass [8,9,33,34] [35,36,12] as well as the nuclear 'bounce-off' predicted by hydrodynamics [4,37] and microscopic models [4,38,37] and discovered first at the Bevalac [39].

Predictions for rapidity distributions and transverse momentum spectra of p, d, t and ^{4}He cluster are shown in Fig. 2 for central reactions Au+Au at 11.6AGeV. The figure contains cascade (solid histograms) and potential calculations (bold solid histograms). The solid lines show Boltzmann-distributions with temperature parameters adjusted to fit the transverse spectra for p_t >2GeV/c as calculated with baryon potential-interaction. All rapidity distributions peak at midrapidity indicating strong stopping in accordance

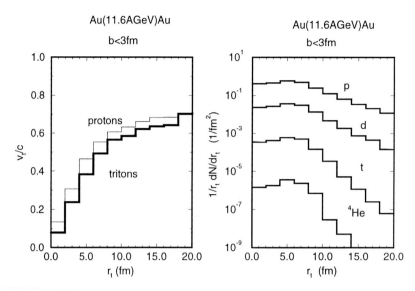

Figure 3: Left: Transverse velocity profiles for protons and tritons in RQMD+coalescence. Right: Transverse freeze-out densities for of p, d, t and ^{4}He integrated over time.

with earlier predictions and preliminary data for protons[38,12]. The transverse spectra have a strong shoulder-arm shape which deviates markedly from distributions expected from a purely thermalized fireball. The shoulder-arm shape becomes most prominent for heavy clusters. For ^{4}He clusters a peak even appears at finite p_t. The high momentum tail of the transverse spectra exhibit different 'apparent' temperatures for clusters with different mass while a thermal system would predict similar slope parameters[40]. Note that the extracted temperature values depend strongly on the p_t-cut choosen. The absolute values extracted by exponential parametrizations always lead to additional systematic errors in the absolute yields according to our calculation overestimating the cluster yields substantially.

The characteristic deviations from thermal distributions are caused by strong transverse expansion and collective flow particularly in massive reactions like Au+Au. The flow-correlations at the microscopic freeze-out are shown in Fig. 3 which contains calculations for the freeze-out velocities and density profiles of protons, deuterons, tritons and ^{4}He. The velocity profiles for all clusters are similar. They exhibit a convex shape and saturate at $\approx 0.7c$. The freeze-out densities have a complicated shape which peakes around 5fm. Most of the nucleons freeze out at larger distance. This is indicated by an average freeze-out radius of $\approx$ 10fm. The strong transverse expansion is caused by the considerable baryon stopping and the pile up of high particle densities near to the reaction center. Note that the region of highest compression ($\rho/\rho_0 > 3$) is large (V≃several hundred fm^3) and contains up to 60% baryons in resonance states [41]. During the expansion phase comoving particles undergo frequent collisions transporting the system collectively sidewards until the flow-induced pressure pushes them into the vacuum. Hence, the suppression of particle emission at $r_t \to 0$ is basically caused by the dynamical expansion: many nucleons are transported through the medium before they reach the 'surface' and freeze out. On the other hand cluster formation is relatively suppressed at the 'surface' ($r_t > 6$fm) which contradicts simple fireball analyses that assume a common density and velocity profile for all particles.

Clusters are clearly dominated by the collective flow components ($\approx 80\%$). Therefore, the freeze-out density and collective velocity profiles determine the final spectra almost exclusively. The different apparent temperatures at high p_t-values are caused by the strong weight of large flow velocities for $r_t > 6$fm. The peak/shoulder in the transverse spectra, however, appears approximately at $p_t/A \approx < \beta_t >$ and directly measures the strength of the transverse flow at the position where most of the clusters at central rapidities freeze out ($r_t \approx 5 - 7$fm). Note that it is not possible to describe the transverse spectra with one single temperature and collective flow velocity in contrast to what has been claimed for reactions in the 1GeV/n regime [34]. Light systems like Si+Si do not provide a comparable transverse expansion. In fact, the 'surface suppression' acts more strongly in the case of the smaller system. Transverse flow turns out to be nearly invisible due to large 'local' momentum components.

The difference between potential and cascade calculations is largest in the low-p_t part of the spectra (Fig. 2). For heavier clusters the distributions close to $p_t = 0$ change by up to a factor of three. Fig. 4 shows the rapidity dependence of the proton and deuteron yields at low transverse momenta – here defined by a cut in transverse momentum $p_t/A < 0.5$GeV/c – and the average transverse momenta of p, d and ^{4}He. While cascade calculations exhibit a clear peak in the dN/dy spectra, the calculations including potentials show a dip even for central events. The potentials also change the average transverse momenta by <20%. The 'transverse communication' is in light systems like Si+Si much smaller and does not allow for a considerable transverse push due to the mean fields. Cascade calculations, however, exhibit concave shaped rapidity spectra which turn to convex distributions if potentials are included.

The role of the shapes of collective velocity and density profiles has been the subject of much previous debate [9,12,34,35,36]. E.g. in Ref. [35] a quadratic r_t-dependence for the velocity profile in combination with a box-shaped density profile was used to explain the low-pt pion enhancement, i.e. a concavely curved p_t-spectrum. As a consequence of these assumptions proton and deuteron spectra show the same behaviour, in particular for $p_t \to 0$. The

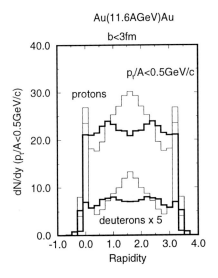
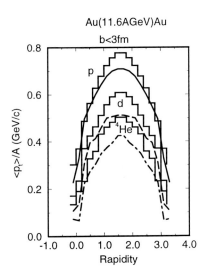

Figure 4: Left: Rapidity distributions of deuterons and protons in central Au(11.6AGeV)Au reactions for low transverse momenta ($p_t/A < 0.5$GeV/c). Right: Average transverse momenta for p, d and ^{4}He as a function of rapidity.

main reason for these misleading results in [35] is probably the misinterpretation of concavely shaped pion spectra. Pions are strongly influenced by the final decays of resonances such as Δ, ρ, B^* (see [41] and Refs. therein). The alternative prediction that the low-p_t pion excess at AGS energy comes from Δ-resonances [42] has been confirmed by experimental reconstruction of the pπ invariant masses which show a strong Δ-signal, in agreement with RQMD [3]. Furthermore, the early preliminary data for protons used in [35] were limited in acceptance ($m_t - m_0 > 200$MeV) and excluded those regions where most of the shoulder-arm effect appears.

Besides the characteristic signals in the inclusive spectra, the correlation between rapidity and directed transverse momentum $p_x(y)$ in Fig. 5 is another indicator of a non-trivial event geometry. Clusters exhibit larger $p_x(y)/A$ values than nucleons although the division by A excludes the trivial effect of the momentum scaling with mass $p_A/A \simeq p_N$ at equal velocity. This stronger correlation for cluster states is well known from Au+Au reactions in the 1GeV/n energy regime [43,5]. Here it is caused by the suppression of cluster formation near to the original beam axis due to higher relative momenta for nucleons. Consequently, the high transverse in-plane velocities are more strongly weighted for clusters.

The value of the flow-correlation $p_x(y)$ in central Au(11.6AGeV)Au collisions is roughly a factor 1.5-2 higher due to the additional sideward push of

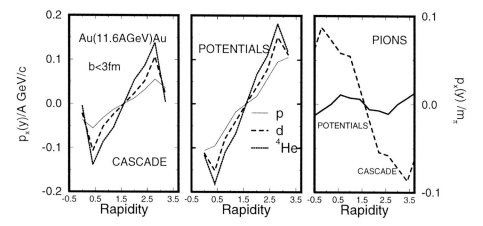

Figure 5: Directed flow observable $p_x(y)$ in central (b¡3fm) reactions Au(11.6AGeV)Au (not rotated). Left: protons, deuterons and ^{4}He in the cascade mode. Middle: protons, deuterons and ^{4}He including potentials. Right: pions with and without baryon-potential interaction.

the mean fields. Fig. 5 includes pionic anti-flow (see also [44,45]), which appears to be sensitive to the baryonic mean fields, too. While cascade calculations show sizeable p_x/m values for πs, the inclusion of baryonic potentials leads to almost vanishing $p_x(y)$-values in the laboratory frame. In the principal axis system, however, the strong anticorrelation of pions to baryons is conserved.

4 Summary and Outlook

We presented a phase space coalescence description for cluster formation in relativistic nucleus-nucleus collisions using the Wigner-function method. The application to the relativistic transport approach RQMD allows the prediction of momentum distributions which have been calculated here for central Au(11.6AGeV)+Au and Si(14.6AGeV)+A reactions.

The microscopic model shows that the strong stopping power recently discovered in nucleus-nucleus collisions at 10-15 AGeV results in observable collective behaviour of the stopped baryon-rich matter. Considerable flow ($< \beta > \approx 0.5$ c) develops due to the internal pressure of the dense matter. The transverse expansion is most visible in the momentum spectra of nuclear clusters. In central Au(11.6AGeV)Au collisions the transverse spectra exhibit a strong shoulder-arm shape, most prominent for heavier mass clusters, apparent temperatures which increase with cluster mass and a clear "bounce-off" event shape. The averaged transverse flow velocities in the reaction plane are for clusters markedly larger than for protons (factor 2). Both the shoulder-arm shape and the large bounce-off signal for nuclear clusters are strongly related to the freeze-out geometry and flow correlations. The cluster spectra and the in-plane flow change markedly if baryon potential-interaction is included: In the reaction Au(11.6AGeV)Au the yields of nuclear clusters change up to a factor of three mostly at low p_t and central rapidities. The $< p_x > /A$-correlation for nucleons and nuclear clusters increases by 1.5-2 while the anticorrelated in-plane flow of pions vanishes.

In general, calculations including baryon potential-interaction agree better with present measurements of various experimental groups at the AGS. Though, the low amount of stopping and the strong transverse momentum production in central Si(14.6AGeV)Au reactions – as discussed in [26,32] – is not yet understood. The sensitivity of nuclear clusters as well as pions and antimatter-cluster [46] to the baryonic phase space distributions are very encouraging: The 'tool-box' of observables provided by pion and (anti-,strange-)cluster flow opens the doorway to study such in-medium properties with nucleus-nucleus collisions at the AGS.

Acknowledgments

This contribution is dedicated to the 60th birthday of Prof. Dr. Dr. h.c. mult. Walter Greiner. This work has been supported by the BMFT, GSI, DFG and the Humboldt foundation.

References

1. U. Vogl and W.Weise: Prog. Part. Nucl. Phys. 27(1991)195; W. Weise: Nucl. Phys. A553(1993)59
2. X. Jin, T.D. Cohen, R.J. Furnstahl and D.K. Kriegel: Phys. Rev. C47(1993)2882 and refs. therein
3. Proceedings contributions of E814,E877,E810,E859,E866 to Quark Matter 1993, published in Nucl. Phys. A566(1993)
4. H.Stöcker and W. Greiner: Phys. Rep. 137(1986)278; H. Kruse, B.V. Jacak, and H. Stöcker: Phys. Rev. Lett. 54(1985)289; J.J. Molitoris, J.B. Hoffer, H. Kruse and H. Stöcker: Phys. Rev. Lett. 53(1984)899; G. Buchwald, G. Graebner, J. Theis, J. Maruhn, and W. Greiner: Phys. Rev. Lett. 52(1984)1594; Ch. Hartnack, M. Berenguer, A. Jahns, A. v. Keitz, R. Mattiello, A. Rosenhauer, J. Schaffner, T. Schönfeldt, H. Sorge, L. Winckelmann, H. Stöcker, W. Greiner: Nucl. Phys. A538(1992)53c;

5. K.-H. Kampert: J.Phys. G15(1989)691; H.H. Gutbrod, K.H. Kampert, B.W. Kolb, A.M. Poskanzer, H.G. Ritter and H.R. Schmidt: Phys. Lett. B216(1989)267

6. J. Barrette for the E877 collaboration: Nucl. Phys. A590(1995)259c

7. T. Abott for the E802 collaboration: Phys. Rev. Lett. 70(1993)1393

8. P.J. Siemens and J.O. Rasmussen: Phys. Rev. Lett. 42(1978)880

9. H. Stöcker, A. Ogloblin, W. Greiner: Z. Phys. A303(1981)259; S. Nagamiya, M.-C. Lemaire, E. Moeller, S. Schnetzer, G. Shapiro, H. Steiner, and I. Tanihata: Phys. Rev. C24(1981)971

10. M. Gyulassy, K. Frankel and E.A. Remler: Nucl. Phys. A402(1983)596

11. J. Aichelin, A. Rosenhauer, G. Peilert, H. Stöcker, and W. Greiner: Phys. Rev. Lett. 58(1987) 1926; J. Aichelin, E.A. Remler: Phys. Rev. C35(1987) 1291

12. R. Mattiello, A. Jahns, H. Sorge, H. Stöcker, W. Greiner: UFTP-preprint 362/1994; Phys. Rev. Lett. 74(1995)2180

13. J.L. Nagle, S. Kumar, D. Kusnezov, H. Sorge, R. Mattiello: Yale-preprint 40609-1160, July 1995; Phys. Rev. C in print

14. E.A. Remler and A.P. Sathe: Ann. of Phys. 91(1975)295 ;E.A. Remler: Ann. of Phys. 95 (1975) 455 ; E.A. Remler: Ann. of Phys. (NY) 136(1981)293

15. R. Mattiello, H. Sorge, H. Stöcker, W. Greiner: BNL-preprint 1996, BNL-63137

16. R.G. Sachs and M. Goeppert-Mayer: Phys. Rev. 53(1938)991

17. L. Hulthen: Ark. Mat. Ast. Fys. 28, No. 5

18. R. Hofstätter: Rev. Mod. Phys. 28(1956)214

19. J. Carlson: Phys. Rev. C38 (1988) 1879

20. C.R. Chen, G.L. Payne, J.L. Friar, B.F. Gibson: Phys. Rev. C33 (1986)1740

21. B.D. Serot and J.D. Walecka: Adv. Nucl. Phys. 15 (1986); J. Theis et al.: Phys. Rev. D28 (1983)2286

22. E.D. Cooper, B.C. Clark, R. Kozack, S. Shim, S. Hama, J.I. Johansson, H.S. Sherif, R.L. Mercer, B.D. Serot: Phys. Rev. C36(1987)2170

23. F. de Jong and R. Malfliet: Phys. Rev. C46(1992)2567

24. J. Ellis, J. Kapusta, K. Olive: Phys. Lett. B273(1991)122

25. H.Sorge, H. Stöcker and W. Greiner: Ann. Phys. (NY) 192(1989)266; Nucl. Phys. A498(1989)567c; H. Sorge, A. v. Keitz, R. Mattiello, H. Stöcker and W. Greiner: Z. Phys. C47 (1990)629

26. H. Sorge, R. Mattiello, H. Stöcker and W. Greiner: Phys. Rev. Lett. 68(1992)286.

27. J.L. Nagle, B.S. Kumar, M.J. Bennett, G.E. Diebold, J.K. Pope, H. Sorge, J.P. Sullivan: Phys. Rev. Lett. 73 (1994) 1219

28. T. Abbott for the E802 collaboration: Phys. Rev. C50 (1994) 1024

29. D. Beavis for the E878 collaboration: Phys. Rev. Lett. 75(1995)3078

30. J. Gillo et al.: Nucl. Phys. A590(1995),483c

31. Ziping Chen for the E802 collaboration: Contrib. to the First Int. Conf. on Frontiers of Physics, Shantou, China, August 1995

32. B. Moscowitz, M. Gonin, F. Videbaek, H. Sorge, R. Mattiello: Phys. Rev. C 51(1995)310

33. P. Danielewicz and Q. Pan: Phys. Rev. C46(1992)2002; Q. Pan and P. Danielewicz: Phys. Rev. Lett. 70(1993)2062,3523

34. M.A. Lisa for the EOS collaboration: Phys. Rev. Lett. 75(1995) 2662

35. K.S. Lee and U. Heinz: Z. Phys. C48(1990)525; K.S. Lee, U. Heinz and E. Schnedermann: Z. Phys. C48 (1990) 525

36. E. Schnedermann and U. Heinz: Phys. Rev. Lett. 69 (1992) 2908

37. N.S. Amelin, E.F. Staubo, L.P. Csernai, V.D. Toneev, K.K. Gudima, D. Strottman: Phys. Rev. Lett. 67(1991)1523

38. H. Sorge, A.v. Keitz, R. Mattiello, H. Stöcker, W. Greiner: Phys. Lett. B 243(1990)7

39. H.A. Gustafsson, H.H. Gutbrod, J. Harris, B.V. Jacak, K.H. Kampert, B.Kolb, A.M. Poskanzer, H.G. Ritter and H.R. Schmidt: Mod. Phys. Lett. A3(1988)1323

40. A. Mekijan: Phys. Rev. Lett. 38(1977)640; Phys. Rev. C17(1978)1051

41. M. Hofmann, R. Mattiello, H. Sorge, H. Stöcker and W. Greiner: Phys. Rev. C51(1995) 2095

42. H. Sorge, R. Mattiello, H. Stöcker and W. Greiner: Phys. Lett. B271(1991)37

43. M. D. Partlan for the EOS collaboration: Phys. Rev. Lett. 75(1995)2100

44. A. Jahns, Chr. Spieles, H. Sorge, H. Stöcker and W. Greiner: Phys. Rev. Lett. 72(1994) 3464

45. S.A. Bass, R. Mattiello, H. Stöcker, W. Greiner and Ch. Hartnack: Phys. Lett. B302 (1993) 381; S.A. Bass, C. Hartnack, H. Stöcker, W.Greiner: Phys. Rev. C51(1995)3343

46. Ch. Spieles, M. Bleicher, A. Dumitru, C. Greiner, M. Hofmann, A. Jahns, U. Katscher, R. Mattiello, J. Schaffner, H. Sorge, L. Winckelmann, J. Maruhn, H. Stöcker, W. Greiner: Nucl. Phys. A590(1995)271 ; M. Bleicher, C. Spieles, A. Jahns, R. Mattiello, H. Sorge, H. Stöcker, W. Greiner: Phys. Lett. B361(1995)10

Composite Particle Probes of Nuclear and Quark Matter

J.L. Nagle and B.S. Kumar

A.W. Wright Nuclear Structure Laboratory, Yale University, New Haven, Connecticut 06520

H. Sorge

New York State University of Stony Brook, Stony Brook, NY 11974

We will discuss how the abundances and kinematic distributions of nuclei and anti-nuclei can be used to probe the spatial distributions at freeze-out and the dynamics of high energy heavy ion collisions. Model predictions will be compared with composite (anti) particle data from both the AGS and CERN.

1 Introduction

There is considerable interest in using high energy heavy ion collisions to probe the structure of nuclear matter at high energy densities and also possibly high baryon densities. It has been proposed that in these collision environments, a phase transition may occur converting hadronic matter into a quark gluon plasma. In order to determine whether the conditions for such a transition are met, we would like to understand the time evolution during these collisions of the energy and baryon densities.

The degree of strong coupling of the system, either through production of baryon resonances or a transition to the plasma state, can be measured by the strength of outward collective expansion. Also, a strongly coupled system should have a longer lifetime and thus a larger volume at freeze-out (when particles cease to interact). The production of light nuclei via the coalescence process is sensitive to the lifetime and volume of the system at freeze-out and to collective flow and should reveal important information about the colliding system [1].

At AGS energies (10-15 A GeV), anti-proton production is near threshold and is thus extremely sensitive in heavy ion collisions to the energy density achieved. Both through the production of "resonance matter" and owing to a phase transition with the restoration of chiral symmetry (quark masses approaching zero) [2,3], one expects an enhancement in anti-baryon production. Hence, in principle, the production of antiprotons and anti-nuclei are an excellent probe of the early stages of the collision.

However, anti-baryons are expected to have a very large annihilation cross in baryon-rich matter. Thus, the anti-baryon losses are a very sensitive probe of the baryon density at all stages of the collision. The two competing effects of possible enhanced production and large annihilation make the interpretation of experimental data difficult. We believe that the additional constraints provided by the combined measurement of antiprotons and anti-nuclei should significantly aid in disentangling these competing effects [4] and give us access to interesting physics.

2 Coalescence of Light Nuclei

2.1 Determining the Collision Volume

Light nuclei produced in heavy ion collisions in the mid-rapidity region are not likely to be fragments from the initial colliding ions. Instead, the light nuclei are mostly the result of individual nucleons which at the point of their final strong interaction (freeze-out) are close enough in both configuration and momentum space and in the right quantum states to fuse together into composite objects. Thus, the yields of composite objects (like the deuteron) will be sensitive to the density of ingredients (nucleons) in phase space at freeze-out.

If the coalescence process were only dependent on the relative momentum of the nucleons, then the experimentally determined coalescence scale factor B_A defined below, which relates the density of composites (of mass A) to the density of ingredients to the Ath power , should be constant (at equal momentum per nucleon $K = A \times k$) [5,6].

$$B_A = \left(\gamma \frac{d^3 N_A}{dK^3}\right) / \left(\gamma \frac{d^3 N_p}{dk^3}\right)^A \tag{1}$$

However, in small impact parameter heavy ion collisions at the AGS and the SPS, the collision volume at freeze-out is considerably larger than the volume of the composite particles formed, and the scale factor B_A becomes sensitive to the spatial extent of this larger source. Data from experiment 814 in Si + Pb collisions at 14.6 A GeV/c, show convincingly that the scale factor decreases as the centrality of the collision increases [7]. This is a direct indication that the size of the collision volume is growing with centrality, as expected.

While simple analytic models have been used with varying degrees of success to interpret B_A, one needs a detailed simulation of nucleus-nucleus collisions to extract information about source dimensions as we have done previously [1].

2.2 Transport Model RQMD as a Tool

In the following, we will use the transport model RQMD (relativistic quantum molecular dynamic) [8] followed by a coalescence extension. The model RQMD

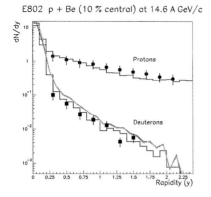

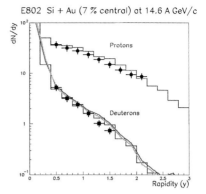

Figure 1: Predictions of the RQMD + Coalescence model for p+Be→p,d. The data are from E802.

Figure 2: Predictions of the RQMD + Coalescence model for Si+Au→p,d. The data are from E802.

has been used extensively to describe the spectra of particles produced in heavy ion reactions at AGS (10-15 A GeV) and CERN (160-200 A GeV) energies. However, RQMD does not include the production of light (anti) nuclei, and thus, such a calculation must be performed separately.

The phase space output of RQMD gives the final momenta and locations at which particles suffer their last interactions (defined for an energy threshold of 2 MeV). We consider neutrons and protons in pairs and if in their center of mass frame they are within a set of cutoff parameters in relative momentum and position we assume they will form a deuteron[9]. Alternatively, we calculate the deuteron formation probability by projecting the nucleon pair phase space on the deuteron wavefunction via the Wigner-function method as described elsewhere[9,10]. The deuteron Wigner density (ρ_d^W) is approximated to be that of the ground-state harmonic oscillator wavefunction (although we have also done the calculation with the Hulthen wavefunction solution for the Yukawa potential).

In figure 1, we show the RQMD prediction for protons and three calculations for deuterons (using the cutoff parameters and the harmonic osciallator and Hulthen wavefunction in the Wigner formalism). We see little sensitivity to the coalescence formalism and good agreement with the data from E802 for p + Be collisions where the system size is expected to be small. In figure 2 there is good agreement in Si + Au central collision data from E802 at 14.6 A

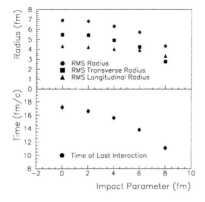

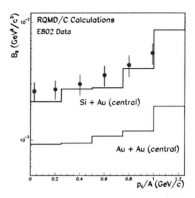

Figure 3: Source dimensions and lifetime from RQMD in Si + Pb collisions
versus collision centrality.

Figure 4: Scale factor B_2 versus transverse momentum per nucleon. The
data are from E802.

GeV/c, and thus we conclude that RQMD must have a reasonable description
of the lifetime and spatial extent of the source at freeze-out. One can then
look at the actual space-time positions of the nucleons which contribute to the
composite particle formation in RQMD.

Shown in figure 3 are the RMS radii and lifetimes of the nucleon source
at freeze-out from RQMD as a function of collision centrality. The numbers
indicate a source radius and lifetime increasing with centrality and significant expansion beyond the size of the initial volume of geometric overlap. We
emphasize that RQMD does not explicitly assume a quark-hadron phase transition. The significant expansion observed is a result of strong coupling of the
system through the large excitation of delta resonances in agreement with data
from E814[11].

2.3 Collective Flow

The above picture of coalescence is complicated by the presence of strong collective expansion of the interaction volume. Collective expansion or flow produces a strong space-momentum correlation for particles. Since the deuteron

state is a result of a correlation in the wavefunction between two nucleons, collective expansion will enhance this correlation and thus increase the number of composite objects. This enhancement can be understood qualitatively in a hydrodynamic plus thermal model as detailed elsewhere [13]. All particles have the same initial temperature in the thermal model and are given the same velocity boost (fluid motion gives the same velocity, not momentum, boost). Heavier particles will get a larger momentum boost and therefore have an apparently larger temperature. The result is an increased number of composites at larger transverse momentum. Data from E802 show that the relative yields of deuterons increases as one goes to higher transverse momentum (as seen in figure 4). The RQMD + Coalesence calculation is in excellent agreement with the data. Data for Au + Au at 11 A GeV/c near $p_t = 0$ is already available from AGS experiments 878 and 886 and should soon be available over a range in transverse momentum from experiments 864, 866, and 877. These data will be compared with our predictions shown in figure 4.

3 Coalescence of Anti-Nuclei

3.1 Anti-proton Results

Experiments measure only the final distribution in momentum space of anti-baryons and as mentioned previously, it is difficult to disentangle possible enhanced production and large annihilation. Experiment 814 published results from Si + Pb collisions and noted that the anti-proton yields appeared to scale with the number of first nucleon-nucleon collisions [14,15]. Their interpretation was that in heavier systems and smaller impact parameters, the expected increased production is balanced by the increased annihilation. Recent data from experiment 878 in Au + Au central collisions show that at $p_t = 0$ the anti-proton yields no longer scale with first collisions and are much reduced [16]. This result could be indicative of increasing annihilation dominating over any possible enhanced production.

In figure 5, anti-proton data from E878 are shown with the predictions of RQMD version 1.07 (lower grey curve) and ARC version 1.9.5 (upper black curve). The dramatic difference between model predictions results mostly from a difference in modeling of the annihilation process. RQMD uses the full free space annihilation cross section. In nature, the annihilation process is actually one of capture and then subsequent annihilation. Thus, it is thought that the capture process may be disturbed in a dense nuclear medium. ARC invokes a screening mechanism whereby the effective annihilation is reduced [17]. RQMD shows a valley in the anti-protons at mid-rapidity (at $p_t = 0$) where there is

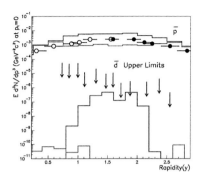

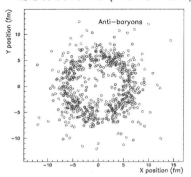

Figure 5: The data and upper limits are from E878 for central Au + Au collisions. Predictions using transport models ARC (top curve) and RQMD (bottom curve) are shown.

Figure 6: The spatial distribution for a slice along the longitudinal direction for anti-baryons produced in RQMD Au + Au central collisions.

the corresponding highest baryon density.

3.2 Shape Sensitivity

Although the difference between the two model momentum distributions for anti-protons is quite dramatic, the spatial distribution is even more extreme. As shown in figure 6, in RQMD the anti-baryons only escape the collision volume at the surface. However, in ARC, because of the reduced annihilation, the anti-baryons escape from all points in the volume. This shape difference results in a four order of magnitude difference in the predicted yields of anti-deuterons at $p_t = 0$ in central Au + Au collisions at 10.8 A GeV/c, as shown in figure 5. (It should be noted that the anti-deuteron calculation was done on the output file of the ARC code using an identical coalescence formalism to that applied using RQMD.) Experiment 878 has set upper limits on the production of anti-deuterons, but does not reach sensitivities sufficient to test these predictions. However, experiment 864 with data taken in the fall of 1995 should be able to either measure anti-deuterons or place significantly improved limits on their production cross sections.

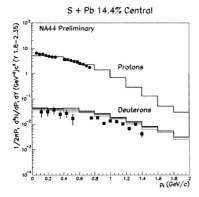
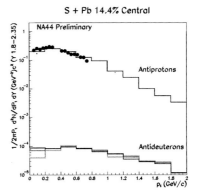

Figure 7: NA44 preliminary data for protons and deuterons compared with RQMD+C calculations.

Figure 8: NA44 preliminary data for anti-protons compared with RQMD+C calculations.

3.3 CERN Results

Recently experiment NA44 at CERN has measured the yields of deuterons and anti-deuterons in S + Pb central collisions at 200 A GeV/c. In figure 7, preliminary data from NA44 [18] are shown with RQMD (version 2.1) + Coalescence predictions. The agreement is reasonable. As seen in figure 8, RQMD results give reasonable agreement with the NA44 anti-protons. Similar comparisons have been made with the preliminary NA44 anti-deuteron data (not shown in figure) and we find that the RQMD + Coalescence predictions (figure 8) are a factor of 5 lower than the data. RQMD at CERN energies incorporates the fusion of color strings into color ropes. These color ropes can "be viewed as locally deconfined quark matter." [19] The ropes lead to a significant enhancement of anti-matter, particularly for strange anti-baryons. Despite the inclusion of this source of enhancment, the anti-deuteron predictions are still below the experimental data. New data from NA44 and NA52 in Pb + Pb collisions will provide more information to understand the anti-deuteron results.

4 Conclusions

The RQMD + Coalescence model reproduces data at the AGS and CERN for lighter systems. Clear evidence is seen for overall expansion and collective flow.

Further studies with heavier composites ($A > 2$) should provide additional information. Already there is exciting new data from CERN on anti-clusters with more data to come in Pb + Pb collisions. The anti-deuteron search at the AGS is underway and should also yield results soon. Both sets of results will be necessary to fully understand the role of enhanced antimatter production and annihilation in heavy ion collisions.

Acknowledgments

We are pleased to acknowledge useful and stimulating discussions with R. Mattiello and J. Simon-Gillo. We thank T. Schlagel for providing ARC events (version 1.9.5). This work was supported in part by grant DE-FG02-91ER-40609 with the U.S. Department of Energy.

References

1. J.L. Nagle et al., Phys. Rev. Lett. **73** , 1219 (1994).
2. S. Gavin and M. Gyulassy, M. Plumer, R. Venugopalan, Phys. Lett. **B234**, 175 (1990).
3. U. Heinz et al., Nucl. Phys. **12** (1966) 1237-1263.
4. J.L. Nagle et al., Phys. Rev. Lett. **73** , 2417 (1994).
5. S.T. Butler and C.A. Pearson, Phys. Rev. Lett. **7** (1961) 69.
6. A. Schwarzschild and C. Zupancic, Phys. Rev. **129**, 854 (1963).
7. E814 Collaboration, J. Barrette et al., Phys. Rev. C **50**, 1077 (1994).
8. H. Sorge, H. Stöcker, and W. Greiner, Ann. Phys. **192**, 266 (1989).
9. J.L. Nagle et al., Phys. Rev. C **53** , 367 (1995).
10. M. Gyulassy, K. Frankel, and E.A. Remler, Nuclear Physics **A402**, 596 (1983).
11. E814, T.K. Hemmick et al., Nucl. Phys. **A566**, 435c (1994).
12. S. Mrowczynski, Phys. Lett. **B308**, 216 (1993).
13. J. Kapusta, Phys. Rev. c **21**, 1301 (1979).
14. J. Barrette et al., Phys. Rev. Lett. **70**, 1763 (1993)
15. B. Shiva Kumar, S.V. Greene, J.T. Mitchell, Phys. Rev. C **50**, 2512 (1994).
16. D. Beavis et al., Phys. Rev. Lett. **75**, 3633 (1995).
17. S. Kahana, Y. Pang, T. Schlagel, and C. Dover, Phys. Rev. C **47**, R1356 (1993).
18. NA44 Collaboration, J. Simon-Gillo et al., Nucl. Phys **A590**, 483c (1995).
19. H. Sorge, Phys. Rev. C, 3291 (1995).

1. Paul Kienle and Walter Greiner.

2. Johanna Stachel, Peter Braun-Munzinger discussing, Fridolin Weber and Norman Glendenning listening.

3. Violeta Sandulescu, Aurel Sandulescu and Krishna Ramayya.

1. Rainer Dreizler and Walter Greiner.

2. Dirk Rischke, Joerg Aichelin and Eckart Grosse on the train trip.

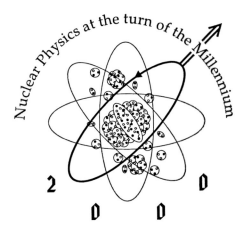

Nuclear Physics at the turn of the Millennium

2 0 0 0

5. Special Topics

DENSITY FUNCTIONAL APPROACH TO QHD

REINER M. DREIZLER, EBERHARD ENGEL AND REINER N. SCHMID
Institut für Theoretische Physik, JW Goethe-Universität,
Robert-Mayer-Str. 8-10, D-60054 Frankfurt (Main),
Germany

A review of presently available applications of density functional concepts to nuclear systems characterised by QHD models is given. It is demonstrated that exchange-only Kohn-Sham results obtained with the local density approximation are essentially equivalent to Hartree-Fock results. In addition, the application of extended Thomas-Fermi models for cold and thermal nuclei is outlined.

Around 1980 the field theoretical model of quantum hadrodynamics (QHD) became a popular tool for the discussion of nuclear properties[1,2]. In this model, the strong interaction between nucleons is mediated by the exchange of mesons,

$$(J^\pi, T): \quad \begin{matrix} \sigma, & \omega, & \rho, & \pi, & \dots \\ (0^+, 0) & (1^-, 0) & (1^-, 1) & (0^-, 1) & \dots \end{matrix} \quad .$$

This model is not expected to be applicable in the regime of very high energies, where the quark substructure of the hadrons becomes apparent. It is expected to be of relevance as an effective field theory for questions of nuclear structure and for collisions up to intermediate energies. In this contribution we will address the question of calculating nuclear structure data on the basis of QHD, in particular the calculation of ground state properties.

Initially, applications of the QHD model were restricted to the Hartree approximation (or relativistic mean field approximation for the meson degrees of freedom). In the meantime, Hartree-Fock results have become available [3,4,5]. It is well known, though, that correlation effects are important in strongly interacting systems. These can be addressed eg. via a Dirac-Brueckner scheme. First results of this type are available for nuclear matter and an application of this scheme for finite nuclei has been given[5].

Another avenue to the discussion of the many body problem at hand is density functional theory[6] (DFT). In order to summarise this approach, we shall rely on the simplest QHD model, QHD-I — the linear σ-ω-model[7]. This model is characterised by the Lagrangian density

$$
\mathcal{L} = \bar{\hat{\psi}}(x)[i\partial\!\!\!/ - M + g_s\hat{\phi}(x) - g_v\hat{V}\!\!\!/(x)]\hat{\psi}(x) + \delta\mathcal{L}_{CTC}
$$
$$
+ \frac{1}{2}[\partial_\mu\hat{\phi}(x)\partial^\mu\hat{\phi}(x) - m_s^2\hat{\phi}(x)^2] \tag{1}
$$

$$-\frac{1}{4}[\partial^\mu \hat{V}^\nu(x) - \partial^\nu \hat{V}^\mu(x)][\partial_\mu \hat{V}_\nu(x) - \partial_\nu \hat{V}_\mu(x)] + \frac{1}{2}m_v^2 \hat{V}_\mu(x)\hat{V}^\mu(x).$$

The nucleons, characterised by a single field $\hat{\psi}$, interact via exchange of massive scalar $(\hat{\phi})$ and vector $(\hat{V}_\nu)$ mesons. The scalar σ-meson, which is responsible for the attractive part of the N-N interaction, couples to the scalar density

$$\hat{\rho}_s(x) = \hat{\bar{\psi}}(x)\hat{\psi}(x). \tag{2}$$

The ω-mesons, which generate the short range repulsion, couple to the fermion four current

$$\hat{j}^\mu(x) = \hat{\bar{\psi}}(x)\gamma^\mu\hat{\psi}(x). \tag{3}$$

The quantity $\delta\mathcal{L}_{CTC}$ contains counterterms for purposes of renormalisation.

The basis statement of DFT is then: The ground state expectation value A_0 of any observable $\hat{A}$ for the many nucleon problem can be expressed rigorously as a functional of the ground state scalar density $\rho_s(\mathbf{x})$ and the ground state four current $j^\mu(\mathbf{x})$,

$$A_0[\rho_s, j^\mu] = <\Psi_0[\rho_s, j^\mu] \mid \hat{A} \mid \Psi_0[\rho_s, j^\mu]> \quad . \tag{4}$$

This extension of the Hohenberg-Kohn theorem of nonrelativistic DFT can be proven explicitly [8,9], provided the Ritz minimum principle is valid for interacting relativistic systems (after renormalisation).

The first task to be faced is the derivation of the functionals in questions, in particular for the ground state energy. We shall sidestep this problem for the moment and first ask the question: What options would be open if the functional for the ground state energy $E_0[\rho_s, j^\mu]$ (or a suitable approximation) would be available?

The Ritz principle leads directly to the variational equations

$$\frac{\delta E_0[\rho_s, j^\mu]}{\delta j^\lambda(\mathbf{x})} = \mu g_{\lambda 0} \quad , \qquad \frac{\delta E_0[\rho_s, j^\mu]}{\delta \rho_s(\mathbf{x})} = 0. \tag{5}$$

These equations are the starting point for the application of extended Thomas-Fermi (ETF) models, in which the ground state energy is fully written as a functional of ρ_s and j^μ.

More accurate results are usually obtained with the Kohn-Sham (KS) scheme [9,10]. In this scheme one represents the density variables in terms of auxiliary spinor orbitals,

$$\rho_s(\mathbf{x}) = \rho_{s,vac}(\mathbf{x}) + \rho_{s,D}(\mathbf{x}) \quad , \tag{6}$$

with

$$\rho_{s,D}(\mathbf{x}) = \sum_{-M < \epsilon_k \le \epsilon_F} \bar{\varphi}_k(\mathbf{x})\varphi_k(\mathbf{x})$$

$$\rho_{s,vac}(\mathbf{x}) = \frac{1}{2}\left[\sum_{\epsilon_k \le -M} \bar{\varphi}_k(\mathbf{x})\varphi_k(\mathbf{x}) - \sum_{\epsilon_k > -M} \bar{\varphi}_k(\mathbf{x})\varphi_k(\mathbf{x}) \right],$$

and corresponding expressions for the components of the four current

$$j^\mu(\mathbf{x}) = j^\mu_{vac}(\mathbf{x}) + j^\mu_D(\mathbf{x}) . \tag{7}$$

One then rearranges the expression for the ground state energy by addition and subtraction of the noninteracting kinetic energy

$$T_s[j^\nu, \rho_s] = T_{s,vac}[j^\nu, \rho_s] + T_{s,D}[j^\nu, \rho_s] \tag{8}$$

with

$$T_{s,D} = \int d^3x \sum_{-M < \epsilon_k \le \epsilon_F} \bar{\varphi}_k(\mathbf{x}) \left(-i\boldsymbol{\gamma} \cdot \boldsymbol{\nabla} + M \right) \varphi_k(\mathbf{x})$$

(and an analogous expression for $T_{s,vac}$) and the Hartree energy

$$\begin{aligned}
E_H &= -\frac{g_s^2}{2} \int d^3x \, d^3y \, \frac{e^{-m_s|\mathbf{x}-\mathbf{y}|}}{4\pi|\mathbf{x}-\mathbf{y}|} \, \rho_s(\mathbf{x}) \, \rho_s(\mathbf{y}) \\
&\quad + \frac{g_v^2}{2} \int d^3x \, d^3y \, \frac{e^{-m_v|\mathbf{x}-\mathbf{y}|}}{4\pi|\mathbf{x}-\mathbf{y}|} \, j^\mu(\mathbf{x}) \, j_\mu(\mathbf{y}),
\end{aligned} \tag{9}$$

so that

$$E_0[\rho_s, j^\mu] = T_s[\rho_s, j^\mu] + E_H[\rho_s, j^\mu] + E_{xc}[\rho_s, j^\mu]. \tag{10}$$

The exchange-correlation (xc) energy is the term, that remains after extraction of the dominant terms,

$$E_{xc} = T - T_s + W - E_H , \tag{11}$$

where W is the ground state expectation value of the complete N-N interaction energy. Exploitation of the variational principle (variation with respect to the spinor orbitals) leads to the selfconsistent Dirac-KS equations,

$$\left\{ -i\boldsymbol{\alpha} \cdot \boldsymbol{\nabla} + \beta\left[M - \phi_H - \phi_{xc} + \gamma_\mu(V_H^\mu + V_{xc}^\mu) \right] \right\} \varphi_k = \epsilon_k \varphi_k , \tag{12}$$

with the standard Hartree potentials ϕ_H, V_H^μ and the xc-potentials

$$V_{\mu,xc}(\mathbf{x}) = \frac{\delta}{\delta j^\mu(\mathbf{x})} E_{xc}[\rho_s, j^\mu] \tag{13}$$

$$\phi_{xc}(\mathbf{x}) = -\frac{\delta}{\delta \rho_s(\mathbf{x})} E_{xc}[\rho_s, j^\mu]. \tag{14}$$

The following direct remarks apply: The solution of the Dirac KS problem including all vacuum corrections (as indicated above) would present a rather tremendous task, as negative energy and positive energy solutions have to be determined at each of the selfconsistency cycles. One of the standard approximations is therefore the no-sea approximation,

$$\rho_{s,vac}, \quad j^\mu_{s,vac}, \quad T_{s,vac}, \quad E_{xc,vac} \quad \longrightarrow \quad 0 \,.$$

If, in addition, all xc-effects are neglected, $E_{xc} \to 0$, one recovers the Hartree approximation. It is important to note, that xc-effects are represented in terms of local (i.e. multiplicative) potentials. This means for instance: Provided one can derive a reasonable functional for exchange (x) effects, then the KS approach will be much easier to handle than the (nonlocal) HF approach.

After setting the stage, we have to face the main task, the derivation of functionals. If one is interested in ETF applications one needs a representation of T_s in terms of the densities rather than orbitals $T_s = T_s[\rho_s, j^\mu]$. This can be obtained with gradient expansion techniques [11,9,12], which are not quite standard as renormalisation (emphasised by the counterterm implied above) is required. Results, to second order in the gradient terms (establishing the ETF2 model), are (for brevity $T_s^* = T_s - \int d^3r \rho_s \phi$ is shown)

$$T_s^*[k, M^*] = T_{s,vac}^*[k, M^*] + T_{s,D}^*[k, M^*] \tag{15}$$

$$T_{s,vac/D}^*[k, M^*] = T_{s,vac/D}^{*,[0]}[k, M^*] + T_{s,vac/D}^{*,[2]}[k, M^*]$$

with the kinetic energy densities

$$t_{s,vac}^{*,[0]}[k, M^*] = -\frac{M^{*4}}{8\pi^2} \ln\left|\frac{M^*}{M}\right| + \frac{1}{32\pi^2}(M^{*4} - M^4)$$

$$+ \frac{M^2}{4\pi^2}(M^* - M)^2 + \frac{5M}{12\pi^2}(M^* - M)^3 + \frac{11}{48\pi^2}(M^* - M)^4$$

$$t_{s,vac}^{*,[2]}[k, M^*] = \frac{1}{12\pi^2}\frac{k^2}{E^2} \ln\left|\frac{M^*}{M}\right|(\nabla k)^2 + \frac{1}{6\pi^2}\frac{kM^*}{E^2} \ln\left|\frac{M^*}{M}\right|(\nabla k \cdot \nabla M^*)$$

$$- \frac{1}{24\pi^2}\left(1 + 2\frac{k^2}{E^2}\right) \ln\left|\frac{M^*}{M}\right|(\nabla M^*)^2$$

$$t_{s,D}^{*,[0]}[k, M^*] = \frac{1}{8\pi^2}\left[kE^3 + k^3 E - M^{*4}\operatorname{arcsinh}\left(\frac{k}{M^*}\right)\right]$$

$$t_{s,D}^{*,[2]}[k, M^*] = \frac{1}{24\pi^2}\frac{k}{E}\left[1 + 2\frac{k}{E}\operatorname{arcsinh}\left(\frac{k}{M^*}\right)\right](\nabla k)^2$$

$$+\frac{1}{6\pi^2}\frac{kM^*}{E^2}\operatorname{arcsinh}\left(\frac{k}{M^*}\right)(\nabla k \cdot \nabla M^*)$$

$$+\frac{1}{24\pi^2}\left[\frac{k}{E} - \left(1 + 2\frac{k^2}{E^2}\right)\operatorname{arcsinh}\left(\frac{k}{M^*}\right)\right](\nabla M^*)^2 \quad,$$

where $k(\mathbf{x}) = [3\pi^2 j^0(\mathbf{x})]^{\frac{1}{3}}$, $M^*(\mathbf{x}) = M - g_s\,\phi(\mathbf{x})$ and $E = \sqrt{M^{*2} + k^2}$. The no-sea approximation has not been evoked, so we explicitly see vacuum corrections. They are due to the scalar meson (they vanish for $\phi = 0$), vacuum corrections due to vector mesons only occur in the fourth order gradient terms.

The local density approximation (LDA) for exchange[10] is obtained by the evaluation of the nuclear matter exchange energy density e_x as a function of the Fermi momentum k_F (again renormalisation is involved) and the subsequent replacement $k_F \rightarrow [3\pi^2 j^0(\mathbf{x})]^{\frac{1}{3}}$. Results can actually be given analytically[10], although in terms of a messy expression finally involving Eulers dilogarithm,

$$e_{x,s}(\beta, M^*) = \frac{g_s^2(M^*)^4}{(2\pi)^4}\left\{\frac{1}{4}(\beta\eta - \ln\xi)^2 + (1 - \frac{w_s}{4})I(w_s, \xi, \xi)\right\} \quad (16)$$

$$e_{x,v}(\beta, M^*) = \frac{g_v^2(M^*)^4}{(2\pi)^4}\left\{\frac{1}{2}(\beta\eta - \ln\xi)^2 - (1 + \frac{w_v}{2})I(w_v, \xi, \xi)\right\}, \quad (17)$$

where $\beta = k_F/M^*$, $\eta = (1 + \beta^2)^{1/2}$, $\xi = \beta + \eta$, $w_{s,v} = m_{s,v}^2/(M^*)^2$ and

$$I(w, \xi_1, \xi_2) = \frac{w-2}{2}(\ln\xi_1 \ln\xi_2 - \beta_1\eta_1 \ln\xi_2 - \beta_2\eta_2 \ln\xi_1) - \beta_1\beta_2 \quad (18)$$

$$+\frac{1}{4}\left[w(\eta_1^2 + \eta_2^2) - 2(\eta_1 - \eta_2)^2\right]\ln\frac{(\xi_1\xi_2 - 1)^2 + w\xi_1\xi_2}{(\xi_1 - \xi_2)^2 + w\xi_1\xi_2}$$

$$+F(\xi_1, \xi_2) - F\left(\xi_1, \frac{1}{\xi_2}\right) - F\left(\frac{1}{\xi_1}, \xi_2\right) + F\left(\frac{1}{\xi_1}, \frac{1}{\xi_2}\right) \quad,$$

with

$$F(x_1, x_2) = \frac{s}{4}\left(\frac{\eta_1}{x_1} + \frac{\eta_2}{x_2}\right)\arctan\left(\frac{2(x_1 x_2 - 1) + w}{s}\right)$$

$$+\frac{i}{8}s\operatorname{Li}_2\left(x_1 x_2 \frac{2 - w - is}{2}\right) - \frac{i}{8}s\operatorname{Li}_2\left(x_1 x_2 \frac{2 - w + is}{2}\right)$$

$$\operatorname{Li}_2(x) = -\int_0^x \frac{\ln(1-z)}{z}dz \quad ; \quad s = \sqrt{w(4-w)} \quad.$$

In this case M^* has to be calculated from ρ_s via the standard nuclear matter relation [10]. Exchange corrections beyond the LDA are not known.

The situation with respect to correlation contributions is even more restricted: Two sets of nuclear matter results have been calculated.

1. The correlation contribution in the RPA limit [13] has been evaluated numerically for QHD-I in the no-sea approximation for the density range characterised by $0.6 fm^{-1} \leq k_F \leq 1.8 fm^{-1}$.

2. The correlation energy due to the sum of individual meson ladder contributions to the selfenergy has been evaluated [14] for the density range $1.0 fm^{-1} \leq k_F \leq 1.8 fm^{-1}$ in the no-pair approximation. In addition to contributions due to σ, ω, ρ and π mesons η and δ-meson contributions have been considered.

As a corollary I note that the DFT-approach can be extended to thermal systems [15,16]. For instance, in order to obtain an expression for the noninteracting free energy

$$F_s = T_s - T S_s$$

the starting point is the thermal Green's function

$$G(x,y) = \mathrm{Tr}\left\{ \frac{e^{-\beta(\hat{H}-\mu\hat{N})}}{\mathrm{Tr}(e^{-\beta(\hat{H}-\mu\hat{N})})} T_\tau \left[\hat{\psi}(x)\hat{\bar{\psi}}(y) \right] \right\} \tag{19}$$

where $x_0 = -i\tau_x$, $y = -i\tau_y$ and T_τ stands for "time ordering" of the real variables τ_x, τ_y. Again all quantities of interest can be extracted from $G(x,y)$. Questions of renormalisation arise, but do not differ from the situation at $T = 0$.

We finally present an indication of DFT results on the basis of the QHD model that have been obtained so far.

1. We begin with a comparison of HF and x-only LDA results for a QHD-I model. The model includes the exchange of photons, so that slight differences between proton and neutron data are observed. Fig.1 shows the proton densities obtained in the two approaches for ^{114}Sn, using the same set of model parameters. The closeness of the results is confirmed for neutron and baryon densities, for binding energies and mean radii of all spherical nuclei that have been considered [10].

The conclusion, that one extracts from these results, is: The LDA for exchange (or, expressed differently, a multiplicative x-potential of rather simple extraction) is adequate in the case of short range interactions.

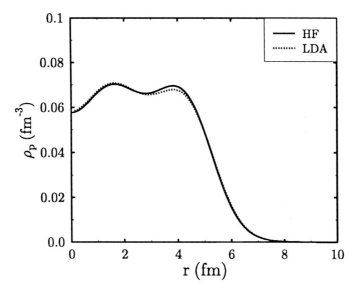

Figure 1: Proton point densities for ^{114}Sn from HF-[17] and LDA-calculations[10] (for the parameter set given in Ref.[3]).

2. Quite similar statements can be made if one compares HF and LDA x-only results for QHD-II (involving the exchange of σ, ω, ρ and π-mesons as well as photons). This time the differences for the charge and baryon densities, obtained with the two approaches, are slightly larger. This is a consequence of the contribution of the π-mesons, which is of longer range. On the other hand, results for binding energies and radii are still very close[18]. This is illustrated in Table 1, where results obtained with the parameter sets HF2 (without centre of mass corrections) and ZJO[4] (with cm corrections) are given.

3. If one compares Hartree results with the corresponding ETF results (available for the TF and ETF2 models[12,20]), one finds once more that these standard ETF functionals do not reproduce shell effects. The gradient corrected ETF2-functional yields, however, a more realistic surface structure for the nuclei.

4. Shell effects are known to be less important if one considers thermal systems. It can thus be expected that temperature dependent TF or ETF models provide reasonable results. A first investigation of various thermal nuclear systems (nuclear slabs, symmetric and asymmetric nuclei) addressing the questions of stability, appropriate equations of state

Table 1: Binding energies per nucleon (in MeV) and charge radii (in fm) from LDA-KS-calculations [18] for several spherical nuclei using two different parameters sets (HF2 [19] — without center of mass correction, ZJO [4] — including center of mass correction) in comparison with HF-calculations (using parameter set HF2 [19] — without center of mass correction) and experimental data (taken from Ref. [3]).

	$-E/A$			R_c			
	HF	LDA	EXP	HF	LDA		EXP
	HF2	HF2	ZJO		HF2	HF2	ZJO	
^{16}O	5.11	4.97	7.63	7.98	2.74	2.76	2.74	2.73
^{40}Ca	6.46	6.35	8.26	8.55	3.46	3.49	3.52	3.46
^{48}Ca	6.72	6.71	8.53	8.67	3.45	3.49	3.53	3.45
^{90}Zr	7.11	7.02	8.73	8.71	4.23	4.27	4.33	4.23
^{208}Pb	6.49	6.52	7.87	7.87	5.47	5.48	5.60	5.47
^{298}114			7.10				6.32	

etc. in terms of a thermal TF model for QHD-II (without photons) is available [16].

In summary we may then state that the application of DFT methods to QHD systems looks quite promising. The obvious next step is the investigation of correlation contributions. First results for the nuclear matter problem indicate that they are quite large (as expected), so that their incorporation is essential. The hope is that nuclear matter correlation results can be used in the LDA form with the same success as the exchange contributions. If this is the case, a number of questions concerning the QHD model itself can be raised and answered, as eg.

• Can more involved models (with nonlinear terms and a larger number of mesons), that are applied with restrictive approximations, be replaced by simpler models, that are applied with full inclusion of correlations?

• Can one find one "realistic" parameter set, that is derived from nuclear structure data and applied to low/intermediate energy collision problems?

Acknowledgements

We gratefully acknowledge the contributions of Dr. C. Speicher and Dr. H. Müller.

References

1. B. D. Serot and J. D. Walecka, in *Advances in Nuclear Physics*, edited by J. W. Negele and E. Vogt (Plenum, New York, 1986), Vol. 16.
2. Y. K. Gambhir, P. Ring, and A. Thimet, Ann. Phys. (N.Y.) **198**, 132 (1990).
3. A. Bouyssy, J.-F. Mathiot, Nguyen Van Giai, and S. Marcos, Phys. Rev. C **36**, 380 (1987).
4. J.-K. Zhang, Y. Jin, and D. S. Onley, Phys. Rev C **48**, 2697 (1993).
5. H. F. Boersma and R. Malfliet, Phys. Rev. C **49**, 1495 (1994).
6. R. M. Dreizler and E. K. U. Gross, *Density Functional Theory*, (Springer, Berlin, 1990).
7. J. D. Walecka, Ann. Phys. (N.Y.) **83**, 491 (1974).
8. E. Engel, H. Müller, C. Speicher, and R. M. Dreizler, in: *Density Functional Theory* by E. K. U. Gross and R. M. Dreizler (eds.), (Plenum, New York, 1995).
9. C. Speicher, R. M. Dreizler, and E. Engel, Ann. Phys. (N.Y.) **213**, 312 (1992).
10. R. N. Schmid, E. Engel and R. M. Dreizler, Phys. Rev. C **52**, 164 (1995).
11. M. Centelles, X. Viñas, M. Barranco and P. Schuck, Nucl. Phys. **A519**, 73c (1990).
12. M. Centelles, X. Viñas, M. Barranco and P. Schuck, Ann. of Phys. **221**, 165 (1993).
13. X. Ji, Phys. Lett. **B258**, 19 (1988).
14. H. F. Boersma and R. Malfliet, Phys. Rev. C **49**, 233 (1994).
15. H. Müller and R. M. Dreizler, Z. Phys. A **341**, 417 (1992).
16. H. Müller and R. M. Dreizler, Nucl. Phys. A **563**, 649 (1993).
17. H. F. Boersma, Ph.D. thesis, Groningen University, 1992.
18. R. N. Schmid, E. Engel and R. M. Dreizler, Phys. Rev. C **52**, 2804 (1995).
19. J.-K. Zhang and D. S. Onley, Phys. Rev. C **44**, 1915 (1991).
20. C. Speicher, E. Engel, and R. M. Dreizler, Nucl. Phys. **A562**, 569 (1993).

THE NUCLEON IN THE NUCLEUS AS GIVEN BY SKYRMIONS

J.M. EISENBERG

School of Physics and Astronomy
Raymond and Beverly Sackler Faculty of Exact Sciences
Tel Aviv University, 69978 Tel Aviv, Israel

G. KÄLBERMANN

Faculty of Agriculture
and Racah Institute of Physics
Hebrew University, 91904 Jerusalem, Israel

After some ten years or more of effort towards understanding the nucleon-nucleon force in terms of skyrmions, the way may now be open to consider its application to various problems involving the manifestations of quantum chromodynamics in nuclear physics. We here consider two such studies: (i) aspects of color transparency as given by skyrmions and (ii) the treatment of nucleon swelling and shrinking in nuclei using skyrmions.

1 Introductory Remarks

1.1 Historical Perspective

During the last ten years or so, roughly ten groups around the world have pursued the issue of correctly describing the nucleon-nucleon force and systems with small numbers of nucleons using skyrmions.[a] In offering this modest contribution in honor of Professor Walter Greiner it was tempting to review the impressive progress that has recently been made in that direction. However, it is one of Professor Greiner's engaging qualities to be more oriented towards the future than the past, and at the opening of this conference he asked that the speakers not stint in addressing the future in their areas—a request which it is a pleasure to comply with here.

Thus we shall take as our point of departure the supposition that the skyrmion has become, or very soon will become, an adequate tool for qualitative and semiquantitative descriptions of few-baryon systems, so that the way is open to begin considering its use for insight into hadronic behavior in nuclei, a topic of increasing importance in a variety of studies of quantum chromodynamics (QCD) effects in nuclei. The first such topic to be considered here is

---

[a]This work has been extensively surveyed, practically every group involved having written a summary of the topic from its own perspective. We here note one review [1] which is close to the usages and approaches of this talk.

color transparency, reviewed, for example, in ref. [2], which involves very-high-momentum transfer processes within nuclei in a test of QCD dynamical issues. The second question relates to nucleon swelling and shrinking in nuclei, a matter that has received increasing attention over the last ten years or so [3,4,5,6,7]. More detailed discussions of both these topics are given in the review noted previously [1] and in the original papers [8,9].

2 Generalities on $B = 1$ Skyrmions

2.1 The Skyrme Lagrangian

It was pointed out a number of years ago by 't Hooft that in the limit of a large number of colors, $N_c \to \infty$, QCD reduces to a theory involving meson fields as the relevant degrees of freedom [10], while Witten pointed out that baryons can then be constructed out of these degrees of freedom as topological solitons [11]. This provided underpinnings within QCD for a very old and prescient idea of Skyrme [12] that baryons might be built up out of topological solutions of a lagrangian embodying chiral symmetry, namely,

$$\mathcal{L} = \mathcal{L}_2 + \mathcal{L}_4 = -\frac{F_\pi^2}{16}\text{tr}(L_\mu L^\mu) + \frac{1}{32e^2}\text{tr}[L_\mu, L_\nu]^2, \tag{1}$$

where F_π is the pion decay constant, with experimental value 186 MeV, and e is the Skyrme parameter which accompanies his stabilizing term with four derivatives. Here

$$L_\mu \equiv U^\dagger \partial_\mu U, \tag{2}$$

where $U(\mathbf{r}, t)$ is the chiral field, here taken in SU(2). For the $B = 1$ static problem we take a time-independent hedgehog solution,

$$U(\mathbf{r}, t) = \exp[i\boldsymbol{\tau} \cdot \hat{\mathbf{r}} F(r)], \tag{3}$$

where $F(r)$ is referred to as the profile function. Substituting eq. (3) into eq. (1) yields eventually an expression for the particle energy or mass in terms of the profile function,

$$\begin{aligned}
M &= 4\pi \int_0^\infty r^2\,dr \left\{ \frac{F_\pi^2}{8}\left[\left(\frac{dF}{dr}\right)^2 + 2\frac{\sin^2 F}{r^2} \right] \right. \\
&\quad + \left. \frac{1}{2e^2}\frac{\sin^2 F}{r^2}\left[\frac{\sin^2 F}{r^2} + 2\left(\frac{dF}{dr}\right)^2 \right] \right\}.
\end{aligned} \tag{4}$$

From this one sees that in order for the mass to remain finite the profile function must approach $n\pi$ as $r \to 0$ and $m\pi$ as $r \to \infty$, where n and m are integers whose difference $n - m$ is a topologically conserved quantity that Skyrme identified as baryon number. Moreover the minimum reached by varying F in eq. (4) yields the Euler-Lagrange equation for the profile function,

$$\left(\frac{1}{4} \tilde{r}^2 + 2 \sin^2 F \right) F'' + \frac{1}{2} \tilde{r} F' + \sin 2F F'^2 - \frac{1}{4} \sin 2F$$
$$- \frac{\sin^2 F \sin 2F}{\tilde{r}^2} = 0, \tag{5}$$

in terms of dimensionless variables for which lengths are scaled by $e F_\pi$ and energies by F_π / e. This dimensionless form will be particularly handy for discussing color transparency below.

2.2 Projection onto N and Δ States

The hedgehog form of eq. (3) obviously mixes the SU(2) space with the space of three-dimensional rotations, and Adkins, Nappi, and Witten [13], in an important paper for modern applications of skyrmions, offered a projection method for generating states of good spin and isospin. This method is very reminiscent of projection techniques long used in nuclear physics [14] to reach states of good quantum numbers from various sorts of mixed states. The general idea is to rotate the hedgehog solution using a time-dependent unitary SU(2) matrix $A(t)$ according to

$$U(\mathbf{r}, t) = A(t) U_0(\mathbf{r}) A^\dagger(t). \tag{6}$$

This matrix $A(t)$ is then converted into a dynamical variable and subjected to first-quantization, leading to a dynamic lagrangian, upon substitution into eq. (1), of the form

$$L = \int \mathcal{L} \, d\mathbf{r} = -M + I \, \mathrm{tr} \left[\frac{dA}{dt} \frac{dA^\dagger}{dt} \right], \tag{7}$$

where M is the mass defined in eq. (4) and

$$I = \frac{2\pi}{3 e^3 F_\pi} \int_0^\infty \sin^2 F \left\{ 1 + 4 \left[F'^2 + \frac{\sin^2 F}{\tilde{r}^2} \right] \right\} \tilde{r}^2 \, d\tilde{r}, \tag{8}$$

with the numerical result $I = 50.9 \, (2\pi / 3 e^3 F_\pi)$. The quantization procedure allows the construction of operators for spin and isospin and of wave functions in the A-space for nucleon and Δ states. Ultimately the nucleon and delta masses are given by

$$M_N = M + \frac{3}{8I}, \qquad M_\Delta = M + \frac{15}{8I}. \tag{9}$$

Table 1: Skyrmion results for $B = 1$.

Physical quantity	Theory	Experiment
F_π	108 MeV	186 MeV
$r_{\text{rms}}(\text{ch}, T = 0)$	0.68 fm	0.72 fm
$r_{\text{rms}}(\text{ch}, T = 1)$	1.04 fm	0.88 fm
$r_{\text{rms}}(\text{mag}, T = 0)$	0.95 fm	0.81 fm
$r_{\text{rms}}(\text{mag}, T = 1)$	1.04 fm	0.81 fm
μ_p	1.97 μ_0	2.79 μ_0
μ_n	-1.24 μ_0	-1.91 μ_0
g_A	0.65	1.25
$g_{\pi NN}$	11.9	13.5
$g_{\pi N\Delta}$	17.8	20.3
$\mu_{N\Delta}$	2.3	3.3

The two parameters in eq. (1) may then be fit to these two values. Higher states with $J = T$ are assumed to be nonphysical (or very broad), as in nuclear rotational bands that are truncated at the maximally allowed spin values for the number of nucleons present [14]; that role is here played by the three valence quarks. All relevant observables for the nucleon and the Δ and transitions between them can then be constructed and calculated, [13] with the rather satisfactory results (to within $1/N_c \sim 33\%$) shown in Table 1.

3   The Product Ansatz for $B = 2$ Systems

In approaching the two-nucleon problem, the earliest [12] and simplest line of attack was based on London–Heitler methods and uses the so-called product ansatz in which the two-nucleon skyrmion is assumed to be well approximated by the product of two one-nucleon forms,

$$U_{B=2}(\mathbf{r}, \mathbf{R}) = U_{B=1}(\mathbf{r} + \frac{1}{2}\mathbf{R}) \, U_{B=1}(\mathbf{r} - \frac{1}{2}\mathbf{R}), \qquad (10)$$

with profile functions known from the $B = 1$ problem. Here $\mathbf{r}$ is the general skyrmion variable and $\mathbf{R}$ is the separation between the two baryon centers. This ansatz automatically fulfills the correct baryon number condition $B = 2$.

It is crucial for the study of the NN system that we project the product ansatz onto nucleon states. This is accomplished by rotating each member of the product ansatz separately according to

$$U_{B=2}(\mathbf{r}, \mathbf{R}) = \left[A\, U_{B=1}\left(\mathbf{r} + \frac{1}{2}\mathbf{R}\right) A^\dagger \right] \left[B\, U_{B=1}\left(\mathbf{r} - \frac{1}{2}\mathbf{R}\right) B^\dagger \right], \quad (11)$$

where A and B are the rotation matrices of eq. (6).

4   Skyrmion Applications in Nuclei: Color Transparency

Color transparency is the phenomenon in which a nucleon is ejected from the nucleus with very high momentum transfer and experiences little final-state interaction as it exits: The high momentum transfer "catches" the nucleon in a small configuration, and, because the color charges of a small QCD object partially neutralize each other, the nucleon emerges while interacting weakly with the other nucleons. Thus color dynamics makes the nucleus transparent to the "small" nucleon (referred to here as a minisculon m). The crucial QCD ingredient in this process is precisely the mutual color neutralization, and thus it is of interest to examine this in QCD-based models. The skyrmion allows one to study this[8] by artificially shrinking the skyrmion and examining the consequent mN potential and total mN cross section. The shrinking is accomplished by using the scaling noted below eq. (5) and taking values for the minisculon such that its parameter $(eF_\pi)_m$ becomes large. The ratio of minisculon size to the size of a normal nucleon is then

$$x \equiv \frac{r_m}{r_N} = \frac{eF_\pi}{(eF_\pi)_m}. \quad (12)$$

The result of the skyrmion calculation using the product ansatz is that the central potential tends to zero more or less linearly with x for $x \le 0.3$. On the other hand, the range of the potential tends to fall very gradually until $x \sim 0.05$ and then plunges rapidly to zero. The total NN cross section, in a rough high-energy approximation, falls linearly to zero as x drops from unity to zero, showing quadratic behavior only when x is below 0.3. This last feature given by the skyrmion estimate is somewhat surprising in that most discussions of color transparency have based themselves on an assumption of quadratic dependence on minisculon size through the whole scale from normal radii to very small ones.

5   Skyrmion Applications in Nuclei: Shrinking and Swelling Nucleons in Nuclei

In order to make any sort of sensible estimate of nucleon swelling in nuclei, one needs to have an NN force that provides for central attraction. Among the many suggestions[1] for achieving this within the product ansatz, two seem to have provided central attraction based on physically plausible mechanisms. These are (i) the admixture of a nucleon vibrational state[15] (the Roper resonance), which appears to be the equivalent for the product ansatz of skyrmion distortions seen within more complete approaches[16]; and (ii) the inclusion of the dilaton,[17] which mocks up the trace anomaly of QCD for theories based on meson degrees of freedom.

Attraction from the Roper is usually discussed in terms of a lagrangian involving terms beyond the simple for of eq. (1). One often adds[1] terms involving four derivatives, for enhanced attraction, and terms with six derivatives, simulating ω-exchange, which are repulsive. The two-baryon states are then taken as

$$
\begin{aligned}
|\widetilde{NN}(R)\rangle &= \alpha(R)|NN(R)\rangle + \beta(R)|N\Delta(R)\rangle + \gamma(R)|\Delta\Delta(R)\rangle \\
&+ \epsilon(R)|NN^*(R)\rangle + \zeta(R)|N^*N^*(R)\rangle.
\end{aligned}
\tag{13}
$$

The Roper states R^* are handled by first-quantizing a collective, breathing-mode, radial-scaling variable[1,15].

The dilaton yields attraction by sharpening the edge of the skyrmion, a reasonable physical expectation here. The dilaton itself is motivated by the need to make the lagrangian obey the same breaking of scale invariance as found in QCD in the context of the trace anomaly[17],

$$
T^\mu_\mu = \partial_\mu D^\mu = -\frac{9\alpha_s}{8\pi} G^a_{\mu\nu} G^{a\mu\nu} \equiv \sigma^4,
\tag{14}
$$

where $D^\mu (= T^{\mu\nu} x_\nu)$ is the dilatation current, α_s is the QCD coupling constant, $G^a_{\mu\nu}$ is the gluon field, and σ is an order-parameter field—the dilaton—which represents the scalar glueball formed from the contraction of the two gluon fields. The modified skyrmion lagrangian that satisfies the trace anomaly is then

$$
\begin{aligned}
\mathcal{L} &= e^{2\sigma}\left[\frac{1}{2}\Gamma_0^2 \partial_\mu \sigma \partial^\mu \sigma - \frac{F_\pi^2}{16}\mathrm{tr}(L_\mu L^\mu)\right] + \frac{1}{32e^2}\mathrm{tr}[L_\mu, L_\nu]^2 \\
&- \frac{C_G}{4}[1 + e^{4\sigma}(4\sigma - 1)].
\end{aligned}
\tag{15}
$$

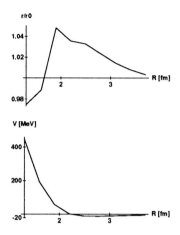

Figure 1: The NN potential and ratio of the radius for the interacting nucleon to that of the free nucleon r/r_0 for the usual skyrmion with product ansatz and baryon-resonance admixtures.

The two parameters Γ_0 and C_G are related to the glueball—or scalar field—mass through $m_\sigma^2 = 4C_G/\Gamma_0^2$. In using the dilaton for $B = 2$ systems, one must augment the product ansatz for the skyrmion by an assumption of additivity for the dilaton, $\sigma_{B=2} = \sigma_1 + \sigma_2$.

Both mechanisms are capable of yielding central attraction on the scale of 10 MeV or more with reasonable parameters. This in turn leads to nucleon swelling of some 3 or 4 percent when the separation between the nucleons places them in the attractive region. This is illustrated for the case of the Roper mechanism in fig. 1, the case for dilaton-induced attraction being quite similar[9].

6 Outlook

The two studies we have discussed represent an early phase of the applications of skyrmions to nucleon behavior and changes in the nuclear medium (see also ref.[18] in this regard). Much more can clearly be done. For example, in the context of heavy-ion collisions there is a great deal to be learned about baryons in a hot medium and here again the skyrmion may offer useful insights. We look forward to learning about these matters from experimental and theoretical work over the next decade.

Acknowledgments

It is a pleasure to acknowledge helpful discussions with Leonid Frankfurt on the subject matter of this report. This work was supported in part by the Israel Science Foundation and in part by the Yuval Ne'eman Chair at Tel Aviv University.

References

1. J.M. Eisenberg and G. Kälbermann, *Int. J. Mod. Phys. E,* to be published.
2. L.L. Frankfurt, G.A. Miller, and M. Strikman, *Annu. Rev. Nucl. Part. Sci.* **45**, 501 (1994).
3. J.V. Noble, *Phys. Rev. Lett.* **46**, 412 (1981) and *Phys. Lett.* B **178**, 285 (1986).
4. L.S. Celenza, A. Rosenthal, and C.M. Shakin, *Phys. Rev. Lett.* **53**, 892 (1984) and *Phys. Rev.* C **31**, 232 (1985).
5. P.J. Mulders, *Phys. Rev. Lett.* **54**, 2560 (1985) and *Nucl. Phys.* A **459**, 525 (1986).
6. L.L. Frankfurt and M. Strikman, *Nucl. Phys.* B **250**, 123 (1985).
7. M. Oka and R.D. Amado, *Phys. Rev.* C **35**, 1586 (1987).
8. J.M. Eisenberg and G. Kälbermann, *Phys. Lett.* B **286**, 24 (1992).
9. G. Kälbermann, L.L. Frankfurt, and J.M. Eisenberg, *Phys. Lett.* B **329**, 164 (1994).
10. G. 't Hooft, NPB **72**, 461 (1974) and *Nucl. Phys.* B **75**, 461 (1974).
11. E. Witten, *Nucl. Phys.* B **160**, 57 (1979).
12. T.H.R. Skyrme, *Proc. Roy. Soc. London, Series A,* **260**, 127 (1960) and **262**, 237 (1961) and *Nucl. Phys.* **31**, 556 (1962).
13. G.S. Adkins, C.R. Nappi, and E. Witten, *Nucl. Phys.* B **228**, 552 (1983).
14. J.M. Eisenberg and W. Greiner, *Microscopic theory of the nucleus* (North-Holland, Amsterdam, 1972).
15. L.C. Biedenharn, Y. Dothan, and M. Tarlini, *Phys. Rev.* D **31**, 649 (1985).
16. T.S. Walhout and J. Wambach, *Int. J. Mod. Phys. E* **1**, 665 (1992).
17. J. Schechter, *Phys. Rev.* D **21**, 3393 (1980).
18. I. Mishustin, *Sov. Phys. JETP* **71**, 21 (1990).

WAVELET-CORRELATIONS IN HIERARCHICAL CASCADE PROCESSES – THE QUESTION OF SCALING AND CLUSTERING IN COMPLEX REACTIONS

MARTIN GREINER

Institut für Theoretische Physik, Technische Universität,
Mommsenstraße 13, D-01062 Dresden, Germany

PETER LIPA

Institut für Hochenergiephysik der Österreichischen Akademie der Wissenschaften,
Nikolsdorfergasse 18, A-1050 Wien, Austria

JENS GIESEMANN

Institut für Theoretische Physik, Justus Liebig Universität,
Heinrich-Buff-Ring 16, D-35392 Gießen, Germany

The wavelet transformation borrowed from signal analysis compresses and analyses the correlation information obtained from complex reaction processes in a new way. It reveals information on scaling and clustering.

1 Introduction

Relativistic heavy-ion collisions represent very complex reactions. A theoretician imagines these reactions to occur in various steps. An experimentalist, on the other hand, only measures the hundreds or thousands of particles which come out as the final state of the reaction. Here, a simple question comes to our minds: Given only the outcome, is it possible to analyse and reconstruct the underlying evolution process? Surely an extensive statistical analysis would be necessary for such a demanding enterprise and this means correlation functions to all orders. With conventional representations of the correlation functions this will not work in practice as we have to cope with an enormous amount of redundant information. New techniques have to be employed, which discard redundancy and concentrate on the really important information.

Such new compression and analysing techniques can be borrowed from signal analysis. Over the past ten years it has been the *wavelet transformation*, which has revolutionized this field.[1-4] It is able to compress signals and video pictures having lots of contrasts to an enormous amount without loosing information; with the relevant structures discovered the reconstruction of the signal out of the compression is hard to distinguish from the original. There is more the wavelet transformation can do: it allows to analyse a signal very efficiently in conjugate variables, say time and frequency, as it reveals what frequencies are present at what time. These issues are hard to address with standard Fourier techniques: whereas the wavelet transformation dissects a signal in very local pieces with compact support, the Fourier transformation expands it into harmonic functions with infinite support.

In order to test the many promising features of wavelets we pick a simple toy-model cascade from fully developed turbulence and investigate the compression and analysing power of wavelets for the statistical analysis. In particular we concentrate on aspects like scaling (selfsimilarity) and clustering.

2   Wavelet-correlations of the p-model

The energy dissipation in one-dimensional fully developed turbulent flow shows pronounced anomalous fluctuations known as intermittency. So far a derivation of this phenomenon from first principles (Navier-Stokes equation) has not been possible. Hence, phenomenology runs the show at the moment. The p-model cascade [5] represents one of these phenomenological, so-called random multiplicative models. It is described as follows: A uniform energy $E_{00} = 1$ on an interval [0,1] splits into a part $E_{10} = pE_{00}$ on the subinterval [0,1/2] and into another part $E_{11} = (1-p)E_{00}$ on [1/2,1]. The random number p can take on the values $(1+\alpha)/2$ or $(1-\alpha)/2$ with equal probability. This is already the cascade prescription for the first step. It is repeated over and over again for the next cascade steps; in each splitting (branching) the random number p is tossed independently from all the other branchings. After J cascade steps we arrive at 2^J subintervals (bins) of length $1/2^J$, each bin $0 \leq k < 2^J$ going with an energy density $\epsilon_{Jk} = E_{Jk}2^J$. Any one p-model configuration obtained in this way shows striking similarity to the intermittent spatial fluctuations of the energy dissipation measured in fully developed turbulence. From the perspective of statistics a more solid comparison is necessary and this means to look on *true* spatial correlation functions.

To determine the spatial bin correlation densities it is most convenient to use the corresponding generating function

$$Z[\vec{\lambda}] = \left\langle \exp\left(i \sum_{k=0}^{2^J-1} \lambda_k \epsilon_{Jk} \right) \right\rangle \quad , \tag{1}$$

where the bracket $\langle \ldots \rangle$ indicates an averaging over all possible p-model configurations. The correlation density functions follow by taking adequate derivatives with respect to the generating coordinates λ_k:

$$\rho_{k_1 k_2 \ldots k_q} = \left\langle \epsilon_{Jk_1} \epsilon_{Jk_2} \cdot \ldots \cdot \epsilon_{Jk_q} \right\rangle$$

$$= \frac{1}{i^q} \frac{\partial^q Z[\vec{\lambda}]}{\partial \lambda_{k_1} \partial \lambda_{k_2} \cdot \ldots \cdot \partial \lambda_{k_q}} \bigg|_{\vec{\lambda}=0} \quad ; \tag{2}$$

q represents the order of the correlation function. Still, we have to construct the generating function for the p-model. This can be done with a forward evolution equation [6]:

$$Z^{(j+1)}[\vec{\lambda}^{(j+1)}] = \tag{3}$$

$$= \int p(q_0^{(j)}) dq_0^{(j)} \ldots \int p(q_{2^j-1}^{(j)}) dq_{2^j-1}^{(j)} \ Z^{(j)}[\vec{\lambda}^{(j)}(q_0^{(j)}, \ldots, q_{2^j-1}^{(j)}; \vec{\lambda}^{(j+1)})] \tag{4}$$

with

$$\vec{\lambda}^{(j)} = (\lambda_0^{(j)}, \lambda_1^{(j)}, \ldots, \lambda_{2^j-1}^{(j)}) \quad ,$$

$$\lambda_n^{(j)} = (1 + q_n^{(j)}) \lambda_{2n}^{(j+1)} + (1 - q_n^{(j)}) \lambda_{2n+1}^{(j+1)}, \tag{5}$$

the splitting function

$$p(q) = \frac{1}{2} \delta(q - \alpha) + \frac{1}{2} \delta(q + \alpha) \tag{6}$$

and

$$Z^{(0)}[\vec{\lambda}^{(0)}] = \exp(i \lambda_0^{(0)}) \quad . \tag{7}$$

Once this evolution equation is inserted into Eq. (2), recursive relations for the correlation densities $\rho_{k_1 \ldots k_q}$ can be derived and solved. A similar procedure has been performed in Refs. 7 and 8, except that a backward evolution equation has been employed for the generating function.

Fig. 1a depicts the two-bin correlation density $\rho_{k_1 k_2}$ after 6 p-model cascade steps; hence $0 \leq k_1, k_2 < 2^6 = 64$. The power-law rise $(1 + \alpha^2)^j$ towards the diagonal is a clear indication of the selfsimilarity of the hierarchical p-model cascade: the closer two bins are together, the more they share a common (cascade) history and the stronger they are correlated.

This representation of the correlation density is based on a monoscale expansion of the p-model configurations:

$$\epsilon(x) = \sum_{0 \leq k < 2^J} \epsilon_{Jk} \phi_{Jk}^H(x) \tag{8}$$

with the Haar-scaling function (box function)

$$\phi_{Jk}^H(x) = \phi^H(2^J x - k) = \begin{cases} 1 & \text{for } k2^{-J} \leq x < (k+1)2^{-J} \\ 0 & \text{otherwise}; \end{cases} \tag{9}$$

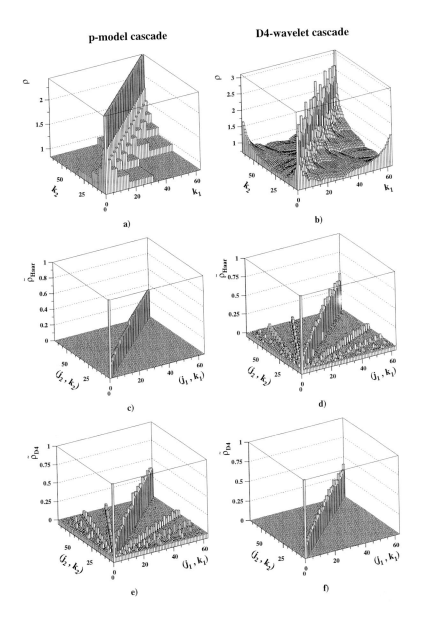

p-model cascade

D4-wavelet cascade

Figure 1: Two-bin correlation density $\rho_{k_1 k_2}$ and its Haar-wavelet and D4-wavelet transformed correlation densities of the p-model cascade (a,c,e) and of the D4-wavelet cacade (b,d,e) with $J = 6$ and $\alpha = 0.4$.

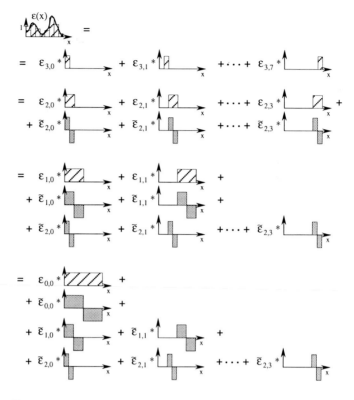

Figure 2: Multiresolution analysis with the Haar-wavelet.

confer the first row of Fig. 2. The bin correlation densities can be thought of as the correlations between the amplitudes ϵ_{Jk} of this expansion. Clearly, this monoscale representation of the correlation functions is not an optimal choice for a hierarchically organized process like the p-model cascade. A multiscale representation is certainly a better choice.

Such a multiresolution expansion can be obtained by applying successive smoothing and "differentiation" operations as illustrated in Fig. 2. We arrive at the decomposition

$$\epsilon(x) = \epsilon_{00}\phi_{00}^H(x) + \sum_{j=0}^{J-1}\sum_{k=0}^{2^j-1} \tilde{\epsilon}_{jk}\psi_{jk}^H(x) \tag{10}$$

with the Haar-wavelet

$$\psi_{jk}^H(x) = \psi^H(2^j x - k) = \begin{cases} 1 & \text{for } k2^{-j} \leq x < (k+1/2)2^{-j} \\ -1 & \text{for } (k+1/2)2^{-j} \leq x < (k+1)2^{-j} \\ 0 & \text{otherwise.} \end{cases} \tag{11}$$

This is not the only orthogonal and compact multiresolution analysis. With the help of the fundamental dilation equations

$$\phi(x) = \sum_m c_m \phi(2x - m) \quad , \qquad \psi(x) = \sum_m (-1)^m c_{1-m}\phi(2x - m) \tag{12}$$

generalisations are possible: the choice $c_0 = c_1 = 1$ leads to (9) and (11), whereas, for example, $c_0 = (1+\sqrt{3})/4$, $c_1 = (3+\sqrt{3})/4$, $c_2 = (3-\sqrt{3})/4$, $c_3 = (1-\sqrt{3})/4$ leads to Daubechies' orthogonal and compactly supported D4-wavelet.

The two representations (8) and (10) each completely characterize a p-model configuration; again, the former is a monoscale representation whereas the latter is a multiscale representation. The wavelet amplitudes $\tilde{\epsilon}_{jk}$ are related to the bin amplitudes ϵ_{Jk} by a linear transformation $\mathbf{W}(c_m)$, which depends on the c_m-coefficients entering the dilation equations (12). We now ask for the correlations

$$\tilde{\rho}_{(j_1k_1)(j_2k_2)\ldots(j_qk_q)} = \left\langle \tilde{\epsilon}_{j_1k_1}\tilde{\epsilon}_{j_2k_2}\cdot\ldots\cdot\tilde{\epsilon}_{j_qk_q} \right\rangle \tag{13}$$

between the wavelet amplitudes $\tilde{\epsilon}_{jk}$. These wavelet correlations are determined by employing the wavelet transformation $\mathbf{W}(c_m)$ either within the generating function (1) or directly to the bin correlation densities (2); for more details see Refs. 7 and 8.

Fig. 1c depicts the second order Haar-wavelet correlations of the p-model cascade. The wavelet amplitudes have been ordered according to

$$\tilde{\vec{\epsilon}} = (\epsilon_{00}, \tilde{\epsilon}_{00}, \tilde{\epsilon}_{10}, \tilde{\epsilon}_{11}, \tilde{\epsilon}_{20}, \ldots, \tilde{\epsilon}_{J-1,2^{J-1}-1}) \quad . \tag{14}$$

It is completely diagonal. In other words, the Haar-wavelet transformation completely "compresses" the second order correlation information to the diagonal. Moreover, the diagonal elements $\tilde{\rho}_{(jk)^2} = \alpha^2(1 + \alpha^2)^j$ reveal the same powerlaw scaling as observed in the conventional two-bin correlation density towards the diagonal and in this way signal selfsimilarity of the underlying process. The Haar-wavelets represent the perfect building blocks for the p-model cascade as they are adapted to the energy splitting rules. – Other choices of wavelets will not be optimal building blocks for the p-model cascade. However, due to the inherent multiresolution character a quasidiagonalisation is expected for the correlations of second order. This is demonstrated in Fig. 1e, where the D4-wavelet correlations of the p-model are depicted.

There is more to learn from wavelet correlations than only the compression aspect. Higher order wavelet correlations benefit from the analysing aspect of wavelets as they depend sensitively on the evolution dynamics. For example, the only nonvanishing contributions to the third order Haar-wavelet correlations $\tilde{\rho}_{(j_1 k_1)(j_2 k_2)(j_3 k_3)}$ are when a small structure $(j_2 k_2) = (j_3 k_3)$ lives inside a larger structure $(j_1 k_1)$. These contributions reveal a double scaling: one scaling for the common history of the small and large structure, i.e. $(1+3\alpha^2)^{j_1}$, and a different scaling, i.e. $(1 + \alpha^2)^{j_2 - j_1}$, from the moment on the small structure evolves from the large structure. These interscale correlations can be naturally interpreted as cluster correlations.

3   D4-wavelet cascade

So far we have discussed the p-model cascade as the prototype for a random multiplicative process. Its spatial tree structure is organized in such a way that branches do not overlap. We now present an example of a cascade process with overlapping branches. The cascade prescription is illustrated in Fig. 3: Given an energy E_{jk} in the subbin k obtained after j cascade steps, E_{jk} splits into $E_{j+1,2k} = (c_0 + q_k^{(j)} c_3) E_{jk}/2$ in the subbin (j+1,2k), $E_{j+1,2k+1} = (c_1 - q_k^{(j)} c_2) E_{jk}/2$ in the subbin (j+1,2k+1), $E_{j+1,2k+2} = (c_2 + q_k^{(j)} c_1) E_{jk}/2$ in (j+1,2k+2) and $E_{j+1,2k+3} = (c_3 - q_k^{(j)} c_0) E_{jk}/2$ in (j+1,2k+3). The coefficients $c_0, \ldots, c_3$ represent the D4-wavelet filter coefficients mentioned after Eq. (12). The variable $q_k^{(j)}$ represents a random multiplier, which is randomly distributed according to the splitting function (6); it describes the random fluctuation introduced in the cascade prescription.

The correlation densities of this cascade are again determined according to Eq. (2) with the help of the evolution equation (3) for the generating function. Also the splitting function (6) remains the same; only Eq. (5) needs to be modified for the new cascade prescription:

$$
\begin{aligned}
\lambda_n^{(j)} = {} & (c_0 + c_3 q_n^{(j)}) \lambda_{2n}^{(j+1)} + (c_1 - c_2 q_n^{(j)}) \lambda_{2n+1}^{(j+1)} \\
& + (c_2 + c_1 q_n^{(j)}) \lambda_{2n+2}^{(j+1)} + (c_3 - c_0 q_n^{(j)}) \lambda_{2n+3}^{(j+1)} \; .
\end{aligned}
\tag{15}
$$

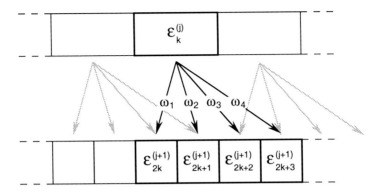

Figure 3: Prescription of the D4-wavelet cascade; $\omega_1 = c_0 + q_k^{(j)} c_3$, $\omega_2 = c_1 - q_k^{(j)} c_2$, $\omega_3 = c_2 + q_k^{(j)} c_1$, $\omega_4 = c_3 - q_k^{(j)} c_0$.

The second order bin correlations, the second order Haar-wavelet correlations and the second order D4-wavelet correlations are shown in Fig. 1b,d,f, respectively. The Haar-wavelet transformation quasidiagonalizes the second order bin correlations; still the Haar-wavelet does not represent the perfect building block for this cascade. However, Daubechies' D4-wavelet does: the D4-wavelet transformation leads to a perfect "compression" of the second order bin correlations to the diagonal. This is the reason why we call the underlying process the D4-wavelet cascade.

4   Conclusions and outlook

Wavelets act like a mathematical microscope, which is able to dissect structures on multiple scales. For hierarchically organized processes wavelets represent "quasi" eigenfunctions as they nearly diagonalize the second order correlation function. Higher order wavelet correlations can be naturally interpreted as multiscale cluster correlations. In this respect the wavelet representation of correlation functions represents a new approach to learn more about the evolution dynamics of the complex process under investigation.

For the ultimate goal to reconstruct the underlying evolution process from correlation functions of (in principle) all orders the wavelet representation can not be more than a tiny step. For the analysis of complex processes in general the wavelet scheme is to rigid. Adaptive schemes are in order: promising candidates are for example wavelet packets or the wavelet modulus maxima method; also a Karhunen-Lòeve expansion is worth considering. These advanced techniques have to go hand in hand with more analytical model studies in order to appreciate their full strengths. Once completed fruitful applications are to be found in the fields of turbulence, multiparticle dynamics, disordered (solid state) systems[9], quantum algebras[10] and even relativistic heavy-ion collisions. For the latter the study of fluctuations in large multiplicity single events will definitely reveal interesting features about the heavy-ion reaction dynamics, maybe even signals from a quark-gluon plasma.

Acknowledgments

M.G. gratefully acknowledges support from DFG, GSI and BMBF. P.L. is indebted to the Österreichische Akademie der Wissenschaften for its support with an APART fellowship.

References

1. I. Daubechies, *Ten Lectures on Wavelets* (SIAM, Philadelphia, 1992).
2. C.K. Chui, *An Introduction to Wavelets* (Academic Press, Boston, 1992); *Wavelets: A Tutorial in Theory and Application* (Academic Press, Boston, 1992).
3. Y. Meyer, *Wavelets: Algorithms and Apllications* (SIAM, Philadelphia, 1993).
4. G. Kaiser, *A Friendly Guide to Wavelets* (Birkhäuser, Boston, 1994).
5. C. Meneveau and K.R. Sreenivasan, *Phys. Rev. Lett.* **59**, 1424 (1987).
6. J. Giesemann, M. Greiner and P. Lipa, *in preparation*.
7. M. Greiner, P. Lipa and P. Carruthers, *Phys. Rev.* E **51**, 1948 (1995).
8. M. Greiner, J. Giesemann, P. Lipa and P. Carruthers, *Z. Phys.* C **69**, 305 (1996).
9. J. Kantelhardt, M. Greiner and E. Roman, *Physica* A **220**, 219 (1995).
10. A. Ludu and M. Greiner, *ICTP-Internal Report* IC/95/214.

INSTANTONS AND THE QCD PHASE TRANSITION

S. SCHRAMM

Gesellschaft für Schwerionenforschung, Darmstadt, Germany

M.-C. CHU

Department of Physics, Chinese University of Hongkong, Hongkong

We study the instanton content of gluon fields on the lattice as function of temperature. We find strong suppression of instantons for temperatures above the deconfinement transition. For temperatures up to about 1.3 T_c, the topological charge correlation agrees well with the profile of a single instanton.

1 Introduction

It is generally assumed, in agreement with lattice QCD calculations[1], that at a critical temperature T_c a phase transition takes place at which quarks and gluons are deconfined and form a new phase of matter, the so-called *quark-gluon plasma*. This possibility stimulated numerous theoretical and phenomenological works on the subject, many of which focus on the possible experimental signatures of the quark-gluon plasma[2]. Especially in view of the experiments at AGS, CERN, and RHIC, theoretical studies of the properties of finite temperature QCD matter are important. Lattice gauge theory provides a useful tool to calculate QCD effects from first principle, and it therefore enables us to study static finite temperature properties of QCD matter.

A particularly important and interesting question regarding QCD matter at finite temperature is what happens to the hadrons. Most of the phenomenological works on quark-gluon plasma signatures are based on a picture of weakly interacting quarks and gluons at temperatures above T_c. However, such a simple picture is not completely supported by lattice calculations. For example, lattice calculations of the so-called screening wavefunctions indicate strong correlations among the quarks in hadronic channels[3].

In lack of a method to calculate the finite-temperature hadron wavefunctions directly, we turn our attention to a class of gauge configurations known to have important effects on the structure and masses of light hadrons at zero temperature[5], namely, large, isolated instantons.

It is well known that instantons break the $U_A(1)$ chiral symmetry, resulting in the famous axial anomalies. At high temperature, Debye screening gives rise to an electric mass for the gluons and thereby suppresses instanton amplitudes; the $U_A(1)$ symmetry should therefore be restored. Whether the $U_A(1)$ restoration, at T_U, occurs at the same temperature as the $SU(N_f)_A$ restoration at T_c

bears important consequences to the hadronic physics at finite temperature: if $T_U = T_c$, the phase transition is of first order[6], and there are drastic changes to hadrons at T_c; if on the other hand, $T_U > T_c$, the phase transition is of second order. The two scenarios give rise to very different hadron physics at temperatures around T_c[7]. One of our goals is to locate T_U via lattice calculation of the topological susceptibility, which is a measure of the instanton content, and thereby differentiate the two scenarios.

There have been calculations of the temperature dependence of the topological susceptibility in pure SU(2)[8,9], pure SU(3)[10], as well as unquenched SU(3) with two light quark flavors[11]. However, these authors vary the temperature by adjusting the coupling constant, making it tricky to compare the susceptibility at different temperatures because of the different lattice length scales. At larger β finite size effects may plague the calculations using small (eg.,$8^3 \times 4$) lattices. There are also some confusions on the behavior of the susceptibility in the SU(2) theory. Whereas Teper et al. observed a sudden suppression of the topological susceptibility at the deconfinement temperature for both SU(2) and SU(3) fields[8,10], Di Giacomo et al. reported that the cooling method gave ambiguous results for SU(2) theory, and the susceptibility determined by their field theoretical method stays almost constant across the phase transition[9].

In the following we present calculations of the temperature dependence of the instanton content of quenched SU(3) fields using the cooling method. We find a suppression of the topological susceptibility starting already at temperatures below T_c. The topological charge correlation function agrees well with the continuum instanton profile, although there seems to be a change in the instanton size parameter across the phase transition.

We first introduce the set of observables we calculated in Section 2. After a brief outline of the cooling method in Section 3, we present our main results in Section 4. We close with a discussion in Section 5.

2 Lattice Quantities

The instanton content of the gauge fields can be monitored by the the topological charge density, which can be defined on the lattice as

$$Q(x_n) = -\frac{1}{32\pi^2}\epsilon_{\alpha\beta\gamma\delta}\mathrm{Re}\ \mathrm{Tr}\left[U_{\alpha\beta}(x_n)U_{\gamma\delta}(x_n)\right] , \tag{1}$$

where $U_{\alpha\beta}$ is the product of the link variables around a plaquette in the $\alpha - \beta$ plane, after some cooling steps were applied to remove lattice artifacts associated with discretization[10]. The topological susceptibility is then a measure of the fluctuations of the topological charge:

$$\chi_t \equiv \frac{1}{N_t N_s^3 a^4} \langle \left(\sum_n Q(x_n) \right)^2 \rangle \ .$$ (2)

$\langle ... \rangle$ indicates configuration averaging, and N_t and N_s are the number of sites in the temporal and spatial direction. The lattice spacing a changes as a function of the lattice inverse coupling β, which makes it tricky to compare χ_t calculated with different β, especially when β is not large enough to be in the asymptotic scaling regime. To study the temperature dependence of the topological susceptibility, we chose to change the temperature by varying the number of time slices

$$T = \frac{1}{N_t a} \ ,$$ (3)

but keeping β, and hence a, fixed. This way the uncertainty in a enters only as an overall constant factor which does not affect the shape of the χ_t vs. T curve.

Due to Euclidean symmetry the space-like and time-like plaquettes (P_x, P_t) have equal expectation values at zero temperature. At finite temperature, however, the Euclidean symmetry is broken, and an asymmetry between P_x and P_t develops at the phase transition temperature proportional to the entropy s of the system [12]:

$$Ts = \epsilon + P = 4\beta \left[1 - 0.16675g^2 + O(g^4) \right] (< P_t > - < P_x >) \ ,$$ (4)

where

$$< P_t > \equiv \frac{1}{3N_s^3 N_t} \sum_{n,i} \frac{1}{3} TrU_{0,i}(x_n) \ ,$$ (5)

and

$$< P_x > \equiv \frac{1}{3N_s^3 N_t} \sum_{n,i<j} \frac{1}{3} TrU_{i,j}(x_n) \ .$$ (6)

Since the pressure p is continuous across the phase transition, the discontinuity of $\epsilon + p$ is the latent heat $\Delta\epsilon$. Understanding the physics origin of the latent heat is important to developing models for finite temperature hadronic matter. For example, it was postulated in Ref. [13] that an instanton liquid reorganizes itself into molecules near T_c and this process contributes to a jump in the energy density at T_c. Here we calculate the latent heat from both cooled and uncooled configurations in order to monitor the contribution of the instantons to this quantity.

3 Cooling Procedure

The method of lattice cooling has been extensively studied and discussed in the literature [10] and can be applied to finite temperature configurations without modifications. We use the same procedure as in ref. [5].

Briefly speaking, the cooling method smooths out the gauge fields locally, by minimizing the action density. After some cooling steps the short-range fluctuations in the gauge fields are suppressed, leaving the long-range "bumps" more or less unchanged. In particular, large isolated instantons, which are protected by topology, dominate the gauge fields after many cooling steps, while the smaller instantons annihilate with nearby anti-instantons. The topological charge approaches a plateau corresponding to an integral number of instantons, when plotted against the number of cooling steps. The topological susceptibility then becomes constant, since the annihilation of instanton pairs does not change its value. Only then one can extract an unambiguous value for the susceptibility using the cooling method.

After many cooling steps, the instantons are well separated in the gauge fields, and the probability distribution for the topological charge is then given by a convolution of Poissonian distributions as one would expect for a dilute gas of instantons [4].

4 Numerical Results

For the generation of the gauge field configurations we used the Metropolis algorithm. We studied a lattice with spatial volume $V_s = 16^3$. Following Eq. (3) we varied the temperature by varying the number of time slices using values $N_t = 4, 6, 8, 10, 12, 14$, and 16, respectively. The coupling constant is chosen to be $\beta = 6.0$ which corresponds to a physical lattice spacing $a \sim 0.1$fm. Thus the calculation covers a temperature range between 125 and 500 MeV. In the case of quenched QCD the critical temperature is $T_c \sim 250$ MeV. We generated 100 configurations each for $N_t = 4, 6, 8, 10, 12$ and 40 configurations for $N_t = 14$ and 16, which were separated by 300 lattice sweeps. The configurations were cooled using a heat bath algorithm in the limit of infinite coupling strength $b = \infty$ as discussed in [5]. We verified the stability of our results by varying the number of cooling steps.

The numerical values for the topological susceptibility are shown in Fig. 1. The plot also shows the approximate phenomenological value for χ_t at zero temperature as estimated by Witten and Veneziano [15]

$$\chi_t = \frac{f_\pi^2}{4N_f}(m_{\eta'}^2 - 2m_K^2 + m_\eta^2) \sim (180 \text{ MeV})^4 \ . \tag{7}$$

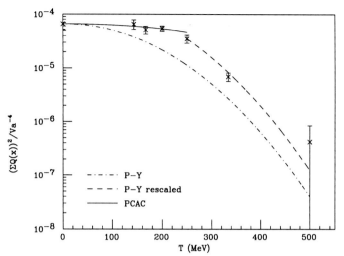

Figure 1: Topological susceptibility as a function of temperature. Lattice data, shown with error bars, are compared with: 1) the PCAC expectation assuming soft pion gas as the heat bath (see [14]), indicated by the solid line; 2) the Pisarski-Yaffe formula (Eq. 8, using $\rho = 0.26$ fm), which assumes instanton suppression via Debye screening (dotdashed line); 3) the Pisarski-Yaffe formula rescaled to match with the value of the topological susceptibility at $T_c = 250$ MeV (dashed line). The phenomenological value, Eq. 7, is also plotted at T=0.

As can be seen the phenomenological value and the numerical result for the $V = 16^4$ lattice agree quite well. The temperature behavior of χ_t exhibits a gentle decrease for $T < T_c$, turning into a relatively sharp decay of the susceptibility around the critical temperature T_c, similar to phenomenological low-temperature results where one models the heat bath as a soft pion gas [14]. The instanton density is already reduced to half of its value for $T = T_c$, with some stronger suppression mechanism operative at temperatures between $T = 200$ MeV and T_c. A rather sharp decrease of the topological susceptibility continues above the phase transition. At temperature $T \sim 334$ MeV topological effects are already strongly suppressed; χ_t is practically zero at $T \sim 500$ MeV. The dotdashed curve in Fig. 1 shows the temperature dependence for χ_t as predicted by Pisarski and Yaffe[16] calculating the Debye-screening suppression of large scale instantons:

$$\chi_t(T) = \chi_t(T = 0) \left(1 + \lambda^2/3\right)^{\frac{3}{2}} \exp\left[-2\lambda^2 - 18\alpha \left(1 + \gamma\lambda^{\frac{-3}{2}}\right)^{-8}\right] , \qquad (8)$$

for quenched $SU(3)$ fields, with $\lambda \equiv \pi\rho T$, $\alpha = 0.01289764$, $\gamma = 0.15858$, and $\rho = 0.26$ fm is the size parameter of instantons we obtained by fitting our

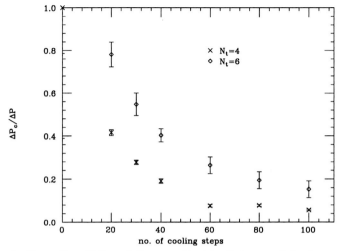

Figure 2: Plaquette splittings ΔP_c, normalized to their uncooled values, as a function of cooling steps for $N_t = 4$ (crosses) and $N_t = 6$ (diamonds). The values at the 100th cooling step represent upper bounds to the fraction of latent heat contributed by the instantons.

$N_t = 6$ data, which differs slightly from a previous lattice calculation at zero temperature[5]. The perturbative result in Eq. 8 is supposed to be valid only at very high temperature. In the range of temperatures we are considering, the perturbative formula gives too large a suppression. The dashed curve in Fig. 1 was computed by rescaling the Pisarski-Yaffe expression to match with the susceptibility at $T_c = 250$ MeV, which then seems to be in agreement with the data, though more data are needed between T_c and $2T_c$ for a more meaningful comparison. The phenomenological consequences of the restoration of $U_A(1)$, such as changes in the η' mass, should be an exciting subject for experimental works at AGS, CERN, and RHIC.

We monitor the contribution of the large, isolated instantons to the latent heat by calculating the splittings of the spatial and temporal plaquettes for both cooled and uncooled configurations. The splitting $\Delta P \equiv < P_t > - < P_x >$, which is proportional to the latent heat, is clearly a function of cooling, and it approaches zero in the limit of infinite cooling steps. We estimate the fraction of latent heat contributed by instantons, f, by

$$f \approx \frac{\Delta P_{\text{cooled}}}{\Delta P_{\text{uncooled}}} \quad . \tag{9}$$

In Fig. 2 we show f as a function of cooling steps for both $N_t = 4$ and 6. In both cases, f drops fast during the first 40 cooling steps, showing that most of the splitting is due to short range fluctuations that got eliminated by cooling.

After about 50 cooling steps, the $N_t = 4$ splitting levels off to a plateau, where its value represents an upper bound to the contribution of the instantons at $T = 500$ MeV. The splitting at $N_t = 6$ has not yet reached a plateau even at 100 cooling steps; we just quote the fraction f at the 100th cooling step as an upper bound at $T = 334$ MeV. The results for f are 0.15 ± 0.04 at $T = 334$ MeV and 0.056 ± 0.004 at $T = 500$ MeV.

5 Conclusion

We discussed the content and the role of instantons in quenched QCD matter at finite temperature. Specifically, we calculated the density of instantons and found that it is suppressed as temperature increases. While at low temperatures the mild dependence of χ_t resembles results from a soft pion gas, the high temperature $(T > T_c)$ behavior deviates significantly from the perturbative formula expressing the Debye-screening suppression of instantons. Our results indicate that the restoration of $U_A(1)$ symmetry is a continuous process starting below T_c; at T_c, about half of the axial anomalies is already removed. Some non-perturbative mechanism enhances the rate of $U_A(1)$-restoration in the temperature range $T_c \leq T \leq 2T_c$. The topological charge correlation function agrees well with the continuum instanton profile, though our data suggests a rapid change in the size parameter from $\rho = 0.33$ fm at and below 250 MeV to 0.26 fm at 334 MeV[4]. By comparing the entropy in the cooled and uncooled gauge fields, we obtain upper bounds - of about 5% at $2T_c$ and 15% at $1.5T_c$- on the instanton contribution to the latent heat. We are working on repeating our calculations within an unquenched simulation. First results indicate the same qualitative behaviour of the instanton suppression around T_c.

References

1. Larry D. McLerran and Benjamin Svetitsky, Phys. Lett. **98B** (1981)195; J. Kuti, J. Polónyi, and K. Szlachányi, Phys. Lett. **98B** (1981)199; J. Engels, F. Karsch, H. Satz, and I. Montvay, Phys. Lett. **101B**(1981)89.
2. See for example, Nucl. Phys. **A566** (1994) 115c, and references therein.
3. M.-C. Chu and S. Schramm, Phys. Rev. **D51**, 4580 (1995).
4. M.-C. Chu, J. M. Grandy, S. Huang, and J. W. Negele, Phys. Rev. **D49**, 6039(1994); Phys. Rev. Lett. **70**, 225(1993); M.-C. Chu and S. Huang, Phys. Rev. **D45**, 2446(1992).
5. R. D. Pisarski and F. Wilczek, Phys. Rev. **D29**, 338(1984).
6. E. Shuryak, Comm. Nucl. Part. Phys. **21**, 235(1994).
7. M. Teper, Phys. Lett. **B171** (1986)81.

620

8. A. DiGiacomo, E. Meggiolaro, and H. Panagopoulos, Phys. Lett. **B277** (1992)491.
9. Jaap Hoek, M. Teper, and J. Waterhouse, Nucl. Phys. **B288** (1987)589.
10. Steven Gottlieb *et al.*, Phys. Rev. **D47**, 3619(1993).
11. J. Engels, F. Karsch, H. Satz, and I. Montvay, Nucl. Phys. **B205** (1982)545; F. Brown, *et al.*, Phys. Rev. Lett. **61**, 2058(1988).
12. E. M. Ilgenfritz and E. Shuryak, Phys. Lett. **B325**, 263(1994); T. Schäfer, E. Shuryak, and J. J. M. Verbaarschot, Report No. SUNY-NTG-93-39.
13. C. Bernard *et al.*, Phys. Rev. Lett. **68**, 2125 (1992); S. Schramm and M.-C. Chu, Phys. Rev. D **48**, 2279 (1993).
14. E. Shuryak and M. Velkovsky, Phys. Rev. **D50**, 3323(1994).
15. E. Witten, Nucl. Phys. **B156**, 269 (1979); G. Veneziano, Nucl. Phys. **B156**, 213 (1979); Phys. Lett. **95B**, 90 (1980).
16. R. D. Pisarski and L. G. Yaffe, Phys. Lett. **B97**, 110(1980).

RENORMALON CONTRIBUTION TO NUCLEON STRUCTURE FUNCTIONS

E. Stein, M. Maul, and M. Meyer-Hermann

Institut für Theoretische Physik, J. W. Goethe Universität Frankfurt,
Postfach 11 19 32, D-60054 Frankfurt am Main, Germany

Data on QCD-observables now have reached such precision that corrections that are powerlike suppressed might be confronted with theoretical predictions. While a rigorous calculation of higher twist corrections to nucleon structure functions is not yet available the calculation of the renormalon ambiguity can be used to estimate the contribution of "genuine" twist-4 corrections as a function of Bjorken-x. The predictions turn out to be in surprisingly good agreement with the experimental data for the longitudinal structure function F_L

One of the interesting quantities that can be measured in deep inelastic lepton-nucleon scattering is the ratio

$$R(x, Q^2) = \frac{\sigma_{\rm L}(x, Q^2)}{\sigma_{\rm T}(x, Q^2)} = \frac{F_2(x, Q^2)}{F_2(x, Q^2) - F_L(x, Q^2)} \left(1 + \frac{4M^2 x^2}{Q^2}\right) - 1 \quad (1)$$

of the total cross-sections for the scattering of longitudinal respectively transversely polarized photons and a nucleon, where M is the nucleon mass and $F_L = F_2 - 2xF_1$. This ratio provides a clean test of the QCD interaction since it vanishes identically in the naive parton model. The experimental information on $R(x, Q^2)$ is still limited [1], but much better data should be available in a few years from now. Phenomenological fits to the existing data [2] suggest surprisingly large higher-twist corrections. While for inclusive processes such as deep inelastic scattering those are well defined in the framework of the operator product expansion the situation is less clear for exclusive processes. Even if there is an operator product expansion the calculation of the matrix elements that determine the power corrections requires non perturbative techniques, such as lattice calculations or QCD sum rules. To get a complete picture of those corrections the calculation of arbitrarily many matrix elements is required and therefore well beyond today's possibilities.

The renormalon approach to power corrections uses the fact that the necessity to include those corrections can be already read off from the leading term. The perturbative expansion of the short distance coefficient diverges. This renormalon divergence occurs because certain classes of Feynman diagrams are sensitive to large distances when calculated up to high orders. In this framework one refers to the large N_f-expansion where fermion bubble-chains are resummed to all orders yielding the coefficient of the $\alpha_S^n N_f^{n-1}$ - term exactly.

IR-renormalons have recently received much attention because of their potential to generate power-like corrections. For a physical quantity like F_L the perturbative QCD series is not summable, even in the Borel sense, due to the appearance of fixed sign factorial growth of its coefficients. It results in a power-suppressed ambiguity of the magnitude $\sim \Lambda_{QCD}^2/Q^2$ [3]. Such terms show the need to include higher twist (non perturbative) corrections to give a meaning to a summed perturbation series [4,5]. On the other hand also the higher twist corrections themselves are ill-defined. The ambiguity in their definition, due to the UV-renormalon, cancels exactly the IR-renormalon ambiguity in the perturbative series which describes the twist-2 term. In turn, the investigation of the ambiguities in the definition of the perturbation series of leading twist shows which higher twist corrections are needed for an unambiguous definition of a physical quantity.

In practice, one has observed the empirical fact that in cases where the perturbative series was studied in parallel with the higher twist corrections, such as the polarized Bjorken sum rule and the Gross Llewellyn-Smith sum rule, the ambiguities produced by IR renormalons in the leading twist contribution were roughly of the same order of magnitude as the best available theoretical estimates of the higher twist corrections [6]. Thus, for phenomenological purposes one may use IR renormalons as a guide for the magnitude of higher-twist corrections [7]. The obvious advantage of such an approach is that the IR calculation can be done for all moments, and hence the result can be extended to the full x-dependence of the higher-twist contributions. One has to keep in mind, however, that the last step is even less justified, as the order of magnitude correspondence between IR ambiguities and higher twist corrections has been tested only for sum rules for first moments of structure functions.

We focus on the flavor non-singlet part of the longitudinal structure function

$$F_L^{p-n}(x, Q^2) \tag{2}$$

i.e. on the difference between the proton and neutron structure function, and calculate the infrared renormalon contribution [8]. This will also provide the exact coefficients of the perturbative series of F_L in the large N_f approximation [9,10,11]. In the framework of the 'Naive Non Abelianization' [12,13] this can be used to approximate the non-leading N_f terms.

We start with the well known hadronic scattering tensor of unpolarized deep inelastic lepton nucleon scattering parameterized in terms of two structure functions F_2 and F_L.

$$
\begin{aligned}
W_{\mu\nu}(p,q) &= \frac{1}{2\pi} \int d^4 z\, e^{iqz} \langle p | J_\mu(z) J_\nu(0) | p \rangle \\
&= \left(g_{\mu\nu} - \frac{q_\mu q_\nu}{q^2} \right) \frac{1}{2x} F_L(x, Q^2) \\
&\quad - \left(g_{\mu\nu} + p_\mu p_\nu \frac{q^2}{(p \cdot q)^2} - \frac{p_\mu q_\nu + p_\nu q_\mu}{p \cdot q} \right) \frac{1}{2x} F_2(x, Q^2)
\end{aligned}
\tag{3}
$$

Here J_μ is the electromagnetic quark current, $x = Q^2/(2p \cdot q)$ and $q^2 = -Q^2$. The nucleon state $|p\rangle$ has momentum p (averaging over the polarizations of the nucleon is understood). The non-singlet moments of the structure functions F_k^{p-n} ($k = 2, L$) can be expressed through operator product expansion [14] in the following form:

$$
\begin{aligned}
M_{k,N}(Q^2) &= \int_0^1 dx\, x^{N-2} F_k^{p-n}(x, Q^2) \\
&= C_{k,N}\left(\frac{Q^2}{\mu^2}, a_s \right) \left[A_N(\mu^2) \right]^{p-n} + \text{higher twist}
\end{aligned}
\tag{4}
$$

where a_s stands for $a_s = \frac{\alpha_s}{4\pi}$ and the A_N are the spin-averaged matrix elements of the spin-N twist-2 operator

$$
\langle p | \bar\psi \gamma^{\{\mu_1} iD^{\mu_2} \ldots iD^{\mu_N\}} \psi | p \rangle^{p-n} = p^{\{\mu_1 \cdots \mu_N\}} \left[A_N(\mu^2) \right]^{p-n}
\tag{5}
$$

The inclusion of quark charges is implicitly understood. The flavors of the quark-operators ψ are combined to yield the proton minus neutron matrix element. $\{\mu\nu\}$ indicates symmetric and traceless combinations. The higher twist corrections are given by matrix elements of twist-4 operators and were derived for the second moments of F_L and F_2 in ref. [15].

$$
\begin{aligned}
M_{k,2}^{twist-4}(Q^2) &= -\frac{C_{L,2}^{(a)}\left(Q^2/\mu^2, a_s \right)}{4Q^2} \langle p | \bar\psi \left(D^\alpha g G_{\alpha\{\mu} \right) \gamma_{\nu\}} \psi | p \rangle^{p-n} \\
&\quad - 3 \frac{C_{L,2}^{(b)}\left(Q^2/\mu^2 \right), a_s)}{8Q^2} \langle p | \bar\psi \left\{ g\tilde{G}_{\alpha\{\mu}, iD_{\nu\}} \right\}_+ \gamma^\alpha \gamma_5 \psi | p \rangle^{p-n}
\end{aligned}
\tag{6}
$$

Due to the high dimension of operators it is at present not possible to calculate the matrixelements reliably in the framework of lattice QCD or QCD sum rules.

An additional problem of such a calculation is that a renormalization scheme has to be found in which quadratic divergences in twist-4 matrix elements do not produce mixing with lower dimension twist-2 operators. That is still an unsolved problem in lattice calculations [16,17]. On the other hand, state of the art calculations of higher twist-corrections never claim an accuracy better than 30-50%. Therefore we claim that calculating the renormalon ambiguity in the coefficient function $C_{L,N}(Q^2/\mu^2, a_s)$ instead of the true higher-twist corrections to the longitudinal structure function is an legitimate procedure. The advantage is that such a calculation can be done for all N therefore allowing to estimate the twist-4 corrections as a function of Bjorken-x. Note that a renormalon ambiguity in the coefficient function of the twist-2 spin-2 operator will only account for twist-4, spin-2, twist-6, spin-2 etc. operators and not for power suppressed twist-2 operators. This implies that target mass effects can not be traced by IR-renormalons.

The truncated perturbative expansion of the coefficient functions of the moments of F_L and F_2 can be written as

$$C_{k,N}(1, a_s) = \sum_{n=0}^{m_0(N)} B_{k,N}^{(n)} a_s^n + C_{k,N}^{(1)} \frac{\Lambda_C^2 e^{-C}}{Q^2} + C_{k,N}^{(2)} \frac{\Lambda_C^4 e^{-2C}}{Q^4} \tag{7}$$

where we have accounted for the asymptotic behaviour of the perturbation series which makes only sense up to a maximal order $m_0(N)$ depending on the magnitude of the expansion parameter a_s. With the standard normalization one finds $B_{L,N}^{(0)} = 0$, $B_{2,N}^{(0)} = 1$ and $B_{L,N}^{(1)} = 4C_F/(1 + N)$ [18], $C_F = 4/3$. We have indicated the ambiguity of the asymptotic expansion by including power suppressed terms. $\Lambda_C^2 e^{-C}$ is a renormalization scheme independent quantity. $C = -5/3$ corresponds to $\Lambda_{\overline{MS}}$. We will show that only ambiguities up to order $1/Q^4$ will appear in the $1/N_f$ approximation for the longitudinal coefficient function. Since non-singlet F_L and F_2 are to leading twist accuracy determined by the same operators we obtain from Eq. (4) with $Q^2 = \mu^2$, $a_s = a_s(Q^2)$

$$M_{L,N}(Q^2) = \left[B_{L,N}^{(1)} a_s + \left(B_{L,N}^{(2)} - B_{L,N}^{(1)} B_{2,N}^{(1)} \right) a_s^2 + \mathcal{O}(a_s^3) \right.$$
$$\left. + \frac{\Lambda_C^2 e^{-C}}{Q^2} \left(C_{L,N}^{(1)} - \left(C_{L,N}^{(1)} B_{2,N}^{(1)} + C_{2,N}^{(1)} B_{L,N}^{(1)} \right) a_s + \mathcal{O}(a_s^2) \right) \right] M_{2,N}(Q^2) \tag{8}$$

To extract the renormalon contribution to $C_{L,N}$ we calculate the coefficients to all orders in a_s in the $1/N_f$ expansion where N_f refers to the number of active flavours [9,10,11]. For that we split the exact coefficient into

$$B_{L,N}^{(m+1)} = \tilde{B}_{L,N}^{(m)} + \delta_{L,N}^{(m)} . \tag{9}$$

While $\tilde{B}_{L,N}^{(m)}$ contains only the effects of one-loop running of the coupling to order m, only $\delta_{L,N}^{(m)}$ requires a true m-loop calculation. It will be checked a posteriori by comparison with those coefficient that are known exactly whether the neglection of $\delta_{L,N}^{(m)}$ is justified (see Table (1)). Note that $B_{L,N}^{(1)} = \tilde{B}_{L,N}^{(0)} = 4C_F/(N+1)$. In the following the NNA approximation to the coefficient function $C_{L,N}(a_s)$ will be written as $\tilde{C}_{L,N}(a_s)$.

A convenient way to calculate $\tilde{C}_{L,N}(a_s)$ is to deal with its Borel transform

$$BT[\tilde{C}_{L,N}(a_s)](s) = \sum_{m=0} s^m \beta_0^{-m} \frac{\tilde{B}_{L,N}^{(m)}}{m!} . \tag{10}$$

The advantage of that representation is manifold. The Borel transform can be used as generating function for the fixed order coefficients

$$\tilde{B}_{L,N}^{(m)} = \beta_0^m \frac{d^m}{ds^m} BT[\tilde{C}_{L,N}(a_s)](s) \Big|_{s=0} \tag{11}$$

and the sum of all diagrams can be defined by the integral representation

$$\tilde{C}_{L,N}(a_s) = \frac{1}{\beta_0} \int_0^\infty ds\, e^{-s/(\beta_0 a_s)} BT[\tilde{C}_{L,N}(a_s)](s) . \tag{12}$$

Technically the most important point is the simplification of the calculation of the Borel transform of diagrams with only one fermion bubble chain. In that case the Borel transform can be applied directly to the effective gluon propagator which resums the fermion bubble chain. The effective (Borel-transformed) gluon propagator is [19]

$$BT[a_s D^{AB}(k)](s) = i\delta^{AB} \frac{k_\mu k_\nu - k^2 g_{\mu\nu}}{(-k^2)^2} \left(\frac{\mu^2 e^{-C}}{-k^2} \right)^s . \tag{13}$$

In fact one only has to calculate the leading-order diagram with the usual gluon propagator substituted by the above one in which only the usual denominator of $(-k^2)^{-2}$ is changed to $(-k^2)^{-(2+s)}$.

To obtain the coefficient function $\tilde{C}_{L,N}(1, a_s)$ we have to calculate the $\mathcal{O}(a_s)$ correction to the Compton forward scattering amplitude with the effective propagator Eq. (13). We get

$$BT[\tilde{C}_{L,N}(a_s)](s) = C_F \left(\frac{\mu^2 e^{-C}}{Q^2} \right)^s \frac{8}{(2-s)(1-s)(1+s+N)} \frac{\Gamma(s+N)}{\Gamma(1+s)\Gamma(N)} \tag{14}$$

		Exact results [20]	NNA approximants
$N = 2$	$N_F = 3$	$43.1254\,a_s^2 + 1386.59\,a_s^3$	$61.3333\,a_s^2 + 2168.\,a_s^3$
	$N_F = 4$	$38.5822\,a_s^2 + 1032.7\,a_s^3$	$56.7901\,a_s^2 + 1858.71\,a_s^3$
$N = 4$	$N_F = 3$	$37.75\,a_s^2 + 1472.58\,a_s^3$	$46.08\,a_s^2 + 1984.51\,a_s^3$
	$N_F = 4$	$34.3367\,a_s^2 + 1155.64\,a_s^3$	$42.6667\,a_s^2 + 1701.4\,a_s^3$
$N = 6$	$N_F = 3$	$32.9091\,a_s^2 + 1433.24\,a_s^3$	$36.3918\,a_s^2 + 1726.31\,a_s^3$
	$N_F = 4$	$30.2134\,a_s^2 + 1152.07\,a_s^3$	$33.6961\,a_s^2 + 1480.03\,a_s^3$
$N = 8$	$N_F = 3$	$29.1822\,a_s^2 + 1373.83\,a_s^3$	$30.1249\,a_s^2 + 1519.45\,a_s^3$
	$N_F = 4$	$26.9507\,a_s^2 + 1117.97\,a_s^3$	$27.8934\,a_s^2 + 1302.68\,a_s^3$

Table 1: Comparison of the NNA approximants to the exact results obtained in [20] for the coefficient function $C_{L,N}(1, a_s)$ up to order $\mathcal{O}(a_s^3)$. We have omitted the $\mathcal{O}(a_s)$ corrections which agree exactly.

The Borel transform exhibits IR-renormalons at $s = 1$ and $s = 2$. The position of the UV-renormalon $s = -1 - N$ depends on the moment N one is dealing with.

The NNA approximants to the coefficient function in all orders in a_s can be derived setting $\mu^2 = Q^2$ and $C = -5/3$ in equation Eq. (14). It is interesting to compare the approximation of the NNA procedure with the exact results derived by Larin et al. [20] for the non singlet moments $N = 2$, 4, 6, 8, denoted by $C_{L,N}$. The subleading N_f coefficients approximate those of the exact expression in sign and magnitude. The leading $N_f a_s^2$ and $N_f^2 a_s^3$ coefficients of course agree exactly. The numerically important cases $N_f = 3$ and $N_f = 4$ are given in table 1. It is interesting to observe that NNA approximates the higher moments consistently better than the lower ones. The most problematic property of NNA is the neglect of multiple gluon emission. As such processes are important for small x we cannot expect our NNA structure functions to be correct in this region. Ever higher moments of the structure functions are less and less sensitive to their small-x behaviour and therefore the NNA should systematically improve.

As can be seen from Eq. (14) the perturbative expansion of $\tilde{C}_{L,N}$ is not Borel summable. The poles in the Borel representation at $s = 1$ and $s = 2$ destroy a reconstruction of the summed series via Eq. (12). Asymptotically the first IR-renormalon, i.e. the pole at $s = 1$ will dominate the perturbative expansion giving rise to a factorial growth of the coefficient

$$\lim_{m \to \infty} \tilde{B}_{L,N}^{(m)} \sim \beta_0^m \frac{d^m}{ds^m} \frac{1}{1-s}\bigg|_{s=0} = \beta_0^m m! \tag{15}$$

This means that a perturbative expansion at best can be regarded as an asymptotic expansion and the expansion makes sense only up to a maximal value $m = m_0 \sim \log(Q^2/\Lambda^2)$. For higher values of $m > m_0$ the fixed order contributions will increase and finally diverge. The general uncertainty in the perturbative prediction is then of the order of the minimal term in the expansion. It can be estimated either directly or by taking the imaginary part $\Im/\pi$ (divided by π) of the Borel transform [6]. From Eq. (12) we get for the function Eq. (14)

$$\frac{\Im}{\pi} \frac{1}{\beta_0} \int_0^\infty ds \, e^{-s/(\beta_0 a_s)} BT[\tilde{C}_{L,N}(a_s)](s) = \pm \frac{8 C_F}{\beta_0} \frac{\Lambda_C^2 e^{-C}}{Q^2} \frac{N}{N+2} \pm \mathcal{O}(1/Q^4)$$

$$\tag{16}$$

The ambiguity in the sign of the IR-renormalon contributions is due to the two possible contour deformations above or below the pole at $s = 1$ and $s = 2$. Transforming to Bjorken-x space we finally arrive at

$$F_L^{p-n}(x, Q^2) + \frac{4x^2 M^2}{Q^2} F_2^{p-n}(x, Q^2)$$

$$= 4 C_F a_s(Q^2) \int_x^1 \frac{dy}{y} \left(\frac{x}{y}\right)^2 F_2^{p-n}(y, Q^2) + \mathcal{O}(a_s^2)$$

$$\pm \frac{8 C_F}{\beta_0} \frac{\Lambda_{\overline{MS}}^2 e^{5/3}}{Q^2} \left[F_2^{p-n}(x, Q^2) - 2 \int_x^1 \frac{dy}{y} \left(\frac{x}{y}\right)^3 F_2^{p-n}(y, Q^2) + \mathcal{O}(a_s) \right]$$

$$+ 4x^3 \frac{M^2}{Q^2} \int_x^1 \frac{dy}{y} F_2^{p-n}(y, Q^2) \tag{17}$$

We have neglected the contribution of the second IR-renormalon since it is of the order of $1/(Q^2)^2$ while there is a contribution of order a_s/Q^2 related to the ambiguity in the coefficient function of F_2 which we have not included. We have included the kinematical and target mass corrections to the order we are working.

In connection to the IR renormalon, it is interesting to investigate the corresponding ambiguity in the definition of the twist-4 matrix elements. This can be done for the second moment of F_L where the contributing twist-4 operators are known, see Eq. (6). Composite operators have to be renormalized individually and have their own renormalization scale dependence. Operators of a higher twist, and therefore of a higher dimension, exhibit power-like UV divergences in addition to the usual logarithmic ones. In particular, a quadratic divergence of a twist-4 matrix element contributing to F_L, results in a mixing with the lower-dimension twist-2 matrix element. When the calculation of the matrix element is done in the framework of dimensional regularization, the quadratic divergence does not appear explicitly. It manifests itself as $\Gamma(1 - d/2)$ factor, singular at $d = 4$, and at $d = 2$, where it corresponds to usual logarithmic divergence. Evaluating the one loop contribution to the quark matrix elements of twist-4 operators, with the Borel transformed propagator (Eq. (13)), in $d = 4$ dimensions we obtain an expression singular at $s = 1$ and observe that indeed the ambiguities cancel [8].

The appearance of an IR-renormalon ambiguity in a perturbative calculation thus indicates the need to include higher twist corrections to interpret the perturbative expansion to all orders. Of course one can argue that the numerical value of the IR-renormalon uncertainty has no physical significance since it has to cancel in a complete calculation. As we explained above, in some case a higher twist estimate based on IR renormalon ambiguity has proven to be a fairly good guess, at least for low moments of the structure functions. In the present case the IR calculation can be easily done for all moments, so that the result can be extended to produce a model of the the full x-dependence of higher twist effects. We keep in mind that such a model cannot have significance beyond phenomenological level for the following reason. The renormalons ambiguity in the coefficient function $C_{L,N}(Q^2)$ is a target independent quantity of pure perturbative nature, while "genuine" higher twist matrix elements are a measure of multiparticle correlations in the target and are process dependent.

In Figure 1 we compare the experimental fit of the higher-twist coefficient with the QCD-calculation, where we have shown target mass and twist-4 contributions separately. We observe a rather large contribution coming from the IR renormalon estimate for the twist-4 part, which accounts for more than half of the discrepancy between the experimental fit and a prediction which takes into account the target mass correction only. The sign of the IR renormalon contribution, which cannot be determined theoretically, should be choosen positive. This leads to an astonishing agreement (possible somewhat fortuitous) with the experimental fit. Our estimate based on the calculation of the IR ambiguity has proven to be phenomenologically surprisingly successful, predicting a high twist-4 contribution to F_L in accordance with experimental results. It further supports the idea that, while the rigorous QCD calculations of higher twist contribution to F_L are not yet available, calculations like the one presented in this paper can be used to predict the order of magnitude of power suppressed corrections. This idea was successfully applied in the calculation of the higher twist corrections to the remaining structure functions F_2, F_3 and g_1 [22,23].

Acknowledgments

We acknowledge the collaboration with L. Mankiewicz and A. Schäfer. This work is dedicated to Walter Greiner at the occasion of his 60th birthday.

References

1. S. Dasu et al., Phys.Rev.Lett. **61**, 1061 (1988)
2. L.W. Whitlow, S. Rock, A. Bodek, S. Dasu, and E.M. Riordan, Phys. Lett. **B250**, 193 (1990)
3. A.H. Mueller, *The QCD perturbation series.*, in "QCD - Twenty Years Later", edited by P.M. Zerwas and H.A. Kastrup, 162, (World Scientific 1992)
4. V. I. Zakharov, Nucl. Phys. **B385**, 452 (1992)
5. A.H. Mueller, Phys. Lett. **B 308**, 355 (1993)
6. V. M. Braun, *QCD renormalons and higher twist effects.* hep-ph/9505317 (1995)
7. M. Beneke and V.M. Braun, Nucl. Phys. **B454**, 253 (1995)
8. E. Stein, M. Meyer-Hermann, L. Mankievicz, and A. Schäfer Phys.Lett.**376**, 177 (1996).
9. D.J.Broadhurst, Z.Phys. **C58**, 339 (1993)
10. D.J.Broadhurst and A.L. Kataev, Phys.Lett. **B315**, 179 (1993)
11. C.N. Lovett-Turner, C.J. Maxwell, Nucl. Phys. **B452** 188 (1995)
12. M. Beneke and V.M. Braun, Phys.Lett. **B348**, 513 (1995)
13. P. Ball, M. Beneke and V.M. Braun, Nucl. Phys. **B452**, 563 (1995)
14. H.D. Politzer, Phys.Rev.Lett. **30**, 1346 (1973)
    D.J. Gross and F. Wilczek, Phys.Rev.Lett. **30**, 1323 (1973)
15. E.V. Shuryak and A.I. Vainshtein, Nucl.Phys. **B199**, 451, (1982)
    R.L. Jaffe and M. Soldate, Phys.Rev. **D26**, 49 (1982)
16. X. Ji, Nucl.Phys. **B448** 51 (1995)
17. G. Martinelli, C.T. Sachrajda, Phys. Lett. **B354** 423 (1995)
18. W.A. Bardeen, A.J. Buras, D.W. Duke, and T. Muta, Phys. Rev. **D18**, 3998 (1978)

19. M. Beneke Nucl.Phys. **B405** 424 (1993)

20. S.A. Larin, T. van Ritbergen, J.A.M. Vermaseren, Nucl.Phys. **B427**, 41 (1994)

21. NMC, M. Arneodo et al., Measurement of the proton and the deuteron structure functions F_2^p and F_2^d, CERN-PPE/95-138, hep-ph/9509406

22. M. Dasgupta, B. R. Webber Phys.Lett.**B382**, 273 (1996).

23. M. Meyer-Hermann, M. Maul L. Mankiewicz, E. Stein, A. Schäfer Phys.Lett.**B 383**, 463 (1996).

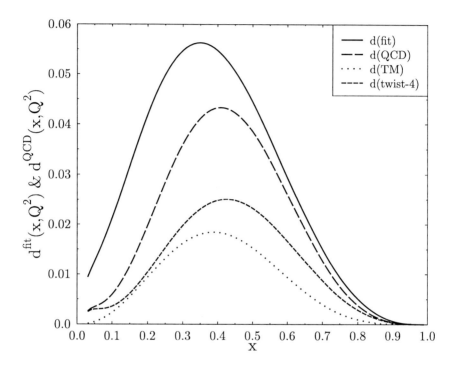

Figure 1: Comparison of $d^{QCD}(x, Q^2)$ (long dashed line), the power suppressed contribution on the right hand side of Eq. (17), with the phenomenological fit $d^{\text{fit}}(x, Q^2)$ (full line). We have also plotted the IR-renormalon (twist-4) part (short dashed line) and the target mass corrections (dotted line) separately. The agreement with experiment is much better including twist-4 corrections than without them (dotted line). We have chosen $\Lambda_{\overline{MS}} = 250$ MeV, $Q^2 = 5$ GeV and $N_f = 4$.

IS SCALAR FIELD THEORY TRIVIAL?

K. SAILER

Department of Theoretical Physics, Kossuth Lajos University,
Poroszlay út 6/c, H-4032 Debrecen, Hungary

The phase structure of the scalar field theory including terms with arbitrary powers of the gradient operator and a local non-analytic potential is investigated by the help of the RG in Euclidean space. Infinitely many non-trivial fixed points of the RG transformations are found. Rather probably the corresponding effective theories, however, do not exhibit any particle content.

1 Introduction

Scalar fields with particular internal symmetries are important ingredients of field theoretical models with spontaneous symmetry breaking. It is well-known that scalar field theory with a local potential providing spontaneous symmetry breaking at the tree level, retains this feature if quantum fluctuations are taken into account perturbatively. It is, however, also well-known that scalar field theory with local polynomial potentials at the UV cut-off is equivalent to a free theory at low momentum scale [1], [2]. That means that scalar field theories treating them non-perturbatively do not provide a non-trivial vacuum structure.

To be more precise let us consider the bare action defined at the UV cut-off Λ in dimension 4:

$$S_\Lambda \;\; = \;\; \int d^4 x \left(\frac{1}{2} (\partial \phi)^2 + V(\phi) \right) \tag{1}$$

with an analytic potential $V(\phi) = \sum_{n=2}^{\infty} c_n \phi^n(x)$. Here cut-off regularization is used. It is the general question how does physics change if the scale of observations changes, i.e. the observables considered are not sensitive to the fluctuations above some momentum scale k. The effective action at the momentum scale k can then be defined by integrating out the Fourier modes ϕ_q of the field for $k < |q| \leq \Lambda$. A lot of efforts to find the IR effective action of scalar field theory with analytic potential lead to the conclusion that the coupling constants of the potential vanish, i.e. $c_n(k) \to 0$ $(n > 2)$ in the limit $k \to 0$. No other than this Gaussian fixed point was found [2],[3] This fact is called as the triviality of the scalar field theory. In the parameter space spanned by the coupling constants c_n any point represents a possible action S_Λ. Starting with any of them and integrating out the high-frequency modes the point moves along the so called RG trajectories towards the point $c_n = 0$ $(n > 2)$ reached for $k \to 0$. This is an IR fixed point on these RG trajectories and corresponds to a free theory.

More recently it has been observed [4] that there are particular directions in the parameter space given by a recursion formula of the type $c_{n+1} = f_a(c_n)$ and parametrized by the continuous parameter $a \in [-1, 0]$. If we start in the neighbourhood of the point $c_n = 0$ with a bare action S_Λ with the couplings satisfying this recursion formula, we obtain RG trajectories running away from that fixed point. For such RG trajectories the fixed point is an UV fixed point and the model is asymptotically free. The corresponding potentials are local, analytic, but non-polynomial. One expects that outside the linearization region of the fixed point those RG trajectories decline from the lines given by the recursion formula. It is an intriguing question where do flow the corresponding RG trajectories for $k \to 0$. Therefore the renewed search for fixed points other than the Gaussian one gets actuality once again.

Our main goal is to perform such a search in the space of actions extended by including terms with arbitrary powers of the gradient operator and non-analytic local potentials. Earlier investigations of the phase structure of the scalar field theory used a parameter space restricted to the usual derivative term and local analytic potentials.

We show that the one-component scalar field theory has infinitely many non-trivial fixed points in any dimension $d > 2$ and determine the fixed point actions and the irrelevant operators in the neighbourhood of the non-trivial fixed points. It is crucial for finding the non-trivial fixed points the enlargement of the space of the scale dependent actions (Hamiltonians) considered.

2 Renormalization Group Equations

The phase structure of the scalar field theory is investigated with the help of the Wegner-Houghton equation [5]. As it is well-known [5,6,7,8], the approach of Wegner and Houghton enables one to carry out renormalization by integrating out the high-frequency modes step by step in infinitesimally thin momentum shells, $|p| \in I_k \equiv (k - \delta k, k]$.

The action at the momentum scale $k - \delta k$ is obtained from the action at the momentum scale k by integrating out the Fourier modes ϕ_p in the momentum shell I_k:

$$\exp\{-S_k[\phi_q||q| \leq k - \delta k]\} = \left(\int \prod_{|p| \in I_k} d\phi_p \right) \exp\{-S_\Lambda[\phi_q||q| \leq k]\}. \ (2)$$

This leads to the Wegner-Houghton equation [5,2,7] for the action S_k at the scale k:

$$k\partial_k S_k = \frac{Vk^d}{2(2\pi)^d} \int d\omega \left[F_p K^{-1}_{p,-p} F_{-p} - \ln \frac{K_{p,-p}}{const.} \right]_{|p|=k}, \qquad (3)$$

where $F_p = \partial S_k/\partial\phi_p|_{\phi_p=0}$, $K_{p,-p} = \partial^2 S_k/\partial\phi_p\partial\phi_{-p}|_{\phi_p=0}$, the volume V and the integral over the solid angle $\int d\omega$ in the d-dimensional momentum space.

In order to derive partial differential equations for functions instead of the functional S_k, we make the following Ansatz: $S_k = \mathcal{G} + \mathcal{U}$ with the local potential

$$\mathcal{U} = \sum_{r=2}^{\infty} u_r(k)V^{-r} \sum_{q_1,\ldots,q_r}^{\leq k} \phi_{q_1}\cdots\phi_{q_r} V\delta_{q_1+\ldots+q_r}, \qquad (4)$$

and with the derivative part

$$\mathcal{G} = \sum_{n=0}^{\infty}\sum_{r=0}^{\infty} g_{nr}(k)V^{-(r+2)} \sum_{q_1,\ldots,q_{r+2}} q_1^{2n}\phi_{q_1}\cdots\phi_{q_{r+2}} V\delta_{q_1+\ldots+q_{r+2}}, \qquad (5)$$

$u_r(k)$ and $g_{nr}(k)$ the running coupling constants at the scale k. The term with $n = 0$ describes a local potential, the term with $n = 1$ represents the usual kinetic term with field dependent wave-function renormalization and the terms with $n > 1$ contain higher order derivatives of the field.

Inserting the Ansatz above in the Wegner-Houghton equation it can be reduced by projection technique to a system of coupled partial differential equations for the generating functions, defined as

$$V(z,k) = \sum_{r=2}^{\infty} u_r(k)z^r, \qquad G(Q^2, z, k) = 2\sum_{n=0}^{\infty}\sum_{r=0}^{\infty} g_{nr}(k)Q^{2n}z^{r+1}. \qquad (6)$$

The projector (acting on the arbitrary functional F)

$$\mathcal{P}F = \left(\exp\left\{ z\frac{\partial}{\partial\phi_0} \right\} F \right)_{\phi=0} \qquad (7)$$

is used [2]. We apply it on both sides of Eq. (3) and on those of the equation obtained from Eq. (3) by taking the second derivatives $\partial^2/(\partial\phi_Q\partial\phi_{-Q})$ of its sides at first, $|Q| \neq 0$ and k. Rewriting the resulting equations in terms of dimensionless quantities (with except of the scale k) we find the following coupled set of partial differential equations for the generating functions:

$$\left(k\partial_k - \frac{d-2}{2} z\partial_z + d \right) V(z,k)$$

$$= -\alpha \ln \frac{\partial_z G(1,z,k) + \frac{1}{2} z^2 \partial_z^3 G(0,z,k) + \partial_z^2 V(z,k)}{[\partial_z G(1,z,k) + \frac{1}{2} z^2 \partial_z^3 G(0,z,k) + \partial_z^2 V(z,k)]_{z_c}}, \tag{8}$$

$$\left(k\partial_k - 2Q^2 \partial_{Q^2} - \frac{d-2}{2} z\partial_z + 2 \right) \partial_z G(Q^2,z,k)$$

$$+ \left(k\partial_k - \frac{d-2}{2} z\partial_z + d \right) \frac{1}{2} z^2 \partial_z^3 G(0,z,k)$$

$$+ \left(k\partial_k - \frac{d-2}{2} z\partial_z + 2 \right) \partial_z^2 V(z,k)$$

$$= -\alpha \frac{\partial_z^3 G(Q^2,z,k) + \partial_z^3 G(1,z,k) + \frac{1}{2} z^2 \partial_z^5 G(0,z,k) + \partial_z^4 V(z,k)}{\partial_z G(1,z,k) + \frac{1}{2} z^2 \partial_z^3 G(0,z,k) + \partial_z^2 V(z,k)} \tag{9}$$

with $\alpha = [2(2\pi)^d]^{-1} \int d\omega$. It is worthwhile mentioning that the tree level term on the r.h.s. of these equations does not appear due to $\mathcal{P} f_p = 0$, and $\mathcal{P} \partial f_p / \partial \phi_Q = 0$. The choice of the constant z_c is connected with the choice of the field independent additive constant of the potential.

Although Eqs. (8) and (9) have beeen derived for generating functions analytic in variables Q^2 and z, we will consider them valid for generating functions non-analytic in variable z as well. Having obtained the generating functions, the reconstruction of the action at scale k can be performed as follows:

$$\mathcal{U}_k = \int d^d x V(\phi(x); k),$$

$$\mathcal{G}_k = \frac{1}{2} \sum_s \int d^d x \phi(x)(-\partial_x^2)^s \phi(x) G_s(\phi(x); k) \tag{10}$$

for $G(Q^2, z; k) = \sum_s (Q^2)^s z G_s(z; k)$.

Furthermore we shall only consider the case with dimensions $d > 2$.

3 Fixed Point Solutions

The equations for the fixed points are obtained by setting zero the derivatives of the generating functions with respect of the scale k in Eqs. (8), (9). We shall seek the fixed point solutions $V^*(z)$, $G^*(Q^2, z)$ by making the Ansatz that the field dependent wave function renormalization can be separated in the derivative part, i.e. $\partial_z G^*(Q^2, z) = H^*(Q^2)h^*(z)$.

3.1 Gaussian Fixed Point

Assuming $V^* = const.$ and $h^* \equiv 1$, one easily finds the Gaussian fixed point: $H^*(Q^2) = H_0^* Q^2$, i.e. $G^*(Q^2, z) = H_0^* Q^2 z$ and the corresponding fixed point action $S^* = -\frac{1}{2} H_0^* \int d^d x \phi(x) \partial_x^2 \phi(x)$. The choice $H_0^* = 1$ can be made without loss of generality (rescaling of the field).

3.2 Non-trivial Fixed Points

We recognize that $V^* = \frac{1}{2} C_V \ln z^2 + V_0^*$, $h^*(z) = z^{-2}$ is the solution of the fixed point equation obtained from Eq. (8) with $C_V = 2\alpha/d$ and $V_0^* = (\alpha/d)\left[(d-2)/d - \ln z_c^2\right]$. Here we see that the choice of z_c is connected with the choice of the additive constant of the potential.

The logarithm on the r.h.s. of Eq. (8) is only well-defined for $b \equiv H^*(1) + 3H^*(0) - C_V \neq 0$. Then we obtain an ordinary differential equation for the function $H^*(Q^2)$. Let us write $H^*(Q^2) = H^*(0) + F(Q^2)$ with $F(0) = 0$ and $H^*(0)$ finite. Then we separate the Q^2 dependent and the constant terms of Eq. (9) and find:

$$Q^2 \frac{dF(Q^2)}{dQ^2} = \kappa F(Q^2) \tag{11}$$

with $2\kappa = d + (6\alpha/b)$. It has the solution:

$$F(Q^2) = F(1) \, (Q^2)^\kappa. \tag{12}$$

From the equation for the constant pieces and the definitions of b and κ we find:

$$H^*(0) = \frac{4C_V d}{14\kappa - 6}, \qquad F(1) = C_V \left(1 + \frac{3d}{2\kappa - d} - \frac{16d}{14\kappa - 6}\right). \tag{13}$$

The non-trivial fixed point solution is then given by

$$G^*(Q^2, z) = -\left[H^*(0) + F(1)(Q^2)^\kappa\right] z^{-1},$$

$$V^*(z) = \frac{1}{2} C_V \ln z^2 + V_0^*. \tag{14}$$

The first term of G^* leads to an additive constant of the potential that can be neglected.

At the first sight Eq. (11) might seem linear. However the constant κ is defined via $F(1)$ and linear combinations of solutions with various values of κ are not solutions.

The powers $\kappa = 3/7$, $d/2$ must be excluded due to $F(1) \rightarrow \infty$. Analyticity and $F(0) = 0$ are only satisfied for $0 < \kappa$ integer. Therefore there are infinitely many non-trivial fixed points characterized by the positive integer $\kappa \neq d/2$.

4 Linearized RG Transformation

As to the next we investigate the linearized RG transformations in the neighbourhood of the fixed points in order to determine the scaling operators. Theories with the action being the sum of the fixed point action and some irrelevant operators belong to the same universality class associated with the given fixed point. Let us insert $V(z, k) = V^*(z) + \delta V(z, k)$, and $G(Q^2, z, k) = G^*(Q^2, z) + \delta G(Q^2, z, k)$ in Eqs. (8) and (9) and linearize them with respect to δV and δG in the neighbourhood of a given fixed point. Let us seek the eigenfunctions of the operator $k\partial_k$ in the separated form:

$$\delta V(z, k) = (k/\Lambda)^{-\lambda} \varphi(z), \qquad \partial_z \delta G(Q^2, z, k) = (k/\Lambda)^{-\lambda} \phi(Q^2)\psi(z) \qquad (15)$$

with $\phi(0) = 0$. The positive, vanishing and negative eigenvalues λ correspond to relevant, marginal, and irrelevant directions, respectively, at the given fixed point in the parameter space.

4.1 Eigensolutions at the Gaussian Fixed Point

The equation for $\phi(Q^2)$ decouples and has the complete set of analytic solutions $\phi(Q^2) = \phi_0 Q^{2n}$ with $n = 1, 2, \ldots$. In the case of field independent wave function renormalization $\psi(z) \equiv 1$, δG and δV are completely decoupled. and $\delta G \sim \phi(Q^2)$ belongs to the eigenvalue $\lambda_n = 2 - 2n$. Therefore the term $\sim Q^2$ is marginal, the higher order derivative terms are irrelevant, as expected. All the known relevant and irrelevant eigenpotentials are recovered.

4.2 Eigensolutions at the Non-trivial Fixed Points

The real eigensolutions of the linearized RG equations satisfying $\phi(0) = 0$ can be characterized by the power $0 < \rho$, $\delta G \sim (Q^2)^\rho$. There are two different eigenvalues belonging to the same value ρ:

$$\lambda_\mp = d - s_\mp(d - 2), \qquad s_\mp = \frac{1}{4}\left(5 \mp \sqrt{1 - \frac{\rho}{\rho_c}}\right) \qquad (16)$$

with $\rho_c = (d-2\kappa)/48$. For $\kappa < d/2$ only the values $0 < \rho < \rho_c$ are allowed. For $\kappa > d/2$ all positive values of ρ lead to real eigenvalues. Due to the requirements δG analytic in variable Q^2 and $\delta G(0, z, k) \equiv 0$ it must be $\rho = L > 0$ integer.

We are interested in the irrelevant directions at the non-trivial fixed points. Those are the directions along which the RG trajectories run into them. For $\kappa < d/2$ and $d < 10$ the only eigenvalues are λ_- that are negative for $L/\rho_c < 1 - (10-d)^2/(d-2)^2$. Positive powers L satisfying this inequality do only exist for $d > 6$. Therefore in dimension $d = 4$ the fixed points with $\kappa < 2$ cannot be IR fixed points.

It is easy to show that for $\kappa > d/2$ and $d > 10$ the eigenvalue $\lambda_- \leq 0$ for any $0 < L$ and these are the only non-positive eigenvalues. For $\kappa > d/2$ and $6 < d < 10$ there are no irrelevant eigenvalues. For $\kappa > d/2$ and $2 < d \leq 6$ the only irrelevant (marginal) eigenvalues are $\lambda_- \leq 0$ for $L \geq \frac{1}{3}(d-6)(d-2\kappa)(d-2)^{-2}$.

Therefore the fixed points with $\kappa > 2$ for dimension $d = 4$ possess irrelevant scaling operators and can be considered as candidates for IR fixed points, where the RG trajectories starting at the Gaussian fixed point along a relevant direction could go in. The corresponding scaling operators are given as:

$$
\begin{aligned}
\delta V^{(L)}(z, k) &= c_L \, (k/k_0)^{-\lambda_-} \, (z^2)^{s_-}, \\
\delta G^{(L)}(Q^2, z, k) &= -c_L \, (k/k_0)^{-\lambda_-} \, 2s_- \left(Q^2\right)^L z^{2s_- - 1}
\end{aligned}
\tag{17}
$$

with the coupling constants c_L.

Summarizing, infinitely many non-trivial fixed points characterized by the positive integer $\kappa \neq d/2$ are found. These fixed points cannot be interpreted as UV fixed points in any cases due to the presence of higher order derivative terms in the fixed point action. Only those with $\kappa < d/2$ for $6 < d < 10$ and $\kappa > d/2$ for $2 < d \leq 6$ or $d > 10$ can be given the physical meaning of IR fixed points.

The following general features of the non-trivial fixed points are worthwhile mentioning. (i) All of these non-trivial fixed points possess infinitely many relevant scaling operators (corresponding to the eigenvalues λ_+). Therefore it seems to be rather unlikely that RG trajectories moving off the Gaussian fixed point can reach them. (ii) For dimension $d = 4$ the only IR fixed points are those with $\kappa > 2$. It is worthwhile to notice that there is a marginal eigenvalue $\lambda_-(L) = 0$ at the fixed point $\kappa = 3L + 2$ for any positive integer L and the corresponding marginal potential is just ϕ^4 due to $2s_-(\kappa = 3L + 2, L) = 4$. For $L = 1$ the derivative part of the marginal operator is given as $\phi\partial_x^2\phi \cdot \phi^2$, i.e. by the usual kinetic term with field dependent (analytic) wave function renormalization. (iii) For $\kappa > d/2$ we obtain $F(1) > 0$. Then the sign of the derivative term in the IR effective theory is just the opposite of that in the original theory defined at the UV cut-off scale. Therefore these fixed point actions could be analogues of anti-ferromagnetic theories. The possibility of the existence of such fixed points has already been argued in [9]. The effective action is minimized by the periodic field configuration $\phi_P(x) = f\cos(Px)$ rather than by a constant one:

$$S_{eff}(P^2, f) \equiv S^*[\phi_P] = -\frac{1}{2}VF(1)(P^2)^\kappa + \frac{1}{2}C_V V \ln f^2 + P^{-d} \cdot \text{const.} \quad (18)$$

Due to the presence of the cut-off $|P| \leq 1$ (P is dimensionless) S_{eff} has an absolute minimum for $|P| = 1$, but the amplitude of the condensate f is vanishing. Therefore no anti-ferromagnetic vacuum is formed. (iv) The logarithmic effective potential and all its derivatives are infinite for the stable field configuration $\phi(x) \equiv 0$ and no particle excitations do exist in the field ϕ. (v) The higher order derivative terms lead to non-localized interactions. Causality is, however, not violated due to the analytic dependence of the Lagrangian on the gradient operator [10]. Unitarity depends on whether the real energy eigenvalue states are all of positive norm [11]. It is, however, not a necessary requirement for an effective theory to be unitary.

Acknowledgments

The author is indebted to J. Alexandre, W. Greiner, I. Lovas, J. Polónyi, J. Rau, and A. Schäfer for the useful discussions. This work has been supported by the Hungarian Research Fund (OTKA, T 017311) and the EC Project ERB-CIPA-CT92-4023 (Prop. 3293).

References

1. K.G. Wilson and J. Kogut, *Phys. Rep.* **12**, 75 (1974).
2. A. Hasenfratz and P. Hasenfratz, *Nucl. Phys.* B **270**, 687 (1986).
3. P. Hasenfratz and J. Nager, *Z. Phys.* C **37**, 477 (1988).
4. K. Halpern and K. Huang, hep-th/9510240;
   Vipul Periwal, hep-th/9512108.
5. F.J. Wegner and A. Houghton, *Phys. Rev.* A **8**, 401 (1973).
6. J. Polónyi, hep-ph/9511243.
7. Sen-Ben Liao and J. Polónyi, *Phys. Rev.* D **51**, 4474 (1995).
8. Sen-Ben Liao and J. Polónyi, *Ann. Phys.* **222**, 122 (1993).
9. G. Gallavotti and V. Rivasseau, *Phys. Lett.* B **122**, 268 (1983).
10. A. Pais and G.E. Uhlenbeck, *Phys. Rev.* **79**, 145 (1950).
11. T.D. Lee and G.C. Wick, *Nucl. Phys.* B **9**, 209 (1969).

Boson Mapping in Field Theory

P.O.Hess[a], J.C.López[b], C.R.Stephens[c]

Instituto de Ciencias Nucleares, UNAM, Circuito Exterior,C.U., A.P. 70-543, Del. Coyoacán, 04510 México, D.F., Mexico

A boson mapping of pair field operators is presented. The mapping preserves all hermiticity properties and the Poisson bracket relations between fields and momenta. The most practical application of the boson mapping is to field theories which exhibit bound states of pairs of fields. As a concrete application we consider, in the low energy limit, the Wick-Cutkosky model with equal mass for the charged fields.

I. Introduction

Boson mapping techniques play an important role in many areas of physics whenever the low lying states in energy are dominated by pair correlations of the fundamental particles. A typical example is the seniority model[1] where the dominant contribution to the ground state of a nuclear system is given by pairs of nucleons of total spin zero. In condensed matter physics a famous example is superconductivity where pairs of electrons are coupled via an electron-phonon interaction to form Cooper pairs.

A pair of creation operators of two fundamental particles can be used to represent the creation of a composite bosonic field if one can find a conjugate annihilation operator which annihilates the field. A pair of annihilation operators of the fundamental particles is not sufficient. However, the boson mapping[1,2] allows one to find just such an operator. ap

It is natural to ask if such a mapping can be applied directly at the level of field operators rather than annihilation and creation operators, the motivation being to map a Hamiltonian (Lagrangian) density to an effective one which depends only on pair operators. For example, if the Hamiltonian density of the gauge field of QCD can be mapped to one which depends only on composite operators representing gluon pairs of total spin and colour zero, then one would end up with a new effective Hamiltonian density which depends only on a scalar field. The simplification in structure is obvious. So, if at low energies the ground state is dominated by pair correlations it is very probable that a boson mapping not only simplifies the structure but also takes into account the physics. This is the case for QCD where perturbatively a pair of gluons of

---
[a] HESS@ROXANNE.NUCLECU.UNAM.MX
[b] VIEYRA@ROXANNE.NUCLECU.UNAM.MX
[c] STEPHENS@ROXANNE.NUCLECU.UNAM.MX

spin and colour zero is the lowest energy state. Additionally, in the quark-anti-quark sector we have ample experimental evidence that the low energy physics is dominated by quark-antiquark pairs of spin and colour zero (e.g., the pion) and by di-quarks. However, up to now no explicit mapping at the level of field operators, depending on the coordinate $x = (\vec{x}, t_o)$, where t_o refers to a fixed time, has been given. All applications have been restricted to creation and annihilation operators[1,2]. Recently, the boson mapping has been extended to pairs of coordinate operators $Q_{ai} = \frac{1}{\sqrt{2}}(b_{ai}^\dagger + b_{ai})$, where the indices i and a refer to a set quantum numbers, and their derivatives $\frac{\partial}{\partial Q_{ai}}$[3]. The main reason to consider such a mapping is that the Hamiltonian densities of field theories have a much simpler structure in terms of coordinate operators Q_{ai} which are the expansion coefficients of the fields $\Phi_a(x)$ in terms of an orthonormal set of functions $f_i(x)$, i.e., $\Phi_a(x) = \sum_i \frac{1}{\sqrt{\omega_i}} f_i(x) Q_{ai}$.

In this contribution we will show how to implement a boson mapping at the level of classical pair fields, e.g., $\left\{ \frac{1}{\sqrt{N}} \sum_a \Phi_a(x)\Phi^a(x) \right\} = q(x)$ with a denoting some quantum numbers, and discuss some problems related to it. Although the mapping is presented at the classical level it is relatively straightforward to check that it goes thr ough in the quantum case where Poisson brackets are replaced by commutators. As a specific example, we will investigate the Wick-Cutkosky model[4] (WCM) which is known to have bound states. In the WCM we integrate out the scalar field which mediates the interaction between two charged fields. For equal masses of the latter and in the limit where the mass of the scalar field is much greater than any relevant energy involved, we will obtain a model which, after boson mapping, will be equivalent to a scalar field theory with only one scalar neutral field. The classical equation of motion is that of an anharmonic oscillator with an attractive fourth order term in the potential. Consequently, the classical system is unstable as $q \to \infty$ when all other classical degrees of freedoms are neglected. It is important to emphasize that in the full quantum theory this unphysical feature is expected to disappear.

We will not discuss here the construction of the ground state and excited states of the WCM. This will be done in another publication where we will compare results obtained from the boson mapping with those derived from a more fundamental renormalization group analysis of the model and investigate the limit of applicability of the boson mapping. We will also restrict our attention here to boson fields as fundamental fields, the extension to fermion fields being straightforward.

II. A Boson Mapping for Fields:

Let us denote a general bosonic field by $\Phi_{ia}(x)$ with $i = 1, ..., M$ and $a = 1, ..., N$. The conjugate momenta to these fields are denoted by $\Pi^{ia}(x)$ and satisfy with $\Phi_{ia}(x)$ the Poisson bracket relation $\{\Phi_{jb}(\vec{y}, t_o), \Pi^{ia}(\vec{x}, t_o)\} = \delta^{ia}_{jb}\delta(\vec{x} - \vec{y})$.[d] We use co- and contravariant indices in order to allow non-cartesian components. Associated to these fields and their conjugate momenta we introduce pair fields $(q_{ij}(x))$ and their conjugate momenta $(p^{ij}(x))$ satisfying the Poisson bracket relation $\{q_{nm}(\vec{y}, t_o), p^{ij}(\vec{x}, t_o)\} = (\delta^{ij}_{nm} + \delta^{ji}_{nm})\delta(\vec{x} - \vec{y})$, where instead of the usual definition of the Poisson bracket we multiplied for convenience the right hand side by an extra factor of one half. (This comes first by noting that $\frac{\delta q_{nm}(x)}{\delta q_{ij}(y)} = (\delta^{ij}_{nm} + \delta^{ji}_{nm})\delta(x - y)$, which reflects the non-normalized nature of $q_{ij}(x)$, and that we require the above mentioned Poisson bracket relation between $q_{nm}(y)$ and $p^{ij}(x)$ be satisfied in analogy to the usual boson pair relation[2].) The $q_{ij}(x)$ is called a pair field because it will be related to a pair of the fundamental fields $\Phi_{ia}(x)$. It is still not normalized, however, as can be seen from the Poisson bracket when $(nm) = (ij)$. This is also for convenience and can later on be corrected by an appropriate normalization of the fields. Furthermore, the pair fields and their conjugate momenta are symmetric with repect to interchange of their discrete indices $(q_{ij}(x) = q_{ji}(x)$, etc.).

We now propose a mapping from fundamental to composite fields such that the Poisson bracket relations and hermiticity properties are conserved. Both conditions are important: the former to ensure that the dynamics is preserved (for at least a physically relevant subspace of the space of states) and the latter to ensure the invariance of matrix elements under the mapping. The explicit mapping is

$$\left\{\frac{1}{\sqrt{N}}\sum_a \Phi_{ia}(x)\Phi^a_j(x)\right\} = q_{ij}(x)$$

$$\left\{\frac{1}{\sqrt{N}}\sum_a \Pi^i_a(x)\Pi^{ja}(x)\right\} = \frac{1}{N}\sum_{k_1 k_2} p^{ik_1}(x)q_{k_1 k_2}(x)p^{k_2 j}(x) \qquad (1)$$

$$\left\{\sum_a \Phi_{ia}(x)\Pi^{ja}(x)\right\} = \sum_k q_{ik}(x)p^{kj}(x)$$

where the curly bracket on the left hand side denotes the mapping, i.e., the left hand side gives the operator in the original space of states while the mapped

---

[d]The form of the Poisson bracket is given in the chapter about continous fields of the book written by H.Goldstein[5].

expression gives the operator in the new space of states, defined by the pair fields. In eq.2 both sides satisfy exactly the same Poisson bracket relations as can be seen by direct calculation. The hermiticity properties are obviously satisfied, i.e., the first two expressions on the left are hermitian and so they are on the right. The last is anti-hermitian and so is the mapped expression on the right. The $\frac{1}{\sqrt{N}}$ on the left hand side is the normalization of the pair and has its origin in the coupling coefficient which couples the field to a scalar with respect to the property characterized by the index $a = 1, 2, ..., N$. For example, in the case of QCD this index refers to colour and the above objects have colour zero. The mapping of eq.2 can easily be generalized to a quantized theory, the only change being that the last operator has to be written symmetrically.

Note that a kinetic energy, quadratic in the momenta of the original fields, is mapped to a kinetic energy quadratic in the pair fields but which now depends also on the field itself and not only on the conjugate momenta. This corresponds to a "mass" parameter which depends on the strength of the field. This situation appears quite often in physics. As an example see ref.[6] which describes the moment of inertia of a nucleus as a function of deformation. It can also be given the interpretation of working in a non-cartesian coordinate system in the configuration space of the fields as will be seen in a less general setting shortly. The appearance of the composite field explicitly in its own kinetic term indicates the high degree of non-linearity of the mapping , something that makes the explicit construction of the inverse mapping quite diffi cult.

We have still to consider the mapping of expressions containing derivatives such as $\sum_a (\nabla \Phi_{ia}) \cdot (\nabla \Phi_j^a)$, which appear in the original Lagrangian density. The problem here is that the field derivative is a limit of differences of its value at two neighbouring points. Thus, the above product contains products of two fields at neighbouring points. The mapping in eq.2 is only constructed for products of fields at the same point (a generalization of it with different vectors x and y can also be given but it leads to non-local field theories, which we try to avoid for the moment). For the case $M = 1$, i.e., $i = j = 1$ the result is

$$\left\{ \frac{1}{\sqrt{N}} \sum_a (\nabla \Phi_a(x)) \cdot (\nabla \Phi^a(x)) \right\} = \frac{(\nabla q(x)) \cdot (\nabla q(x))}{4q(x)} \qquad . \tag{2}$$

(For the general case it is more involved.)

We can also see this mapping of the derivative term, and indeed the entire boson mapping itself, by considering a change of variables in the measure of the functional integral. For instance, in the case of a complex scalar field $\Phi(x)$

we change variable $\Phi(x) = \sqrt{q(x)}exp(i\theta(x))$. The volume element $d\Phi d\Phi^*$ changes to $dqd\theta$, apart from a constant. The expression $(\partial_\mu \Phi)(\partial^\mu \Phi^*)$ becomes $\frac{(\partial_\mu q)(\partial^\mu q)}{4q} + q(\partial_\mu \theta)(\partial^\mu \theta)$. For the spatial part $(\nabla \Phi) \cdot (\nabla \Phi^*)$ the first term corresponds to the mapping, when restricted to pairs $q(x)$ only. For the part $|\partial_o \Phi|^2$ we have to take the expression for the classical conjugate momentum $p(x)$ (given in the next section) whereupon we will arrive at the mapping proposed in eq.2 for the square of the momenta $\Pi(x)$. Note, that in this example we have an extra contribution $q(\partial_\mu \theta)(\partial^\mu \theta)$. Generically this type of term becomes important when the contribution of the fields not coupled to pairs start to dominate. For that case the mapping in eq.2 has to be generalized along the same line as done in the examples of nuclear physics[2] when broken pairs are introduced. The mapping in eq.2 is, therefore, restricted (projected) to states of the space of states which are dominated by the pairs $q(x)$. The simple change of variables, used in the above example, is generalized in eq.2.

In the future we intend to extend the mapping in eq.2 such that pairs of fields at different space-time points are considered. As mentioned above, this leads to non-local field theories, however, by using operator product expansion methods this yields local field theories. The work is still in progress.

III. The Wick-Cutkosky model (WCM):

In the WCM[1] the Langragian, with equal masses of the charged fields, is given by

$$
\begin{aligned}
\mathcal{L}(x) = {} & \frac{1}{2}\sum_{b=1}^{2}(\partial_\mu \Phi_b^\dagger(x))(\partial^\mu \Phi^b(x)) + \frac{1}{2}((\partial_\mu \phi(x))(\partial^\mu \phi(x)) \\
& - \frac{M^2}{2}\sum_{b=1}^{2}\Phi_b^\dagger(x)\Phi^b(x) - \frac{\mu^2}{2}\phi^2(x) \\
& - g(\sum_{b=1}^{2}\Phi_b^\dagger(x)\Phi^b(x))\phi(x) \quad .
\end{aligned}
\tag{3}
$$

In eq.3 the charged field is described by a complex field which in terms of real and imaginary parts implies that the sum over a involves four real fields. In this case it manifestly has the structure of eq.2[e].

---

[e] In fact the mapping in eq.2 is the same for complex fields in which one of the fields has to be taken as the complex conjugate and the value N is twice the number of complex fields.

Using functional integration with Euclidian measure we can integrate over the real field $\phi(x)$ using gaussian integration. We obtain finally for an effective Lagrangian density

$$\mathcal{L}_{eff}(x) = \frac{1}{2}\sum_{b=1}^{2}\partial_o\Phi_b^\dagger(x)\partial_o\Phi^b(x) - \frac{1}{2}\sum_{b=1}^{2}(\nabla\Phi_b^\dagger(x))\cdot(\nabla\Phi^b(x)) +$$
$$-\frac{M^2}{2}\sum_{b=1}^{2}\Phi_b^\dagger(x)\Phi^b(x)$$
$$+\frac{g^2}{2}(\sum_{b=1}^{2}\Phi_b^\dagger(x)\Phi^b(x))\int d^4y\Delta_F(x-y)(\sum_{c=1}^{2}\Phi_c^\dagger(y)\Phi^c(y)) \quad (4)$$

where $\Delta_F(x-y)$ is the Feynman propagator for scalar fields.

In the limit of μ^2 much greater than the momenta involved which corresponds for a given ultraviolet cutoff Λ to $(\frac{\mu}{\Lambda})^2 >> 1$, i.e. that the momenta involved are within the low energy limit, the interaction part of the effective Lagrangian density in eq.4 acquires the form

$$\frac{g^2}{2\mu^2}(\sum_{b=1}^{2}\Phi_b^\dagger(x)\Phi^b(x))^2 \quad . \qquad (5)$$

We now express the complex fields $\Phi_b(x)$ in terms of their real and imaginary part, i.e., $\Phi_b(x) = \Phi_{1,b}(x) + \Phi_{2,b}(x)$ which yields four different fields $\Phi_a(x)$ with a linearized index $a = 1, ..., 4$ ($a = 1$: $(1,b)$; $a = 2$: $(2,b)$; etc.). Then we change to the Hamiltonian formulation because the mapping involves the conjugate momenta. This yields for the kinetic energy the form $\frac{1}{2}\sum_{a=1}^{4}\Pi_a(x)\Pi^a(x)$ and the potential part in eq.4 (including the term which involves the square of the gradient of the fields) changes its sign. With the mapping of eq.2 (with $i,j = 1$) and the normalization of the pair fields $q(x)$ by multiplying them and their conjugate momenta by $\frac{1}{\sqrt{2}}$ (denoting them afterwards with the same letters) this Hamiltonian density is mapped to

$$\mathcal{H}_{bos}(x) = \frac{1}{\sqrt{2}}p(x)q(x)p(x) + \sqrt{2}M^2q(x) - \frac{4g^2}{\mu^2}q^2(x)$$
$$+\frac{\sqrt{2}}{4}\frac{(\nabla q(x))\cdot(\nabla q(x))}{q(x)} \quad . \qquad (6)$$

The kinetic energy of the Hamiltonian density contains a dependence on the field $q(x)$ and the potential has a term $\sim q(x)$ which can be interpreted as a magnetic type of interaction. ¿From eq.6 the momentum as a function of $q(x)$ can be deduced:

$$p(x) = \frac{1}{\sqrt{2}q(x)} \frac{\partial q(x)}{\partial t} \tag{7}$$

The equation of motion is derived from the new Hamiltonian density and is of the form

$$\Box q(x) - \frac{1}{2q(x)}(\nabla^\mu q(x)) \cdot (\nabla^\mu q(x))$$

$$+2M^2 q(x) - \frac{8\sqrt{2}g^2}{\mu^2}q^2(x) = 0 \quad . \tag{8}$$

The second term is a connection term associated with, as mentioned previously, the fact that we are using non-cartesian coordinates. Making the substitution $q(x) = \chi^2(x)$ we arrive at the equation

$$\Box \chi(x) + M^2 \chi(x) - \frac{4\sqrt{2}g^2}{\mu^2}\chi^3(x) = 0 \quad . \tag{9}$$

This is the classical equation of motion of an anharmonic oscillator where the anharmonic term in the potential is attractive. This term destabilizes the system for large coupling constant g and produces a transition to a state with infinite expectation value $<q>$. It, therefore, indicates that the system is unstable under the formation of a q-condensate. Higher, repulsive terms, not included in the Hamiltonian density, should make it stable.

The approximation we have made in the above is restricting the space of states to functionals which depend on the pair fields only. It remains to be seen that at low energy this assumption is justified, i.e., that the lowest lying states are dominated by bound states which are comprised of a coupling of two fields. This is under investigation[7].

IV. Conclusion

In this contribution we proposed a mapping of fundamental pair fields and their conjugate momenta to new composite fields and their corresponding conjugate momenta. We restricted attention to the case of classical fields, with the proviso that the mapping can be directly extended to a quantized picture without undue difficulty. The mapping will be particularly useful whenever pair correlations dominate the low energy structure, as is the case for QCD. Only boson fields as fundamental fields were considered, the extension to fermionic fields being direct. (There is some relation of the boson mapping, as presented here on the operator level, to the formulation with Feynman path integrals[8]. We thank F.J.W.Hahne for pointing this out during the conference.)

The mapping was applied to the Wick-Cutkosky model, where in the limit of low energy and equal masses for the charged fields we obtained a field theory equivalent to that of a neutral self-interacting scalar field. The classical equation of motion was that of an anharmonic oscillator with an attractive anharmonic term of fourth order in the potential. As a result, the system is unstable to small fluctuations.

What still has to be shown in a fundamental (basic) calculation is that at low energies the Wick-Cutkosky model exhibits the pair structure used. Also under investigation is how far the boson model gives correct results compared to the exact description. This is important in order to probe the range of application of the boson model and deduce from there possible applications to QCD.

1. P.Ring, P.Schuck, *The Nuclear Many Body Problem*, (Springer-Verlag, New York/Heidelberg, 1980)
2. A.Klein, E.R.Marshalek, Rev. Mod. Phys. **63**, 375 (1991)
3. P.O.Hess, J.C.López, Jour. Math. Phys. **36**, 1123 (1995)
4. N.Nakanishi, Prog. Theor. Phys. **Suppl. No. 43**, 1 (1969)
5. H.Goldstein, *Classical Mechanics*, (Addison-Wesley, 1969)
6. J.M.Eisenberg, W.Greiner, *Nuclear Theory I: Nuclear Models*, 3rd edition, (North-Holland Physics Publishing, Holland, 1987)
7. P.O.Hess, J.C.López, C.R.Stephens, in preparation
8. F.G.Scholtz, A.N.Theron, H.B.Geyer, Phys. Lett.**B345**, 242 (1995)

NONLINEAR SYMMETRIES FOR A COLLECTIVE MODEL IN CLUSTER PHYSICS

A. LUDU[a], A. SANDULESCU[b]

Institut für Theoretische Physik der J. W. Goethe - Universität,
Robert-Mayer-Strasse 8-10, Frankfurt am Main,
D-60054 Germany

We consider a collective model for the cluster preformation. The discrete non-Lie symmetries of this model are investigated in the framework of q-groups and "bubble"-states, degenerated in energy with the spherical symmetric initial state, are obtained through a spontaneous symmetry breaking.

1 Introduction

Generally the continuous symmetries are described in terms of Lie groups and algebras, and their unitary irreducible representations act on the Hilbert space of the states. The special advantage of continuous symmetries over the discrete ones appears clearly in the Lagrangean field theories, where due to the Noether theorem, one can define currents and conserved quantities associated with the time evolution of the dynamical system. This is possible essentially due to the existence of infinitesimal transformations of the Lie group (the basis vectors of the Lie algebra), which, in the case of discrete symmetries, are absent.

The main point in this contribution is to introduce a continuous mapping of some discrete symmetries (inversions) towards unity, in the frame of the q-deformation theory of Lie algebras (for a recent review see Chang[1]). We investigate the discrete symmetries of a spontaneous breaking of symmetry for a nonrelativistic field Lagrangian with soliton and instanton solutions.

The inversion transformations are related with the theory of non-linear deformations, braids and knots[2]. The problem of smoothing of the inversion transformation is similar with the problem of mirroring a knot. The right-handed and the left-handed systems of coordinates in the plane (x, y) are equivalent with two knots which keep the same topological informations as the original image, modulo non-singular moves in the plane. Following this construction, the right-handed coordinate frame is equivalent with a knot K and the left-handed coordinate frame is equivalent with its mirror image. So,

---

[a] Permanent Adress: Bucharest University, Department of Theoretical Physics, Bucharest-Măgurele P.O. Box MG-5211, Romania, e-mail: ludu@th.physik.uni-frankfurt.de

[b] Permanent address: Romanian Academy, Calea Victoriei 125, Bucharest, Romania, e-mail: sandulescu@roifa.ifa.ro

the discrete transformation of mirroring the y axis of the coordinate frame, is realised as a mirror transformation of a knot. There are examples of knots which can be continuously deformed into their mirror image (e.g. the Eight knot), so-called achiral knots, and examples of knots which are not allowed to do this (e.g. the Trifoil knot), so-called chiral knots. This can be done by using the Reidemeister moves of ambiental isotopy. Consequently, an algebraic treatment of the Reidemeister moves can lead our research to the realisation of the continuous embedding of discrete mirrorings of the space. Since the quantum group $gl_q(2)$ provides representations of such moves, we construct special q-deformations in relation with knots and braids theory.

2 q-Discrete symmetries and the q-plane

In the following we introduce a set of continuous transformations of the coordinates, depending on one complex parameter q, which have no Lie algebraic equivalent and which, for certain fixed values of q, fall into some usual discrete transformations (space inversions, etc.). These general structures are nonlocal, non-Lie and non-linear transformations, i.e. they do not form a Lie group (they are not of the form $e^{\epsilon v}$ with v a vector field of a Lie algebra not depending on ϵ). Such transformations are generaly named nongeometric or "hidden". We define the infinitesimal generator for dilation in the x_i direction as $d_i = x_i \partial_i$ and we use the special q-deformation with $q = e^{is}, s \in [0, 2\pi]$. A q-deformed operator integer function $Q(d_i, s) = \sum_k C_k(s) d_i^k$ has the action on real integer functions $f(x_i) = \sum_j f_j x_i^j$, $Q(d_i, s) f(x_i) = \sum_j f_j Q(j, s) x_i^j \to f(x_i)$, for $Q(d_i, 1) = 1$. This operator should be linear and should have the following limits: $Q(d_i, 1) = 1, Q(d_i, -1) = e^{i\pi d_i} = I_i$. The Hamiltonian is invariant to the symmetry introduced by the operator Q if $[H, Q] = 0$. The simplest example is given by a dilation in the x_i direction $Q = e^{isd_i}$. We choose a general one-dimensional Hamiltonian in the form $(x_i = x)$

$$H = -\partial_x^2 + V(x) + W(x\partial_x),$$ (1)

and the corresponding Schrödinger equation $Hf(x) = Ef(x)$. The Hamiltonian consists in the sum of the kinetic energy term and a classical potential $V(x) = \sum_j V_j x^j$. Since Q commutes with any function of $x\partial_x$ we can add to the Hamiltonian an arbitrary integer function $W(x\partial_x) = \sum W_j(x\partial_x)^j$. The condition of symmetry of H under the transformation Q applied on an arbitrary integer function $f(x) = \sum f_k x^k$, after identification of the corresponding powers of x, reads

$$f_{k+2}(k+1)(k+2)(Q(k) - Q(k+2)) = \sum_{j=0}^{k-1} V_{k-j} f_j(Q(k) - Q(j)). \quad (2)$$

Equation (2) must be solved together with the Schrödinger equation for f which, expanded in power of x, reads:

$$f_{k+2}(k+1)(k+2) = \sum_{l=0}^{k} f_l V_{k-l} - E f_k + f_k W(k). \quad (3)$$

Equations (2-3) give the conditions for the existence of solutions with continuous symmetry described by the operator Q. Both these equations are identities for $Q = 1$. In the case of the inversion $x \rightarrow -x$, $Q = I_x$, i.e. $Q(k) = (-1)^k$, eq.(2) asks for $V_k = f_k = 0$ for k odd ($V(-x) = V(x), f(x) = f(-x)$) like in the traditional case. In the simplest case of a free particle we have $V(x) = 0$ and from eq.(1) we get $Q(k) = Q(k+2)$, which results in the action $Qf(x) = \frac{1}{2}\left(Q(0)(f(x) + f(-x)) + Q(1)(f(x) - f(-x))\right)$. Taking into account the limiting restrictions at $q = \pm 1$, the unique solutions for Q are $Q(k_{odd}) = \pm 1$ and $Q(k_{even}) = 1$. This restricts the allowed symmetries for the free particle to inversion only. The free particle does not admit continuous symmetries between identity and the mirroring of the x-axis. In the case of the harmonic oscillator, $V_k = V_2 \delta_{k,2}$, we obtain the condition $q^4 = 1$. One can generalise this for any potential of the form $V(x) = x^n$ and we obtain the corresponding condition $q^n = 1$. In this case the number of admissible discrete symmetries between 1 and I_x is larger (n roots of the unity) but still there is no way of continuous mapping of the identity into the mirroring. The transformations of the potential, corresponding to eqs.(1,2), can be written explicitly in the form of a nonlocal operator applied on $V^0(x)$ and results in

$$V(x,q) = \frac{q(q+1)}{2} \int \frac{dV^0(x)}{dx} d_q x, \quad (4)$$

that is the q-primitive[1] of the derivative of V^0

$$\int f(x) d_q x = (q^{-1} - q) \sum_{n=0}^{\infty} q^{2n+1} f(q^{2n+1} x) + constant. \quad (5)$$

For $q = 1$ the q-primitive tends to the normal integration and eq.(4) describes the action of the identity operator on $V^0(x)$. The transformation of the potential is a deformation of a nilpotent operation: one acts first with an operator (the derivative) and then with a q-deformation of the inverse of this operator (q-integration). The result is not the indentity but a sort of a "defect" of the identity: an infinitesimal derivative followed by a finite difference integration.

In the following we use these above results in a model based on a non-commutative q-plane [3]. Contrary to the commutative plane where the generators: $P_x = \partial_x$, $P_y = \partial_y$ (translations or $-i$ times the momentum operators), and $R = y\partial_x - x\partial_y$ (rotation or the L_z component of the angular momentum operator) fulfil the commutator relations:

$$[R, P_x] = P_y, \quad [R, P_y] = -P_x, \quad [P_x, P_y] = 0, \tag{6}$$

in the non-commutatice plane, eqs.(6) do not close under the commutator relations. In this q-deformed non-commutative realisation we have not a closed q-algebra, unlike the case of $E_q(2)$, which reduces to eqs.(6) when $q \to 1$. The corresponding commutator relations are quadratic, and in order to close this q-deformed algebra one needs to introduce the coordinate operators, too. We solve this in a particular situation, $q = -1$. In this case eqs.(6) become:

$$[p_x, p_y] = 2p_x p_y, \quad [p_x, L_z] = -ip_y - 2L_z p_x, \quad [p_y, L_z] = ip_x + 2L_z p_y \tag{7}$$

and the q-algebra closes. We shall concentrate on the following example: in the commutative case, from the second commutator relation in eqs.(6), the RHS is zero on a subspace of the (x, y)−representation of the Hilbert space, given by wave functions $\Psi(x, y) = \Psi(y)$. In this case both L_z and p_y, due to the above uncertainty relation, can be measured with the same precision, i.e. there is no macroscopic motion in the y direction. We have full delocalisation in the x direction and $p_x \Psi = 0$. In this case $\Psi(y)$ can only be an eigenvector for p_y, a plane wave in y direction, and it results $< L_z >= 0$, which gives a trivial situation. In the non-commutative case the situation is different and by looking for such subspaces, which annihilate the RHS in the third commutator relation in eqs.(7), we get a nontrivial partial differential equation for $\Psi(x, y)$ and $\Psi(x, -y)$

$$(\partial_x + 2y\partial_x\partial_y + 2x\partial_y^2)\Psi(x, y) = 0. \tag{8}$$

We search solutions in the form $\Psi(x, y) = g(x)f(y)$. By introduction this form in eq.(8), performing the derivations by taking care of the order of the operation in x and y, and with the additional restriction $f(y) = -f(-y)$, we obtain one bounded $L_2(R)$ exact solution, in the form

$$\Psi(x, y) = e^{-\alpha x^2} \sqrt{y} I_{1/4}\left(\frac{\alpha y^2}{2} e^{-\frac{-\alpha y^2}{2}}\right) \tag{9}$$

where $I_{1/4}$ is the Bessel function of imaginary argument and α is an arbitrary real parameter. This solution represents a bounded function at ∞ but has one pole for $y = 0$. Its asympthotic behaviour for $y \to \infty$ is given by: $\Psi(x, y) \simeq e^{-\alpha x^2}$ and describe a wavefunction which is constant with respect to y. The wave function is localised in the x direction and $< L_z > |_\Psi = < p_x > |_\Psi = < p_y > |_\Psi = 0$. Consequently, the effect of the non-commutativity of the plane provides a behaviour of the particle similar with a potential valley in x direction. Taking into account these results, we note that for such non-commutative plane and for a Schrödinger equation q-invariant for $q = -1$, a free particle behaves like a particle in a potential valley directed in the x axis.

3 Spontaneous symmetry breaking by q-discrete transformations

In the following we discuss other examples of such symmetries given by a free scalar, real field $\phi(x^\mu)$, $(x^4 = it)$, in the classical case, described by a Lagrangian density $L(\phi, \partial_\nu \phi)$. For continuous transformations in coordinates and in the field these invariants are given by the well-known Noether's theorem. We define a set of n continuous transformations, each depending on a parameter ϵ_a, by $5n$ differentiable functions $X_{\mu a}, \psi_a, a = 1, 2, ..., n$ and $\mu = 1, ..., 4$ through the infinitesimal transformations

$$x_\mu \to x'_\mu = x_\mu + X_{\mu a} \delta \epsilon_a \tag{10}$$

$$\phi(x) \to \phi'(x') = \phi(x) + \psi_a \delta \epsilon_a \tag{11}$$

In the following we apply another approach in order to investigate the discrete symmetries for a problem of spontaneous symmetry-breaking. Let us consider the system described by some Lagrangian and Hamiltonian and the transformations in eqs.(10-11). The states of minimum energy describe the vacuum wave functions of the system. One vacuum state is invariant under the action of one of the transformations if it transforms into itself and noninvariant otherwise. In the case of local relativistic quantum field theory, this problem is managed by the Coleman theorem [5], i.e. if the vacuum state is invariant the Lagrangian must necessarily be invariant, too (exact symmetries). If only the vacuum state is non-invariant whereas the Lagrangian is invariant, the symmetry breaking is called spontaneous. In our case, the Lagrangian has an arbitrary form and we have to investigate its invariance by using the group transformation methods.

We want to investigate the problem of spontaneous symmetry breaking in the case of the mirroring discrete transformation of the space coordinate x or of the field ϕ. That is, to find general forms of Lagrangians which are invariant under this discrete transformations together with the corresponding vacuum states. We introduce the infinitesimal transformations associated with the Lie group defined in eqs.(10,11), in the form

$$v = a(x, \phi)\partial_x + b(x, \phi)\partial_\phi \qquad (12)$$

where v is the infinitesimal generator, a and b are its components which are to be identificated for each symmetry, and we do not introduce transformations concerning the time. In the determination of the symmetries of the Euler partial differential equation $\triangle(x, t, \phi, \phi_{tt}, \phi_x, \phi_{xx}) = 0$ we proceed by identifying this equation with a manifold in the appropriate jet space [4]. We then look for a vector field v such as its second order prolongation pr^2v (Euler equation is of second order) is tangent to this manifold: $pr^2v\triangle|_{\triangle=0} = 0$. In the 1+1 dimesional case, $\phi(x, t)$, we choose a Lagrangian in the form

$$L = \frac{1}{2}(\phi_t^2 + \phi_x^2 + F^2(\phi, \phi_x)) \qquad (13)$$

where F is an arbitrary function. The Euler equation has the form

$$F\frac{\partial F}{\partial \phi} - \partial_x\left(F\frac{\partial F}{\partial \phi_x}\right) = 0 \qquad (14)$$

By application of the second prolongation pr^2v on eq.(14) we obtain a partial differential equation which represents the condition of invariance of the Euler equation against the transformations given by eqs.(10,11). These transformations are continuous, but for certain values of the parameter they have the same action like the discrete mirroring. We investigate only the stationary solutions ($\phi_t = 0$) which are relevant to the vacuum states. We use only the special dilation of the space coordinate x, which is responsible for the mirroring, i.e. $v = x\partial_x$. Under these restrictions the Euler equation, eq.(14) becomes

$$\phi_{xx} = f(x, \phi, \phi_x), \qquad (15)$$

and we obtain the most general form for the Euler eq.(14), invariant to the special dilation, and consequently invariant to the mirror transformation $x \rightarrow -x$

$$\phi_{xx} = f(x, \phi, \phi_x) = \frac{f_0(\ln(x) - \epsilon, \ln(\phi_x, \phi))}{x^2} \qquad (16)$$

For example, if we take $f_0(x_1, x_2, x_3) = x_3$ we obtain the Euler equation in the form

$$\phi_{xx} = \frac{\phi}{x^2} \qquad (17)$$

which arrises from the Lagrangian

$$L = \frac{1}{2}(\phi_t^2 - \phi_x^2 + \frac{\phi\phi_x}{x^2})$$ (18)

As it is easy to check this Lagrangian is not only invariant to mirroring in x but also to any dilation in x. It is the most general 1+1 dimension Lagrangian for a scalar field with this property. Concerning the vacuum states of this Lagrangean eq.(13), we have the corresponding Hamiltonian

$$H = \frac{1}{2}(\phi_t^2 + \phi_x^2 - \frac{\phi\phi_x}{x^2})$$ (19)

which reaches its minimum for two possible solutions: $\phi_0 = C =$const. and for $\phi_0 = Ce^x$. The system has two different vacuum states, each of them corresponding to the same minimum energy, hence we have the spontaneous symmetry breaking. Here we should like to stress the following general result: the condition of invariance of the Lagrangian and, simultaneous, of the vacuum solutions, at dilation transformations are impossible to be fulfiled simultaneously.

In the following we present a direct application of the above result, concerning a localised field model. In literature such models have the common feature that external corections (e.g. shell corections for the nucleus) are needed in order to make the energy of a localised state equal to the energy of normal (non-localised) state. By localised we mean that the field goes asymptotically to 0 when $x \to \infty$ and the Hamiltonian density is localised in space. Due to the spontaneous breaking of the discrete symmetry the formation of localised states could be spontaneous. We choose a real scalar field in 1+1 dimensions $\Phi(t, \theta)$, defined on a circle, described by the Lagrangean density:

$$\mathcal{L} = \frac{1}{2}\left\{ \Phi_t^2 - (\Phi_x^2 - U(\Phi))^2 \right\}$$ (20)

where the function U is arbitrary and restricted by the periodicity conditions: $\Phi(t, -\pi) = \Phi(t, \pi)$

The Euler-Lagrange equation of Lagrangean eq.(20) is:

$$\Phi_{tt} - \Phi_{xx} + U(\Phi)U'(\Phi) = 0$$ (21)

and the coresponding conserved quantity given by Noether's theorem (the energy) is:

$$E[\Phi] = \int_{-\pi}^{\pi} dx \frac{1}{2}\left(\Phi_t^2 + (\Phi_x - U(\Phi))^2 \right)$$ (22)

The energy functional reaches its minimmum at field configurations wich satisfy:

$$\begin{cases} \Phi_t & = & 0 \\ \Phi_x & = & U(\Phi) \end{cases} \tag{23}$$

The configuration wich satisfy eq.(23) also satisfy the equation of motion eq.(21). We have to classes of solutions: singular solutions of the form $\Phi(x) = \Phi_0$ =constant,where $U(\Phi_0) = 0$ and regular solutions given by the equation:

$$\int \frac{d\Phi}{U(\Phi)} = x - x_0 \tag{24}$$

Both these solutions are classical vacuum solutions, satisfying the boundary condition. A soliton-like solution satisfies eq.(24). If $\Phi(x)$ is a regular solution then $\Phi(-x)$ it is also a solution of the Euler-Lagrange equations. Obviously, $\Phi(-x)$ it is not a vacuum solution and have a nonvanishing energy. Consequently, the quantum fluctuations built around $\Phi(-x)$ will have higher energies than the excitation states of the regular solution, though the "shape" of $\Phi(-x)$ is identical with the shape of $\Phi(-x)$. So, we have two classical configurations wich are identical in shapes but different in energies. The first one is a vacuum configuration, therefore it mihgt be attainable in the coresponding quantum theory by spontaneous transitions from other vacuum configurations and the second is a classically excited configuration. This feature clearly support the interpretation that the non-trivial solution is at least related with a separate object wich may exist alone, separate from the object who created it. The non-trivial vacuum configuration ($\Phi(x)$) would be, in this interpretation, the configuration wich consist in a preformated quantum extended particle plus wathever remains if this is emitted, and the non-trivial excited configuration ($\Phi(-x)$) would be a bound-state configuration of descendant and the emmited particle.

References

1. Z. Chang, *Phys. Rep.* **262**, 137 (1995).
2. H. L. Kauffman *Knots and Physics*, (World scientific, Singapore, 1991).
3. J. Wess and B. Zumino, *Covariant differential calculus on the quantum hyperplane*, preprints: CERN-TH-5697/90, LAPP-TH-284/90.
4. G. Gaeta and M. A. Rodrigues, *J. Ph. A: Math. Gen.* **29**, 859 (1996).
5. C. G. Callan, S. Coleman, *Phys. Rev.* D **16**, 1762 (1977) and S. Coleman, *The Uses of Instantons* in *The Whys of Subnuclear Physics*, ed. A. Zichichi, (Plenum Press, New York, 1979).

CURRENT AND FUTURE RESEARCH AT THE SOUTH AFRICAN NATIONAL ACCELERATOR CENTRE

J.F. SHARPEY-SCHAFER

National Accelerator Centre, P O Box 72, Faure, South Africa

The main facility at the National Accelerator Centre (NAC) is a K=200 Separated Sector Cyclotron (SSC). This cyclotron produces beams for radiotherapy, isotope production and nuclear physics. Current studies in nuclear physics utilize light ion beams and involve both reaction mechanisms and spectroscopy. A new AFRican Omnipurpose Detector for Innovative Techniques and Experiments (AFRODITE) is currently being constructed to take advantage of the wide range of heavy ion beams from the SSC. Current and planned physics will be described.

1 Introduction

I was very pleased to discover that Walter Greiner's 60th birthday fest was being held in South Africa. I admit to being somewhat surprised. I had fondly imagined that a physicist as productive, famous and as senior as Walter must be more than three years older than myself! Indeed, I seem to remember thinking of Walter of being about 60 for the past twenty years? But, to be serious, I thought you would like to know something about South Africa's biggest accelerator laboratory. The National Accelerator Centre (NAC) is situated on the Cape Flats between Khayelitsha and Eerste River, which is about 35 km east of down-town Cape Town. My topic has nothing to do with the title of this meeting "Structure of the Vacuum and Elementary Matter", except perhaps to consider what positive and constructive things can come out of the vacuum created by over 30 years of the appalling apartheid system.

I was appointed the new Director of NAC from 1st July 1995, but only managed to arrive here full-time from the beginning of October. As NAC is a large multidisciplinary and cross-disciplinary science laboratory I have been through an intensive learning process over the last few months. Research at NAC is based around two accelerators. The first is a 6 MV HVEC CN single ended Van de Graaff which originally formed the Southern Universities Nuclear Institute (SUNI) to service the research needs of the Universities of Cape Town and Stellenbosch. The second is a K=200 separated-sector cyclotron (SSC) which first started producing beams in 1987.

The cyclotron facilities at NAC are, by international standards, more extensive than is normal for a country where only 5 million of its inhabitants had access to reasonable educational standards and personal freedoms. It is however of the size needed to cater for the research and technology development of a population of over 40 million in the several areas of science that NAC addresses. Hence NAC is fortunately positioned, in that it has the opportunity to expand its User base and play a full and significant role in the reconstruction and development of South African science and technology. Very high standards of technical competence are demonstrated at NAC. It is just the vehicle needed for skills transfer and science and technology building in a country suddenly readmitted as a full member of the community of nations.

The three pillars of NAC are:
1. A wide range of research from nuclear reactions and structure through radiobiology, materials science to the use of the Van de Graaff microprobe in applications to geology, zoology and botany.
2. The production of proton-rich radioisotopes for use in medical diagnosis and general research.
3. Radiotherapy using both neutrons and protons.

The current throughput of young people taking M.Sc. and Ph.D. degrees can be almost doubled in most areas. However, in South Africa it is currently only too common for postgraduates to take more than the 2 or 3 years supposedly required for completion of their course. NAC is no exception to this problem which needs to be solved as it represents a loss of, and inefficient use of, resources. One area in which data collection can be made much more effective is by exploiting heavy-ion beams from the SSC. To this end a new spectrometer, AFRODITE, is being designed and constructed to exploit the very nice characteristics of these heavy-ion beams.

In this talk I will give an overview of the current facilities at NAC followed by a brief description of the AFRODITE spectrometer and the physics it can address.

2   The Separated-Sector Cyclotron (SSC)

A plan of the cyclotron facilities [1,2,3] at NAC is shown in figure 1. There are two solid pole injector cyclotrons SPC1 and SPC2. SPC1 injects high currents of protons and other very light ions into the SSC. It was the original injector cyclotron. The second injector, SPC2, was designed to pre-accelerate beams from the ECR heavy-ion source and the polarized proton source. SPC2 was commissioned in [4] in 1994.

The SSC is a four-sector cyclotron with a sector angle of $34°$. The two $\lambda/2$ resonators give an energy gain per turn of 1 MeV for protons. The main operations are:
(i)   85 or 28 μA protons at 66 MeV giving up to 5.6 kW of power in the beam. These beams are used for isotope production and neutron therapy respectively. When neutron therapy is taking place the beam is switched to isotope production if a patient is not in the process of being irradiated. This switching of the beam gives a very efficient use of the facility.

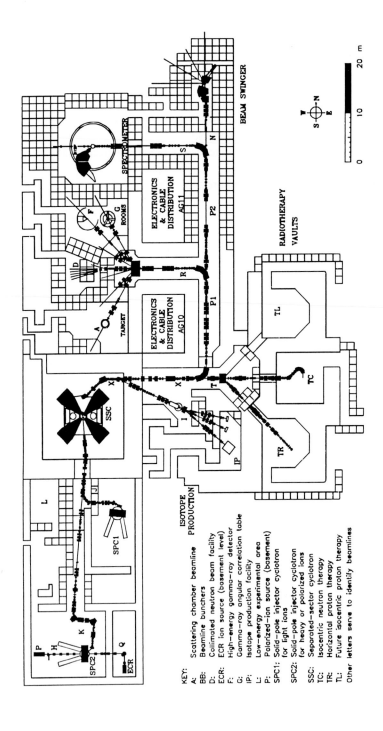

KEY:

A: Scattering chamber beamline
BB: Beamline bunchers
D: Collimated neutron beam facility
ECR: ECR ion source (basement level)
F: High−energy gamma−ray detector
G: Gamma−ray angular correlation table
IP: Isotope production facility
L: Low−energy experimental area
P: Polarized−ion source (basement)
SPC1: Solid−pole injector cyclotron
      for light ions
SPC2: Solid−pole injector cyclotron
      for heavy or polarized ions
SSC: Separated−sector cyclotron
TC: Isocentric neutron therapy
TR: Horizontal proton therapy
TL: Future isocentric proton therapy
Other letters serve to identify beamlines

Fig. 1: Layout of the injector cyclotrons SPC1 and SPC2, the separated sector cyclotron SSC and the experimental areas etc. at the National Accelerator Centre near Cape Town.

(ii) 1 µA protons at 200 MeV. This beam is used for proton therapy . There is no parallel isotope production.

(iii) Other beams for physics research, such as 60-200 MeV polarized protons, deuterons, ^{3}He, ^{40}Ar, ^{84}Kr etc.

In figure 2 the weekly scheduling of the SSC beams is shown. Currently proton therapy takes place on Mondays and Fridays with neutron therapy and isotope production interleaved in between. Physics starts on Friday afternoons and continues over the weekend. This schedule minimises the number of beam changes at 4 per week (fig. 2a). Due to the demand for more proton therapy sessions per week a new schedule (fig. 2b) is to be introduced in August 1996. This will require 9 changes of beam per week and will be very challenging for both the operators and the equipment! In 1995 the SSC beam was available for 6389 hours which is 72% of the complete year.

Table 1: Medical radioisotopes produced at NAC and delivered to South African hospitals for the year ending 31/3/96.

Number of Consignments	Compound	Main Use
5	^{18}F-deoxyglucose)	PET and SPET studies of the brain and heart
5	^{18}F-solution)	
503	^{67}Ga-citrate	Localization of certain tumors and inflammatory regions
250	^{81}Rb/^{81m}Kr-generators	Lung ventilation studies
88	^{123}I-sodium iodide	Thyroid studies and localization of certain tumours
183	^{123}I-meta-iodobenzylguanidine (mIBG)	Localization of tumours such as neuroblastoma and phaeochro-mocytoma; heart studies
12	^{123}I-isopropyl-amphetamine	Brain studies
29	^{123}I-fatty acids (IPPA and BMIPP)	Heart metabolism studies
7	^{123}I-Vasoactive intestinal peptide	Localization of VIP-receptor bearing tumours
6	^{123}I-IBZM	Brain dopamine receptor studies
24	^{201}Tl-chloride	Heart studies, brain tumours

NEW AND PREVIOUS OPERATING SCHEDULES FOR THE NAC CYCLOTRONS

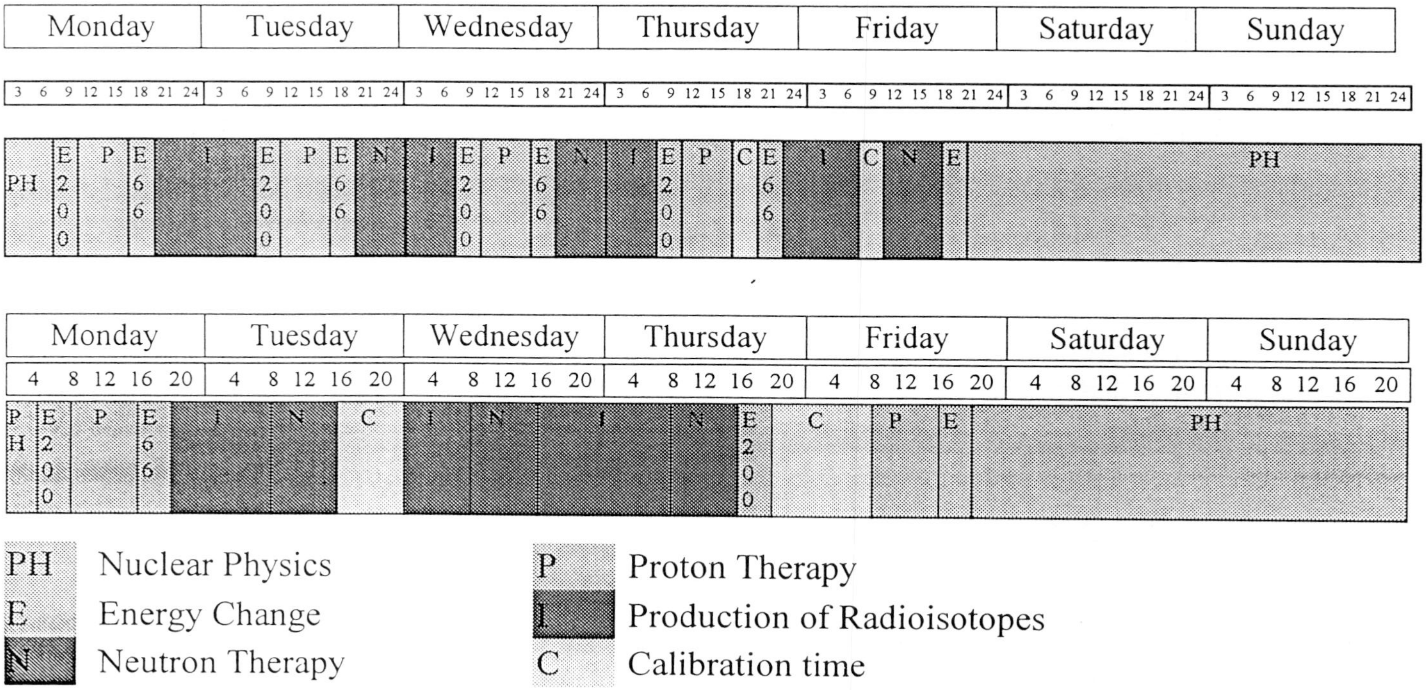

Fig. 2   Weekly schedules for the use of the SSC beams (a) present scheduled (b) new schedule planned for August 1996.

3. Radioisotope Production

Radioisotope production uses the same 66 MeV, high current proton beam as neutron therapy . Remote target handling is used and careful design [5] minimises activation in the vicinity of the target. Extensive hot cell facilities are used for chemical separation and purification. Emphasis is placed on developing and using closed chemical processes [6,7]. There are three types of radioisotopes produced:

(i) radioisotopes of short half-life used in nuclear medicine departments in 32 hospitals throughout South Africa. These radioisotopes are used in the diagnosis of about 7,000 patients per year.
(ii) radioisotopes developed for use by medical research institutes in South Africa.
(iii) radioisotopes for non-medical use and supplied to local users and also sold for export.

The medical radioisotopes produced at NAC in 1995/6 are given in table 1 with an indication of their use and the number of batches produced during the 1994/5 financial year. Radioisotopes for non-medical use are given in table 2.

Table 2: Radioisotopes for non-medical use produced by NAC and supplied to local users or exported for the year ending 31/3/96.

Radioisotope	Number of Consignments	Activity (GBq)
^{22}Na	8	13.7
^{111}In	10	1.7

4. Radiotherapy

About 30% of people contract cancer, of which about half die of it. Radiation therapy and surgery are used as the main forms of localised treatments to attempt to eliminate the primary tumour. Secondary tumours, metastases, originating from the primary, are usually treated by chemotherapy. Most radiotherapy uses photons from ^{60}Co($\sim$1 MeV) or from the bremsstrahlung from electron linacs ($\sim$8 MeV). By using isocentric geometries the radiation can be focused on the cancerous growth giving lower radiation doses to healthy tissue. The total radiation dose is usually divided into small daily "fractions" to give better tumour control. Each "fraction" may be divided into several fields, the radiation entering the patient from a different direction.

I must say I always regarded radiotherapy in the same light as trying to mend a TV by kicking it! However, if you think about it, most treatments, i.e. surgery or

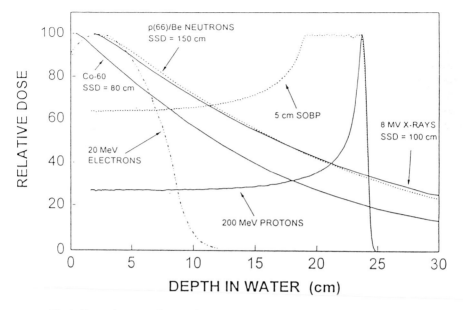

Fig. 3: Energy loss curves in water for 8 MeV γ-rays, 60 MeV neutrons and 200 MeV protons

medicines, attack the root problem and then the final mending is done by the body's own defence and repair mechanisms. Hence, although radiation therapy machines can be quite expensive and require high technical expertise, there are many advantages:

(a) the patient is not anaesthetised, suffers no damage from invasive surgery, does not require an expensive sterile environment and suffers no chemical side effects;
(b) few people are involved in patient treatment which does not usually require the daily presence of a clinician;
(c) most patients are treated as out-patients and do not have to occupy a hospital bed or require intensive care. The after-care is uncomplicated and mostly consists of monitoring the progress of the patient.

NAC is carrying out research [8,9,10] into the use of neutron and proton radiation, as alternatives to the normal photontherapy, in cases where the latter is not very effective or has major difficulties. In fig. 3 the energy loss in water of 8 MeV photons, 60 MeV neutrons and 200 MeV protons is shown. The neutrons have the same energy deposition rate as the photons and therefore require an isocentric treatment facility. The neutrons lose energy mostly by colliding with the hydrogen protons in the water. These have a short recoil path compared with the electrons Compton scattered by the photons. Hence neutrons have a higher biological effect than the photons and can be used to treat tumours insensitive to photons. Examples of tumours which are effectively treated by neutrons include salivary gland tumours, large breast tumours and certain tumours of other soft tissues. NAC has an isocentric gantry which produces a beam of neutrons from 66 MeV protons impinging on a Be target. The neutrons are collimated appropriately for each tumour. Treatment has been carried out [11] on over 750 patients since 1989. Typically each patient receives three fractions per week over a period of four weeks A summary of the neutron therapy statistics is given in table 3.

Proton Therapy at NAC uses a 200 MeV beam of protons with an initial beam current of about 1 µA. This beam [8,10] is shaped, and has its energy adjusted by absorbers, to suit each individual field for a given tumour. The beam is delivered horizontally at the patient who sits in a special chair [12] which holds the patient and locates her/him, and hence the tumour, to an accuracy of ± 1.0 mm. This

Table 3: Neutron Therapy Protocol Accrual up to 31/3/96

Trial Identification	Accrued 95/96	Total since activation
B1 Head and neck	12	148
B2 Salivary gland	42	209
B4 Soft tissue sarcoma	10	74
B5 Breast	0	74
B7 Uterine cervix	0	5
B8 Bronchus	0	6
B9 Non-study	18	198
B10 Uterine sarcoma	2	45
B11 Mesothelioma	0	21
B12 Paranasal sinuses	3	28
Total	**87**	**808**

accuracy is required by the nature of the treatment. As the protons are charged particles their energy loss is given by the Bragg curve, shown in fig. 3, and is quite different to the curves for photons and neutrons. Because the Bragg curve has a sharp cut-off at the maximum penetration and the maximum energy deposition occurs just before this cut-off, protons can be used to treat tumours or other abnormalities near sensitive structures. Examples of such structures are the optic nerve, spinal cord, kidneys etc. where other forms of radiation would do too much damage to healthy tissue. Protons can, with accurate planning systems, be steered, focused and be given exactly the right energy to stop at any particular point within the body, thus completely protecting any organs beyond this range.

The isodoses for a four-field plan for an intercranial treatment are shown in fig. 4. In the past 2.5 years 130 patients have received proton beam irradiations at NAC. The programme has concentrated on arteriovenous malformations (AVM), meningiomas, acoustic neuroms and pituitary adenomas (table 4), all conditions for which no further or alternative treatments were available.

Most proton therapy treatments have involved only 3 to 4 fractions. The number of fractions has been limited by the availability of the proton therapy beam being restricted to Mondays and Fridays (fig. 2a). A new schedule is planned (2b) to start in August 1996 giving a proton beam every morning in the first 4 days of the week. This will allow patients to be enrolled on international protocols involving several fractions per week.

5. Light-ion Physics

The physics programme with the SSC has mainly concentrated on the use of high energy light-ions. The main apparatus for experiments consists of:

(i) a large 0,75 m radius scattering chamber;
(ii) a K=600 kinematically corrected magnetic spectrometer;
(iii) a beam swinger to be used in conjunction with a neutron time-of-flight facility with a path length up to 100 m. The 200 MeV proton beam pulse has a time fwhm of less than 0.5 ns on target.

A programme of knock-out reactions has been carried out using the (p,2p) reaction [13,14], (p,α) reaction [15] and (α,2α) reaction [16]. Proton continuum scattering, between 100 and 200 MeV incident energies, has been studied on both heavy [17] and light [18] nuclei. The data on ^{12}C, ^{14}N and ^{16}O is also a very useful practical input into the planning calculations used in the NAC proton therapy programme.

The ^{4}He(α,α)^{4}He; ^{4}He(α,^{3}He)^{5}He and ^{4}He(α,^{3}H)^{5}Li reactions have been studied [19]. Clear neutron and proton stripping transitions are seen to unbound ground states in ^{5}He and ^{5}Li. These data are being analysed with modified DWBA codes.

The K=600 spectrometer has been used to study the ^{55}Mn(d,^{3}He)^{54}Cr reaction to investigate the Clement-Perez non-energy-weighted sum rules [20].

Other programmes have been carried out which I do not have space to mention here.

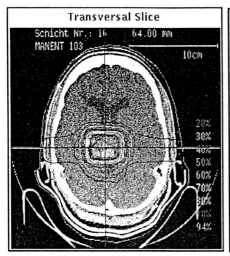
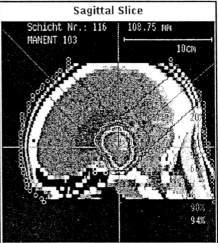

Fig 4:   Isodoses for a four-field treatment plan for an intercranial lesion using 200 MeV protons. Left: a
transversal plane through the isocentre. Right: A sagitally reconstructed plane through the isocentre.

Table 4: Summary of proton therapy statistics to 31/3/96

Diagnosis	Stereotactic intracranial lesions	Non-stereotactic extracranial lesions
AVM	31	-
Acoustic neuroma	20	-
Meningioma	16	-
Brain metastasis	14	-
Pituitary adenoma	13	-
Glioma (low grade)	6	-
Craniopharyngioma	5	-
Carcinoma of the antra	-	3
Oropharyngeal cancer	-	4
Carcinoma of the orbit	-	2
Prostate cancer (boost)	-	2
Glioma (high grade)	2	-
Skull base tumours	-	5
Spinal cord tumours	-	3
Choroidal melanoma	-	1
Haemangiopericytoma	-	1
Paraspinal metastasis	-	2
Subtotals	107	23

6.   Research with the 6MV Van de Graaff

This research falls into two main areas: Ion-Beam Analysis and Materials Research. The main tools in the ion-beam analysis are Rutherford back-scattering (RBS) and PIXE analysis of samples using a nuclear microprobe [21,22]. This latter technique consumes about 75% of the running time on the Van de Graaff which runs 12 hours a day, 7 days a week. A wide range of science is addressed including applications to geology [23], archaeology [24], biology [25] and materials technology [26]. The beam spot size achieved with the microprobe is less than 1.0 μm.

RBS measurements are made to investigate the phases of thin layers of metal-semiconductors. The results are compared with the effective heat of formation model [27]. This is a very broad programme which also extends to studies of high T_c thin-film superconductors [28].

7. AFRODITE

The ability of ECR ion sources [29] to produce high currents of very highly charged heavy-ions, has not only proved to be a boon to accelerators like the UNILAC + SIS at GSI but has also enabled cyclotrons [30] to become very useful providers of heavy-ions. The SSC at NAC is well capable of producing beams with energies E>5MeV/u as heavy as Xe. This makes it an ideal facility for spectroscopic studies of nuclei using heavy-ion reactions near the Coulomb barrier. During the past 15 years the most spectacular advances in nuclear spectroscopy have been made using arrays of escape suppressed germanium γ-ray detectors [31-33]. Clearly such a device is needed to properly exploit the beams of heavy-ions from the NAC SSC.

There are many advantages to using such arrays:
(a) they can be used to address a very wide range of physics;
(b) they subtend a large solid angle at the target and are therefore very efficient in collecting relevant information in an experiment This is needed as the SSC is only available at the weekends for physics experiments;
(c) this efficiency allows more projects to be run - i.e. more Ph.Ds per weekend!

In designing an γ-ray spectrometer array to use with the SSC there are several self-evident criteria to be considered:

(i) this array can not be nearly as large and powerful as the Gammasphere [32] and Euroball [33, 34] arrays built by extensive collaborations in the USA and Europe respectively. Therefore this array must exploit advantages of the SSC and put other ancillary detectors in the gaps which we cannot afford to fill with γ-ray detectors!

(ii) there is no point in reinventing the wheel, hence current and proven designs of γ-ray detectors should be used that give the maximum performance-for-cost advantage;

(iii) as wide a range of ancillary detectors as possible should be available for use in coincidence with the γ-ray detectors. Usually these detectors will be cheaper than Ge detectors per solid angle, giving us a cost saving. Ancillary detectors can often be used to give extra selection onevents which improves the achievable signal to noise;

(iv) an instantly identifiable name for the array needs to be invented in order to ensure favourable publicity. The best I have been able to come up with is "AFRODITE" which I claim stands for AFRican Omnipurpose Detector for Innovative Techniques and Experiments! At least the name locates the devices and is a joke for my more serious Greek friends about my chronic inability to spell.

Our design response to the above criteria is as follows:

(i) we will use a simple geometric design, fig. 5; a rhombicuboctahedron, which has 18 square faces and 9 triangular faces. In order to maintain a simple symmetry, the beam will enter and exit two of the square faces. This gives 4 square faces at 45°, 8 square faces at 90° and 4 square faces at 135° to the beam direction. The 8 triangular faces are at 54.7° or 125.3° to the beam direction. The frame will hold up to 16 Ge γ-ray detectors in the square faces and ancillary detectors in the triangular faces.

The frame will be designed in two parts, as shown in fig. 4. Each part will be movable away from the beam-line. The larger part of the frame will have 11 square and 4 triangular faces. The smaller part will have 5 square and 4 triangular faces. The target chamber will reflect the geometry of the detector frame.

(ii) the design of Ge detectors and BGO shields used will be of the "clover" type [35,36] which are currently used in the Eurogam2 and Euroball arrays [32,33]. Four Ge crystals, of individual relative efficiency of 20-25%, are shaped to mount compactly in a single cryostat. The front face of the Ge crystals will be 195 mm from the target. The design allows a compromise between granularity and efficiency. The smaller size of the individual crystals also gives better timing resolution, with which to exploit the excellent timing of the pulsed beams, and shorter time constants to be used in the amplifiers with little loss in energy resolution. The clover detectors can also be used to measure the linear polarization of reasonably intense γ-rays [35]. The BGO shields differ moderately from the original Eurogam2 design [32,33] in that they have a square symmetry for AFRODITE and slightly thicker BGO. Currently we have 8 clover detectors and 6 BGO shields on order.

(iii) Some ancillary detectors are already available at NAC. These are; a large number of 75 mm diameter NaI(Tl) γ-ray detectors that can be used as a multiplicity filter [31]; a large 240 x 360 mm NaI(Tl) detector to detect high energy photons; and some plastic position sensitive neutron detectors that can be used to form a wall downstream of the target. The large NaI(Tl) detector may be used to observe high energy γ-rays from GDR decay. My own feeling is that this mode will be most useful as a time scale to measure the fusion lifetimes of collisions between very heavy-ions [36].

In order to be able to study γ-rays produced in reactions other than fusion-evaporation reactions, we will require a very large solid angle position and energy sensitive heavy ion detector that is reasonably transparent to 100 keV γ-rays. We are therefore building a mosaic of solar cells which will give both position and energy information [37-39] on heavy-ions and fission fragments while being insensitive to evaporation protons and alpha particles. The shape of the mosaic will reflect the rhombicuboctahedron shape of the AFRODITE frame and target chamber (fig. 5).

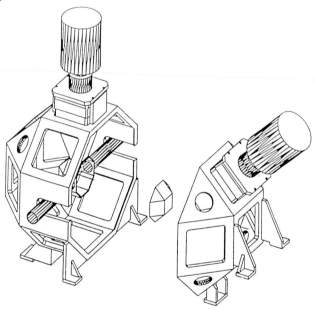

Fig. 5: Sketch of the framework for the AFRODITE spectrometer. The shape is a rhombicuboctohedron. Two hyperpure Ge "clover" γ-ray detector and BGO escape suppression shield assemblies are shown in place.

As we would like to study the spectroscopy of nuclei with Z≥80, it would be sensible to construct a conversion electron transport system that accepts a large solid angle for electrons from the target in the energy range $50 \leq E_{ce} \leq 300$ keV. The electron detector should be segmented so that e-e-γ and e-γ-γ coincidences can be made. The power of such a device has been demonstrated by Peter Butler and his colleagues [40] at the Jyväskylä cyclotron. Design studies for such a spectrometer are in hand.

Lastly it would be very expedient to augment the array with LEPS X-ray detectors to give separation of elements by their characteristic X-rays. This is especially useful for fission-fragment spectroscopy [41].

The physics we can address using AFRODITE is very varied, as I mentioned before. We will not attempt initially to make any effort to study boring old superdeformed nuclei! We will go back even further in time and look at the spectroscopy of fission fragments formed after fusion-fission and at the detailed states populated by particle transfer and deep inelastic reactions. These reactions will also be studied across the Coulomb barrier and examined in the light of Niel Rowley's barrier distribution picture [42-44].

The power of γ-ray arrays to study the structure of neutron-rich nuclei populated by fusion-evaporation reactions has been demonstrated graphically by Porquet et al. [45]. Using the mosaic of solar cells to detect one or more fission fragment, extra selection can be obtained. If a thick target is used and one escaping fission fragment is detected, the DSAM technique can be used to measure level lifetimes in the other fragment which has to recoil into the target. Reactions can be chosen to populate neutron-rich nuclei that are not accessible with spontaneous fission sources.

Nuclei populated in particle transfer and deep inelastic reactions have been investigated by Rafael Broda et al. [46] and Peter Butler et al. [47] in Europe and by Pat Daly et al. [48] in the USA. Their data show that the nuclei populated are dominated by the mass and charge equilibration process. This means that a very wide range of nuclei can be populated by making suitable target and projectile choices.

Our first experiments will be to explore the strengths and weaknesses of the AFRODITE array in order to gain experience in this relatively new field. I hope that we will be able to make some impact on both nuclear spectroscopy and reaction studies using the combination of excellent beams from the SSC and the flexibility of the AFRODITE detector.

Acknowledgements

I would like to thank all those at NAC who have been very generous to me with their considerable and patient help and also for their welcoming kindness. Also many thanks to all of my old colleagues for their valiant efforts to keep me up-to-date! Finally many thanks to Walter himself for his enthusiastic encouragement and effective promotion over many years of the fascinating and beautiful subject of Nuclear Physics.

References:

1. A H Botha, H N Jungwirth, J J Kritzinger, D Reitmann and S Schneider; Proc. 11th Int. Conf. on Cyclotrons and their Applications, Tokyo (1987) World Scientific, p9.
2. A H Botha et al.; Proc. 12th Int. Conf. on Cyclotrons and their Applications, (1990) World Scientific, p80.
3. J L Conradie et al.; Proc. 13th Int. Conf. on Cyclotrons and their Applications, Vancouver (1992) World Scientific, p95.
4. Z B du Toit et al.; Proc. 14th Int. Conf. on Cyclotrons and their Applications, Cape Town (1995) World Scientific, ed. J C Cornell, p28.
5. F M Nortier, N R Stevenson and W Z Gelbart; Nucl. Instr. Meth. A355 (1995) 236.
6. D du T Rossouw and J H Langenhoven; Appl. Radiat. Isot. 45 (1994) 902.
7. T N van der Walt and G J Haasbroek; Proc. 5th Int. Symp. on the Synthesis and Aopplications of Isotopically Labelled Compounds; Strasbourg, France 1994. ed J Allen and R Voges John Wiley and Sons (1994) p211.
8. D T L Jones et al; in "Hadron Therapy Oncology", eds. U Amaldi and B Larsson, Elsevier Sci. BV (1994) 307.
9. L Böhm and B J Smit; S A Med. J. 85 (1995) 116.
10. D T L Jones; in "Ion Beams in Tumour Therapy", ed. U Linz, Chapman and Hall (1995) p350.
11. C E Stannard et al; Radiother. Oncol. Invest. 2 (1995) 245.
12. D T L Jones et al.; Proc. 1st Research Co-Ordination Meeting of IAEA Prog. on Appl. of Heavy Charged Particles in Cancer Radiotherapy, Vienna (1995); in press.
13. A A Cowley; Proc. 7th Int. Conf. on Nuclear Reaction Mechanisms, Varenna, 1994, ed. E Gadioli; Univ. Degli Studi di Milano #100, p122.
14. A A Cowley et al.; Phys. Lett. B359 (1995) 645.
15. A A Cowley et al.; Phys. Rev. C54 (1996) 778.
16. A A Cowley et al.; Phys. Rev. C50 (1994) 2449.
17. W A Richter et al.; Phys. Rev. C49 (1994) 1001.
18. G J Arendse et al; to be published.
19. J J Lawrie et al.; to be published.
20. C F Clement and S M Perez; Rep. Prog. Phys. 53 (1990) 127.
21. V M Prozesky et al.; Proc. 4th Int. Conf. on Nucl. Microprobe Technology and Applications, Shanghai 1994, Nucl,. Instr. Meth. B104 (1995) 36.
22. C G Ryan et al.; Nucl. Instr. Meth. B47 (1990) 55.
23. F M Meyer et al.; Exploration and Mining Geology 3 (1994) 207.
24. C A Bollong et al.; J Archaeol. Sci. (in press).
25. J Mesjasz-Przybylowicz et al.; Nucl. Instr. & Meth. B89 (1994) 208.
26. J L Campbell et al.; Nucl. Instr. Meth. B85 (1994) 108.

27. R Pretorius et al.; J. Appl. Phys. **70** (1991) 3636.
28. R Naidoo, R Pretorius and F Saris; Supercond. Sci and Tech. **8** (1995) 25.
29. R Geller; IEEE. Trans. Nucl. Sci. <u>NS-23</u> (1976) 904. and <u>NS-26</u> (1979) 2120.
30. C M Lyneis, Z Q Xie, D J Clarke; Proc. 14th Int. Conf. on Cyclotrons and their Applications, Cape Town (1995) World Scientific, ed J Cornell, p173.
31. J F Sharpey-Schafer and J Simpson; Prog. in Part. Nucl. Phys., ed. A Faessler, **21** (1988) 293.
32. P J Nolan, F A Beck and D B Fossan; Ann. Rev. Nucl. Part. Sci. **45** (1994) 561.
33. C W Beausang and J Simpson; J Phys. **G22** (1966) 527.
34. R M Lieder; Expt. Tech. in Nucl. Phys., eds. D N Poenaru and W Greiner (1995) de Gruyter.
35. P M Jones et al.; Nucl. Instr. Meth. Phys. Res. **A362** (1995) 556.
36. P Paul; Proc. Int. Conf. on Nucl. Physics, Beijing (1995) in press.
37. G Siegert; Nucl. Instr. Meth. **164** (1979) 437.
38. E Liatard et al.; Nucl. Instr. Meth. in Phys. Res. **A267** (1988) 231.
39. N N Ajianand et al.; ibid **A300** (1991) 354.
40. P A Butler et al.; Acta. Phys. Pol. **B27** (1996) 463.
41. I Ahmad and W R Phillips; Rep. Prog. Phys. **58** (1995) 1415.
42. N Rowley, G R Satchler and P H Stelson; Phys. Lett. B254 (1991) 25.
43. A M Stefanini et al.; Phys. Rev. Lett. **74** (1995) 864.
44. D J Hinde; Phys. Rev. Lett. **74** (1995) 1294.
45. M-G Porquet; Acta Phys. Pol. **B27** (1996) 179.
46. R Broda et al.; Phys. Rev. **C49** (1994) 575.
47. J F C Cocks et al.; Acta. Phys. Pol. **B27** (1996) 213.
48. P J Daly et al.; Proc. 5th Int. Spring Seminar on Nucl. Phys. "New Perspectives in Nucl. Str." Ravello Italy (1995), World Sci., ed. A Covello p91.

1. Andre Gallmann, Walter Greiner, Friedel Sellschop, and Amand Faessler.

2. Bärbel Greiner and B. Fricke, J. Konopka, S. Bass, H. Oeschler and M. Weigel in the background.

1. Friedel Sellschop, Bärbel Greiner, Walter Greiner, and Marylin Bilpuch.

2. Edward Bilpuch, Friedel Sellschop, Sue Sellschop, Walter Greiner, and Marylin Bilpuch.

1. World's greatest Grandpa and A. V. Ramayya.

2. Christopher, Johanna, Annkatrin, and Josephine Greiner.

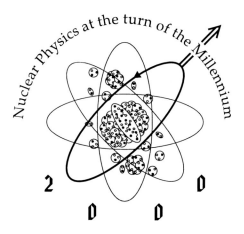

Nuclear Physics at the turn of the Millennium

2000

Epilogue

Greeting and grilling Greiner at the conference dinner

A. Gallmann

*Université Louis Pasteur et Centre de Recherches Nucléaires 67 037   Strasbourg
(France)*

J. H. Hamilton

Vanderbilt University, Physics Department, Nashville, TN 37235

1   Reflections on the distinguished career of Walter Greiner

As Master of Ceremonies for the "Roast" for Walter Greiner at the conference held in honor of his sixtieth birthday, I was most delighted to make it a special evening for a truly special colleague. Over thirty scientists came forward tonight to pay tribute to Walter. To give a flavor of the occasion, I have summarized my remarks that evening.

Ladies and Gentlemen, never in the history of science has one person had so many crazy ideas - and gotten them all published!. Let me recount just a few of his crazy ideas to show just how schizophrenic he has been.

1. Among his first crazy ideas were his efforts with Amand Faessler to improve the highly successful Bohr-Mottelson Model that was soon to win them the Nobel Prize. With Faessler he gave the first solution, the so-called Bohr-Mottelson Hamiltonian, and proposed a Rotation-Vibration Interaction Model – unfortunately, experimentalists found their model was an improvement, so this encouraged even crazier ideas.

2. Not content to solve the B-M collective Hamiltonian to give the first consistent solutions in the model for axially symmetric nuclei, he set out to advance the other Nobel Prize winning nuclear model, the spherical shell model of Mayer and Jensen. Not contented with then a one-center shell model, he had to have a two-centered shell model! Surely, nature could never be described by such a model. But, unfortunately, again experiments revealed many important uses for this model, too – nuclear-molecules, fission, fusion, to name just a few.

3. Then, being schizophrenic, he thought nuclei can be schizophrenic too. Also, why should neutrons and protons have the same deformation? They could be different and vibrate against each other. Moreover, he proposed nuclei can have multiple personalities; one nucleus could have levels characteristic of the Bohr-Mottelson Model and, overlapping these, levels

677

characteristic of a spherical type shape in the Mayer-Jensen Model. The Gneuss-Greiner Model giving solutions of arbitrarily shaped potential energy surfaces was derived. This time I was one of the unfortunate ones, almost a decade later, to discover his crazy ideas about nuclear shape coexistence were right.

4. Would anyone believe nuclei could kiss and stick – nuclear molecules. Not deterred by the laughter of others, he calculated nuclear and atomic properties of nuclear molecules, including K-shell binding energies all the way to $Z = 184$ –impossible to test, of course, since the heaviest element then known was 106! Yet again, nature let our sensibilities down and experiments confirmed that quasi-nuclear molecules can be found in U up to Cm collisions, and their K-shell binding energies to such high Z were correct.

5. In another foolish direction, he thought to develop the collective model still further into a Generalized Collective Model; where nuclear potential energy surfaces (PES) would give clear, vivid pictures of how the nuclear landscape changes along long N and Z chains. Such PES's turned out to be so useful that we see such calculations in all sorts of papers to encourage Walter to greater heights of folly.

6. Not content with challenging today's theories, he decided to challenge ideas that go back to the ancient Greeks – the concept of the vacuum as empty space. He and his colleagues predicted that the vacuum could decay with the spontaneous creation of particles in over-critical fields. This work has created tremendous excitement, but this time the experiments have been too difficult to unravel.

7. His fantastic follies reached new heights when he and his Romanian colleagues predicted a new form of radioactive decay, cluster radioactivities, were not α, β and γ decay enough. Moreover, it was bad enough to have α particles running around in a nucleus, but clusters like ^{14}C, ^{24,26}Ne, and ^{28,30}Mg running around and escaping – nonsense, and these foolish predictions were rightly ignored until ^{223}Ra was found to emit their predicted ^{14}C and other nuclei to emit the heavier ones predicted.

8. Since every madman has a "Theory of Everything", in this spirit Walter and his Romanian colleagues followed these ideas of the two-center shell model and cluster radioactivities with their "Theory to Explain Everything" from α decay to spontaneous fission. As a theory of everything, they predicted cold fusion and cold fission valleys, new ways to make

new elements – really now? It was foolishness, until elements 108-111 were discovered by using target-projectile combinations to take advantage of their predicted cold fusion valleys. Then, in spontaneous fission, our group found every case of cold fission; zero neutron emission spontaneous fission (a new extreme of cluster radioactivity), they predicted. It was really shameless of nature, the way nature cooperated with the predictions of Greiner and his colleagues; urging them onward.

9. You know about Mach I and Mach II shock waves when objects travel faster than the speed of sound in air, but how about Greiner I and Greiner II – nuclear shock waves – never! Nuclear matter is incompressible, so most certainly nature will not cooperate in the perpetration of this hoax which he made with Werner Scheid. Well, who said we could trust nature? Nuclear shock waves are for real, as a new and only way to compress and heat nuclear matter to investigate the nuclear matter equation state and to possibly reach the quark-gluon plasma. And so, armed with Greiner's predictions, experimentalists have run off to build more and more heavy ion accelerators with higher and higher energies, up to RHIC with 100 GeV gold on 100 GeV gold and LHC with 2.7 TeV ^{208}Pb on 2.7 TeV ^{208}Pb, 1100 TeV total collision energy!!

10. As if all these crazy ideas (and others time did not permit to be included) were not enough, Walter got out his crystal ball and, gazing into it, said, "What is the craziest thing I could urge experimentalists to do in the future?" So, while falling asleep he dreamed of new periodic charts of the elements: why not a periodic chart of antimatter with all antiprotons, antineutrons, and positrons with shell gaps, magic numbers and all, and in flights of pure fantasy dreamed of a strange matter periodic chart with Memo's (Memo's?) and strangelets and all sorts of other exotic intensities. He got Horst Stöcker and even his own son, Carsten, to join him in these proposals which could help bring about a new synthesis of particle and nuclear physics he dreamed. How really crazy can you get, because these two groups are always fighting each other over who controls what experiments and which theories. Surely experimentalists will not be fooled and nature cooperate this time. But then, a number of hypernuclei and even double Λ^0 hypernuclei are observed to exist. Worse, recently an antihydrogen atom was produced! So, in the next century we can expect to see equally crazy scientists breathing new life into our field by eagerly pursuing new antimatter and strange periodic charts.

11. Finally, I came to what Walter considered would be his crowning achievement. In a dream he typed out this work on his home computer and sent

off the disk with the only record of the work. He has kept this work so secret that only the journal Editor-In-Chief knew about it (and myself, because he left at Vanderbilt by mistake a written copy of the Editor-In-Chief's remarks). Why was he so secretive? Why?,... well Walter had discovered the key to the Universe – the True God Particle which can unite all forces, known and future unknown forces. So complete was his theory that it would bring an end to all our physics research. What is this True God Particle? No less than GREINERONIUM? The key to the Universe. Fortunately, fellow scientists, there was one Editor-In-Chief, who had backbone. He called his Board of Editors together and said, Enough is enough. Walter Greiner has bombarded the world of science with far too many crazy ideas already. GREINERONIUM – this is true nonsense. (I cannot repeat what the editor really said.) But he told his board, this will never be published and burned the only existing copy before their eyes. So, you and I can continue in our individual quests for the ultimate nature of reality knowing that the true secrets of the universe are still hidden.

Walter, everyone had great fun celebrating in humorous ways your truly amazing career. No one has made so many significant contributions in so many different areas of nuclear, atomic, and particle physics as you. My list above encompasses only some selected highlights. Your many seminal ideas have had enormous impacts on the directions and progress of physics in the last thirty years. You have set new directions that will carry us many years into the next century. Moreover, you have extended your influence and impacts through an equally remarkable career in training young scientists in their Ph.D. research. These talented scientists now fill important positions in universities and industries around the world and, through their own outstanding contributions, multiplied many times over your remarkable legacy. For me personally, Walter, it has been fantastic to be engaged in research, teaching, conferences and more with you. You have stimulated, challenged, and enhanced my research in so many, many ways. Equally as important, you have been a colleague and true friend in the finest sense. I am reminded of the words of George Eliot who wrote, "Friendship is the inexpressible comfort of feeling safe with a person, having neither to weigh thoughts nor measure words". Ours has been such a marvelous friendship. Congratulations, bester Freund, on a fantastic career and best wishes for many, many more years.

Joseph H. Hamilton

2   A toast to Prof. Dr. Walter Greiner

The common thread running through this conference is friendship – built on loyalty and sharing our mutual interest in good physics. I first met Walter in 1963 when he visited Larry Biedenharn at Duke, and I readily admit that my life has been enriched by his friendship. So I thank you, Walter, for this and for your support of my physics. After all, heavy ion physics is about nucleon-nucleon interactions, nucleon-nucleus and nucleus-nucleus interactions. So I am in the center. Walter likes to see excitation functions, so I can't resist showing the excitation function that is probably the most complex ever recorded. And, Walter it was taken in small steps of 100 eV. Such data lend themselves to the fine structure of analogue states, level density measurements, signatures for chaotic systems and possible tests of charged particle parity violation.

In conclusion, a toast:

> Here's to Walter,
> Who never falters.
> He's always ahead of the curve.

> Theorists three–
> We meet with glee;
> Since Greiners are never finer
> Than when in the Carolinas
> Whether 60 years old or fewer than three.

Edward G. Bilpuch

3   Grilling Walter Greiner

I do not believe in the barbarian Anglo-Saxon custom of chosing celebration days to tell the celebree all the unpleasant things you never told him before - even when using wit as an accessory. It makes me think of the story of the rabbi in Galitzia, known for his wisdom and sincerity, who would openly tell anyone his qualities and defects. So once the local prince invited the rabbi and asked him to enumerate all his - the prince's - qualities and defects. The rabbi started with the qualities and, as he was telling, two servants were stuffing them into a huge sack. But, regardless of how many the rabbi enumerated, there was always room in the sack. Finally, the turn came for the defects. However, as soon as the servants started to stuff them into a sack, the prince called "Full, full!" and quickly tied the sack.

So you see why I'll do the grilling my way.

Strangely enough, neither Walter nor I can remember the occasion (time, venue) where we first met. The obvious explanation is that we knew each other from time immemorial: like in the early Universe, the time and space of our first encounter were totally fuzzed. There is, however, a less grand explanation: neither of us was overly impressed by the other, so that our first interaction went unnoticed.

My strongest interaction with Walter resulted in the application of the anharmonic vibrator - rotator model to the classification of resonances in light heavy-ion collisions. It was at the time of a DPG conference in Bad Honnef, that Franco Iachello came out with a classification scheme of these resonances, based on group theory symmetries. This scheme (essentially the first terms of the Dunham expression) was successfully applied by Karl Erb and Allan Bromley to the resonances discovered in $12^C + 12^C$ collisions, resonances which many of us cherished under the name molecular or quasi-molecular.

Walter immediately realized that he could obtain an equally good or - as he said - better result, by assuming that, together with the rotation, the "molecule" undergoes an anharmonic vibration. In other words, an anschauliche geometrical picture of the collision would be as efficient as the more abstract group theory one. The only addition would be the reasonable supposition, that at these energies, collisions engender anharmonic surface vibrations. The idea was so cute, that, in his final talk, Walter told me:"Nikola, go home, do your homework and come back with results." I saluted, said "Jawohl, Walter!" and went home.

After a few months of hard work, I realized that Walter was right - as in most cases. I did come back with results. They are still in J. Phys. G, 1983, and in many other articles.

I owe much to Walter, in physics, in private life and in the combination of the two. To enumerate all of it would be tedious - and it would hurt Walter's proverbial delicacy. Yet, I'll mention something, for which I shall always be grateful. In 1972 Walter did not come to the Plitvice Lakes Conference on intermediate phenomena. Instead, he sent Werner Scheid. That's how I came to know Werner.

You see, life with Walter is this and that, but it is never dull.

Nikola Cindro

4   Lieber Walter, dear friends,

I speak here as an old friend and a stationary, nearest neighbour. Everybody who knows our Institute, knows that visitors come and go, students come and go and sometimes Walter comes and goes, so there has to be somebody who is relatively stationary. That is me.

But let me explain in more detail. First, I claim to be the oldest friend of Walter in this room, with the exception naturally of his wife Bärbel. The story starts during the end of 1959, when I was the one and only student of Professor Marschall in Freiburg, working on a Diploma thesis. I was very pleased, when, sometime in the early part of 1960 Marschall told me that another person would be coming. This person was Walter Greiner, who was working towards a doctorate under Marschall. He arrived, started to work but nothing more happened. The next one to arrive was Amand Faessler, who, I think, joined us during the Summer of 1960. At that time Walter thought "now we have a group" and we started to work together. The suggestion was that we learn about the Rot-Vib model, that we learn about the collective model of the Copenhagen group. So we each got a copy of the relevant issues of this Danish journal - Dan. Fys. Math. med. - and set to work day, after day. However, in the meantime I managed to finish my thesis, married and set off to Australia. The rest, who were joined by Florian Scheck, Max Huber and others, worked on and, as you know, were very successful. They were so successful that people started to talk about the Freiburg Mafia. As I left too early, I can at best qualify as an associate member of the Freiburg Mafia. But actually, I don't think that they are a Mafia because they were never organised enough.

Let me then address, after having established my relative seniority over nearly everybody in this room, the question of the next neighbour more fully. You might know, and certainly Walter knows, that there is a saying in Germany which goes "Willst Du Rat vom Teufel haben, mußt Du den Kollegen fragen". I provide a translation "If you wish to have advice from the devil, ask one of your colleagues". - I have to address a side remark to Fritz Hahne: This quotation is not from Goethe's Faust, so I do not qualify for the competition - I am glad to say, that Walter and I did not conform to this saying, but rather, by a gentleman's agreement, tried to coordinate our efforts and tried to pull on the same proverbial rope. In this fashion we were able to fight off constant suggestions of financial reductions by our Ministry and our experimental colleagues. We even were able to acquire some computing equipment and I think that the unspoken gentleman's agreement was able to get us around most difficulties. What really aided this, was probably the fact that we are quite different characters. Thus, for example, Walter gives his lectures in a suit and tie, while I wear jeans. Or, Walter likes to talk to experimentalists, I (if I talk at all) rather talk to mathematicians. In this and other ways, we presented a rather complementary picture to the students and I think this complementarity is the reason why we got on well. In conclusion I say it then in public: You were a very good colleague, Walter, and this for the better part of 24 years.

Well, that was history. Looking ahead, I am afraid to say that we will not be colleagues for another 24 years. But there is still time to do something for the Institute, there is enough time to do some very good physics. So I suggest that we just go ahead and do it, and I wish you luck, Walter!

Reiner M. Dreizler

5 Liebe Frau Greiner, lieber Walter, ladies and gentlemen,

This month it is 36 years ago since I met in March 1960 for the first time Walter Greiner. Walter just had come from Darmstadt to join the group of Hans Marschall to do his PhD in Freiburg. I came back from Munich where I had studied physics in the winter semester 1959 - 1960. When I came to Freiburg, other students told me that Hans Marschall got a new very excellent PhD student from Darmstadt. They and also I were very much impressed that Walter had received from Siemens - without applying for the job - an offer for working in the plasma physics theory group. In Freiburg the new physics building was not yet finished and Walter was sitting in a room with Rainer Dreizler in the old building in the second floor. When I asked Hans Marschall if I could get a diploma thesis, he told me that I have first to give a seminar so that he can convince himself that I am good enough to work in his group. He gave me a copy of the review article on Coulomb excitation in "Reviews of Modern Physics" from 1958 of Alder, Bohr, Huus, Mottelson and Winter. In March 1960 I started with the seminar and Walter always helped me to prepare my talks. We soon realized that it is not enough to speak about the review article alone but I had to present material from the book "Multipole fields" of M. E. Rose and give an introduction into "Angular momentum algebra". In addition , we realized that we have to learn the Bohr-Mottelson-model. Walter was always helping me to understand the material and preparing my talks. The seminars where on each Wednesday at three o'clock and lasted often until after eight o'clock in the evening. Apart of Hans Marschall also Theo Schmidt (He detected the nuclear quadrupole moments in 1935 and derived for the nuclear magnetic moments of the odd nuclei the "Schmidt-lines".) was present in these seminars. From the people at this symposium there was apart of Walter and myself also Rainer Dreizler participating. This sequence of seminars lasted until July 1960 and then finally Hans Marschall told me that I could start with the diploma thesis in his group working on an improved description of the "Sternheimer-Effect". Walter was working on his PhD calculating the polarization of nuclei in muonic atoms. During the seminars from March until July 1960, Walter and I realized that the solution which Bohr and Mottelson had found for describing the vibrations and rotations of nuclei was not a very optimal one.

Fall 1960 was a very important time: Walter got married, we both moved into the new building into one room in the eighth floor with an excellent view to the court of the Freiburg jail and we both started to work on an improved version of the Bohr-Mottelson-model, which we called the rotational-vibrational-model. In spring 1961 we went both to the meeting of the nuclear physics section of the German Physical Society to Bad Pyrmont. Among others we met there Max Born. Walter spoke about the rotational-vibrational model and I about the Sternheimer effect. It was the first time that we were speaking at a meeting. But there we also realized that what we had done in relative isolation in Freiburg for the description of the rotational and vibrational spectra in deformed and transitional nuclei was an interesting contribution which was respected also by more experienced colleagues.

Somewhere in 1961 Walter finished his PhD and I my diploma degree. He left for the States to the University of Maryland and I started extending the rotational-vibrational model to odd nuclei as my PhD thesis. Walter worked with Mike Danos from the National Bureau of Standards in Washington on applying the rotational-vibrational model to giant multipole resonances. He described the splitting of the giant resonances due to the deformations and their coupling to quadrupole vibrations. In addition he started there to collaborate with Judah Eisenberg who is also at this symposium. At a Gordon Research Conference Walter got to know Ray Sheline from the Florida State University in Tallahassee. Ray had at that time the most complete collection of data on rare earth and actinide nuclei for excitation energies, branching ratios and transition probabilities. Partially these were his own data from (d, p) reactions at the FSU Tandem, partially they were from the literature, but a large part was unpublished. Most of this unpublished data he had collected by his ability to copy from projected slides in the dark all information in a speed which seemed to me almost a miracle. Ray realized that all these data could be explained by the rotational-vibrational model and invited Walter Greiner and also me in summer and fall 1963 to Tallahassee. I came to Tallahassee with my computer codes, written in the internal language for the "Siemens 2002". FSU had the most advanced computer at that time, an IBM 709, so I first had to learn FORTRAN and rewrite the programs. I arrived on Friday evening and I worked until Monday morning with almost no sleep and then the programs were running again on the IBM 709, which still had electronic tubes. In intervalls of about two hours one of the many tubes was dying and had to be replaced. So one had to stay the whole night watching the computer. Within two months Walter and I in collaboration with Ray Sheline explained all these data in three papers: one for the rare earth and actinide nuclei, a letter for the transitional nuclei in the Os region and one paper giving a tabulation of the results of the rotational-vibrational model as a function of three parameters, which allowed to everyone to use the results just by interpolation in this table.

Before I left at the beginning of August 1963 to the States, I had written up my thesis on the extension of the rotational-vibrational model to odd mass nuclei. But when I returned somewhere in October, Hans Marschall had not yet looked into my thesis, so it took another month of discussions with Hans Marschall and of corrections and rewriting the text until he was finally satisfied and I got my PhD in the middle of December 1963. Somewhere in 1964 Walter had returned from the states and got an offer for a full professorship in the Institute of Theoretical Physics of the University of Frankfurt. After a big celebration he left Freiburg to Frankfurt and shortly afterwards I left Freiburg to go for a short time in spring 1965 to Florida State University to Ray Sheline as a research associate and from there in August 1965 as an assistant professor to the University of California at Los Angeles.

The activity of Walter Greiner at the University of Frankfurt is an unparalleled success story. He built up very fast a large group of young and smart scientists. He almost monopolized the description of collective excitations in nuclei and he initiated a new research field in nuclear physics "Heavy ion physics". But this success story of the Frankfurt school of theoretical nuclear physics of Walter Greiner will be told here by people who directly participated in it.

Since our common work in Freiburg, Walter and I are good friends. I am very thankful to him for this friendship and I am sure that it will continue far beyond his 60th birthday.

Amand Faessler

6 Dear Walter, dear friends, ladies and gentlemen,

At the reception given by Barbara and Walter GREINER, last November in Frankfurt, for Walter's 60 th birthday, I mainly talked about the relationship between his Institute and the Centre de Recherches Nucléaires at Strasbourg.

So, when it had been suggested that I give a short address this evening, I did not know what I could say in addition to my Frankfurt speech. But on Sunday evening, Fritz HAHNE, in his welcome address, used the words "friendship" and "loyalty". This gave me the idea to elaborate on these two qualities of Walter, and add another one, namely "reliability".

We, a few true friends of Walter, thought a long time ago, that for your sixties birthday, Walter, you would deserve a ceremony worth of your huge contribution to physics in general and theoretical physics in the Heavy Ion domain in particular. An International Conference on the STRUCTURE OF VACUUM AND ELEMENTARY MATTER , one of your many fields of interest, would be the right answer to this challenge.

Horst Stöcker and myself had first thought to organize a NASI (Nato Advanced Study Institute), like the one Walter, Horst and myself had held two and a half years ago in Bodrum, Turkey. However, such an Institute has to last for two weeks, a too long a period for many participants. But it would have had the big advantage of carrying with it a non negligible amount of financial aid, which can be used, and has to be used, to pay part of the expenses for "students" and the lecturers.

But when you agreed to the idea of a conference in your honor, Walter, you expressed three wishes:

1.) The conference should not last for more than ONE week,

2.) You would be pleased to be surrounded by your family at large and by your old and younger friends.

3.) For the location you desired a warm place where you can enjoy swimming.

Walter, once one has been accepted in your inner circle of friends, one realizes that you are a truly, loyal and reliable friend. And I think that it is quite normal that you expect from your friends and collaborators. reciprocity. This, of course, is not always the case, and disloyalties do not happen only to you. It's human, and if you cannot forgive them, so at least forget them.

Now all of your wishes have been fulfilled and we hope that you are pleased with the overall organization of this conference: you are surrounded by people who really want to honor you, because they came without any support from the Conference Committee, as this would have been the case with a NASI.

Your wonderful family is here: Barbara, your wife; your sons and their wives: Martin and Claudia, Carsten and Daniela and last, but not least, your adored and beloved grand children: Johanna, Josephine, Anne Catherine and Christopher.

LET US APPLAUD THEM !

Concerning your friends, I will come back to them in a moment.

But first: why are we here in South Africa ?

Among many places for holding this conference, our final choices where the Caribics, Sanibel Island off the west coast of Florida and South Africa. Walter decided for South Africa. Although he never expressed it, I think that one of the main reasons was a sentimental one. It is in South Africa, at the University of the Witwatersrand that you, Walter, received, your first doctor *honoris causa* degree, following a proposal of Friedel SELLSCHOP, a true and reliable friend. And the first *honoris causa* counts a great deal.

Fritz Hahne and Friedel accepted with enthusiasm to take care of the local organization of this symposium, and Sue SELLSCHOP, with ardor and competency, made sure that the ladies could visit a maximum of interesting spots in her beloved native country.

The second and third doctor *honoris causa* degrees you received almost at the same time. Judah Eisenberg, another reliable and old friend, who is here with his wife Neilly, was your sponsor for the distinction you received from the University of Tel Aviv. The fourth degree, namely the *honoris causa* from the University of Bucharest, and a honorary membership of the Romanian Academy of Science, were proposed by our colleagues POENARU and SANDULESCU.

So, Walter, you can see that all of your sponsors are here. A proof of real friendship. You have received other honors before and after the ones I have mentioned. Because of lack of time, I will skip them. Instead, I want to point to the presence of Joe HAMILTON, our master of ceremony, doctor *honoris causa* of your University, and Ed BILPUCH, also doctor *honoris causa* of the Johann Wolfgang Goethe Universität and Marilyn, his wife.

If I go on to cite names, I would certainly forget many of your friends, which are here, and came from all over the world, from 14 different countries. Many of the participants are former students of you; others are loyal and reliable friends like Amand FAESSLER and Werner SCHEID.

Concerning your students, not only did you train them, but you helped them to get a job after their degree. Many of them are now in industry or are professors at Universities or senior physicists in Research Centers. I hasten to mention that your intercession was not limited to your own students. You often put your intervening weight into the balance to turn it in favor of good physicists, from other Institutes than yours, when they were in competition for a job.

At this point it is time to recall that physics, Walter, is not your whole world. You enjoy life, good food and good wines. One of your favored dishes in Alsace is *CHOUCROUTE*. You must really love it, because I remember that, at one of your numerous visits to Strasbourg, it happened that I had organized a lunch with *Choucroute*. For dinner of this same day, we went to the *Winstub "Chez Yvonne"*. And there you had the choice between many alsatian specialties. What did you order ? .... *CHOUCROUTE* !!

Incidentally, *"Chez Yvonne"* has been your favored *Winstub* long before your chancellor Helmuth KOHL and Jacques CHIRAC had dinner there, a few days after CHIRAC had been elected president of France.

In Alsace, you also learned to appreciate Alsatian wines, your favored one being *Gewurztraminer* and *Riesling*, and, with the years, you have improved your taste. The proof? nowadays, nothing less than a *"Grand Cru"* or a wine which carries the label *"Médaille d'or"* would do it.

Walter, forgive me if I interrupt here my address for a short publicity: Alsace does not sum up with *Choucroute* and wine. It's a beautiful region, with charming villages along the *"route du vin"*; nice cities like Strasbourg and Colmar; a lot of cultural, scientific and folklore manifestations and many more. I recommend all of you to visit Alsace at your first opportunity.

End of the publicity.

Before I finish my address, I would like to point out another of your qualities, Walter. Not only are you a good teacher, but you do not hesitate to adjust your explanations to the level of pedestrians, like me, so that they can understand, or at least have the feeling to understand, your answers to questions like: Why is the vacuum not empty? Is there something which is empty? By the way, on this subject I got a few years ago, also a one hour private lesson by Professor Bernd Müller at Duke University.

Walter, you have ensured that the name GREINER will be on scientific papers for the next decades, with the initials "M" or "C", and this is certainly not the least of your satisfactions, today, at the celebration of your 60 th birthday.

Happy birthday, Walter!

Happy birthday for the first sixty years. For the next sixty, our friend Gabriel SIMONOFF, professor at the University of BORDEAUX, would like to say a few words. Gabriel is a nuclear physicist who, for many years now, has transferred nuclear technologies to biology and medicine, domains in which he is now very much interested and involved.

Gabriel .....

Dear Professor Walter GREINER,

You are now 60 years old; that means you are just in the middle of your path to reach the normal human longevity, which is 120 years.

A little more than 2 years ago, I wrote a book based on our research results. It is a sort of a guide on how to reach the human longevity in good physical shape and good intellectual state. The book is in French, but recently it was translated into German. So it is a great pleasure for me to offer you this german version.

I want to add:

Everybody knows you as an outstanding theoretician, but I dare to suggest you to make the following experiment.

1. Try to read the book in a very near future.

2. Try to follow the advices which it contains.

3. If you reopen this book in 60 years from now, you will be a quite exceptional experimenter.

Ich wünsche Ihnen ein langes, glückliches und gesundes Leben.

Walter, I did not bring my gift for you to South Africa. I have given it to you at your last visit to Strasbourg.

A song of Jacques BREL says: "I did not bring you flowers, but candies". I will say it the other way around: "I did not bring candies. I have flowers....." but not for you... for Barbara.

Wilderness *(South Africa)* march 14th, 1996

André Gallmann

7   Inspirations in the Wilderness

Ten years ago the family and close relatives joined to celebrate my father's 50th birthday. Some of the relatives expected me to toast to my father. I was not prepared and at once became very nervous about this spontaneous emission of words. Before completely diving into the lost sea, my father somehow detected the excited vacuum surrounding his oldest son and immediately took care of the strong disturbing fields.

What worked out nicely ten years ago, won't work again. Still, I am not a man of many words; this has not changed. But this time I do not want to have an excuse. Let me take the opportunity to share with you some of my lifelong experiences with my father as head of the family and my half lifelong encounters with him as a physicist. Certainly my mother would do better in the first part and some of his early collaborators and friends would do better in the second part. But only my brother and myself can combine both aspects!

60 years! This is not the time to only look back, but also to look forward. I think this very much reflects the spirit of my father's philosophy. Let me make this point clear.

It was the time when I was a small boy. My first grandfather just died. I asked my father: "Why do we not live forever?" I will never forget his answer: "When we grow up we experience good and bad things. It is our task to pass our good experiences to the younger and warn them about the bad ones. If we are no longer capable of doing so, we are of no use anymore. That is what we call evolution." That was the answer!

You can call this my father's principle! He taught this to his sons and to his students. He showed us how to develop fantasy and visions. Devotion and hard work is necessary to stay focused on what can be achieved and he prepared us to build up confidence. Yet another advice from him, maybe most important: Do things in such a way to respect yourself!

Times during this education have not always been easy. Being young, sometimes impatient and believing in one's own embryonic strength, often led to conflicts with the master. It is not these "relativistic collisions", which let you feel like a dwarfish coward facing a monstrous hurricane, it is the longterm benefit from him, which opens the door for such noble qualities as there is friendship, loyalty and the care for each other. With his black and white philosophy, allowing no "inbetweens", he sculptured me in such a way, and this certainly also holds for my brother and for his students, as we had to learn to find our own way, knowing the two extremes.

60 years! Isn't it like the question a child could pose: "Grandpa, who determines how old I am?" You see there is no reason for getting old! There are still noble goals to aim for in the future. Take this advice from your grown-up son: step back from the bureaucratic cancer, grasp a small group of young and bright students and teach them the good old way!

60 years. And many more to come!

Martin Greiner

8 Ladies and gentlemen, dear colleagues and friends, dear Walter Greiner

This is my first visit to South Africa, to this wonderful country. I'm from former East Germany and I am living in Dresden. For a long time, more than 30 years, I worked at the Nuclear Physics Department of the Institute for Nuclear Research of East Germany in Rossendorf near Dresden, for some years also at the Laboratory of Nuclear Reactions of the International Dubna-Institute.

This conference here is dedicated to Walter Greiner, to honour him at the occasion of his 60th-birthday. Thank you very much, members of the organizing commitee, for your kind invitation to this conference. I accepted with pleasure.

I have known Walter Greiner for more than 30 years, however, for a long time by name and his publications only. In 1962 I read two publications by Walter Greiner and Amand Faessler about a new model to describe the coupling of vibration and rotation in deformed nuclei in the Zeitschrift für Physik and later in Nuclear Physics. At that time our group in Rossendorf investigated the level structure of medium mass strongly deformed nuclei in beta-decay experiments and later on by means of in-beam gamma-spectroscopy. The Greiner Faessler-model was very important and helpful for the interpretation of our experimental results. - This model was my first encounter with Walter in Physics. A crossing point or a first level crossing between the experimental and theoretical investigations in Rossendorf (later also in Dubna) and the theoretical activities of Walter Greiner and his Theoretical Physics Institute in Frankfurt.

I myself and my coworkers in Rossendorf and in Dubna had many more such encounters in various fields of nuclear and heavy ion physics in the 70th and 80th. Here are some typical examples: Nuclear structure investigations, two-centre-shell-model for fusion and fission, investigation of quasimolecular X-rays in heavy ion collisions, investigation of extreme states of nuclear matter in high energy nucleus-nucleus collisions.

You know very well the creative potential of Walter Greiner in all fields and problems of modern nuclear and microscopic matter physics. For a long time Walter Greiner has stimulated all our investigations in nuclear physics in East Germany and in Dubna. I believe that it would be difficult to find a teacher better than Walter Greiner, especially for our young people in East Germany.

Of course, until 1989 our possibilities to communicate and cooperate with West German physicist were always a complicated matter. In Germany we lived in the shadow of the wall in Berlin, built in 1961 and destroyed in 1989.

And this wall was high and difficult to overcome. You know quite well what the fission barrier of heavy nuclei is. Different from the fission barrier for the wall in Berlin the probability of tunneling was strongly forbidden by strong selection rules. For a long time we lived behind this wall in a ground state of a very closed system. "Excited level states" we reached through international conferences and through the work at the international Dubna-Institute [as an isomeric state in the second minimum by exchange with our western colleagues], but these occasions were rare. To meet Walter Greiner at such circumstances and to discuss with him physical problems was always an extremely exciting event for us.

The existence of the wall in Berlin lead also to strange situations and curious stories. For instance, let me tell you a curious story that happened in 1968. Walter had invited me to come to Frankfurt and to give a seminar about the level structure of strongly deformed odd mass nuclei. I was happy in expecting my first visit and the seminar on the other side of the iron curtain. The administration of the Institute in Rossendorf told me: It is okay! Only two days before the seminar was scheduled the same man of the administration told me: Sorry, you cannot go to the seminar in Frankfurt, the political situation excludes this, a.s.o. Oh God! In my trouble I asked him: Can you tell me what I can say Professor Greiner about the reasons for such nonsense to cancel this seminar, for me it would be a shame? He said: You can send a telegram that the traffic to West Germany is closed because of foot-and-mouth disease! I followed his advice without comments hoping that Walter Greiner and his coworkers will take this nonsens in good humor!

Let me close at this point. Now we have a better time and a much better situation. Of course, the time is going on and we became 30 years older. I mean that 60 years is also a good time.

Dear Walter it is my hope that you will stimulate the physical investigations also in the next 30 years! Thank you.

Karl-Heinz Kaun

9 Mathematical theory of the happiness

My father was a poor man. He lived in this century surviving several wars, revolutions, contrerevolutions and dictators. In spite of these circumstances he was a happy man. I wanted to follow him and I decided that I will be also happy. Time to time I meditated about the secret of happiness. Once upon a time I decided to work out the mathematical theory of happiness.

First of all it is necessary to define the measure of happiness by the help of a real function having negative and positive values as well. Of course it is the function of the time t. Let us call it "my happiness" and denote it by $mh(t)$. In addition to this let us introduce two other functions $h \geq 0$ and $u \geq 0$ as a measure for the happiness and for the unhappiness of other peoples, respectively.

Now the task is to give an algorithm to determine the value of $mh(t)$. We will do it by succesive approximation.

I am ashamed, nevertheless I must confess, that I am happy, when one of my enemies e is unhappy, consequently $u(e)$ gives a positive contribution to my happiness:

$$mh \sim u(e).$$

Since I have more than one enemy I must integrate over all of them.

$$mh \sim \int u(e)de.$$

On the other hand if one of my enemies is happy then my happiness gets a negative contribution $-h(e)$:

$$mh \sim \int u(e)de - \int h(e)de.$$

Similarly if my friends (f) are unhappy the contribution is also negative:

$$mh \sim \int u(e)de - \int h(e)de - \int u(f)df.$$

Finally we arrive to the most important contribution: the happiness of my friends increases the value of my happiness:

$$mh \sim \int u(e)de - \int h(e)de + \int h(f)df - \int u(f)df.$$

The functions $mh, u(e), h(e), u(f)$ and $h(f)$ depend also on time, therefore to get my happiness at the time t, I have to average these functions over the time interval $[t - T, t]$ where T is the characteristic time of my memory:

$$mh(t) \sim \int (\overline{u(e)} - \overline{h(e)})de + \int (\overline{h(f)} - \overline{u(f)})df \ .$$

It is easy to see that this average fluctuates around zero: in the average I am neither happy, nor unhappy.

I tried to push my happiness toward positive values by introducing a positive weight factor $\mathbf{W} > \mathbf{1}$, which enhances the contribution to my happiness when my friends are happy:

$$mh(t) \sim \int (\overline{u(e)} - \overline{h(e)})de + \int (\mathbf{W}\overline{h(f)} - \overline{u(f)})df .$$

This trick, however does not help, because, I know that it was me, who introduced this arbitrary weighting factor, therefore a psychological renormalisation takes place and the effect of this factor is cancelled.

What to do?

I began to observe systematically those people who seem to be happy and I discovered the "trick". They do not introduce arbitrarily the weight. Instead of that, they make every efforts to generate real happiness for their friends:

$$mh(t) \sim \int (\overline{u(e)} - \overline{h(e)})de + \int (\mathbf{WG}\overline{h(f)} - \overline{u(f)})df .$$

This is the great trick of genuine happiness.

Istvan Lovas

10 Ladies and gentlemen, dear colleagues and friends, dear Walter Greiner

It is a great honor as well as a pleasure for me to attend this happy occasion in celebration of Walter. First, I need to transmit the "official" congratulations and expressions of appreciation from the Physics Division of the Oak Ridge National Laboratory. We at ORNL are fortunate to have Walter visit us on a yearly basis. Over the years, he has made great contributions to many of our programs, and his insight and guidance are very much appreciated. All my colleagues at ORNL join me in congratulating Walter on the occasion of his 60th birthday.

Second, I need to transmit a very personal greeting from Mike Strayer, the head of our Theory Group. Mike has FAXed a special message which it is my pleasure to read to you.

Finally, I wish to add a personal note. In my view, Walter has made outstanding contributions to physics in three areas. First, there are his own superb intrinsic contributions to the field, many of which are highlighted at this conference. There is little that I can add here, except to say that both the breadth and the depth of his contributions are outstanding. Second, there are the students that have been trained by Walter, and the contributions to physics that they make. It is indeed a distinguished group. My own professional life has been very much influenced by Walter's former students such as Ulrich Mosel, Berndt Müller, and Horst Stöcker. Third, and by no means least, there are Walter's sons, who are making their own major contributions to physics. In the U.S. we are in the midst of a political campaign. The politicians play lip-service to "family values" for purposes of their own political gain. Seeing Walter here, together with his sons and their families, gives a special meaning to the term "family values," and I wish that the hollow politicians were here to learn this meaning.

Walter, please accept the very best wishes from all of us at the Oak Ridge National Laboratory, and very special personal wishes from Mike Strayer, and from Carol and myself.

Frank Plasil

11   A toast at the Banquet

In praise of the many facets that make up my personal images and personal experiences of Walter Greiner, I offer as my paean of praise the following quotations:

1. **Walter the pious**
   The Researcher's Prayer
   (with acknowledgment to Proceedings of the Chemical Society, 1963, pp 8 - 10)

   Grant, oh God. Thy benedictions
   On my theory's predictions,
   Lest the facts, when verified,
   Show Thy servant to have lied.

   May they make me B.Sc.,
           A Ph.D. and then
   A D.Sc. and F R.S.,
           A Times obit. Amen.

   Oh Lord, I pray, forgive me please,
   My unsuccessful syntheses,
   Thou know'st, of course — in Thy position —
   I'm up against such competition.

Let not the hardened Editor,
With referee to quote,
Cut all my explanation out
And print it as a Note.

2. **Walter the scholar:**
   From Shakespeare's King Henry VIII —

   "He was a scholar, and a ripe and good one;
   Exceeding wise, fair-spoken, and persuading;
   Lofty and sour to them that lov'd him not,
   But to those men that sought him, sweet as summer"

3. **Walter the avid mushroom (pfifferling) hunter** in the Sunday afternoon woods ...
   from John Ford's *The Broken Heart* (written in 1633):

   "I am . . . . . . a mushroom
   On whom the dew of heaven drops now and then"

4. **Walter the loyal . . . . . .**
   This is a difficult one to put to appropriate verse, but from my papers I
   found these two:
   From the Engish poet, Joseph Trapp in ca. 1700

   "The King, observing with judicious eyes,
   The state of both his universities.
   To Oxford sent a troop of horse, and why ?
   That learned body wanted loyalty,
   To Cambridge books, as well discerning,
   How much that body wanted learning"

   But perhaps more appropriate is to quote from Winston Churchill ...

   "The loyalties that centre upon (ones friend) are enormous. . . . . .
   if he trips he must be sustained.
   if he makes mistakes, they must be covered,
   if he sleeps he must not be wantonly disturbed"

5. **Walter the visionary ...**
   from the Bible. the book of Joel
   "I shall pour out my spirit upon all flesh,
   and your sons and your daughters shall prophesy,
   your old men shall dream dreams,
   your young men shall see visions"

6. **Walter the creative ...**
   DH Lawrence in commenting on the essay by Leo
   Shostov, spoke of
   "... (to act) spontaneously from the for ever
   incalculable prompting of the creative wellhead
   within him."

7. **Walter the statesman ...**
   from Oliver Goldsmith, early 18th century Irish dramatist:

   "Who, too deep for his hearers, still went on refining,
   And thought of convincing, while they thought of dining,
   Tho' equal to all things, for all things unfit,
   Too nice for a statesman, too proud for a wit".

8. **Walter the dashing driver** (of a sporty red Jaguar) ...
   from Sir William Gilbert:

   "And the brass will crash,
   And the trumpets bray,
   And he'll cut a dash,
   When he drives today ... "

9. **Walter the friend ...**
   there is so much to say, but few words must suffice
   from Alexander Pope, of the late 17th century, we cull the lovely stanza
   "Shall then this verse to future age pretend
   Thou wert my guide, philosopher, and friend ?
   That urg'd by thee, I turn'd the tuneful art
   From sounds to things, from fancy to the heart"

But my closing words are short and simple, repeating once again Shakespeare's King Richard II, that in relation to my friendship to Walter I can but say

"I count myself in nothing else so happy, as in a soul rememb'ring my good friend"

Friedel Sellschop

12 Dear Walter

I am sitting in Oak Ridge trying to think what I can say that would herald this celebration of a remarkable scientist and friend who is half a world away. It is difficult, to say the least. I have never met anyone quite like Walter; he is simply unique. Nuclear physics is a small community in the world today, and we have much to thank him for. He has guided and led so many of us in new and exciting directions that it is hard to imagine what the field would have been like without him. He is indeed a true pioneer and a great educator.

Walter, I am saddened that circumstances have prevented me from being with you on this happy occasion. I know that the depth and breadth of the science being discussed will lead to new and exciting endeavors over the next few years. Rather than discuss such things, I would like to relate a personal story about Walter and myself.

I have never been sure if Walter remembers our first meeting, I was interviewing in Frankfurt, a fresh Ph.D. out of M.I.T. with an enthusiasm and zeal for nuclear structure physics. We spent much of the day together arguing about strong field electrodynamics. I believe at that time I said "that such things were possibly interesting, but did not merit the full-time attention of a serious theorist." This was the period in the early 70s when student demonstrations and unrest were abounding throughout Germany. On the day of my colloquium, Walter had to face about 3,000 noisy and angry students, so he sent me away with the advice to think more about physics and argue less. Needless to say, I did not get the job. However, I did follow his advice with the result that more than half of the past twenty years was devoted to various aspects of strong field QED.

It is fair to say that Walter has many talents, more than any one person can easily accommodate. I have vivid memories of him, as we all do. A single meeting is enough to have a permanent imprint on one's memory, To me, Walter has always been warm, sincere, generous, and occasionally outrageous; in other words, a true friend. Walter, my very best regards on this happy occasion.

Michael Strayer

13   Ladies and Gentlemen.

"This has been a week that would test anyone's stamina. It's late; it's been a very long day at an extremely long conference; we traveled forever to get here; we had a long excursion yesterday and we have had many exceptionally long speeches tonight. There would be every justification to be brief but you can rest assured that I will not flinch from my responsibilities. Not only will I speak, I shall speak at length!"

"I should like to begin by letting you into a little secret. The real reason that I made such a long journey from California to South Africa was not to learn new physics, although I have certainly done that. Nor was it to celebrate Walter Greiner's birthday although I am pleased to be able to do that too. No, like so many of you here this evening, the real reason that I traveled so far was to see the game! However, some of you may perhaps be unaware that you don't have to go all the way to Kruger park to see wonderful animals. In fact, during the past few days, I have been fortunate enough to make some splendid sightings right here in George!"

"The first thing that you need to understand when big game watching is the classification system. The animals come in a variety of sizes from the thin fast moving graduate students, all the way up to the biggest and most impressive of all: the legendary and fearsome C4s. These latter beasts are without doubt at the summit of the evolutionary pyramid, their ruthless natures honed to razor sharpness by the forces of natural selection."

"The first C4 that I was lucky enough to spot was the legendary *Felis Leo Garchingensis*, the African Lion, an awesome symphony of claws and teeth. This particular specimen recently stepped down as king of the Hessian jungle but is still easily recognizable by his fine yellow mane (so much more impressive than that of his successor). Late the same day, as I was climbing into the misty hills behind George, I came upon yet another C4, *Gorilla Gorilla Beringei Tubingensis*, the highland Gorilla. What a sight! This magnificent specimen is well known for his deep wisdom, his deep chest and his very deep voice. Then yesterday, as I was strolling through the woods, I caught a flash of movement as *Felis Leo Pardus Heidelbergensis*, the Leopard, jumped onto her prey from a tree! This svelte C4, known around the parks as the 'new kid on the block' is widely touted as future jungle queen material. However, the best of my C4 spotting was saved for the last. Just today, in a clearing near the hotel, I was lucky enough to encounter five tons of *Loxodonta Africana Frankfurter*, the African Elephant. Without a doubt this is the true king of the jungle, with his elaborate social system, his great knowledge, the kindness he shows in caring for his young and his exceptionally long memory. But, watch out:

although normally a placid and peaceful animal, when necessary he protects himself with his 6 foot tusks!"

"That's enough about big game, but there are a few more things I have to say. In particular, I would like to add my congratulations, Walter, and to convey the very warmest greetings possible from all your friends in Berkeley. Let me end by sharing a little anecdote which was told to me recently at another birthday party given for Norman Glendenning who is also here this evening. He and I were talking with Karsten Pruess, one of your students (and now a geophysicist at LBNL) who gave us something of the flavor of life in Frankfurt in those days. Apparently another new student was sitting at his desk when you passed by and asked him what he was doing. On being informed that he was reading a paper, you responded, 'Young man, you are here to write papers, not to read them!' On that note, I close."

Happy Birthday, Walter!

James Symons

Reinhardt, J.	Goethe Universität, Frankfurt am Main, Germany, jr@th.physik.uni-frankfurt.de
Richter, W.	University of Stellenbosch, Stellenbosch, South Africa, RICHTER@SUNVAX.SUN.AC.ZA
Rischke, D.	Columbia University, New York, NY, USA, drischke@nt2.phys.columbia.edu
Röpke, G.	Universität Rostock, Rostock, Germany, roepke@darss.mpg.uni-rostock.de
Sailer, K.	Kossuth Lajos University, Debrecen, Hungaria, SAILER@DTP.ATOMKI.HU
Sandulescu, A.	Academia Romana, Bucharest, Romania, sandulescu@roifa.bitnet
Satarov, L.M.	IV Kurchatov Institute for Theoretical Physics, Moscow, Russia, satr@kiae.su
Schaffner, J.	Nuclear Science Division, LBL, Berkeley, CA, USA, schaffne@nsdssd.lbl.gov
Scheid, W.	Justuts-Liebig-Universität Gießen, Gießen, Germany, werner.scheid@theo.physik.uni-giessen.de
Schempp, A.	Goethe Universität, Frankfurt am Main, Germany, a.schempp@em.uni-frankfurt.d400.de
Schmidt, R.	TU Dresden, Dresden, Germany, schmidt@ptprs3.phy.tu-dresden.de
Schramm, S.	Goethe Universität, Frankfurt am Main, Germany, schramm@clri6a.gsi.de
Schukraft, J.	Europ. Organisation for Nucl. Research, CERN, Switzerland, SKS@crnvma.cern.ch
Sellschop, J.P.F.	University of the Witwatersrand, Johannesburg, Wits, South Africa, 099sell@witsvma.wits.ac.za
Sharpey-Schafer, J.	National Accelerator Centre, Faure, South Africa, DIRECTOR@srvnac1.nac.ac.za
Simonoff, G.	Legnan, France
Soff, G.	TU Dresden, Dresden, Germany, soff@ptprs7.phy.tu-dresden.de
Soff, S.	Goethe Universität, Frankfurt am Main, Germany, ssoff@th.physik.uni-frankfurt.de
Solov'yov, A.V.	Academy of Sciences of Russia, St. Petersburg, Russia, SOLOVYOV@MATH-ATOM.IOFFE.RSSI.RU
Spieles, C.	Goethe Universität, Frankfurt am Main, Germany, spieles@th.physik.uni-frankfurt.de

Stachel, J.   Universität Heidelberg, Heidelberg, Germany,
                       stachel@nuclear.physics.sunysb.edu
Stein, E.   Goethe Universität, Frankfurt am Main, Germany,
                       stein@th.physik.uni-frankfurt.de
Stöcker, H.   Goethe Universität, Frankfurt am Main, Germany,
                       stoecker@th.physik.uni-frankfurt.de
Strayer, M.R.   Oak Ridge National Laboratory, Oak Ridge, TN, USA,
                       strayer@orph01.phy.ornl.gov
Stroebele, H.   Goethe Universität, Frankfurt am Main, Germany,
                       stroebele@ikf006.ikf.uni-frankfurt.de
Strottmann, D.   Los Alamos National Laboratory, Los Alamos, USA,
                       dds@lanl.gov
Symons, J.   Lawrence Berkeley Laboratory, Berkeley, CA, USA,
                       TJSymons@lbl.bitnet
Uggerhøj, E.   Aarhus University, Aarhus C, Denmark,
                       ISA@DFI.AAU.DK
Vaagen, J.S.   University of Bergen, Bergen, Norway,
                       JanS.Vaagen@fi.uib.no
van der Ventel, B.I.S.University of Stellenbosch, Matieland, South Africa,
                       9020381@sunvax.sun.ac.za
Weber, F.   Universität München, München, Germany,
                       fweber@mfl.sue.physik.uni-muenchen.de
Weigel, M.   Universität München, München, Germany,
                       weigel@mfl.sue.physik.uni-muenchen.de
Wyngaardt, S.M.   University of Stellenbosch, Matieland, South Africa,
                       WYNGAARDT@nacdh4.nac.ac.za

Walter Greiner's Scientific Interests

1961	Nuclear polarization in muonic atoms.	
1961-63	Rotation-Vibration-Model, (first solution of the dynamics contained in Bohr-Hamiltonian).	(with A. Faessler)
1964/65	First to point out that shape isomers in collective potential energy surfaces might exist.	
1962-67	Dynamical Collective Model	(with M. Danos, H.J. Weber, M. Huber, D. Drechsel)
	(Coupling of giant dipole and quadrupole resonances to shape vibrations and rotations of nuclei).	
1964/65	Different proton-neutron-vibrations, lowering of g-factor from Z/A, scissor mode idea.	
1964	First calculation of spreading width of giant resonances.	(with M. Danos)
1965-67	Eigenchannel theory of the S-matrix-nuclear reactions.	(with M. Danos)
1969-72	Dispersion effects in electron scattering (an application of eigenchannel theory).	(with Ch. Toepfer)
1965-72	Superheavy nuclei, structure, stability against fission, α-decay e^--capture.	(with U. Mosel, B. Fink, J. Grumann)
1971	Electronic structure of superheavies.	(with B. Fricke)
1967-79	QED of strong fields (vanishing of energy-gap-) spontanous pair creation, superheavy quasimolecules predicted, quasimolecular two center Dirac equation, K-vacancy production, -new particles? - falsification by comparing consequences with precision experiments. Spectroscopy of superheavy quasimolecular orbitals. Supercritical fields in other areas like gravitation (Hawking-radiation) strong colour fields (pair production in strings).	(with W. Pieper, B. Müller, J. Rafelski, J. Reinhardt, A. Schäfer, S. Schramm)

1968-75	Theory of nuclear molecules. Double-resonance mechanism, nuclear two center shell model, coupled channel formulation of nuclear molecular reactions.	(with W. Scheid H.J. Fink K. Pruess P. Holzer)
1969-85	The "Frankfurt Collective Model" ("Gneuß-Greiner-Model"), Potential energy surfaces with shape isomers, transition from spherical to deformed nuclei, γ-instability, triaxiality, etc.	(with G. Gneuß, J. Maruhn Sedlmayers, P. Hess)
1968-80	Development of the idea of compression and heating of nuclear matter - nuclear shock waves - squeeze out & bounce off, cluster flow, development of nuclear fluid dynamics.	(with W. Scheid, H. Müller, H. Stöcker, J. Hofmann)
1972-77	Development of fragmentation theory.	(with W. Scheid, J. Maruhn, Liran, Sandulescu)
1975/76	Prediction of cold valleys for fusion of superheavy elements. All superheavy elements made so far, i.e. 102 ... 112, were synthesized with the reactions predicted here.	(with Sandulescu, Gupta)
1978-80	Prediction of Cluster radioactivity. Four years later experimental confir–mation by Rose and Jones (Oxford), B. Price (Berkeley), A. Ogloblin (Moskau). These are new modes of nuclear decay, in addition to the conventional ones like γ-decay, β-decay and α-decay.	(with A. Sandulescu, D. Poenaru)
1977-85	Calculation of potential energy surfaces exhibiting the cold valleys for fusion, cold fission, bimodal fission and cluster radioactivity.	(with J. Maruhn, Depta, Hermann) (with Sandulescu)
1985	Antimatter clusters emitted from a quark gluon plasma - a signal for the plasma due to the enhancement of the production rate by several orders of magnitude.	(with Stöcker, Heinz)

1986-88	Meson field theory for hot and dense nuclear matter. Phase transition (chiral restauration, i.e. baryon masses $->$ zero) within meson field theory.	(with Theis, Maruhn, Stöcker)
1988-96	Meson field theory applied to nuclei and exotic objects, e.g. multi-Λ-nuclei	(with Maruhn, Schaffner, Rufa, P.G. Reinhard, Stöcker)
1982-94	Development of relativistic fluid dynamics for high energy heavy ion collisions.	(with Maruhn,Gräbner, Waldhauser, Katscher, Stöcker)
1991-95	Antimatter production in thermal meson field theory.	(with Mishustin, Satarov, Stöcker)
1987-93	Quantum Molecular Dynamics applied to nucleus nucleus collisions -Pauli potentials - detailed flow (shock wave effects) and Cluster production rates.	(with Stöcker, Peilert, Konopka),
1989	Relativistic Quantum Molecular Dynamics. First reaction model allowing secondary, ternary etc. collisions of incoming baryons and produced particles.	(with Sorge, Stöcker)
1993	Antiflow of pions and antimatter.	(with Bass, Spieles, Stöcker)
1991-93	Quark-Gluon Plasma is a Cluster Plasma.	(with Rischke, Stöcker, Gorenstein)
1993-96	Extension of the periodic system into the new directions of strangeness and antimatter, structure of the baryonic and mesonic vacuum of high densities and temperature.	(with Carsten Greiner, Schaffner, Stöcker)

LIST OF PARTICIPANTS

Aichelin, J.	Universite de Nantes, Nantes, France, aichelin@nanvs2.in2p3.fr
Arenhövel, H.	J. Gutenberg-Universität, Mainz, Germany, arenhoevel@vkpmza.kph.uni-mainz.de
Bass, S.	Goethe Universität, Frankfurt am Main, Germany, bass@th.physik.uni-frankfurt.de
Bilpuch, E.	Duke University, Durham, NC, USA, Bilpuch@TUNL.TUNL.Duke.edu
Blann, M.	Lawrence Livermore Lab., Univ. of California, Livermore, CA, USA, BLANN@LLNL.GOV
Bleicher, M.	Goethe Universität, Frankfurt am Main, Germany, bleicher@th.physik.uni-frankfurt.de
Braun-Munziger, P.	GSI, Darmstadt, Germany, P.Braun-Munzinger@gsi.de
Britt, H.C.	Lawrence Livermore Nat. Lab., Livermore, CA, USA, Britt1@llnl.gov
Cindro, N.	Ruder Boskovic Institute, Zagreb, Croatia, nikola@ds5000.IRB.HR
Cleymans, J.	Univ. of Capetown, Rondebosch/Cape, South Africa, cleymans@physci.uct.ac.za
Coffin, J.P.	Centre de Recherches Nucl., Univ. Louis Pasteur, Strasbourg, France, coffin@frcpn11.in2p3.fr
Csernai, L.P.	University of Bergen, Bergen, Norway, csernai@fi.uib.no
Dreizler, R.	Goethe Universität, Frankfurt am Main, Germany, dreizler@th.physik.uni-frankfurt.de
Eggers, H.C.	University of Stellenbosch, 7602 Stellenbosch, South Africa, eggers@bohr.sun.ac.za
Eisenberg, J.	University of Tel Aviv, Tel Aviv, Israel, judah@giulio.tau.ac.il
Faessler, A.	Universität Tübingen, Tübingen, Germany, faessler@mailserv.zdv.uni-tuebingen.de
Fricke , B.	Universität Kassel, Kassel, Germany, fricke@physik.uni-kassel.de
Gal, A.	Hebrew University, Racah Institute of Physics, Jerusalem, Israel, avragal@vms.huji.ac.il

Gallmann, A.   Centre des Recherches Nucl., Univ. Louis Pasteur, Strasbourg, France,
                      kueny@frcpn11.in2p3.fr
Glendenning, N.K.   Lawrence Berkeley Laboratory, Berkeley, CA, USA,
                      nkg@LBL.GOV
Gorenstein, M.   Academy of Ukraine, Kiev, Ukraine,
                      hedpitp@gluk.apc.org
Greiner, C.   Universität Gießen, Gießen, Germany,
                      greiner@theorie.physik.uni-giessen.de
Greiner, M.   Technische Universität Dresden, Dresden, Germany,
                      greiner@Ptprs1.phy.tu-dresden.de
Greiner, W.   Goethe Universität, Frankfurt am Main, Germany,
                      greiner@th.physik.uni-frankfurt.de
Grosse, E.   FZ Rossendorf, Dresden, Germany,
                      E.Grosse@fz-rossendorf.de
Gutbrod, H.H.   Universite de Nantes, Nantes, France,
                      gutbrod@vxwa80.cern.ch
Gyulassy, M.   Columbia University, New York, NY, USA,
                      gyulassy@nt1.phys.columbia.edu
Hahne, F.J.W.   University of Stellenbosch, Stellenbosch, South Africa,
                      FJWH@LAND.SUN.AC.ZA
Hamilton, J.H.   Vanderbilt University, Nashville, TN, USA,
                      hamiltj1@ctrvax.vanderbilt.edu
Harris, J.W.   Lawrence Berkeley Nat. Lab., Univ. of California, Berkeley, CA, USA,
                      harris@Csa5.LBL.Gov
Herrmann, N.   GSI, Darmstadt, Germany,
                      nh@dsaf.gsi.de
Hess, P.   Instittuto de Ciencas Nucleares, UNAM, Mexico,
                      HESS@ROXANNE.NUCLECU.UNAM.MX
Hillhouse, G.C.   University of Stellenbosch, Matieland, South Africa,
                      GCH@LAND.SUN.AC.ZA
Hofmann, S.   GSI, Darmstadt, Germany,
                      S.Hofmann@gsi.de
Jundt, F.   Centre de Recherches Nucl., Univ. Louis Pasteur, Strasbourg, France,
                      JUNDT@FRCPN11.IN2P3.FR
Kämpfer, B.   FZ Rossendorf, Dresden, Germany,
                      kaempfer@gamma.fz-rossendorf.de
Kaun, K.H.   Dresden, Germany,
Kienle, Paul   TU München, Garching, Germany,
                      Paul.Kienle@physik.tu-muenchen.de

Konopka, J.   Goethe Universität, Frankfurt am Main, Germany,
                     konopka@th.physik.uni-frankfurt.de
Lindsay, R.   University of the Western Cape, Bellville, South Africa,
                     ROBBIE@physics.uwc.ac.za
Lovas, I.   Lajos Kossuth University, Debrecen, Hungary,
                     LOVAS@HEAVY-ION.ATOMKI.HU
Ludu, A.   Bucharest University, Bucharest-Magurele, Romania,
                     ludu@fizica.unibuc.ro
Lynen, U.   GSI, Darmstadt, Germany,
                     U.Lynen@gsi.de
Mattiello, R.   State University of New York, Stony Brook, NY, USA,
                     mattiell@nth3.phy.bnl.gov
McLerran, L.   University of Minnesota, Minneapolis, MN, USA,
                     mclerran@physics.spa.umn.edu
Mishustin, I.   Niels-Bohr-Institute, Kopenhagen, Denmark,
                     mishustin@nbivax.nbi.dk
Mosel, U.   Justus-Liebig-Universität, Gießen, Germany,
                     mosel@piggy.physik.uni-giessen.de
Müller, B.   Duke University, Durham, NC, USA,
                     muller@phy.duke.edu
Münzenberg, G.   GSI, Darmstadt, Germany,
                     G.Muenzenberg@gsi.de
Nagle, J.   Yale University, New Heaven, CN, USA,
                     nagle@dunes.physics.yale.edu
Oeschler, H.   TH Darmstadt, Darmstadt, Germany,
                     h.oeschler@gsi.de
Oganessian, Y.   Joint Institute for Nuclear Research, Dubna, Russia,
                     oyuts@ljar9.jinr.dubna.su
Paul, P.   State University of New York at Stony Brook, Stony Brook, NY, USA,
                     paul@nuclear.physics.sunysb.edu
Pelte, D.   Universität Heidelberg, Heidelberg, Germany,
                     pelte@pel2.mpi-hd.mpg.de
Plasil, F.   Oak Ridge National Laboratory, Oak Ridge, TN, USA,
                     plasilf@ornl.gov
Poenaru, D.   Institute of Atomic Physics, Bucharest Magurele, Romania,
                     poenaru@ifa.ro
Pretzl, K.   University of Bern, Bern, Switzerland,
                     pretzl@lhep.unibe.ch
Ramayya, A.V.   Vanderbilt University, Nashville, TN, USA,
                     ramayya1@ctrvax.vanderbilt.edu